TechOne: Automotive Brakes

W9-BFY-080

TechOne: Automotive Brakes

Jack Erjavec

Professor Emeritus,
Columbus State Community College
Columbus, Ohio

THOMSON

DELMAR LEARNING

Australia   Canada   Mexico   Singapore   Spain   United Kingdom   United States

THOMSON
DELMAR LEARNING

TechOne: Automotive Brakes

Jack Erjavec

Vice President, Technology and Trades SBU:
Alar Elken

Editorial Director:
Sandy Clark

Acquisitions Editor:
David Boelio

Developmental Editor:
Matthew Thouin

Marketing Director:
Cyndi Eichelman

Channel Manager:
Fair Huntoon
Beth Lutz

Production Director:
Mary Ellen Black

Production Editor:
Barbara L. Diaz

Art/Design Specialist:
Cheri Plasse

Technology Project Manager:
Kevin Smith

Editorial Assistant:
Kevin Rivenburg

Library of Congress Cataloging-in-Publication Data:

Erjavec, Jack.
   TechOne : automotive brakes / Jack Erjavec.
       p. cm.
   Includes index.
   ISBN 1–40183–526–0 (pbk. : alk. paper)
   ISBN 13: 978-1-4018-3526-2
   ISBN 1–40183–527–9
(instructor's ed. : pbk./CD-ROM)
   1. Automobiles—Brakes. I. Title.
TL269.E75 2004
629.2'46'0288—dc22   2003053380

NOTICE TO THE READER

Contents

vi • Contents

Preface

THE SERIES

Welcome to Delmar Learning's *TechOne*, a state-of-the-art series designed to respond to today's automotive instructor and student needs. *TechOne* offers current, concise information on ASE and other specific subject areas, combining classroom theory, diagnosis, and repair into one easy-to-use volume.

You'll notice several differences from a traditional textbook. First, a large number of short chapters divide complex material into chunks. Instructors can give tight, detailed reading assignments that students will find easier to digest. These shorter chapters can be taught in almost any order, allowing instructors to pick and choose the material that best reflects the depth, direction, and pace of their individual classes.

TechOne also features an art-intensive approach to suit today's visual learners: images drive the chapters. From drawings to photos, you will find more art to better understand the systems, parts, and procedures under discussion. Look also for helpful graphics that draw attention to key points in features like You Should Know and Interesting Fact.

Just as importantly, each *TechOne* volume starts off with a section on safety and communication, which stresses safe work practices, tool competence, and familiarity with workplace "soft skills" such as customer communication and the roles necessary to succeed as an automotive technician. From there, learners are ready to tackle the technical material in successive sections, ultimately leading them to the real test—an ASE practice exam in the Appendix.

THE SUPPLEMENTS

TechOne comes with an **Instructor's Manual** that includes answers to all chapter-end review questions and a complete correlation of the text to NATEF standards. A **CD-ROM**, included with each Instructor's Manual, consists of

PowerPoint Slides for classroom presentations and a **Computerized Testbank** with hundreds of questions to aid in creating tests and quizzes. Chapter-end review questions from the text have also been redesigned into adaptable **Electronic Worksheets**, so instructors can modify questions if desired to create in-class assignments or homework.

Flexibility is the key to *TechOne*. For those who would like to purchase jobsheets, Delmar Learning's NATEF Standards jobsheets are a good match. Topics cover the eight ASE subject areas and include:

- Automatic Transmissions and Transaxles
- Automotive Brakes
- Automotive Electrical and Electronic Systems
- Automotive Engine Repair
- Automotive Engine Performance
- Automotive Heating and Air Conditioning
- Automotive Suspension and Steering
- Manual Drive Trains and Axles

Visit **http://www.autoed.com** for a complete catalog.

OTHER TITLES IN THIS SERIES

TechOne is Delmar Learning's latest automotive series. We are excited to announce these future titles:

- Advanced Automotive Electrical Systems
- Advanced Engine Performance
- Air Conditioning
- Automatic Transmissions
- Automotive Computer Systems
- Automotive Engine Repair
- Basic Automotive Service and Maintenance
- Engine Performance
- Fuels and Emissions
- Steering and Suspension

Check with your sales representative for availability.

A NOTE TO THE STUDENT

There are now more computers on a car than aboard the first spacecraft, and even gifted backyard mechanics long ago turned their cars over to automotive professionals for diagnosis and repair. That's a statement about the nation's need for the knowledge and skills you'll develop as you continue your studies. Whether you eventually choose a career as a certified or licensed technician, service writer or manager, an automotive engineer, or even if you decide to open your own shop, hard work will give you the opportunity to become one of the 840,000 automotive professionals providing and maintaining safe and efficient automobiles on our roads. As a member of a technically proficient, cutting-edge workforce, you'll fill a need, and even better, you'll have a career to feel proud of.

Best of luck in your studies,
The Editors of Delmar Learning

About the Author

Jack Erjavec, a master-certified ASE technician and professor emeritus at Columbus State Community College in Columbus, OH, has become a fixture in the automotive textbook world with more than 30 works to his credit. In addition to his bestselling comprehensive text, *Automotive Technology*, Jack is also series editor of Delmar Learning's popular *Today's Technician* series and the new *TechOne* automotive series.

In addition to assuming these editorships, Jack was the Product Development Manager at Delmar Learning for business, industry, government, and retail automotive materials, as well as a professional consultant for several automotive manufacturers. He is a long-time affiliate of the North American Council of Automotive Teachers (NACAT), having served on the board of directors and as executive vice president. Jack is also associated with a number of professional organizations, including the Society of Automotive Engineers, and remains active in the industry. In his free time, he most cherishes spending time with his family: Rose, Megan, Craig, Hannah, Grace, Moira, Judah, and Erin.

Acknowledgments

I'd like to thank the following reviewers, whose technical expertise was invaluable in creating this text:

David Crowley
Community College of Southern Nevada
North Las Vegas, NV

John Eichelberger
St. Philip's College
San Antonio, TX

Larry Hecker
Motorist Assurance Program
Bethesda, MD

Wally Marciniak
Dana Brake & Chassis
McHenry, IL

Russell Strayline
Lincoln Technical Institute
Philadelphia, PA

Features of the Text

TechOne includes a variety of learning aids designed to encourage student comprehension of complex automotive concepts, diagnostics, and repair. Look for these helpful features:

Section Openers provide students with a **Section Table of Contents** and **Objectives** to focus learners on the section's goals.

Interesting Facts spark student attention with industry trivia or history. Interesting facts appear on the section openers and are then scattered throughout the chapters to maintain reader interest.

Section 2

Basic Theories and Services

SECTION OBJECTIVES

After you have read, studied, and practiced the contents of this section, you should be able to:

- List and describe the operation of the basic parts of a brake system.
- Explain how kinetic energy in a brake system is affected by mass, weight, momentum, and inertia.
- Explain the importance of kinetic and static friction in a brake system.
- Describe the effects of pressure, surface area, and friction material on the performance of a braking system.
- Explain how the coefficient of friction is determined for a material.
- Describe the three basic types of friction materials currently used in brake systems.
- Explain how hydraulics can be used to transmit force and motion.
- Explain the relationships of force, pressure, and motion in a hydraulic system.
- Explain how vacuum and atmospheric pressure can be used to increase force.
- Explain the basic principles of electricity.
- Explain how to use Ohm's Law to determine the amount of current, resistance, or voltage in a circuit.
- Name the various electrical components and their uses in electrical circuits.
- Diagnose electrical problems by logic and symptom description.
- Perform troubleshooting procedures using meters, test lights, and jumper wires.
- Inspect, test, and replace the brake warning lamp, parking brake indicator lamp, brake stop lamp switch, and master cylinder fluid level sensor.
- Repair electrical wiring and connectors.

Interesting Fact Hydraulic brakes were invented in 1918 in California in Malcolm Loughead's shop. The spelling of his name was later changed to Lockheed. The Lockheed brake system first appeared on the custom-made 1920 Dusenberg car. The first mass-produced car to use the Lockheed hydraulic brake system was the 1924 Chrysler.

An **Introduction** orients readers at the beginning of each new chapter. **Technical Terms** are bolded in the text upon first reference and are defined.

You Should Know informs the reader whenever special safety cautions, warnings, or other important points deserve emphasis.

Chapter 4

Power and Special Tools

Introduction

Power tools make a technician's job easier. They operate faster and with more torque than hand tools. However, power tools require greater safety measures. Power tools do not stop unless they are turned off. Power is furnished by air (pneumatic), electricity, or hydraulic fluid. Technicians typically use pneumatic tools because they have more torque, weigh less, and require less maintenance than electric power tools. However, electric power tools tend to cost less than the pneumatics. Electric power tools can be plugged into most electric wall sockets, but to use a pneumatic tool, you must have an air compressor and an air storage tank.

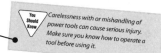

> **You Should Know** *Carelessness with or mishandling of power tools can cause serious injury. Make sure you know how to operate a tool before using it.*

Properly diagnosing and servicing brake systems requires many different special tools. These are also discussed in this chapter. Without these special tools, performing your work properly can become next to impossible.

CLEANING EQUIPMENT AND CONTAINMENT SYSTEMS

The following systems and methods are used to safely contain brake dust while you are doing brake work.

Negative-Pressure Enclosure and HEPA Vacuum Systems

A negative-pressure enclosure is a containment system in which cleaning and the inspection of brake assemblies is performed inside a tightly sealed protective enclosure that covers and contains the entire brake assembly **(Figure 1)**. The enclosure prevents the release of asbestos fibers into the air. It is designed so that you can clearly see the work in progress. It has impermeable sleeves and gloves that let you do what you need to do.

A vacuum pump and a **high-efficiency particulate air (HEPA) filter** keep the enclosure under negative pressure as work is done. Because particles cannot escape the enclosure, compressed air can be used to remove dust, dirt, and asbestos fibers from brake parts. The HEPA vacuum also can be used to loosen the asbestos-containing residue from the brake parts. Once the asbestos is loose, draw it out of the enclosure with the vacuum pump. The dust is then trapped in the vacuum cleaner filter.

Asbestos waste must be collected, recycled, and disposed of in sealed impermeable bags or other closed, impermeable containers. Any spills or release of asbestos-containing waste material from inside the enclosure or vacuum hose or vacuum filter should be immediately cleaned up using vacuuming or wet cleaning methods.

Low-Pressure Wet Cleaning Systems

Low-pressure wet cleaning systems are the most commonly used containment systems. These wash dirt from the brake assembly and catch the contaminated cleaning agent in a basin **(Figure 2)**. The reservoir contains water with an organic solvent or wetting agent. To prevent any

---

Pedal feedback is transmitted through the discharge valve and input pushrod to the brake pedal. If application force increases, the cycle repeats until the booster again reaches the holding position. During a panic stop, the valve opens fully and provides a high-pressure power assist for the brakes.

Brakes Being Released

When the driver releases the brake pedal, the pushrod force no longer holds the apply valve open. The apply valve spring closes the apply valve and prevents additional pressure from entering the booster cavity. Then the discharge valve spring opens the discharge valve, allowing booster cavity pressure to bleed off quickly to the fluid reservoir **(Figure 9)**. Because no pedal force or booster pressure is present, the master cylinder piston springs return the master cylinder pistons to the unapplied position.

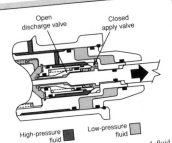

Figure 9. PowerMaster valve positions and fluid flow right after the brake pedal is let up.

A **Summary** concludes each chapter in short, bulleted sentences. **Review Questions** are structured in a variety of formats, including ASE style, challenging students to prove they've mastered the material.

Summary

- The Hydro-Boost system operates with pressure from the power steering pump.
- The PowerMaster system has its own independent hydraulic power source driven by an electric motor.
- Pressure from the power steering pump is routed to the hydraulic booster assembly. The hydraulic booster has a large spool valve that directs pressure to a power

chamber when the brakes are applied. The boost piston in the power chamber reacts to this pressure and provides a force to the master cylinder primary piston.
- The PowerMaster system has apply and discharge valves that control a power piston. The power piston controls the hydraulic boost to the pistons of the master cylinder.

Review Questions

1. Which of the following statements about Hydro-Boost systems is not true?
   A. When the brake pedal is not depressed, pressurized fluid flows through the Hydro-Boost valve to the steering gears.
   B. When the brakes are applied, pressurized fluid flows to the Hydro-Boost power cavity and to the steering gears.
   C. When the brake pedal is released, the spool valve moves fully rearward to the unapplied position and vents pressure from the pressure chamber.
   D. A failure in the power steering system, such as a broken power steering hose, a broken power steering drive belt, or a pump failure could cause a loss of brakes.
2. Technician A says that application pressure can be stored by a spring in an accumulator. Technician B says

that application pressure can be stored by gas under pressure in an accumulator. Who is correct?
   A. Technician A only
   B. Technician B only
   C. Both Technician A and Technician B
   D. Neither Technician A nor Technician B
3. While discussing a Hydro-Boost power brake system, Technician A says that the brake hydraulic system uses DOT 3 or DOT 4 brake fluid as specified by the manufacturer. Technician B says that the power steering pump forces fluid into the master cylinder's reservoir. Who is correct?
   A. Technician A only
   B. Technician B only
   C. Both Technician A and Technician B
   D. Neither Technician A nor Technician B
4. While discussing a PowerMaster power brake system, Technician A says that the brake hydraulic system uses

Appendix A

ASE PRACTICE EXAM FOR BRAKE SYSTEMS

An **ASE Practice Exam** is found in the **Appendix** of every *TechOne* book, followed by a **Bilingual Glossary**, which offers Spanish translations of technical terms alongside their English counterparts.

1. The brake system specifications for a car call for DOT 4 brake fluid. Technician A says that DOT 5 fluid can be used. Technician B says that DOT 3 fluid can be used. Who is correct?
   A. Technician A only
   B. Technician B only
   C. Both Technician A and Technician B
   D. Neither Technician A nor Technician B

2. While servicing a dual reservoir master cylinder, Technician A removes the secondary piston stop bolt in order to remove the primary piston assembly; Technician B cleans the master cylinder with a degreasing solvent. Who is correct?
   A. Technician A only
   B. Technician B only
   C. Both Technician A and Technician B
   D. Neither Technician A nor Technician B

3. When multiple trouble codes are present in an antilock brake system, look for:
   A. a weak connection at a common ground
   B. an open circuit
   C. low-voltage signals
   D. high-voltage signals

4. Which of the following is a true statement?
   A. Excessive amounts of air in the system will cause the brakes to drag.
   B. Dragging brakes are typically caused by a stuck wheel cylinder or caliper piston.
   C. A brake pedal return spring with too much tension can cause the brakes to drag.
   D. All of the statements are true.

5. Technician A says that the master cylinder should be bench bled before being installed on the vehicle. Technician B says that after installing the master cylinder on the vehicle, the entire system should be bled at each wheel. Who is correct?
   A. Technician A only
   B. Technician B only
   C. Both Technician A and Technician B
   D. Neither Technician A nor Technician B

6. Two technicians were discussing brake lines. Technician A says that damaged sections of brake line can be replaced with a piece of tubing and compression fittings. Technician B says that the entire line that is damaged should be replaced and the new one bent to the right shape with a tube bender. Who is correct?
   A. Technician A only
   B. Technician B only
   C. Both Technician A and Technician B
   D. Neither Technician A nor Technician B

7. Which of the following is not likely to cause a pulsating brake pedal?
   A. Loose wheel bearings
   B. Worn brake pad linings
   C. Excessive lateral runout
   D. Nonparallel rotors

A comprehensive **Index** helps instructors and students pinpoint information in the text.

Bilingual Glossary

ABS event A rapid reduction in speed with the brakes applied and when one or more wheels begin to lock up.
Evento ABS *Reducción rápida de velocidad con los frenos aplicados durante la cual una o más ruedas se empiezan a bloquear.*

Acceleration sensor This sensor provides information about the rate of forward or reverse acceleration or deceleration. Also known as a G-switch.
Sensor de aceleración *Este sensor proporciona información acerca de la velocidad de aceleración o desaceleración hacia adelante o en reversa. También se conoce como interruptor G.*

Acceleration slip regulation (ASR) ASR is the name of the traction control system manufactured by Bosch.
Regulación de deslizamiento de aceleración (ASR) *Nombre del sistema de control de tracción fabricado por Bosch.*

Accumulator A container that stores hydraulic fluid under pressure. It can be used as a fluid shock absorber or as an alternate pressure source. A spring or compressed gas behind a sealed diaphragm provides the accumulator pressure.
Acumulador *Recipiente que almacena fluido hidráulico bajo presión. Se puede usar como amortiguador o como una fuente de presión alterna. Un resorte o gas comprimido detrás de un diafragma sellado proporciona la presión del acumulador.*

Actuator Any device that receives an output signal, or command, from a computer and does something in response to the signal.
Actuador *Cualquier dispositivo que recibe un comando o señal de salida de una computadora y que hace algo en respuesta a la señal.*

Air bag system A system that uses impact sensors, a vehicle's on-board computer, an inflation module, and a nylon bag in the steering column and dash to protect the driver and passenger during a head-on collision.
Sistema de bolsa de aire *Sistema que usa sensores de impacto, la computadora a bordo del vehículo, el módulo de*
inflado y una bolsa de fibra sintética en la columna de dirección y el tablero para proteger al conductor y al pasajero durante una colisión de frente. Sistema que usa sensores de impacto, la computadora a bordo, el módulo de inflado y una bolsa de fibra sintética en la columna de dirección y el tablero para proteger al conductor y al pasajero durante una colisión de frente. Sistema que usa sensores de inflado y una bolsa de fibra sintética en la columna de dirección y al pasajero durante una colisión de frente.

Allen wrenches Allen wrenches, or hex-head wrenches, are used to tighten and loosen setscrews and fit into a machined hex-shaped recess in the bolt or screw.
Llaves Allen *Las llaves Allen o de cabeza hexagonal, se usan para apretar y aflojar tornillos de presión y ajustarlos en un hueco maquinado con forma hexagonal en el perno o tornillo.*

Alternating current (AC) Electrical current that changes direction between positive and negative.
Corriente alterna *Corriente eléctrica que cambia de dirección entre positivo y negativo.*

Ammeter The instrument used to measure electrical current flow in a circuit.
Amperímetro *Instrumento que se usa para medir el flujo de corriente eléctrica en un circuito.*

Ampere The unit for measuring electrical current; usually called an amp.
Amperio *Unidad para medir la corriente eléctrica. Generalmente se conoce como amp.*

Amplitude Signal strength, or the maximum measured value of a signal.
Amplitud *Fuerza de señal o el valor medido máximo de una señal.*

Analog Signal A voltage signal that varies within a given range (from high to low, including all points in between).
Señal analógica *Señal de voltaje que varía dentro de un rango (de alto a bajo, incluidos todos los puntos intermedios).*

Index

A

ABS (antilock brake systems), 54, 359
  automatic stability control, 426
  Bendix systems, 387, 398
  bleeding, 398
  Bosch systems, 378, 399
  central-valve cylinders, 124
  components of, 363
  computer systems, 361
  control module operation, 372
  control module replacement, 418
  DaimlerChrysler systems, testing, 406
  Delphi Chassis (Delco Moraine) systems, 383, 399
  Delphi Chassis VI with TCS, 424
  diagnostic trouble codes (DTCs), 395, 404
  Ford systems, testing, 405
  four-channel systems, 370, 372
  General Motors systems, 374, 402
  Honda systems, testing, 409
  hydraulic pressure, 8, 165
  inspection basics, 393
  integral and nonintegral systems, 368
  Kelsey-Hayes systems, 385, 400
  Lucas-Varity systems, 387
  master cylinder removal, 129
  modulator assembly, 373
  one-channel systems, 368, 370
  operating range tests, 395
  pedal feel, 360
  pressure control valves, 52

  pressure modulation, 359
  pump and motor replacement, 417
  safety considerations, 8
  self-diagnosis, 373
  service information, 393
  servicing, 397
  speed sensor circuits, 79
  speed sensor replacement, 416
  speed sensor testing, 413
  Sumitomo systems, 389
  switch testing, 412
  testing basics, 392
  Teves systems, 102, 381, 400
  three-channel systems, 370
  Toyota real-wheel antilock systems, 389
  Toyota TRAC system, 425
  traction control. *See* traction control systems
  warning lamps, 393
  wheel speed sensors, 366
ABX4 systems, 389, 398
AC (alternating current), 68
acceleration sensor, ABC, 382
acceleration slip regulation (ASR), 378, 421, 380, 421. *See also* traction control systems
accident response, 7
accumulators, hydraulic boosters, 172
  depressurizing, 397
  Hydro-Boost boosters, 197, 204
  PowerMaster boosters, 199
  actuators, 72
Acura antilock brake systems, testing, 409

  adjusted tapered roller bearings, 326
  cleaning and inspection, 329
  installation and adjustment, 331
  lubrication, 329
  removing, 327
  servicing, 327
  troubleshooting, 327
adjustment, 136
  auxiliary drum brake shoes, 350
  brake lathes, 312
  brake pedal freeplay, 97, 89. *See also* testing and diagnostics
  brake shoes, manual, 303
  caliper piston clearance, 303
  disc brake self-adjusters, 231
  driveline parking brakes, 354
  drum brake self-adjusters, 215, 220, 225, 276
  drum brakes, 275
  height-sensing proportioning valves, 163
  master cylinder pushrod, 136, 188
  parking brake cables and linkage, 348
  power brake systems, pedal height, 173
  stop lamp switch, 89
  tapered roller bearings, 331
aerosol can cleaners, 42
air
  air bag system, 7
  air ratchets, 29
  air wrenches, 29
  breathing safety, 40
  entrapment tests, 106
  pressure. *See* pressure

Section 1

Safety and Communication

Interesting Fact

For many years, people worked on brake systems without knowing the hazards of brake dust and brake fluid. Many of them suffered from lung and other diseases without knowing the source. Yet some technicians today ignore the hazard warnings and precautions, thereby subjecting themselves to the same health hazards faced by the brake mechanics of yesterday.

SECTION OBJECTIVES

After you have read, studied, and practiced the contents of this section, you should be able to:

- Wear proper clothing for working in a shop and use eye, hand, and hearing protection when needed.
- Identify the type of fire extinguisher for common fires that may occur in the shop.
- Know how to relieve pressure, before service, from high-pressure brake systems.
- Explain the principles of hazard communication and the right-to-know laws.
- List the basic units of measure for length and volume in the metric and United States customary systems.
- Identify and describe the purpose and use of common hand tools.
- Identify and describe the purpose and use of commonly used power tools.
- Operate power and shop tools safely.
- Identify and use the major measuring tools and instruments used in brake service work.
- Identify and describe the purpose and use of special tools used for brake service.
- Identify and use the electrical test tools used in antilock brake system (ABS) diagnostic work.
- Explain the requirements of a brake system performance test as specified by Federal Motor Vehicle Safety Standard (FMVSS) 105.
- Understand the health hazards created by asbestos dust and perform the proper tasks needed to protect yourself from harm.
- List the principal methods, or causes, of chemical poisoning and explain how to avoid them.
- Use cleaning equipment safely and properly and comply with all requirements for handling and disposing of hazardous materials.
- List the basic safety requirements for working with brake fluid.

Chapter 1

Safe Work Practices

Introduction

Safety is an important topic, important to you and those around you. Although the safety issues presented in this chapter are categorized, true safe work practices are based on nothing else but common sense and knowledge of the safety equipment.

SAFE WORK AREAS

Your work area should be kept clean and safe. The floor and bench tops should be kept clean, dry, and orderly. Any oil, coolant, or grease on the floor can make it slippery. Slips can result in serious injuries. To clean up oil, use commercial oil absorbent. Keep all water off the floor. Water is slippery on smooth floors, and electricity flows well through water. Aisles and walkways should be kept clean and wide enough to easily move through. Make sure the work areas around machines are large enough to safely operate the machine. Make sure all drain covers are snugly in place. Open drains or covers that are not flush to the floor can cause toe, ankle, and leg injuries.

Shop safety is the responsibility of everyone in the shop. Everyone must work together to protect the health and welfare of all who work in the shop.

PERSONAL SAFETY

Personal safety simply involves those precautions you take to protect yourself from injury. Your eyes can become infected or permanently damaged by many things in a shop. Eye protection should be worn whenever you are working in the shop. Many types of eye protection are avail-

able. To provide adequate eye protection, safety glasses have lenses made of safety glass. They also offer some sort of side protection. Nearly all brake work is performed on the vehicle, but a few procedures are completed on a workbench. For nearly all services performed on the vehicle or on a bench, eye protection **(Figure 1)** should be worn. This is especially true while working under the vehicle.

Some procedures require that you wear other eye protection in addition to safety glasses. When you are cleaning parts with a pressurized spray, for example, you should wear a face shield. A face shield not only gives added protection to your eyes but also protects the rest of your face.

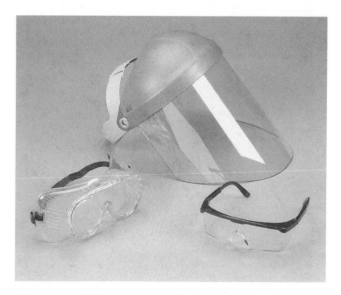

Figure 1. Different types of eye protection worn by automotive technicians; clockwise: face shield, safety glasses, and goggles.

If chemicals such as battery acid, fuel, or solvents get into your eyes, flush them continuously with clean water. Have someone call a doctor and get medical help immediately. Many shops have eyewash stations or safety showers that should be used whenever you or someone else has been sprayed or splashed with a chemical.

Your clothing should be well fitted and comfortable but made with strong material. If you have long hair, tie it back or tuck it under a cap. Never wear rings, watches, bracelets, and neck chains. These can easily get caught in moving parts and cause serious injury.

Automotive work involves the handling of many heavy objects, which can be accidentally dropped on your feet or toes. Always wear shoes or boots of leather or similar material with no-slip soles. Steel-tipped safety shoes can give added protection to your feet.

Good hand protection is often overlooked. A scrape, cut, or burn can limit your effectiveness at work for many days. A well-fitted pair of heavy work gloves should be worn during operations such as grinding and welding or when handling high-temperature components. Always wear approved rubber gloves when handling strong and dangerous caustic chemicals. They can easily burn your skin. Many technicians wear thin, surgical-type latex gloves whenever they are working on vehicles. These offer little protection against cuts but do offer protection against disease and grease buildup under and around the fingernails. These gloves are comfortable and inexpensive.

LIFTING AND CARRYING

When lifting a heavy object such as a transmission, use a hoist or have someone else help you. If you must work alone, *always* lift heavy objects with your legs, not your back. Bend down with your legs, not your back **(Figure 2)**, and securely hold the object you are lifting. Then stand up, keeping the object close to you **(Figure 3)**. Trying to muscle something with your arms or back can result in severe damage to your back and can end your career and limit what you do for the rest of your life.

Figure 2. Lift with your legs, not your back.

Figure 3. Hold the object close to you while lifting or moving it.

FIRE HAZARDS AND PREVENTION

Many items around a typical shop are a potential fire hazard. These include gasoline, diesel fuel, cleaning solvents, and dirty rags. Each of these should be treated as a potential firebomb and handled and stored properly.

In case of a fire, you should know the location of the fire extinguishers and fire alarms in the shop and how to use them. You should also be aware of the different types of fires and the fire extinguishers used to put out each type of fire.

Basically, there are four types of fires: class A fires are those in which wood, paper, and other ordinary materials are burning; class B fires are those involving flammable liquids, such as gasoline, diesel fuel, paint, grease, oil, and other similar liquids; and class C fires are electrical fires. Class D fires are a unique type of fire because the material burning is a metal. The magnesium used in the construction of some wheels and late-model General Motors transfer cases is a flammable metal and will burn fiercely when subjected to high heat.

USING A FIRE EXTINGUISHER

Remember, during a fire never open doors or windows unless it is absolutely necessary; the extra draft will only make the fire worse. Make sure the fire department is contacted before or during your attempt to extinguish a fire. To extinguish a fire, stand six to ten feet from the fire **(Figure 4)**. Hold the extinguisher firmly in an upright position. Aim the nozzle at the base and use a side-to-side motion, sweeping the entire width of the fire. Stay low to avoid inhaling the smoke. If it gets too hot or too smoky, get out. Remember, never go back into a burning building for anything. To help remember how to use an extinguisher, memorize the word "PASS."

Figure 4. Know how to use a fire extinguisher before you need to know.

Pull the pin from the handle of the extinguisher.
Aim the extinguisher's nozzle at the base of the fire.
Squeeze the handle.
Sweep the entire width of the fire with the contents of the extinguisher.

If there is not a fire extinguisher handy, a blanket or fender cover can be used to smother the flames. You must be careful when doing this because the heat of the fire may burn you and the blanket. If the fire is too great to smother, move everyone away from the fire and call the local fire department. A simple under-the-hood fire can cause the total destruction of the car and the building and can take lives. You must be able to respond quickly and precisely to avoid a disaster.

SAFE TOOLS AND EQUIPMENT

Whenever you are using any equipment, make sure you use it properly and that it is set up according to the manufacturer's instructions. All equipment should be properly maintained and periodically inspected for unsafe conditions. Frayed electrical cords or loose mountings can cause serious injuries. All electrical outlets should be equipped to allow for the use of three-pronged electrical cords. All equipment with rotating parts should be equipped with safety guards that reduce the possibility of the parts coming loose and injuring someone. Never use tools and equipment for purposes other than those they are designed for. Using the proper tool in the correct way will not only be safer but will also allow you to do a better job.

Do not depend on someone else to inspect and maintain equipment. Check it out before you use it. If you find the equipment unsafe, put a sign or tag on it to warn others, disconnect its power source, and notify the person in charge.

LIFT SAFETY

Always be careful when raising a vehicle on a lift or a hoist. Adapters and hoist plates must be positioned correctly on twin post-and-rail-type lifts to prevent damage to the underbody of the vehicle. There are specific lift points. These points allow the weight of the vehicle to be evenly supported by the adapters or hoist plates. The correct lift points can be found in the vehicle's service manual. Always follow the manufacturer's instructions. Before operating any lift or hoist, carefully read the operating manual and follow the operating instructions.

Once you feel the lift supports are properly positioned under the vehicle, raise the lift until the supports contact the vehicle. Check the supports to make sure they are in full contact with the vehicle. Shake the vehicle to make sure it is securely balanced on the lift and then raise the lift to the desired working height. Also, make sure that the safety locks for the hoist are fully engaged.

JACK AND JACK STAND SAFETY

A hoist or lift is much safer and easier, but a vehicle can also be raised off the ground with a hydraulic jack. The lifting pad of the jack must be positioned under an area of the vehicle's frame or at one of the manufacturer's recommended lift points. Never place the pad under the floor pan or under steering and suspension components; these are easily damaged by the weight of the vehicle. Always position the jack so the wheels of the vehicle can roll as the vehicle is being raised.

> **You Should Know** *Never use a lift or jack to move something heavier than it is designed for. Always check the rating before using a lift or jack. If a jack is rated for two tons, do not attempt to use it for a job requiring five tons. It is dangerous for you and the vehicle.*

Safety (jack) stands are supports of different heights that sit on the floor. They are placed under a sturdy chassis member, such as the frame or axle housing, to support the vehicle **(Figure 5)**. Once the safety stands are in position, the hydraulic pressure in the jack should be slowly released until the weight of the vehicle is on the stands. Like jacks, jack stands also have a capacity rating. Always use a jack stand with the correct rating.

Never go under a vehicle when it is supported only by a hydraulic jack; rest the vehicle on the safety stands before going under the vehicle. The jack should be removed after the jack stands are set in place. This eliminates a hazard, such as a jack handle sticking out into a walkway. A jack handle that is bumped or kicked can cause a tripping accident or cause the vehicle to fall.

> **You Should Know** *If battery acid gets on your skin, wash it off immediately and flush your skin with water for at least five minutes. If the electrolyte gets into your eyes, immediately flush them out with water and then immediately see a doctor. Never rub your eyes; just flush them well and go to a doctor. While working with and around batteries, it shows common sense to wear safety glasses or goggles.*

BATTERIES

When possible, you should disconnect the battery of a car before you disconnect any electrical wire or component. This prevents the possibility of a fire or electrical shock. It also eliminates the possibility of an accidental short, which can ruin the car's electrical system. This is especially true of newer cars that are equipped with many electronic and computerized controls. Any electrical arcing can cause damage to the components. To properly disconnect the battery, disconnect the negative or ground cable first **(Figure 6)** and then disconnect the positive cable. Because electrical circuits require a ground to be complete, by removing the ground cable you eliminate the possibility of a circuit accidentally becoming completed. When reconnecting the battery, connect the positive cable first and then the negative.

Figure 5. Jack stands should be used to support the vehicle after it has been raised by a jack.

Figure 6. When disconnecting a battery, remove the negative cable first.

Disconnecting the battery also erases the memory of electronic components that require continuous battery power for their memory circuits. These devices include electronic radios and power seats with memory. More importantly, these kinds of devices also include electronic controllers with long-term adaptive memory, such as the engine control module and the transmission control module. It is easy but time consuming to record the stations set on the radio before disconnecting the battery and then reset them and the clock after the battery is reconnected. Unfortunately, you cannot do the same for the adaptive memory circuits in electronic controllers. Therefore, it is a good idea to install a computer memory saver before disconnecting the vehicle battery.

The active chemical in a battery, the **electrolyte**, is basically sulfuric acid. Sulfuric acid can cause severe skin burns and permanent eye damage, including blindness, if it gets in the eyes.

ELECTRICAL SYSTEM REPAIRS

Some electronic replacement parts are very sensitive to **static electricity**. These parts will be labeled as such. Whenever you are handling a part that is sensitive to static, you should follow these guidelines to reduce any possible electrostatic charge buildup on your body and the electronic part:

1. Do not open the package until it is time to install the component.
2. Before removing the part from the package, ground the package to a known good ground on the car.
3. Always touch a known good ground before handling the part. This should be repeated while handling the part and more frequently after sliding across the seat, sitting down from a standing position, or walking a distance.
4. Never touch the electrical terminals of the component.
5. Use grounding straps to prevent damage to components.

ACCIDENTS

Make sure you are aware of the location and contents of the shop's first aid kit. There should be an eyewash station in the shop so that you can rinse your eyes thoroughly should you get acid or some other irritant into them. If there are specific first aid rules in your school or shop, make sure you are aware of them and follow them. Some first aid rules apply to all circumstances and are normally included in everyone's rules. If someone is overcome by carbon monoxide, get that person fresh air immediately. Burns should be cooled immediately by rinsing them with water. Whenever there is severe bleeding from a wound, try to stop the bleeding by applying pressure with clean gauze on or around the wound and get medical help. Never move someone who may have broken bones unless the person's life is otherwise endangered. Moving that person may cause additional injury. Call for medical assistance.

Your instructor or supervisor should be immediately informed of all accidents that occur in the shop. It is a good idea to keep a list of up-to-date emergency telephone numbers posted next to the telephone. The numbers should include a doctor, hospital, and the fire and police departments.

AIR BAG SAFETY AND SERVICE WARNINGS

The dash and steering wheel contain the circuits that control the **air bag system**. Whenever working on or around air bag systems, it is important to follow some safety warnings. There are safety concerns with both **deployed** and live (undeployed) air bag modules.

1. Wear safety glasses when servicing the air bag system.
2. Wear safety glasses when handling an air bag module.
3. Wait at least ten minutes after disconnecting the battery before beginning any service on or around the air bag system. The reserve energy module is capable of storing enough power to deploy the air bag for up to ten minutes after battery voltage is lost.
4. Handle all air bag sensors with care. Do not strike or jar a sensor in such a manner that deployment occurs.
5. When carrying a live air bag module, face the trim and bag away from your body.
6. Do not carry the module by its wires or connector.
7. When placing a live module on a bench, face the trim and air bag up.
8. Deployed air bags may have a powdery residue on them. Sodium hydroxide is produced by the deployment reaction and is converted to sodium carbonate when it comes into contact with atmospheric moisture. Although it is unlikely that sodium hydroxide will still be present, you should not put yourself at risk. This compound is toxic and can cause serious skin irritations. Wear safety glasses and gloves when handling a deployed air bag, and wash your hands immediately after handling the bag.
9. A live air bag module must be deployed before disposal. Because the deployment of an air bag is through an explosive process, improper disposal may result in injury and in fines. A deployed air bag should be disposed of in a manner consistent with Environmental Protection Agency **(EPA)** and manufacturer procedures.
10. The air bag electrical system is identified by its yellow wires, connectors, and wire conduit. Never use a test light, a battery- or AC-powered voltmeter, an ohmmeter, or any other type of test equipment that is not specified for use by the service manual on these circuits.

ANTILOCK BRAKE SYSTEM (ABS) SAFETY

Slightly different braking techniques are required for vehicles with a four-wheel **antilock brake system (ABS)** and those with a rear-wheel-only ABS. Rear-wheel antilock brake systems are found only on light trucks and allow the driver to maintain steering control while preventing the rear wheels from locking. The front wheels can still lock. If this happens, the driver should ease up on the brake pedal just enough to let the front wheels start rolling again and maintain steering control.

A four-wheel ABS keeps all four wheels from locking under heavy braking or on slippery roads. Because the front wheels are kept from locking up, the driver should observe these guidelines when driving a vehicle with a four-wheel ABS:

- Maintain firm and continuous pressure on the pedal during braking. *Do not pump* the brakes.
- Always allow adequate stopping distance (at least two car lengths) from the vehicle in front of you. Do not tailgate. The ABS does not decrease stopping distances for normal braking without wheel lockup.
- Always brake and steer in an emergency stopping situation. Maintain continuous pressure on the brake pedal and steer normally. Releasing the pedal will disengage the antilock action.
- Expect noise and vibration from the brake pedal when the antilock system is operating. These sensations are normal.

ABS Hydraulic Pressure Safety

Many ABSs generate extremely high brake fluid pressures that range from 2000 to 3000 pounds per square inch (psi). Failure to fully depressurize the hydraulic accumulator of an ABS before servicing any part of the system could cause severe injury from high-pressure brake fluid escaping from a service connection. Always follow the manufacturer's procedure for depressurizing the system. The typical procedure includes making sure that the ignition switch is in the off position and disconnecting the negative (–), or ground, battery cable. Pump the brake pedal with medium pressure at least 25–50 times or until you feel a definite increase in pedal pressure. Pump the pedal a few more times to ensure complete relief of hydraulic pressure from the system.

Hydraulic Power-Assisted Brakes

Brake systems with hydraulic power-assist units also have high operating pressures that can range from 700 to 1000 pounds per square inch. The source of this pressure is the power steering pump. Like the ABS, hydraulically assisted power brake systems use an accumulator to store brake fluid under pressure. The method of relieving this pressure is also similar to that for ABS.

HAZARDOUS MATERIALS

Many solvents and other chemicals used in an auto shop have warning and caution labels that should be read and understood by everyone that uses those materials. These products are typically considered hazardous materials. Also, many service procedures generate what are known as **hazardous wastes**. Dirty solvents and liquid cleaners are good examples of these.

Every employee in a shop is protected by **right-to-know laws** concerning hazardous materials and wastes. The general intent of these laws is for employers to provide a safe working place, as it relates to hazardous materials. All employees must be trained about their rights under the legislation, the nature of the hazardous chemicals in their workplace, the labeling of chemicals, and the information about each chemical listed and described on **material safety data sheets (MSDS)**. These sheets are available from the manufacturers and suppliers of the chemicals. They detail the chemical composition and precautionary information for all products that can present health or safety hazards.

Employees must be familiar with the intended purposes of the substance, the recommended protective equipment, accident and spill procedures, and any other information regarding the safe handling of hazardous materials. This training must be given annually to employees and provided to new employees as part of their job orientation. The Canadian equivalents to the MSDS are called workplace hazardous materials information systems (WHMIS).

All hazardous materials should be properly labeled, indicating what health, fire, or reactivity hazard they pose and what protective equipment is necessary when handling them. The manufacturer of hazardous materials must provide all warnings and precautionary information, which must be read and understood by all users before they use them. You should pay great attention to the label information **(Figure 7)**. By doing so, you will use the substances in

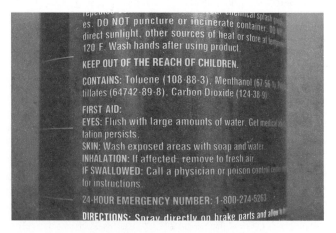

Figure 7. The label on this can of brake fluid lists the hazards, warnings, and first aid information for the fluid.

the proper and safe way, thereby preventing hazardous conditions.

A list of all hazardous materials used in the shop should be posted for the employees to see. Shops must maintain documentation on the hazardous chemicals in the workplace, proof of training programs, records of accidents or spill incidents, satisfaction of employee requests for specific chemical information via the MSDS, and a general right-to-know compliance procedure manual used within the shop.

> **You Should Know** *When handling any hazardous material, always wear the appropriate safety protection. Always follow the correct procedures while using the material and be familiar with the information given on the MSDS for that material.*

Many government agencies are charged with ensuring safe work environments for all workers. These include the Occupational Safety and Health Administration (**OSHA**), the Mine Safety and Health Administration (MSHA), and the National Institute for Occupational Safety and Health (NIOSH). These, in addition to state and local governments, have instituted regulations that must be understood and followed. Everyone in a shop has the responsibility for adhering to these regulations.

OSHA

In 1970, OSHA was formed by the federal government to "assure safe and healthful working conditions for working men and women; by authorizing enforcement of the standards developed under the Act; by assisting and encouraging the States in their efforts to assure safe and healthful working conditions by providing research, information, education, and training in the field of occupational safety and health."

Safety standards have been established that will be consistent across the country. It is the employers' responsibility to provide a place of employment that is free from all recognized hazards and that will be inspected by government agents knowledgeable in the law of working conditions. OSHA controls all safety and health issues of the automotive industry.

Strict rules and regulations of OSHA and the EPA help to promote safety in the auto shop. These are described throughout this text whenever they are applicable. Maintaining a vehicle involves handling and managing a wide variety of materials and wastes. Some of these wastes can be toxic to fish, wildlife, and humans when improperly

managed. No matter the amount of waste produced, it is to the shop's legal and financial advantage to manage the wastes properly and, even more importantly, to prevent pollution.

HANDLING SHOP WASTES

Shop wastes are those things that are left over when a service procedure is completed. Some of these wastes must be treated as potentially hazardous and should be handled and disposed of accordingly.

- Recycle engine oil. Set up equipment, such as a drip table or a screen table with a used oil collection bucket, to collect oils dripping off parts. Place drip pans underneath vehicles that are leaking fluids onto the storage area. Never allow oil to enter into the shop's sewage or drain system.

- Drain used oil filters for at least 24 hours and then crush, and recycle them.

- Recycle old batteries by sending them to a reclaimer or back to the distributor.

- Collect metal filings when machining metal parts. Keep them separate from other dirt and recycle the metal, if possible. Prevent metal filings from falling into a storm sewer drain.

- Recover and/or recycle refrigerants during the service and disposal of motor vehicle air conditioners and refrigeration equipment. It is illegal to knowingly vent refrigerants to the atmosphere. Recovery and/or recycling during service must be performed by an EPA-certified technician using certified equipment and following specified procedures.

- Replace hazardous chemicals with less toxic alternatives that have equal performance. For example, substitute water-based cleaning solvents for petroleum-based solvent degreasers. Hire a hazardous waste management service to clean and recycle solvents. Store solvents in closed containers to prevent evaporation. Properly label spent solvents.

- Store materials such as scrap metal, old machine parts, and worn tires under a roof or tarpaulin to protect them from the elements and to prevent the potential to create contaminated runoff. Consider recycling tires by retreading them.

- Collect and recycle coolants from radiators. Store transmission fluids, brake fluids, and solvents containing chlorinated hydrocarbons separately and recycle or dispose of them properly.

- Keep dirty shop rags in a closed container marked "Contaminated Shop Towels Only." To reduce costs and liabilities associated with disposal of used towels, which can be classified as hazardous wastes, investigate using a laundry service that is able to treat the wastewater generated from cleaning the towels.

Summary

- Shop safety is the responsibility of everyone in the shop, and everyone must work together to protect the health and welfare of everyone in the shop.
- Eye protection should be worn whenever you are working in the shop.
- If chemicals or solvents get into your eyes, flush them continuously with clean water. Have someone call a doctor and get medical help immediately.
- Dress to protect yourself and to avoid accidents. Wear well-fitted and comfortable clothing and foot and hand protection. Never wear anything that may get caught in moving parts or can conduct electricity.
- When lifting heavy objects, use a hoist or have someone help you, and always lift with your legs, not your back.
- Gasoline, diesel fuel, cleaning solvents, and dirty rags should be treated as potential firebombs and handled and stored properly.
- Know where the fire extinguishers and fire alarms are located in the shop and know how to use them. (Think of the word PASS.)
- Inspect all tools and equipment before using them. Check them for unsafe conditions.
- Never use tools and equipment for purposes other than those they are designed for.
- Make sure the contacts of a lift are properly positioned on the vehicle before lifting it.
- Use jack stands to secure a vehicle after it has been raised by a jack.
- When necessary, disconnect the negative cable of the battery before you work around the battery or disconnect any electrical wire or component.
- Follow the proper procedures for eliminating static when handling parts that are static sensitive.
- Know the location and contents of the shop's first aid kits and eyewash stations and the list of emergency telephone numbers.
- Whenever working on or around air bag systems, it is important to follow certain procedures.
- When driving a vehicle with a four-wheel ABS, always maintain firm and continuous pressure on the pedal during braking.
- Handle and dispose of all hazardous wastes according to local laws and your own common sense.
- Know where the MSDS can be found in the shop, and know how to quickly find the appropriate information on them.

Review Questions

1. When should you wear eye protection?
2. What is the correct procedure for putting out a fire with an extinguisher?
3. What do all employees of a shop have the right to know?
4. The general intent of _____-____-_____ laws is for employers to provide a safe working place as it relates to an awareness of hazardous materials.
5. Technician A says that accidents can be prevented by not having anything dangle near rotating equipment and parts. Technician B says that most accidents can be prevented by using common sense. Who is correct?
   A. Technician A only
   B. Technician B only
   C. Both Technician A and Technician B
   D. Neither Technician A nor Technician B
6. Technician A says that unsafe equipment should have its power disconnected and marked with a sign to warn others not to use it. Technician B says that all equipment should be inspected for safety hazards before being used. Who is correct?
   A. Technician A only
   B. Technician B only
   C. Both Technician A and Technician B
   D. Neither Technician A nor Technician B
7. While discussing ways to create an accident-free work environment Technician A says that everyone in the shop should take full responsibility for ensuring safe work areas; Technician B says that the appearance and work habits of technicians can help prevent accidents. Who is correct?
   A. Technician A only
   B. Technician B only
   C. Both Technician A and Technician B
   D. Neither Technician A nor Technician B
8. While discussing a car's electrical system Technician A says that you should always disconnect the negative or ground battery cable first, then disconnect the positive cable. Technician B says that you should always connect the positive battery cable first, then the negative. Who is correct?
   A. Technician A only
   B. Technician B only
   C. Both Technician A and Technician B
   D. Neither Technician A nor Technician B
9. If a chemical gets into someone's eye, what should you do?
10. What is the typical procedure for relieving pressure in ABSs?

Chapter 2

Measuring Systems, Fasteners, and Measuring Tools

Introduction

Servicing automotive brake systems requires the use of various tools. Many of these tools are used to remove and install fasteners. Fasteners of different sizes and shapes are used on today's cars. Basically, the correct tool for the job is the tool that fits the fastener. The size of both tools and fasteners is expressed according to increments defined by a measuring system.

In this chapter, the most common measuring systems are explained. These systems are then used to describe various fasteners. Also covered in this chapter are the commonly used measuring instruments. The proper use of these instruments ensures that parts are installed properly and according to specifications.

MEASURING SYSTEMS

Two different systems of weights and measures are currently used in the United States—the United States Customary System (USCS) and the International System (SI), which is commonly referred to as the metric system. The USCS units of measurement were brought to the United States by the original English settlers and are sometimes referred to as the English System. The United States is slowly changing over to using the metric system. During the changeover, cars produced in the United States are being made with both English and metric fasteners and specifications. Most of the world outside the United States uses the metric system.

In the USCS, linear measurements are measured by inches, feet, and yards. The inch can be broken down into fractions, such as $1/64$, $1/32$, $1/16$, $1/8$, $1/4$, and $1/2$ of an inch. The inch is also commonly broken down by decimals. When an inch is divided into tenths, each part is a tenth of an inch (0.1 inch). Tenths of an inch can be further divided by ten into hundredths of an inch (0.01 inch). The division after hundredths is thousandths (0.001 inch); these are followed by ten-thousandths (0.0001 inch).

In the metric system, the basic measurement unit of length is the meter. For exact measurements, the meter is divided into units of ten. The first division is called a decimeter (dm), the second division is the centimeter (cm), and the third and most commonly used division is the millimeter (mm). One decimeter is equal to 0.1 meter, 1 centimeter equals 0.01 meter, and 1 millimeter equals 0.001 meter.

Because some vehicles have metric fasteners, some have USCS fasteners, and others have both, automotive technicians must have both English and metric tools. Vehicle specifications are normally listed in both meters and inches; therefore, measuring tools are available in both measuring systems.

FASTENERS

Fasteners are those things that are used to secure or hold parts of something together. Many types and sizes of fasteners are commonly used. Each fastener is designed for a specific purpose. One of the most commonly used types of fastener is the threaded fastener. Threaded fasteners include bolts, nuts, screws, and similar items that allow a technician to install or remove parts easily.

Threaded fasteners are available in many sizes, designs, and threads. The threads can be either cut or rolled into the fastener. Rolled threads are 30 percent stronger than cut threads. They also offer better fatigue resistance because there are no sharp notches to create stress points. Fasteners are made to Imperial or metric measurements. The four classifications for the threads of USCS fasteners are: unified national coarse (UNC), unified national fine (UNF), unified national extra fine (UNEF), and unified national pipe thread (UNPT or NPT). Metric fasteners are also available in fine and coarse threads. Coarse threads are used for general-purpose work, especially that in which rapid assembly and disassembly are required. Fine-threaded fasteners are used where greater holding force is necessary. They are also used where greater resistance to vibration is desired.

Bolts have a head on one end and threads on the other. Bolts are identified by their head size, shank diameter, thread pitch, length, and grade **(Figure 1)**. Bolts have a shoulder below the head, and the threads do not travel all the way from the head to the end of the bolt.

Studs are rods with threads on both ends. Most often, the threads on one end are coarse, whereas the other end is fine thread. One end of the stud is screwed into a threaded bore. A hole in the part to be secured is fitted over the stud and held in place with a nut, screwed over the stud. Studs are used when the clamping pressures of a fine thread are needed and a bolt will not work. If the material the stud is being screwed into is soft (such as aluminum) or granular (such as cast iron), fine threads will not withstand a great amount of pulling force on the stud. Therefore, a coarse thread is used to secure the stud in the work piece, and a fine-threaded nut is used to secure the other part to it. Doing this allows the clamping force of fine threads and the holding power of coarse threads.

Nuts are used with other threaded fasteners when the fastener is not threaded into a piece of work. Many different designs of nuts are found on today's cars. The most common one is the hex nut, which is used with studs and bolts and is tightened with a wrench.

BOLT IDENTIFICATION

The **bolt head** is used to loosen and tighten the bolt; a socket or wrench fits over the head and is used to screw the bolt in or out. The size of the bolt head varies with the diameter of the bolt and is available in USCS and metric wrench sizes. Many confuse the size of the head with the size of the bolt. The size of a bolt is determined by the diameter of its shank. The size of the bolt head determines what size wrench is required to screw it. Bolt diameter is the measurement across the major diameter of the threaded area, or across the bolt shank. The length of a bolt is measured from the bottom surface of the head to the end of the threads.

The **thread pitch** of a bolt in the Imperial system is determined by the number of threads that are in 1 inch of the threaded bolt length and is expressed in number of threads per inch. A UNF bolt with a 3/8-inch diameter would be a 3/8 × 24 bolt. It would have 24 threads per inch. Likewise a 3/8-inch UNC bolt would be called a 3/8 × 16. The distance, in millimeters, between two adjacent threads determines the thread pitch in the metric system. This distance will vary between 1.0 and 2.0 and depends on the diameter of the bolt. The lower the number, the closer the threads are placed and the finer the threads are.

The bolt's tensile strength, or grade, is the amount of stress or stretch it is able to withstand before it breaks. The type of material the bolt is made of and the diameter of the bolt determine its grade. In the English system, the tensile strength of a bolt is identified by the number of radial lines **(grade marks)** on the bolt's head. More lines mean higher tensile strength. Count the number of lines and add two to determine the grade of a bolt.

A property class number on the bolt head identifies the grade of metric bolts. This numerical identification is

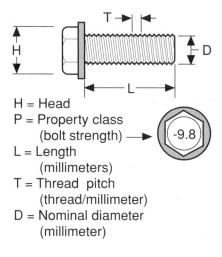

H = Head
G = Grade marking
    (bolt strength) →
L = Length
    (inches)
T = Thread pitch
    (thread/inch)
D = Nominal diameter
    (inches)

A

H = Head
P = Property class
    (bolt strength) →
L = Length
    (millimeters)
T = Thread pitch
    (thread/millimeter)
D = Nominal diameter
    (millimeter)

B

Figure 1. English (A) and metric (B) bolt terminology.

comprised of two numbers. The first number represents the tensile strength of the bolt. The higher the number, the greater the tensile strength. The second number represents the yield strength of the bolt. This number represents how much stress the bolt can take before it is not able to return to its original shape without damage. The second number represents a percentage rating. For example, a 10.9 bolt has a tensile strength of 1000 megapascals (145,000 psi) and a yield strength of 900 megapascals (90 percent of 1000). A 10.9 metric bolt is similar in strength to an SAE grade 8 bolt.

Nuts are graded to match their respective bolts. For example, a grade 8 nut must be used with a grade 8 bolt. If a grade 5 nut were used, a grade 5 connection would result. Grade 8 and critical applications require the use of fully hardened flat washers. These will not dish out when torqued like soft washers will.

Bolt heads can pop off because of **fillet** damage. The fillet is the smooth curve where the shank flows into the bolt head. Scratches in this area introduce stress to the bolt head, causing failure. Removing any burrs around the edges of holes can protect the bolt head. Also, place flat washers with their rounded, punched side against the bolt head and their sharp side to the work surface.

Fatigue breaks are the most common type of bolt failure. A bolt becomes fatigued from working back and forth when it is too loose. Undertightening the bolt causes this problem. Bolts can also be broken or damaged by overtightening, being forced into a nonmatching thread, or bottoming out, which happens when the bolt is too long.

TIGHTENING BOLTS

Any fastener is nearly worthless if it is not as tight as it should be. When a bolt is properly tightened, it will be "spring-loaded" against the part it is holding. This spring effect is caused by the stretch of the bolt when it is tightened. Normally, a properly tightened bolt is stretched to 70 percent of its elastic limit. The elastic limit of a bolt is that point of stretch at which the bolt will not return to its original shape when it is loosened. Not only will an overtightened or stretched bolt not have sufficient clamping force, it will also have distorted threads. The stretched threads will make it more difficult to screw and unscrew the bolt or a nut on the bolt. Always check the service manual to see if there is a torque specification for a bolt before tightening it. If there is, use a torque wrench and tighten the bolt properly.

MEASURING TOOLS

Many of the procedures discussed in this manual require exact measurements of parts and clearances. Accurate measurements require the use of precision measuring devices that are designed to measure things in very small increments. Measuring tools are delicate instruments and should be handled with great care. Never strike, pry, drop, or force these tools. Also, make sure you clean them before and after every use.

Many different measuring devices are used by automotive technicians. This chapter covers only those that are commonly used to service brake system components.

Machinist's Rule

The machinist's rule looks very much like an ordinary ruler. Each edge of this basic measuring tool is divided into increments based on a different scale. A typical machinist's rule based on the Imperial system of measurement can have scales based on $\frac{1}{8}$-, $\frac{1}{16}$-, $\frac{1}{32}$-, and $\frac{1}{64}$-inch intervals. Of course, metric machinist rules are also available. Metric rules are usually divided into 0.5- and 1-millimeter increments.

Some machinist rules are based on decimal intervals. These are typically divided into $\frac{1}{10}$-, $\frac{1}{50}$-, and $\frac{1}{1000}$-inch (0.1, 0.02, and 0.001-inch, respectively) increments. Decimal machinist rules are helpful when measuring dimensions that are specified in decimals.

Feeler Gauge

A **feeler gauge** is a thin strip of metal or plastic of known and closely controlled thickness. Several of these metal strips are often assembled together as a feeler gauge set, which looks like a pocket knife **(Figure 2)**. The desired thickness gauge can be pivoted away from others for convenient use. A steel feeler gauge pack usually contains strips or leaves of 0.002- to 0.010-inch thickness (in steps of 0.001 inch) and leaves of 0.012- to 0.024-inch thickness (in steps of 0.002 inch). A feeler gauge can be used by itself to measure clearances and gaps or it can be used with a precision straightedge to measure alignment and surface warpage.

Figure 2.   A feeler gauge set.

Screw Pitch Gauge

The use of a **screw pitch gauge** provides a quick and accurate method of checking the thread pitch of a fastener. The leaves of this measuring tool are marked with the various pitches. To check the pitch of threads, simply match the teeth of the gauge with the threads of the fastener. Read the pitch from the leaf. Screw pitch gauges are available for the various types of fastener threads used by the automotive industry.

Calipers and Dividers

Calipers and dividers transfer lengths to another measuring instrument. These are used to measure inside and outside diameters. To take a measurement with these tools, adjust the legs of the tool to fit the inside or outside diameter of the object. Then lay the caliper or divider over a rule and read the length on the rule **(Figure 3)**.

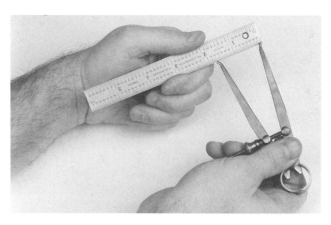

Figure 3. Calipers and dividers transfer lengths to another measuring instrument.

Vernier Caliper

A **vernier caliper** is a measuring tool. It can make inside, outside, or depth measurements. It is marked in both USCS and metric divisions called a vernier scale. A vernier scale consists of a movable scale in parallel with a stationary scale **(Figure 4)**, in this case the vernier bar to the vernier plate.

Each inch of the stationary scale is divided into 10 parts, each equal to 0.100 inch. The area between the 0.100 marks is divided into four, each equaling 0.025 inches. The vernier scale has 25 divisions, each one representing 0.001 inch. Measurement readings are taken by combining the main and vernier scales. At all times, only one division line on the main scale will line up with a line on the vernier scale. This is the basis for accurate measurements.

Dial Caliper

The **dial caliper** is an easier-to-use version of the vernier caliper. USCS calipers commonly measure dimensions from 0 to 6 inches. Metric dial calipers typically measure from 0 to 150 millimeters in increments of 0.02 millimeter. The dial caliper features a depth scale, bar scale, dial indicator, inside measurement jaws, and outside measurement jaws.

The main scale of an Imperial dial caliper is divided into one-tenth (0.1)-inch graduations. The dial indicator is divided into one-thousandth (0.001)-inch graduations. Therefore, one revolution of the dial indicator needle equals one-tenth inch on the bar scale. A metric dial caliper is similar in appearance but the bar scale is divided into 2-mm increments. Additionally, on a metric dial caliper,

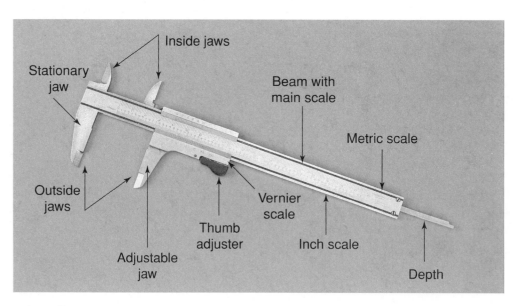

Figure 4. A vernier caliper.

one revolution of the dial indicator needle equals 2 millimeters.

Both English and metric dial calipers use a thumb-operated roll knob for fine adjustment. When you use a dial caliper, always move the measuring jaws backward and forward to center the jaws on the object being measured. Make sure the caliper jaws lie flat on or around the object. If the jaws are tilted in any way, you will not obtain an accurate measurement.

Although dial calipers are precision measuring instruments, they are accurate only to plus or minus two-thousandths (0.002) of an inch. Micrometers are preferred when extremely precise measurements are desired.

Dial Indicator

The **dial indicator** is calibrated in 0.001-inch (one-thousandth inch) increments. Metric dial indicators are also available. Both types are used to measure movement. Common uses of the dial indicator include measuring valve lift, journal concentricity, flywheel or brake rotor runout, gear backlash, and crankshaft endplay. Dial indicators are available with various face markings and measurement ranges to accommodate many measuring tasks.

To use a dial indicator, position the indicator rod against the object to be measured. Push the indicator toward the work until the indicator needle travels far enough around the gauge face to permit movement to be read in either direction **(Figure 5)**. Zero the indicator needle on the gauge. Always be sure that the range of the dial indicator is sufficient to allow the amount of movement required by the measuring procedure. For example, never use a 1-inch indicator on a component that will move 2 inches.

Figure 5.   This dial indicator setup will measure the runout of the brake rotor.

> **You Should Know** *Like all tools, measuring tools should be used only for the purpose they were designed for. Some instruments are not accurate enough for very precise measurements; others are too accurate to be practical for less critical measurements.*

MICROMETERS

The **micrometer** is used to measure the outside diameter of an object or the inside diameter of a bore or drum. Both outside and inside micrometers are calibrated and read in the same manner. Measurements on both are taken with the measuring points in contact with the surfaces being measured.

The major components and markings of a micrometer include the frame, anvil, spindle, locknut, sleeve, sleeve numbers, sleeve long line, thimble marks, thimble, and ratchet. Micrometers are calibrated in either inch or metric graduations and are available in a range of sizes.

To measure small objects with an outside micrometer, open the jaws of the tool and slip the object between the spindle and the anvil **(Figure 6)**. While holding the object against the anvil, turn the thimble, using your thumb and forefinger, until the spindle contacts the object. Use only enough pressure on the thimble to allow the object to just fit between the tips of the anvil and spindle. The object should slip through with only a

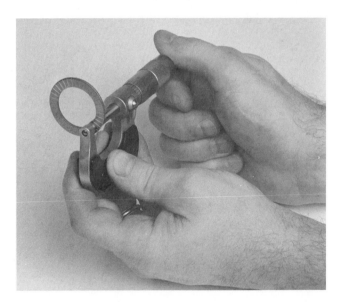

Figure 6.   To measure small objects with an outside micrometer, open the jaws of the tool and slip the object between the spindle and the anvil.

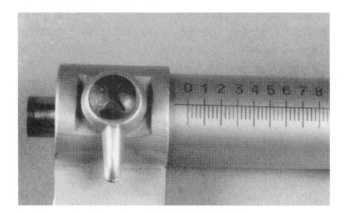

Figure 7. The graduations of a micrometer sleeve represent 0.025 inch.

very slight resistance. When a satisfactory feel is reached, lock the micrometer. Because each graduation on the sleeve represents 0.025 inch **(Figure 7)**, begin reading the measurement by counting the visible lines on the sleeve and multiply that number by 0.025. The graduations on the thimble assembly define the area between the lines on the sleeve; therefore, the number indicated on the thimble should be added to the measurement shown on the sleeve. The sum is the outside diameter of the object.

To measure larger objects, hold the frame of the micrometer and slip it over the object. Turn the thimble while continuing to slip the micrometer over the object until you feel a very slight resistance. Rock the micrometer from side to side while doing this to make sure the spindle cannot be closed any farther. Lock the micrometer and take a measurement reading.

Some technicians use a digital micrometer, which is easier to read. These tools do not have the various scales; rather, the measurement is displayed and read directly off the micrometer.

Outside micrometers come in many different sizes, but the measurement range of the spindle is usually just 1 inch, or 25 millimeters. That is, a 4-inch micrometer measures from 3 to 4 inches. Special micrometers are made for measuring brake rotor thickness. The throats on these micrometers are deeper than those on standard micrometers to allow for the diameter of the rotor. The anvil and spindle of a rotor micrometer are usually pointed to allow measurements at the deepest points of any grooves in the rotor.

Reading a Metric Outside Micrometer

The metric micrometer is read in the same manner as the inch-graduated micrometer, except that the graduations are expressed in the metric system of measurement. Readings are obtained as follows.

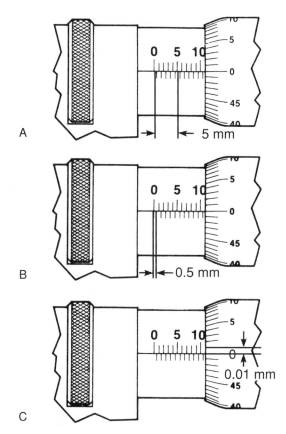

Figure 8. Reading a metric micrometer: 5 millimeters (A) plus 0.5 millimeter (B) plus 0.01 millimeter (C) equals 5.51 millimeters.

1. Each number on the sleeve of the micrometer represents 5 millimeters (mm), or 0.005 meter (m) **(Figure 8A)**.
2. Each of the ten equal spaces between each number, with index lines alternating above and below the horizontal line, represents 0.5 millimeter, or five-tenths of a millimeter. One revolution of the thimble changes the reading one space on the sleeve scale, or 0.5 millimeter **(Figure 8B)**.
3. The beveled edge of the thimble is divided into 50 equal divisions with every fifth line numbered: 0, 5, 10 . . . 45. Because one complete revolution of the thimble advances the spindle 0.5 millimeter, each graduation on the thimble is equal to one-hundredth of a millimeter **(Figure 8C)**.
4. As with the inch-graduated micrometer, the three separate readings are added together to obtain the total reading.

Reading a Depth Micrometer

A depth micrometer is used to measure the distance between two parallel surfaces. The sleeves, thimbles, and ratchet screws operate in the same way as other micrometers. Likewise, depth micrometers are read in the same way

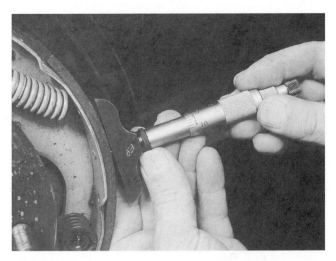

Figure 9.   The thickness of a brake shoe lining can be measured with a depth gauge.

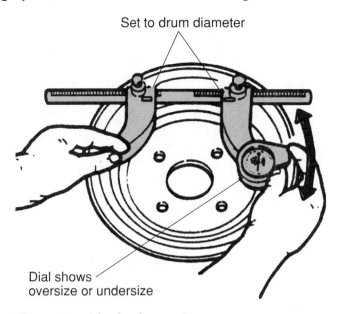

Set to drum diameter

Dial shows oversize or undersize

Figure 10.   A brake drum micrometer.

as other micrometers. A depth gauge is used to check the lining thickness of drum brake shoes **(Figure 9)**. If a depth micrometer is used with a gauge bar, it is important to keep both the bar and the micrometer from rocking. Any movement of either part will result in an inaccurate measurement.

Telescoping Gauge

Telescoping (snap) gauges are used for measuring bore diameters and other clearances. They are available in sizes ranging from fractions of an inch through 6 inches. Each gauge consists of two telescoping plungers, a handle, and a lock screw. Snap gauges are normally used with an outside micrometer. To use the telescoping gauge, insert it into the bore and loosen the lock screw. This will allow the plungers to snap against the bore. Once the plungers have expanded, tighten the lock screw. Remove the gauge and measure the expanse with a micrometer.

Small-Hole Gauge

A small-hole gauge works just like a telescoping gauge; however, it is designed to be used on small bores. After it is placed into the bore and expanded, it is removed and measured with a micrometer. Like the telescoping gauge, the small-hole gauge consists of a lock, handle, and an expanding end. The end expands or retracts by turning the gauge handle.

DRUM MICROMETERS

A drum micrometer is a single-purpose instrument used to measure the inside diameter of a brake drum

(Figure 10). A drum micrometer has two movable arms on a shaft. One arm has a precision dial indicator; the other arm has an outside anvil that fits against the inside of the drum. In use, the arms are secured on the shaft by lock screws that fit into grooves every ⅛ inch (0.125 inch) on the shaft. The dial indicator is graduated in 0.005-inch increments.

To use a drum micrometer, loosen the lock screws and move the arms along the shaft so that the micrometer can fit inside the drum. Extend the arms on the shaft until they align with the increments that indicate the nominal size of the drum. To measure a 10.5-inch drum, set one arm at the 10-inch mark on the shaft and the other arm at the other 10-inch mark, plus four 0.125-inch marks (0.500 inch) for the 10.5-inch drum. Place the micrometer inside the drum and rock it gently on the drum surface until the highest reading is obtained. If the dial indicator reads 0.019 inch, for example, add this reading to the 10.5-inch measurement for a total diameter of 10.519 inches. Always take 6–8 measurements around the drum circumference to check for drum distortion. Metric drum micrometers work the same way except that the shaft is graduated in 1-centimeter major increments, and the lock screws fit in notches every 2 millimeters.

BRAKE SHOE ADJUSTING GAUGES (CALIPERS)

A brake shoe adjusting gauge is an inside-outside measuring device **(Figure 11)**. This gauge is often called a brake shoe caliper. During drum brake service, the inside part of the gauge is placed inside a newly surfaced drum

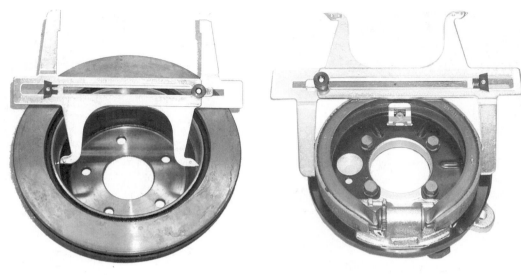

1. Set to drum diameter   2. Adjust correct brake
                                     shoe diameter

Figure 11. A brake shoe adjusting gauge is typically used for preliminary shoe adjustment. In this illustration, the gauge is used to make the only adjustment on this parking brake assembly.

and expanded to fit the drum diameter. The lock screw is then tightened and the gauge moved to the brake shoes installed on the backing plate. The brake shoes are then adjusted until the outside part of the gauge just slips over them. This provides a quick adjustment of the brake shoes.

Summary

- In the USCS, the inch is commonly broken down into fractions or decimals.
- In the metric system, the basic measurement unit of length is the meter, which for exact measurements is divided into units of ten. The most commonly used division is the millimeter (mm).
- The size of a bolt is determined by the diameter of its shank. The size of the bolt head determines what size wrench is required to screw it.
- The length of a bolt is measured from the bottom surface of the head to the end of the threads.

- The thread pitch of a bolt in the Imperial system is determined by the number of threads that are in 1 inch of the threaded bolt length and is expressed in number of threads per inch.
- The bolt's tensile strength, or grade, is the amount of stress or stretch a bolt is able to withstand before it breaks.
- Measuring instruments that are commonly used to service brake systems include the machinist's rule, screw pitch gauge, dial indicator, and micrometer.

Review Questions

1. Technician A says that the USCS measuring system is based on the inch and fine measurements are expressed in fractions of an inch or decimals. Technician B says that the metric system is based on the meter and fine measurements are expressed in tenths, hundredths, and thousandths. Who is correct?
   A. Technician A only
   B. Technician B only
   C. Both Technician A and Technician B
   D. Neither Technician A nor Technician B

2. Technician A says that 1 millimeter is equal to 0.001 meters. Technician B says that 1 meter is equal to 0.03937 inches. Who is correct?
   A. Technician A only
   B. Technician B only
   C. Both Technician A and Technician B
   D. Neither Technician A nor Technician B

3. While discussing automotive fasteners Technician A says that bolt sizes are listed by their appropriate wrench size; Technician B says that whenever bolts are replaced, they should be replaced with exactly the same size and type as the manufacturer installed. Who is correct?
   A. Technician A only
   B. Technician B only
   C. Both Technician A and Technician B
   D. Neither Technician A nor Technician B

4. While discussing the purpose of micrometers Technician A says that micrometers are used to measure the diameter of an object; Technician B says that outside micrometers are used to measure the outside diameter of an object, whereas inside micrometers are used to measure the inside diameter. Who is correct?
   A. Technician A only
   B. Technician B only
   C. Both Technician A and Technician B
   D. Neither Technician A nor Technician B

5. Technician A says that micrometers are commonly used to measure the thickness of a brake rotor. Technician B says that special micrometers are used to measure brake rotors. Who is correct?
   A. Technician A only
   B. Technician B only
   C. Both Technician A and Technician B
   D. Neither Technician A nor Technician B

6. Technician A says that a bolt's tensile strength is the amount of stress or stretch a bolt is able to withstand before it breaks. Technician B says that bolts are identified by their head size, shank diameter, thread pitch, length, and grade. Who is correct?
   A. Technician A only
   B. Technician B only
   C. Both Technician A and Technician B
   D. Neither Technician A nor Technician B

7. Technician A uses a dial caliper to take inside and outside measurements. Technician B uses a dial caliper to take depth measurements. Who is correct?
   A. Technician A only
   B. Technician B only
   C. Both Technician A and Technician B
   D. Neither Technician A nor Technician B

8. List five features of a bolt that are commonly used to identify it.

9. Explain the two primary purposes of a feeler gauge.

10. One millimeter is equal to how many inches?

Chapter 3

Common Hand Tools

Introduction

Although every technician's toolbox contains many different tools, certain hand tools are a must. These are described in this chapter.

WRENCHES

A basic tool set should include a set of box and open-end wrenches. A box-end wrench **(Figure 1)** completely encircles a nut or the head of a bolt and is less apt to slip and cause damage or injury. Box-end wrenches are available in either 6 or 12 points. Twelve-point box-end wrenches allow you to work in tighter areas than do 6-point wrenches. Often, it is difficult to place a box-end wrench around a nut or bolt because of its surroundings. An open-end wrench has an open squared end and often can be used where a box-end wrench will not fit. Open- and box-end wrenches have different sizes at either end.

You should have a complete set of both open- and box-end wrenches in your tool set. However, to reduce cost and storage space, you may want to have a set of combination wrenches, which have an open-end wrench on one end and a box-end wrench on the other. Both ends of these wrenches are sized the same and can be used interchangeably on the same nut or bolt. To be able to work on today's brake systems, you should have complete sets of both USCS and metric wrenches.

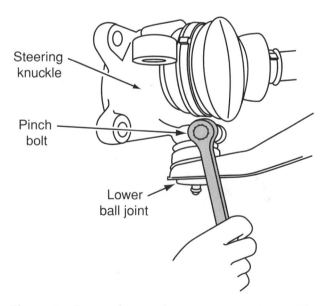

Figure 1. Box-end wrenches are commonly used by driveline technicians.

> **You Should Know** *Metric and USCS wrenches are not interchangeable. For example, a 9/16-inch wrench is 0.02 inch larger than a 14-millimeter nut. If the 9/16-inch wrench is used to turn or hold a 14-millimeter nut, the wrench will probably slip. This may cause the points of the bolt head or nut to round off and can possibly cause skinned knuckles.*

Allen wrenches, or hex-head wrenches, are used to tighten and loosen setscrews. The appropriately sized wrench fits into a machined hex-shaped recess in the bolt or screw.

Flare-Nut (Line) and Bleeder Screw Wrenches

Flare-nut, or line, wrenches **(Figure 2)** should be used to loosen or tighten brake line or tubing fittings. Using open-ended wrenches on these fittings will tend to round the corners of the nut, which are typically made of soft metal and can distort easily. Flare-nut wrenches surround the nut and provide a better grip on the fitting. They have a section cut out so that the wrench can be slipped around the brake line and dropped over the flare nut.

Special bleeder screw wrenches often are used to open bleeder screws **(Figure 3)**. Bleeder screw wrenches are small, 6-point box-end wrenches with strangely offset

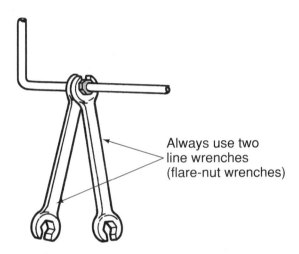

Always use two line wrenches (flare-nut wrenches)

Disconnecting hydraulic lines

Figure 2. Line wrenches surround the nut and provide a better grip on the fitting.

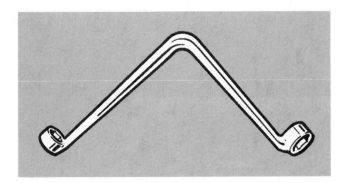

Figure 3. Bleeder screw wrenches are specially made to allow for a good grip on the bleeder screw and a workable position to loosen and tighten the screw.

handles for access to bleeder screws in awkward locations. The 6-point box end grips the screw more securely than a 12-point box-end wrench can and avoids damage to the screw.

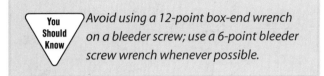

You Should Know *Avoid using a 12-point box-end wrench on a bleeder screw; use a 6-point bleeder screw wrench whenever possible.*

RATCHETS AND SOCKETS

A set of USCS and metric sockets combined with a ratchet handle and a few extensions should also be included in your tool set. These sockets should be ³⁄₈-inch drive, although ¹⁄₄- and ¹⁄₂-inch drive sets are also handy. The ratchet allows you to turn the socket in one direction with force and in the other direction without force. This allows you to tighten or loosen a bolt without removing and resetting the wrench after you have turned it. You may also want a long breaker bar to fit your sockets. Breaker bars offer increased leverage when loosening very tight bolts.

There are many designs of sockets for most sizes. A 6-point socket has stronger walls and improved grip on a bolt compared with a normal 12-point socket. However, 6-point sockets have half the positions of a 12-point socket. Six-point sockets are mostly used on fasteners that are rusted or rounded. Eight-point sockets are available to use on square nuts or square-headed bolts. Sockets are also available as deep-well sockets that are used to reach a nut when it is on a bolt or stud with long threads.

You Should Know *Deep-well sockets are also good for reaching nuts and bolts in limited-access areas. Deep-well sockets should not be used when a regular-size socket will work. The longer socket develops more twist torque and tends to slip off the fastener.*

Extensions allow you to put the handle in the best position while working. They range from 1 inch to 3 feet long. Also available are universal joints that allow a technician to work a bolt or nut at a slight angle.

TORQUE WRENCHES

Torque wrenches (Figure 4) measure how tight a nut or bolt is. Many of the vehicle's nuts and bolts should

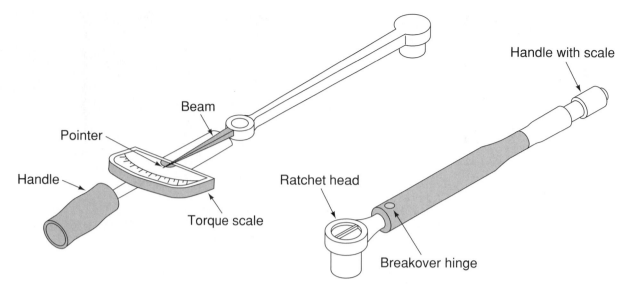

Figure 4. The desired torque is dialed in on a breakover-type torque wrench, and the wrench makes an audible click when you have reached the correct force.

be tightened to a certain amount and have a torque specification, which is expressed in foot-pounds (USCS) or Newton-meters (metric). A foot-pound is the work or pressure accomplished by a force of 1 pound through a distance of 1 foot. A Newton-meter is the work or pressure accomplished by a force of 1 kilogram through a distance of 1 meter.

Torque wrenches come with drives that correspond with sockets: ¼, ⅜, and ½ inch. The four types of torque wrenches are the dial type, the breakover type, the torsion bar type, and the digital readout type. For most brake work, a dial, torsion bar, or digital readout type is recommended. These have a scale that can be read and used to measure turning effort, as well as tightening bolts. With the breakover type, you must dial in the desired torque. The wrench makes an audible click when you have reached the correct force.

SCREWDRIVERS

Many styles of screwdrivers are available. The commonly used styles are the standard tip, or blade style, and Phillips types. A blade style fits into a straight slot in the head of the screw. A Phillips screwdriver has a cross point that allows for more gripping power and stability. The cross point has four surfaces that insert into a like pattern in the head of the screw, making the screwdriver less likely to slip out of the screw. Your tool set should include both blade and Phillips drivers in a variety of lengths from 2-inch "stubbies" to 12-inch screwdrivers. Some vehicles may require special screwdrivers, such as those with a square, clutch, or Torx head design.

PLIERS

The two most commonly used pliers are interlocking jaw pliers that are about eight or nine inches long and diagonal cutters about seven inches long, which can be used to cut wire and remove cotter pins. Other designs are also used while servicing a vehicle's drivetrain. These include: slip-joint, needle-nose, duck-bill, adjustable-joint, and offset needle-nose pliers. It is also recommended that you have a pair of vise-grip pliers to hold parts while grinding or to use as a "third hand."

HAMMERS AND MALLETS

Your tool set should include at least three hammers: two ball-peen hammers (one 8-ounce and one 12- to 16-ounce hammer) and a small sledgehammer. You should also have a plastic and lead or brass-faced mallet. Hammers are used with punches and chisels, and mallets are used for tapping parts apart or aligning parts together. Soft-faced mallets will not harm the part they are hitting against, whereas a hammer will.

PUNCHES AND CHISELS

Your tool set should include a variety of drift punches and starter punches. Drift punches are used to remove drift and roll pins **(Figure 5)**. Some drift punches are made of brass; these should be used whenever you are concerned about possible damage to the pin or to the surface surrounding the pin. Tapered punches are used to line up bolt-holes. Starter or center punches are used to make an indent

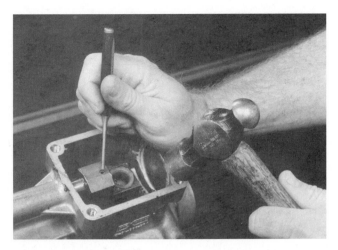

Figure 5.   Drifts and punches have many purposes; one is to drive retaining roll pins out of their bores.

Figure 6.   Service manuals are important tools.

before drilling to prevent the drill bit from wandering. A variety of chisels is also recommended. These should include flat, cape, round-nose cape, and diamond-point chisels.

FILES

A set of files should be included in your tool set. Files are used to remove metal and to deburr parts. Files are available in many different shapes: round, half-round, flat, crossing, knife, square, and triangular. The most commonly used files are the half-round and flat with either single-cut or double-cut designs. A single-cut file has its cutting grooves lined up diagonally across the face of the file. The cutting grooves of a double-cut file run diagonally in both directions across the face. Double-cut files are considered first cut or roughening files because they can remove large amounts of metal. Single-cut files are considered finishing files because they remove small amounts of metal.

TAPS AND DIES

Often, problems are caused by defective fasteners or damaged threads in the bore of an assembly. Fasteners can be replaced or their threads restored with a **die**. A **tap** can cut and restore the threads in a bore. It is recommended that you have two sets of taps and dies: one USCS and one metric.

SERVICE INFORMATION AND MANUALS

Service manuals **(Figure 6)** are essential for safe, complete brake service. They are needed to obtain specifica-

tions on torque values and critical measurements such as drum and rotor discard limits. Manuals also provide drawings and photographs that show where and how to perform service procedures on a particular vehicle. Special tools or instruments also are listed and shown when they are required. Precautions are given to prevent injury or damage to parts. Perhaps the most important tools you will use are service manuals.

Most automobile manufacturers publish a service manual or set of manuals for each model and year of their cars. These manuals provide the best and most complete information for those cars. The most commonly used manuals are comprehensive service manuals published by compilers of information. Various editions are available that cover different ranges of model years for both domestic and imported cars. Although similar to manufacturers' manuals in many ways, they do not provide as much information or detail as do the manufacturers' manuals.

Although the manuals from different publishers vary in presentation and arrangement of topics, all service manuals are easy to use after you become familiar with their organization. Most shop manuals are divided into a number of sections, each covering different aspects of the vehicle. The beginning sections commonly provide vehicle identification and basic maintenance information. The remaining sections deal with each different vehicle system in detail, including diagnostic, service, and overhaul procedures. Each section has an index indicating more specific areas of information.

To use a service manual:
1. Select the appropriate manual for the vehicle being serviced.
2. Use the table of contents and index to locate the section that applies to the work being done.
3. Carefully read the information and study the applicable illustrations and diagrams.

4. Follow all of the required steps and procedures given for that service operation.

5. Adhere to all of the given specifications and perform all measurement and adjustment procedures with accuracy and precision.

Throughout this book, you are told to refer to the appropriate shop manual to find the correct procedures and specifications. Although the various systems of all automobiles function in much the same way, there are many variations in design. Each design has its own set of repair and diagnostic procedures. Therefore, it is important that you always follow the recommendations of the manufacturer to identify and repair problems.

Because many technical changes occur on specific vehicles each year, manufacturers' service manuals need to be constantly updated. Updates are published as service bulletins (often referred to as technical service bulletins, or TSBs) that show the changes in specifications and repair procedures during the model year. The car manufacturer provides these bulletins to dealers and often to repair facilities on a regular basis.

Aftermarket Suppliers' Guides and Catalogs

Many of the larger parts manufacturers have excellent guides on the various parts they manufacture or supply. They also provide updated service bulletins on their products. Other sources for up-to-date technical information are trade magazines and trade associations.

Flat-Rate Manuals

Flat-rate manuals contain standards for the length of time a specific repair is supposed to require. Normally, they also contain a parts list with approximate or exact prices of parts. They are excellent for making cost estimates and are published by the manufacturers and independents.

Computer-based Information

The same information that is available in service manuals and bulletins is also available electronically: on CD-ROMs

Figure 7. The use of CD-ROMs and a computer makes accessing information quick and easy.

(**Figure 7**), DVDs, and the Internet. This is the most common way dealerships and independent repair shops access service information. A single compact disk can hold a quarter million pages of text. This eliminates the need for a huge library to contain all of the printed manuals. Using electronics to find information is also easier and quicker. The disks are normally updated monthly and contain not only the most recent TSBs but also engineering and field service fixes. On-line data can be updated instantly and require no space for physical storage.

Hotline Services

Many companies provide on-line help to technicians. As the complexity of the automobile grows, so does the popularity of these services. Technicians can call and talk to experts who are technicians familiar with the different systems of certain manufacturers. Armed with various factory service manuals, general service manuals, and electronic data sources, these brand experts give information to technicians to help them through diagnostic and repair procedures.

Summary

- A box-end wrench completely encircles a nut or the head of a bolt.
- An open-end wrench has an open squared end.
- A ratchet allows you to turn the socket in one direction with force and in the other direction without force.

- Torque wrenches measure how tight a nut or bolt is. Many of a vehicle's nuts and bolts should be tightened to a certain amount and have a torque specification.
- The most common screwdrivers are the standard tip, or blade style, and Phillips types.

- The two most commonly used pliers are interlocking jaw pliers and diagonal cutters, used to cut wire and remove cotter pins.
- Hammers are used with punches and chisels, and mallets are used for tapping parts apart or aligning parts together.
- Tapered punches are used to line up boltholes. Starter or center punches are used to make an indent before drilling to prevent the drill bit from wandering.

- Fasteners can be replaced or their threads restored with a die. A tap can cut and restore the threads in a bore.
- Service manuals provide the procedures and specifications needed to repair an automobile correctly.
- Flat-rate manuals contain standards for the length of time a specific repair is supposed to require.

Review Questions

1. Which of these wrenches has the least chance of slipping off the head of a bolt while heavy pressure is put on the wrench: open-end, flare nut, or box-end?
2. What is the basic difference between a drift punch and a starter punch?
3. Describe the appearance of a bleeder screw wrench.
4. Why should you not use a deep-well socket when a regular-size socket will work?
5. Describe how most service manuals are divided and how information is organized.
6. Technician A says that a single-cut file has its cutting grooves lined up diagonally across the face of the file. Technician B says that single-cut files are considered first cut or roughening files because they can remove large amounts of metal. Who is correct?
   A. Technician A only
   B. Technician B only
   C. Both Technician A and Technician B
   D. Neither Technician A nor Technician B
7. While discussing the purpose of torque wrenches Technician A says that they are used to tighten fasteners to a specified torque; Technician B says that they are used for added leverage while loosening or tightening a bolt. Who is correct?
   A. Technician A only
   B. Technician B only
   C. Both Technician A and Technician B
   D. Neither Technician A nor Technician B

8. Technician A says that pliers are commonly used to hold items. Technician B says that pliers can be used to tighten or loosen nuts and bolts. Who is correct?
   A. Technician A only
   B. Technician B only
   C. Both Technician A and Technician B
   D. Neither Technician A nor Technician B
9. Technician A says that one of the most important tools for a technician is a service manual. Technician B says that service manuals are not needed if a good selection of TSBs is available. Who is correct?
   A. Technician A only
   B. Technician B only
   C. Both Technician A and Technician B
   D. Neither Technician A nor Technician B
10. Technician A says that an open-end wrench should be used to loosen or tighten brake line or tubing fittings. Technician B says that flare-nut wrenches should only be used on a hardened bolt. Who is correct?
    A. Technician A only
    B. Technician B only
    C. Both Technician A and Technician B
    D. Neither Technician A nor Technician B

Chapter 4

Power and Special Tools

Introduction

Power tools make a technician's job easier. They operate faster and with more torque than hand tools. However, power tools require greater safety measures. Power tools do not stop unless they are turned off. Power is furnished by air (pneumatic), electricity, or hydraulic fluid. Technicians typically use pneumatic tools because they have more torque, weigh less, and require less maintenance than electric power tools. However, electric power tools tend to cost less than the pneumatics. Electric power tools can be plugged into most electric wall sockets, but to use a pneumatic tool, you must have an air compressor and an air storage tank.

Carelessness with or mishandling of power tools can cause serious injury. Make sure you know how to operate a tool before using it.

Properly diagnosing and servicing brake systems requires many different special tools. These are also discussed in this chapter. Without these special tools, performing your work properly can become next to impossible.

CLEANING EQUIPMENT AND CONTAINMENT SYSTEMS

The following systems and methods are used to safely contain brake dust while you are doing brake work.

Negative-Pressure Enclosure and HEPA Vacuum Systems

A negative-pressure enclosure is a containment system in which cleaning and the inspection of brake assemblies is performed inside a tightly sealed protective enclosure that covers and contains the entire brake assembly **(Figure 1)**. The enclosure prevents the release of asbestos fibers into the air. It is designed so that you can clearly see the work in progress. It has impermeable sleeves and gloves that let you do what you need to do.

A vacuum pump and a high-efficiency particulate air filter **(HEPA filter)** keep the enclosure under negative pressure as work is done. Because particles cannot escape the enclosure, compressed air can be used to remove dust, dirt, and asbestos fibers from brake parts. The HEPA vacuum also can be used to loosen the asbestos-containing residue from the brake parts. Once the asbestos is loose, draw it out of the enclosure with the vacuum pump. The dust is then trapped in the vacuum cleaner filter.

Asbestos waste must be collected, recycled, and disposed of in sealed impermeable bags or other closed, impermeable containers. Any spills or release of asbestos-containing waste material from inside the enclosure or vacuum hose or vacuum filter should be immediately cleaned up using vacuuming or wet cleaning methods.

Low-Pressure Wet Cleaning Systems

Low-pressure wet cleaning systems are the most commonly used containment systems. These wash dirt from the brake assembly and catch the contaminated cleaning agent in a basin **(Figure 2)**. The reservoir contains water with an organic solvent or wetting agent. To prevent any

Figure 1. This negative-pressure enclosure is used to contain brake dust during cleaning.

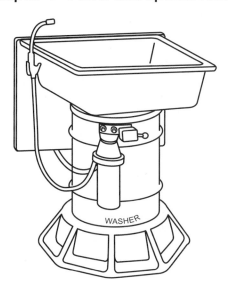

Figure 2. A typical wet, low-pressure parts-cleaning tank.

Wet Cleaning Tools and Equipment

The wet cleaning method of containing asbestos dust is the simplest and easiest to use, but it must be done correctly to provide protection. First, thoroughly wet the brake parts using a spray bottle, hose nozzle, or other implement that creates a fine mist of water or cleaning solution. Once the components are completely wet, wipe them clean with a cloth. Place the cloth in a correctly labeled, impermeable container and properly dispose of it. The cloth can also be professionally laundered by a service equipped to handle asbestos-laden materials and then reused.

Vacuum Cleaning Equipment

Several types of vacuum cleaning systems are available to control asbestos in a shop. The vacuum system must have a HEPA filter to handle asbestos dust. A general-purpose shop vacuum is not an acceptable substitute for a special brake vacuum cleaner with a HEPA filter. After vacuum cleaning, wipe any remaining dust from components with a damp cloth.

Because they contain asbestos fibers, the vacuum cleaner bags and any cloth used for asbestos cleanup are classified as hazardous materials. Such hazardous materials must be disposed of in accordance with OSHA regulations. Always wear your respirator when removing vacuum cleaner bags or handling asbestos-contaminated waste. Seal the cleaner bags and cloths in heavy plastic bags. Label and dispose of the container according to local regulations.

MULTIMETERS

A **multimeter** is a must for diagnosing the individual components of antilock brake systems. Multimeters have

asbestos-containing brake dust from becoming airborne, the flow of liquid should be controlled so that the brake assembly is gently flooded.

You can use the cleaning liquid to wet a brake drum and backing plate before the drum is removed. After the drum is removed, thoroughly wet the wheel hub and the back of the assembly to suppress dust. Wash the brake's backing plate and shoes and the parts used to attach the brake shoes before removing the old shoes.

Some wet cleaning equipment has a filter for the liquid. When the filter is dirty or restrictive, spray it with a fine mist of water; then remove the filter and place it in an impermeable container. Label and dispose of the container according to local regulations.

different names, depending on what they measure and how they function. A volt-ohm-milliamp meter is referred to as a VOM, or DVOM if it is digital. A **DMM** is a **digital multimeter** that can measure many more things than volts, ohms, and low current. Most multimeters measure direct current (DC) and alternating current (AC) in amperes, volts, and ohms. More advanced multimeters can also measure diode continuity, frequency, temperature, engine speed, and dwell and/or duty cycle.

DMMs provide great accuracy by measuring volts, ohms, or amperes in tenths, hundredths, or thousandths of a unit. Several test ranges are usually provided for each of these functions. Some meters have multiple test ranges that must be manually selected; others are auto ranging.

Analog meters use a sweeping needle against a scale to display readings and are not as precise as digital meters. Analog meters have low input impedance and should not be used on sensitive electronic circuits or components. Digital meters have high impedance and can be used on electronic circuits, as well as electrical circuits.

SCAN TOOLS

The introduction of computer-controlled systems brought with it the need for tools capable of troubleshooting electronic control systems. A variety of computer scan tools are available today that do just that. A scan tool is a microprocessor designed to communicate with the vehicle's computer. Connected to the computer through diagnostic connectors, a scan tool can access trouble codes, run tests to check system operations, and monitor the activity of the system. Trouble codes and test results are displayed on a **light-emitting diode (LED)** screen or printed out on the scanner printer.

Scan tools will retrieve fault codes from a computer's memory and digitally display these codes on the tool. Many scan tools also can activate system functions to test individual components. A scan tool can also perform many other diagnostic functions depending on the year and make of the vehicle. Most aftermarket scan tools have removable modules that are updated each year. These modules are designed to test the computer systems on various makes of vehicles. For example, some scan testers have a three-in-one module that tests the computer systems on Chrysler, Ford, and General Motors vehicles. A ten-in-one module is also available to diagnose computer systems on vehicles imported by ten different manufacturers. These modules plug into the scan tool.

Scan tools are capable of testing many onboard computer systems such as transmission controls, engine computers, antilock brake computers **(Figure 3)**, air bag computers, and suspension computers, depending on the year and make of the vehicle and the type of scan tester. In many cases, the technician must select the computer system to be tested with the scanner after it has been connected to the vehicle.

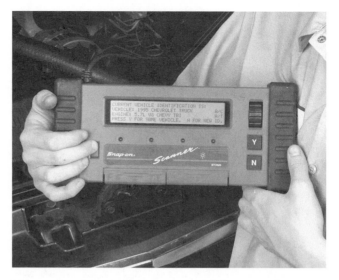

Figure 3. Scan tools are essential for complete testing of ABSs.

The scan tool is connected to specific diagnostic connectors. Most manufacturers have one diagnostic connector. This connects the data wire from each onboard computer to a specific terminal in this connector. Other vehicle manufacturers have several different diagnostic connectors on each vehicle, and each of these connectors can be connected to one or more onboard computers. A set of connectors is supplied with the scanner to allow tester connection to various diagnostic connectors on different vehicles.

The scanner must be programmed for the model year, make of vehicle, and type of engine. With some scan tools, this selection is made by pressing the appropriate buttons on the tester, as directed by the digital tester display. On other scan testers, the appropriate memory card must be installed in the tester for the vehicle being tested. Some scan testers have a built-in printer to print test results, whereas other scan testers might be connected to an external printer.

As automotive computer systems become more complex, the diagnostic capabilities of scan testers continue to expand. Many scan testers now have the capability to store, or "freeze," data into the tester during a road test and then play back this data when the vehicle is returned to the shop. Some scan testers now display diagnostic information based on the fault code in the computer memory. Service bulletins published by the manufacturer of the scan tester may be indexed by the tester after the vehicle information is entered in the tester. Other scan testers will display sensor specifications for the vehicle being tested.

Trouble codes are set by the vehicle's computer only when a voltage signal is entirely out of its normal range. The codes help technicians identify the cause of the problem when this is the case. If a signal is within its normal range but is still not correct, the vehicle's computer will not display a trouble code. However, a problem will still exist. As an aid to identify this type of problem, most manufacturers recommend that the signals to and from the computer be carefully looked at. This is done through the use of a scan tool or breakout box. A breakout box allows the technician to check voltage and resistance readings between specific points within the computer's wiring harness.

All vehicles manufactured from 1996 onward have the second generation of onboard diagnostics (**OBD II**). In these systems, the diagnostic connectors are located in the same place on all vehicles. Also, any scan tools designed for OBD II will work on all OBD II systems; therefore, the need to have designated scan tools or cartridges is eliminated. The OBD II scan tool has the ability to run diagnostic tests on all systems and has "freeze frame" capabilities.

AIR WRENCHES

An impact wrench uses compressed air or electricity to hammer or impact a nut or bolt loose. Light-duty impact wrenches are available in three drive sizes, $\frac{1}{4}$, $\frac{3}{8}$, and $\frac{1}{2}$ inch, and two heavy-duty sizes, $\frac{3}{4}$ and 1 inch.

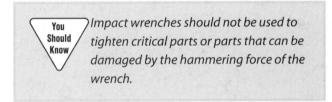

> **You Should Know** *Impact wrenches should not be used to tighten critical parts or parts that can be damaged by the hammering force of the wrench.*

Special thick-walled sockets, called **impact sockets (Figure 4)**, are designed to be used with impact wrenches.

Figure 4. Only specially designed sockets should be used with an impact wrench.

These sockets are designed to withstand the force of the impact. Ordinary sockets must not be used with impact wrenches; they will crack or shatter because of the force and can cause injury. Impact wrenches and sockets are commonly used to remove the wheel assemblies from the vehicle.

> **You Should Know** *It is possible to injure your hand while tightening bolts with an air ratchet. The snap of the ratchet once the bolt or nut is tight may pinch or smash your hand on another object.*

Air ratchets are often used during disassembly or reassembly work to save time. Because the ratchet turns the socket without an impact force, these wrenches can be used on most parts and with ordinary sockets. Air ratchets usually have a $\frac{3}{8}$-inch drive. Air ratchets are not torque sensitive; therefore, a torque wrench should be used on all fasteners after snugging them up with an air ratchet.

BENCH GRINDER

The bench grinder, an electric power tool, is generally bolted to a workbench. The grinder should have safety shields and guards. Always wear face protection when using a grinder. Three types of wheels are available with this bench tool.

- A wire wheel brush, used for general cleaning and buffing; removing rust, scale, and paint; deburring; and so forth.
- A grinding wheel, used for a wide variety of grinding jobs from sharpening cutting tools to deburring.
- A buffing wheel, used for general-purpose buffing, polishing, and light cutting.

TROUBLE LIGHT

Adequate light is necessary when working under and around vehicles. A trouble light can be battery powered (like a flashlight) or need to be plugged into a wall socket. Some shops have trouble lights that pull down from a reel suspended from the ceiling. Trouble lights use either an incandescent bulb or fluorescent tube. Because incandescent bulbs can pop and burn, it is highly recommended that you use only fluorescent bulbs. Take extra care when using a trouble light. Make sure the cord does not get caught in a rotating object. The bulb or tube is surrounded by a cage or enclosed in clear plastic to prevent accidental breaking and burning.

PRESSES

Many automotive jobs require the use of powerful force to assemble or disassemble parts that are **press fit** together. Removing and installing rear-axle bearings and pressing brake drum and rotor wheel studs are common examples. Presses can be hydraulic, electric, air, or hand driven. Capacities range up to 150 tons of pressing force, depending on the size and design of the press. Smaller arbor and C-frame presses can be bench or pedestal mounted, whereas high-capacity units are freestanding or floor mounted.

JACKS AND LIFTS

Jacks are used to raise a vehicle off the ground and are available in two basic designs and in a variety of sizes. The most common jack is the hydraulic floor jack. These are classified by the weights they can lift: 1½, 2, and 2½ tons, etc. These jacks are controlled by moving the handle up and down. The other design of portable floor jack uses compressed air. Pneumatic jacks are operated by controlling air pressure at the jack.

The hydraulic floor lift is the safest lifting tool and is able to raise the vehicle high enough to allow you to walk and work under it. Various safety features prevent a hydraulic lift from dropping if a seal does leak or if air pressure is lost. Before lifting a vehicle, make sure the lift is correctly positioned.

HOLDDOWN SPRING AND RETURN SPRING TOOLS

Brake shoe return springs used on drum brakes are very strong and require special tools for removal and installation **(Figure 5)**. Most return spring tools have special sockets and hooks to release and install the spring ends. Some are built like pliers.

Holddown springs for brake shoes are much lighter than return springs, and many such springs can be released and installed by hand. A holddown spring tool looks like a cross between a screwdriver and a nut driver. A specially shaped end grips and rotates the spring retaining washer.

DRUM BRAKE ADJUSTING TOOLS

Although almost all drum brakes built over the last 30 years have some kind of self-adjuster, the brake shoes still require an initial adjustment after they are installed. The star wheel adjusters of many drum brakes can be adjusted with a flat-blade screwdriver. Brake adjusting spoons (see Figure 5) and wire hooks designed for this specific purpose can make the job faster and easier, however.

BOOT DRIVERS, RINGS, AND PLIERS

Dust boots attach between the caliper bodies and pistons of disc brakes to keep dirt and moisture out of the caliper bores. A special driver **(Figure 6)** is used to install a dust boot with a metal ring that fits tightly on the caliper body. The circular driver is centered on the boot placed against the caliper and then hit with a hammer to drive the boot into place. Other kinds of dust boots fit into a groove in the caliper bore before the piston is installed. Special rings or pliers (see Figure 6) are then needed to expand the opening in the dust boot and let the piston slide through it for installation.

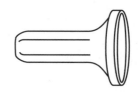

Dust boot installer

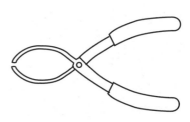

Dust boot pliers

Figure 6. Special tools for installing a dust boot on a disc brake caliper.

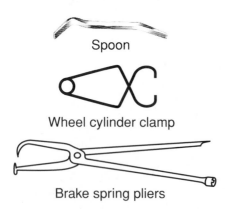

Spoon

Wheel cylinder clamp

Brake spring pliers

Figure 5. Special brake tools include (from top to bottom): adjusting spoons for drum brakes, wheel cylinder clamps, and spring tools.

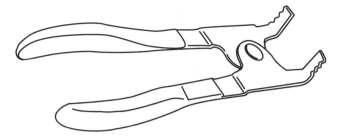

Figure 7. Special pliers are used to remove the pistons from a caliper.

CALIPER PISTON REMOVAL TOOLS

A caliper piston can usually be slid or twisted out of its bore by hand, but rust and corrosion (especially where road salt is used in the winter) can make piston removal difficult. One simple tool that will help with the job is a set of special pliers that grip the inside of the piston and let you move it by hand with more force **(Figure 7)**. These work well on pistons that are only mildly stuck.

For a severely stuck caliper piston, a hydraulic piston remover can be used. This tool requires that the caliper be removed from the car and installed in a holding fixture. A hydraulic line is connected to the caliper inlet and a hand-operated pump is used to apply up to 1000 pounds per square inch (psi) of pressure to loosen the piston. Because of the danger of spraying brake fluid, always wear eye protection when using this equipment.

BRAKE CYLINDER HONES

Cylinder hones (Figure 8) are used to clean light rust, corrosion, pits, and built-up residue from the bores of master cylinders, wheel cylinders, and calipers. A hone can be a very useful—sometimes necessary—tool when you have to overhaul a cylinder. A hone will not, however, save a cylinder with severe rust or corrosion.

The most common cylinder hones have two or three replaceable abrasive stones at the ends of spring-loaded arms. Spring tension usually is adjustable to maintain proper stone pressure against the cylinder walls. The other end of the hone is mounted in a drill motor for use, and the

hone's flexible shaft lets the motor turn the hone properly without being precisely aligned with the cylinder bore.

Another kind of hone is the brush hone. It has abrasive balls attached to flexible metal brushes that are, in turn, mounted on the hone's flexible shaft. In use, centrifugal force moves the abrasive balls outward against the cylinder walls; tension adjustment is not required. A brush hone provides a superior surface finish and is less likely to remove too much metal than a stone hone.

Either kind of hone must be lubricated with brake fluid during use. Do not use a petroleum-based lubricant. After honing a cylinder, flush it thoroughly with denatured alcohol to remove all abrasives and dirt.

BRAKE PEDAL EFFORT GAUGE

The brake pedal effort, or force, gauge **(Figure 9)** is used more and more often to test and service modern brake systems. Some carmakers specify the use of this gauge when checking brake pedal travel and effort or when adjusting the parking brakes. A brake pedal effort gauge is also used to diagnose power brake systems. The gauge assembly is mounted on a steel bracket that can be attached to the brake pedal and has a hydraulic plunger and gauge that measures the force applied by your foot in pounds or Newton-meters (N•M).

TUBING TOOLS

The rigid brake lines, or pipes, of the hydraulic system are made of steel tubing to withstand high pressure and to resist damage from vibration, corrosion, and work hardening. Although it is flexible and easy to work with, copper tubing cannot be used for brake lines. Copper tubing cannot withstand high pressure and is susceptible to breaking due to vibration.

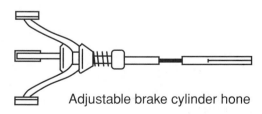

Adjustable brake cylinder hone

Figure 8. A typical brake cylinder hone.

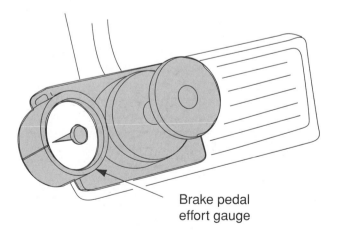

Brake pedal
effort gauge

Figure 9. A brake pedal effort gauge is used to measure the force applied to the brake pedal while applying the brakes.

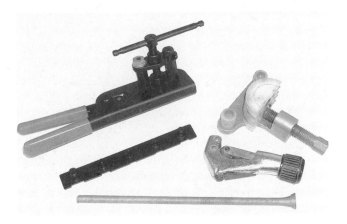

Figure 10. An assortment of tubing tools.

Rigid brake lines often can be purchased in preformed lengths to fit specific locations on specific vehicles. Straight brake lines can be purchased in many lengths and several diameters and bent to fit specific vehicle locations. Even with prefabricated lines available, you probably will have many occasions to cut and bend steel lines and form flared ends for installation. The common tools **(Figure 10)** you should have are:

● A tubing cutter and reamer.
● Tube benders.
● A double flaring tool for SAE flares.
● An ISO flaring tool for European-style **ISO flares**.

POWER STEERING PRESSURE GAUGE

A power steering pressure gauge is used to test the power steering pump pressure. This test is important when checking hydraulic boost brake systems. Because the power steering pump delivers extremely high pressure during this test, the recommended procedure in the vehicle manufacturer's service manual must be followed. However, the typical procedure for using a pressure gauge on power steering systems is given here as an example.

To check the pressure of the pump, a pressure gauge with a shut-off valve is needed. With the engine off, disconnect the pressure hose at the pump. Install the pressure gauge between the pump and the steering gear. Use any adapters that may be necessary to make good connections with the vehicle's system. Open the shut-off valve and bleed the system as described in the service manual.

Start the engine and run it for approximately 2 minutes or until the engine reaches normal operating temperature. Then stop the engine and add fluid to the power steering pump if necessary. Now restart the engine and allow it to idle. Observe the pressure reading. The readings should be about 30–80 psi (200–550 kilopascals). If the pressure is lower than what is specified, the pump may be faulty. If the pressure is greater than specifications, the problem may be restricted hoses.

Now close the shutoff valve, observe the pressure reading, and reopen the valve. Do not keep the valve closed for more than 5 seconds. With the valve closed, the pressure should have increased to 600–1300 psi (4100–8950 kilopascals). Check the pressure reading on the gauge. If the pressure is too high, a faulty pressure relief valve is suggested. If the pressure is too low, the pump may be bad.

BELT TENSION GAUGE

A belt tension gauge is used to measure drive belt tension. The belt tension gauge is installed over the belt, and the gauge indicates the amount of belt tension.

BEARING PULLERS

Many tools are designed for a specific purpose. An example of a special tool with a specific purpose is a bearing puller. Many bearings have a slight interference fit (press fit) when they are installed on a shaft or in a housing. This press fit prevents the parts from moving on each other. The removal of these bearings must be done carefully to prevent damage to the housing, bearings, or shafts. Prying or hammering can break or bind the parts. Using the proper puller, the force required to remove a bearing can be applied with a slight and steady motion.

BUSHING AND SEAL PULLERS AND DRIVERS

Another commonly used group of special tools includes the various designs of **bushing** and seal drivers **(Figure 11)** and pullers. Pullers are either a threaded or

Figure 11. A bushing and seal driver.

Figure 12. A slide hammer–type bushing and seal puller.

Figure 13. Snapring pliers.

slide hammer–type tool **(Figure 12)**. Always make sure you use the correct tool for the job; bushings and seals are easily damaged if the wrong tool or procedure is used. Car manufacturers and specialty tool companies work closely together to design and manufacture special tools required to repair cars. Most of these special tools are listed in the appropriate service manuals.

TIE-ROD END AND BALL JOINT PULLER

Some car manufacturers recommend a tie-rod end and ball joint puller to remove tie-rod ends and pull ball joint studs from the steering knuckle. A tie-rod end remover is a safer and easier way of separating ball joints than a pickle fork.

Ball joint removal and pressing tools are designed to remove and replace pressed-in ball joints on front suspension systems. Often these tools are used in conjunction with a hydraulic press. The size of the removal and pressing tool must match the size of the ball joint. Some ball joints are riveted to the control arm and the rivets are drilled out for removal.

FRONT BEARING HUB TOOL

Front bearing hub tools are designed to remove and install front wheel bearings on front-wheel-drive cars. These bearing hub tools are usually designed for a specific make of vehicle, and the correct tools must be used for each application. Failure to use the correct tools can result in damage to the steering knuckle or hub and will waste quite a bit of your time.

SPECIAL PLIERS

Snapring or lock ring pliers are made with a linkage that allows the movable jaw to stay parallel throughout its

range of opening **(Figure 13)**. The jaw surface is usually notched or toothed to prevent slippage. Often, a technician will run into many different styles and sizes of retaining rings that hold subassemblies together or keep them in a fixed location. Using the correct tool to remove and install these rings is the only safe way to work with them. All technicians should have an assortment of retaining ring pliers.

BRAKE LATHES

Brake lathes are special power tools used only for brake service. They are used to turn (or surface) and resurface brake rotors and drums **(Figure 14)**. Turning involves cutting away very small amounts of metal to restore the surface of the rotor or drum. The traditional brake lathe is an assembly mounted on a stand or workbench. This so-called bench lathe required that the drum or rotor be removed from the vehicle and mounted on the lathe for service. As the drum or rotor is turned on the lathe spindle, a carbide steel cutting bit is passed over the drum or rotor **friction** surface to remove a small amount of metal. The cutting bit is mounted rigidly on a lathe fixture for precise control as it passes across the friction surface.

Figure 14. A typical bench brake lathe used to turn brake drums and discs.

Obviously a brake drum must be removed from its axle or spindle to turn it on a lathe. The friction surface of a rotor is exposed, however, when the wheel and tire are removed. Then it is possible to apply a cutting tool to the rotor friction surface without removing the rotor from the car. With the universal adoption of disc brakes, on-car brake lathes were developed for rotor service.

An on-car lathe is bolted to the vehicle suspension or mounted on a rigid stand to provide a stable mounting point for the cutting tool **(Figure 15)**. The rotor may be turned by either the vehicle's engine and drivetrain (for a front-wheel-drive car) or by an electric motor and drive attachment on the lathe. As the rotor is turned, the lathe cutting tool is moved across both surfaces of the rotor to refinish it. An on-car lathe not only has the obvious advantage of speed, it rotates the rotor on the vehicle wheel bearings and hub so that these sources of runout, or wobble, are compensated for during the refinishing operation.

Most drum and rotor lathes include attachments for applying a final surface finish to the rotor or for grinding hard spots on drums. Some bench lathes also may have a shoe arcing attachment for arc grinding the linings of brake shoes, but such attachments have become rare. Shoe arcing was a necessary operation years ago, but concerns about asbestos dust led brake lining manufacturers to develop preground shoes that do not require arcing. Whenever possible, it is good practice to buy preground shoes. They not only speed up the brake job, they eliminate asbestos dust hazards due to shoe arcing.

PRESSURE BLEEDERS

Removing the air from the closed hydraulic brake system is very important. This is done by bleeding the system. Bleeding can be done manually, with a vacuum pump, or with a pressure bleeder **(Figure 16)**. The latter is preferred because it is quick and efficient: the master cylinder does not have to be refilled several times, and the job can be done by one person.

A pressure bleeder is a tank separated into two sections by a flexible diaphragm. The top section is filled with brake fluid. Compressed air is fed into the bottom section, and as the air pushes on the diaphragm the brake fluid above it also is pressurized. A pressure bleeder is normally pressurized to about 30 psi. Higher pressures should be avoided because fluid can be forced so rapidly through the hydraulic system that it can cause a swirling or surging action. This, in turn, can create air pockets in the lines and valves and make bleeding difficult.

A supply hose runs from the top of the tank to the master cylinder. The hose is connected to the master cylinder by an adapter fitting that fits over the reservoir, taking the place of the reservoir cap. These adapters exist in different shapes for the different types of reservoirs, including the plastic reservoirs on some of the newer vehicles.

The pressurized brake fluid flows into the master cylinder and out through the brake lines, quickly forcing air out of the lines. Because most brake fluids (except for silicone) tend to absorb moisture from the air, always keep containers tightly capped. It is better to buy smaller containers of brake fluid and keep them sealed until needed. Taking

Figure 15.   A typical on-car brake lathe mounted to the vehicle.

Figure 16.   A typical pressure bleeder shown with some adapters.

these steps to minimize water in the brake fluid will help reduce corrosion and keep the brake fluid boiling point high throughout the hydraulic system.

Other tools used in brake bleeding operations include: a large rubber syringe, used to remove fluid from the master cylinder on some systems; master cylinder bleeder tubes, used to return fluid to the master cylinder reservoir from the outlet ports during bench bleeding; and assorted line and port plugs, used to close lines and valves temporarily during service and keep out dirt and moisture.

Summary

- Negative-pressure enclosures, HEPA vacuum systems, and low-pressure wet cleaning systems are used to safely contain asbestos while doing brake work.
- A DMM is a digital multimeter and is a must for diagnosing the individual components of ABSs.
- A scan tool is a microprocessor designed to communicate with the vehicle's computer and is used to check the operation of the entire computer system.
- There are many special tools designed to make brake service easier and safer. These special tools include holddown spring and return spring tools, drum brake adjusting tools, boot drivers, rings, pliers, and caliper piston removal tools.
- Cylinder hones are used to clean light rust, corrosion, pits, and built-up residue from the bores of master cylinders, wheel cylinders, and calipers.
- To repair and replace brake lines, you should have a tubing cutter and reamer, a tube bender, a double flaring tool for SAE flares, and an ISO flaring tool for European-style ISO flares.
- A power steering pressure gauge is used to test the power steering pump pressure, which is an important check on hydraulic boost brake systems.
- Brake lathes are used to turn and resurface brake rotors and drums.
- The traditional brake lathe is a bench lathe that requires the drum or rotor to be removed from the vehicle and mounted on the lathe for service.
- An on-car lathe is bolted to the vehicle's suspension or mounted on a rigid stand to provide a stable mounting point for the cutting tool. The rotor may be turned by either the vehicle's engine and drivetrain (for a front-wheel-drive car) or by an electric motor and drive attachment on the lathe.
- Pressure bleeding is a fast and efficient way to bleed a brake system and uses air pressure to force brake fluid through the system.

Review Questions

1. While discussing on-car brake lathes Technician A says that a disadvantage of this type of lathe is that the rotor is still mounted on its bearings and any bearing runout is not compensated for; Technician B says that the advantage of this type of lathe is that the time required to surface a rotor is drastically less than if using a bench lathe. Who is correct?
   A. Technician A only
   B. Technician B only
   C. Both Technician A and Technician B
   D. Neither Technician A nor Technician B
2. Why should you not use an air impact wrench to tighten critical components?
3. A good flashlight is often the preferred type of trouble light because incandescent bulbs can create a safety hazard. What can happen to incandescent bulbs?
4. Technician A says that a brake pedal effort gauge is used to check brake pedal travel and effort or when adjusting the parking brakes. Technician B says that a brake pedal effort gauge is used to diagnose power brake systems and to check brake pedal travel and effort. Who is correct?
   A. Technician A only
   B. Technician B only
   C. Both Technician A and Technician B
   D. Neither Technician A nor Technician B
5. While discussing the fit of bearings Technician A says that an interference fit allows the parts to be easily disassembled; Technician B says that parts that are press fit have an interference fit. Who is correct?
   A. Technician A only
   B. Technician B only
   C. Both Technician A and Technician B
   D. Neither Technician A nor Technician B

6. Technician A says that using a pressure bleeder is the fastest way to bleed a brake system. Technician B says that with a pressure bleeder, one technician can do the job. Who is correct?
   A. Technician A only
   B. Technician B only
   C. Both Technician A and Technician B
   D. Neither Technician A nor Technician B

7. Why should you not use regular sockets with an air impact wrench?

8. Technician A says that lock ring pliers are made with a linkage that allows the movable jaw to stay parallel throughout its range of opening and that their jaw surface is usually notched or toothed to prevent slippage. Technician B says that retaining ring pliers are identified by their pointed tips that fit into holes at the ends of retaining or snaprings. Who is correct?
   A. Technician A only
   B. Technician B only
   C. Both Technician A and Technician B
   D. Neither Technician A nor Technician B

9. While discussing the use of an air impact wrench Technician A says that impact sockets can be used with an air impact wrench; Technician B says that sockets that have cracks or breaks should never be used with an air impact wrench. Who is correct?
   A. Technician A only
   B. Technician B only
   C. Both Technician A and Technician B
   D. Neither Technician A nor Technician B

10. Which of the following statements about scan tools is *not* true?
   A. Many scan tools also can activate system functions to test individual components on vehicles with bidirectional controls.
   B. OBD II scan tools are connected to many different DLCs to retrieve data from separate electronic control systems.
   C. Trouble codes retrieved by a scan tool are set by the vehicle's computer only when a voltage signal is entirely out of its normal range.
   D. Any scan tools designed for OBD II will work on all OBD II systems; therefore, the need to have designated scan tools or cartridges is eliminated.

Chapter 5

Working As a Brake Technician

Introduction

To be a successful automotive technician you need to have good training, a desire to succeed, and be committed to become a good technician and a good employee. A good employee works well with others and strives to make the business successful. The required training is not just in the automotive field. Good technicians **(Figure 1)** need to have good reading, writing, and math skills. These skills will allow you to better understand and use the material found in service manuals and textbooks, as well as provide you with the basics for good communications with customers and others.

COMPENSATION

Technicians are typically paid according to their abilities. Most often, new or apprentice technicians are paid by the hour. While being paid, they are learning the trade and the business. Time is usually spent working with a master technician or doing low-skilled jobs. As apprentices learn more, they can earn more and take on more complex jobs. Once technicians have demonstrated a satisfactory level of skills, they can go on **flat rate**.

Flat rate is a pay system in which technicians are paid for the amount of work they do. Each job has a flat rate time. Pay is based on that time, regardless of how long it took to complete the job. To explain how this system works, let us look at a technician who is paid $15.00 per flat rate. If a job has a flat-rate time of 3 hours, the technician will be paid $45.00 for the job, regardless of how long it took to complete it. Experienced technicians beat the flat-rate time nearly all of the time. Their weekly pay is based on the time

Figure 1. Good technicians need to have good reading, writing, and math skills.

"turned in," not on the time spent. If the technician turns in 60 hours of work in a 40-hour workweek, he or she actually earned $22.50 each hour worked. However, if the technician turned in only 30 hours in the 40-hour week, the hourly pay is $11.25.

The flat-rate system favors good technicians that work in a shop that has a large volume of work. The use of flat-rate times allows for more accurate repair estimates to the customers. It also rewards skilled and productive technicians.

EMPLOYER-EMPLOYEE RELATIONSHIPS

When you begin a job, you enter into a business agreement with your employer. When you become an employee, you sell your time, skills, and efforts. In return, your employer pays you for these resources.

CUSTOMER RELATIONS

Another responsibility you have is good customer relations. Learn to listen and communicate clearly. Be polite and organized, particularly when dealing with customers. Always be honest.

Look like and present yourself as a professional, which is what automotive technicians are. Professionals are proud of what they do and they show it. Always dress and act appropriately and avoid crude and obscene language and cursing, even when you think no one is near.

Respect the vehicles you work on. They are important to the lives of your customers. Always return the vehicle to the owner in a clean, undamaged condition. Remember, a car is the second largest expense a customer has. Treat it that way. It does not matter if you like the car or not. It belongs to the customer; treat it respectfully.

Explain the repair process to the customer in understandable terms. Whenever you are explaining something to a customer, make sure you do this in a simple way without making the customer feel stupid. Always show customers respect and be courteous to them. Not only is this the right thing to do but it also leads to customer loyalty.

ASE CERTIFICATION

An obvious sign of your knowledge and abilities, as well as your dedication to the trade, is ASE certification. The National Institute for Automotive Service Excellence **(ASE)** has established a voluntary certification program for automotive, heavy-duty truck, auto body repair, and engine machine shop technicians. In addition to these programs, ASE also offers individual testing in some specialty areas. This certification system combines voluntary testing with on-the-job experience to confirm that technicians have

Figure 2.   ASE's certification combines voluntary testing with on-the-job experience to confirm that technicians have the skills needed to work on today's vehicles.

the skills needed to work on today's vehicles **(Figure 2)**. ASE recognizes two distinct levels of service capability—the automotive technician and the master automotive technician **(Figure 3)**.

Figure 3.   A master automotive technician.

To become ASE certified, you must pass one or more tests that stress diagnostic and repair problems. Automotive Brake Systems is one of the eight basic automotive certification areas. After passing at least one exam and providing proof of 2 years of hands-on work experience, you become ASE certified. Retesting is necessary every 5 years to remain certified. A technician who passes one examination receives an automotive technician shoulder patch. The master automotive technician patch is awarded to technicians who pass all eight of the basic automotive certification exams.

Each certification test consists of 40–80 multiple-choice questions. The questions are written by a panel of technical service experts, including domestic and import vehicle manufacturers, repair and test equipment and parts manufacturers, working automotive technicians, and automotive instructors. All questions are pretested and quality-checked on a national sample of technicians before they are included in the actual test. Many test questions force the student to choose between two distinct repair or diagnostic methods. The test questions focus on basic technical knowledge, repair knowledge and skill, and testing and diagnostic knowledge and skill.

DUTIES OF A BRAKE TECHNICIAN

As an automotive brake technician, you will be responsible for servicing, troubleshooting, and repairing all of the components that make up the vehicle's brake system. You are dealing with a very important operation of the vehicle, its ability to stop. Doing a good job is crucial to the safety of others.

Brake work also involves health considerations for those who are working on the vehicles. Some of these considerations are regulated and monitored by government agencies. All of the safety and health guidelines should be followed at all times.

DIAGNOSTICS

The true measure of a good technician is an ability to find and correct the cause of problems. Service manuals and other information sources will guide you through the diagnosis and repair of problems, but those guidelines will not always lead you to the exact cause of the problem. To do this, you must use your knowledge and take a logical approach while troubleshooting. Diagnosis is not guessing, and it is more than following a series of interrelated steps in order to find the solution to a specific problem. Diagnosis is a way of looking at systems that are not functioning the way they should and finding out why. It is knowing how the system should work and deciding if it is working correctly. Through an understanding of the purpose and operation of the system, you can accurately diagnose problems.

Most good technicians use the same basic diagnostic approach. Simply because this is a logical approach, it can quickly lead to the cause of a problem. Logical diagnosis follows these steps:

1. Gather information about the problem.
2. Verify that the problem exists.
3. Thoroughly define what the problem is and when it occurs.
4. Research all available information and knowledge to determine the possible causes of the problem.
5. Isolate the problem by testing.
6. Continue testing to pinpoint the cause of the problem.
7. Locate and repair the problem, then verify the repair.

SPECIFIC BRAKE SYSTEM SERVICE ISSUES

In the United States, brake systems are regulated by Part 571 of the Federal Motor Vehicle Safety Standards **(FMVSS)**. The U.S. Department of Transportation (DOT) has the responsibility for establishing and enforcing these regulations. Many states and Canadian provinces have additional regulations that govern brake safety, condition, and operation. Currently there are seven federal standards that relate to brake systems.

- FMVSS 105 Hydraulic Brake Systems
- FMVSS 106 Brake Hoses
- FMVSS 108 Lamps, Reflective Devices, and Associated Equipment
- FMVSS 116 Motor Vehicle Brake Fluids
- FMVSS 121 Air Brake Systems
- FMVSS 122 Motorcycle Brake Systems
- FMVSS 211 Wheel Nuts, Wheel Discs, and Hub Caps

General performance requirements for service brakes and parking brake systems are governed by FMVSS 105. This standard was set to ensure safe braking performance under normal and emergency conditions. The standard does not dictate how a brake system should be designed or what components it should have; rather it establishes the requirements for brake system performance. The standard regulates four major features of brake systems: instrument-panel warning lamps, the fluid reservoir and its labeling, automatic adjustment, and mechanically operated friction parking brakes.

FMVSS 105 defines the minimum performance requirements for the hydraulic brake system on any vehicle driven on the highway. These minimums are the same for all vehicles. For example, to meet the criteria set by the initial brake effectiveness test, the tested vehicle, with fresh brake linings, makes six stops from 30 mph and six stops from 60 mph. At least one of the stops from 30 mph must be made in 57 feet or less, and one stop from 60 mph must be in 216 feet or less. These stopping distances are absolute requirements for any vehicle of any size and weight **(Figure 4)**.

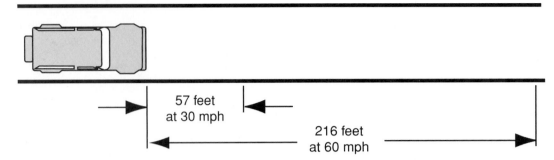

Figure 4. One of the eighteen stages of the brake performance test in FMVSS 105 requires one stop from 30 mph in 57 feet or less and one stop from 60 mph in 216 feet or less.

The performance test of FMVSS 105 includes measuring stopping distances under a number of conditions to ensure that the design of a vehicle's brake system provides a minimal level of safety.

BRAKE SERVICE LAWS AND REGULATIONS

After a new vehicle is sold, the owner of the vehicle is responsible for maintaining the brake system. The owner, in turn, relies on the brake technician to keep the brake system in the proper operating condition. Many states and Canadian provinces have laws that govern brake system operation and brake service.

Some states require periodic vehicle safety inspection, either every year or every 2 years. These inspections usually include at least an inspection of the components of the brake system. Some also include dynamic stopping tests conducted on a brake system analyzer or on a measured course. If a vehicle fails any part of the safety inspection, its registration will not be renewed until all defects are fixed. Some states require that a vehicle that has failed a brake test or inspection or has been cited for unsafe brakes by a police officer can be repaired only at a state-authorized repair facility.

Additionally, some states, provinces, counties, and cities have regulations for the licensing or certification of brake service technicians. Some areas conduct their own certification programs; others rely on ASE certification in brake service.

ASBESTOS HEALTH HAZARDS

One of the greatest safety concerns while doing brake work is exposure to **asbestos** dust. Exposure to asbestos was a greater problem many years ago than it is today; however, the importance of avoiding asbestos exposure and the concern for asbestos safety have not decreased.

Asbestos was used as the primary material for brake and clutch linings. Asbestos is a silicate compound that has excellent heat dissipation abilities and a high coefficient of friction.

It contains millions of small, linked fibers that give it both strength and flexibility. As the asbestos friction materials wear from use, these fibers are released in the dust from the linings. Asbestos fibers as small as 5 microns (0.00020 inch) are a serious health hazard if inhaled in even small quantities.

Asbestos fibers are inert and do not deteriorate or decompose naturally. Once inhaled, these fibers are with you forever. Even moderate quantities of inhaled asbestos fibers can lead to serious diseases, such as asbestosis and lung cancer.

Asbestosis is a progressive lung disease caused by asbestos fibers continually lodging in the lungs and inflaming the lungs' air sacs. The inflammation caused by asbestosis can heal, but it leaves scar tissue in the lungs that thickens the air sacs and makes it increasingly harder for oxygen to enter the bloodstream. Once started, asbestosis is irreversible. Its severity and progress are determined by the amount of asbestos inhaled.

Lung cancer is the most deadly of any asbestos-related disease. Asbestos exposure, combined with other respiratory irritations, such as tobacco smoke, can accelerate the development of cancer and produce more severe effects. It is possible for a person to develop both asbestosis and lung cancer from severe asbestos exposure. Heavy exposure to asbestos also can lead to other cancers of the respiratory and digestive systems.

ASBESTOS CONTROL LAWS AND REGULATIONS

The Occupational Safety and Health Administration (OSHA) has set regulations to control asbestos exposure. Its regulations state that fibers of 5 microns or larger are hazardous. Furthermore, these regulations put a very low limit on the amount of asbestos a worker can be exposed to in an 8-hour period. These extremely low exposure limits can be maintained while performing brake work through the proper use of brake cleaning equipment and respiratory safety devices.

Along with OSHA, the U.S. EPA regulates some aspects of asbestos safety. The EPA regulations are concerned primarily with the handling and disposal of asbestos waste. These regulations state that any waste material containing more than 1 percent asbestos must be disposed of by rigidly controlled methods that do not endanger public health.

Your concern with asbestos safety does not end with prescribed cleaning of brake systems and respiratory safety; you must dispose of the residue from cleaning according to EPA regulations. Because brake dust may contain more than 1 percent asbestos, any vacuum cleaner bags, filters, and cloths used to wipe up brake dust must be sealed in double plastic bags or a similar nonpermeable container. The bag or container must then be labeled with an asbestos exposure warning, similar to "DANGER. Contains Asbestos Fibers. Avoid Creating Dust. Cancer and Lung Disease Hazard."

In most areas of the country, it is acceptable to turn over properly contained asbestos residue to local sanitation agencies for burial in a landfill. This eliminates the hazard of airborne fibers. Local asbestos disposal regulations may vary, however, and some may require additional special handling. It is your responsibility to know the regulations in your area and ensure that they are observed in your workplace.

ASBESTOS PRECAUTIONS

The best way to avoid asbestos dust is not to create it. Never use compressed air or dry brushing for cleaning brake dust from brake drums, backing plates, and brake assemblies. Nor should you ever dry sweep or blow brake dust off shop equipment or the shop floor. Use approved vacuum cleaning equipment or the proper wet-cleaning procedures to remove the dust.

You can also protect yourself and your family by leaving work clothes that could be contaminated with asbestos

at work. If possible, shower at work or shower as soon as you get home to remove any asbestos from your skin. Wash thoroughly before eating and avoid eating, drinking, and smoking in an area where brake work is being performed. During work, wash yourself well before eating, drinking, or smoking, especially if you are doing brake work.

RESPIRATORS

Concern for asbestos exposure has led to a reduction of its use in automotive components. Asbestos has been eliminated from the brake linings of new cars and light trucks sold in North America and from replacement brake linings made in the United States and Canada. These restrictions do not apply, however, to replacement brake linings manufactured outside North America and imported into the United States.

As asbestos content was reduced in brake linings, other materials took its place. Today, many brake linings are made of organic or semimetallic compounds. Like asbestos, however, these materials wear and create brake dust. Semimetallic brake linings, for example, can contain copper or iron compounds, and these materials become part of the brake dust. Although exposure to these metals may not be as hazardous as exposure to asbestos, it cannot be good for you to inhale copper or iron dust. For all of these reasons, proper use of brake cleaning equipment and respiratory safety devices is as important today as it has ever been.

Although all brake work should be done in an enclosure unit that has been certified by OSHA and the EPA, additional breathing protection can be provided by respirators. The respirator can be a simple throwaway mask or a mask with replaceable filters **(Figure 5)**. Put your respirator on when you first start removing the wheels from

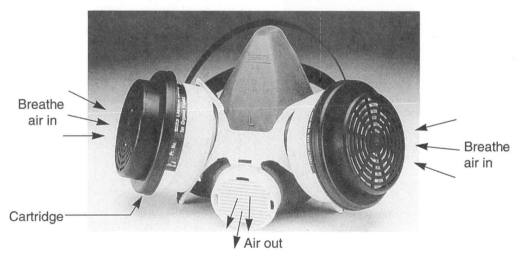

Figure 5.   This approved filter-type respirator is ideal for brake work.

the car and keep it on until the job is finished. Also, wear a respirator if someone in another service bay is doing brake work.

Respirators work properly only if they fit properly and have the correct type of filter asbestos protection. Felt filters use an electrostatic charge to help trap particulates with the fibers. High-efficiency paper filters use a fiberglass paper with fine pores to stop the passage of particulates.

Filter life depends upon the amount of particulates in the air. Change your respirator's filters as soon as you notice an increased resistance to inhalation. Place the filter in the snap-off filter cap. Then install the cap over the cartridge, if used, or on the receptacle in combination with a cartridge. Do not place the filter directly into the receptacle next to the inhalation valve. It is recommended that the respirator's filter be changed on a regular basis.

BRAKE VACUUM CLEANERS

Working around asbestos requires some other special precautions. Never blow or brush the dust off brake parts. A brake dust vacuum cleaner system **(Figure 6)** or a brake washing system should be used during brake disassembly. These will keep asbestos dust within safe limits for the entire shop area. These units mount over the brake system and remove dust from the parts as they are disassembled.

In addition to cleaning brake system components, the vacuum system can be used for general floor cleanup in the service area. Use the recommended attachments. If a vacuum is not available, apply water or other dust suppressants to the floor before any sweeping or mopping to remove asbestos dust. Never sweep the floor with a dry broom or use compressed air to blow away dust.

Figure 6.   A heavy-duty vacuum cleaner with an HEPA filter can handle asbestos cleanup, as well as general shop cleaning.

CLEANING WITH CHEMICALS

Parts cleaning is an important part of any brake repair job. Often you will need to clean parts to find problems and measure for wear. Brake parts cleaning must always be done with approved equipment because of the danger of asbestos exposure.

Be careful when using solvents. Most are toxic, caustic, and flammable. Avoid placing your hands in the solvent; wear protective gloves. Read all manufacturer's precautions and instructions and the appropriate MSDS before using any cleaning solution.

The cleaning of small parts can often be done with aerosol can cleaners. These spray cans contain chemicals that will break down dirt and grease. Wear eye protection, gloves, and a shop coat to prevent exposure to your skin or eyes. Also, always do your cleaning in a well-ventilated area.

> **You Should Know** *Vapors from many brake cleaning solvents may be toxic if inhaled, particularly in large quantities for prolonged periods. Use brake cleaning solvents only in well-ventilated areas. Wear a respirator and avoid inhaling vapors.*

At times, small parts are cleaned with cleaning solvent in a parts cleaner. An electric motor pumps the solvent through a hose that can be used to flush off parts. The solvent is circulated through a filter to keep it clean. Many of the solvents used in cleaning tanks are flammable. Be careful to prevent an open flame around the solvent tank.

Some solvents can be absorbed through the skin and into your body. This is especially true if you have a cut on your hand. Never use compressed air on your hands if they get wet with solvent; this can cause the solvent to penetrate into your skin. Always wear neoprene gloves when washing parts.

Brake Cleaning Solvents

Wetting the dirt and dust residue on a brake assembly with solvent keeps asbestos fibers and other toxic materials out of the air. However, these solvents may present a risk of their own.

Although they are a lesser health hazard than asbestos, various cleaning solvents used on brake systems must be handled with specific precautions **(Figure 7)**. Among the most significant from a safety standpoint are those that contain **chlorinated hydrocarbon solvents**. These are colorless solvents with a strong odor of ether or chloroform. The vapors from these solvents are narcotic and can cause

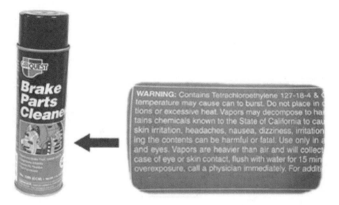

Figure 7. This aerosol can of brake cleaner has a label on it to identify how to safely use it and how to deal with accidents resulting from its use.

drowsiness or loss of consciousness. Very high levels of exposure, even for a short time, can be fatal. Although these hydrocarbon solvents are not flammable, they decompose when exposed to flame and release toxic gases such as **phosgene**, carbon monoxide, and hydrogen chloride.

> You Should Know | Solvent vapors are heavier than air and will collect at floor level. This is particularly dangerous in areas containing gas appliances with open pilot lights.

Chlorinated hydrocarbon solvents react in the atmosphere and deplete the earth's ozone layer. Their manufacture has been restricted since January 1, 1996. Other solvents such as hexane, heptane, and xylene are replacing chlorinated hydrocarbons in brake cleaners. Hexane and heptane are flammable, so make sure all fire safety precautions are observed when using these solvents.

Chemical Precautions

You may be exposed to chemical health hazards in three ways: ingestion, inhalation, and skin contact. The MSDS describe any poisoning hazards and the methods for counteracting poisonous effects.

Solvents can be ingested by a smoker who lights a cigarette while working with the solvent, but contact with solvents occurs most often through inhalation or absorption through the skin. Inhalation probably is the most serious and has the most immediate effect. Absorption can be just as dangerous, however, and its effects may not be noticeable until several days after exposure.

The immediate effect of contact is the removing of natural skin oils and drying of your skin, causing redness and irritation. Prolonged skin contact with solvent can have the same effects as inhalation. Exposure to chlorinated hydrocarbons and other solvents by any means can cause nausea, drowsiness, headache, dizziness, and eventually unconsciousness. Prolonged exposure can lead to liver and kidney damage.

Solvents must always be handled carefully and kept in properly labeled containers. When not in use, the containers must be stored away from untrained personnel or children. Basic common sense will go a long way toward protecting you from the hazards of chemical poisoning.

BRAKE FLUID SAFETY

The EPA and other environmental agencies consider used brake fluid a hazardous waste material. New or used brake fluid also is toxic if ingested, and it can be harmful to vehicles if used incorrectly.

The most common brake fluids are polyglycol liquid compounds. They are identified by the DOT specification numbers DOT 3 and DOT 4 **(Figure 8)**. Polyglycol fluids act as solvents on most automotive paints and will damage the finish of a vehicle. Always handle brake fluid containers carefully and wipe up any spilled fluid immediately.

DOT 3 and DOT 4 brake fluids are **hygroscopic**, which means they will absorb moisture. Even the smallest amount of moisture mixed with brake fluid lowers the fluid boiling point and reduces its effectiveness as a hydraulic fluid. These conditions, in turn, can lead to unsafe brake system operation. Therefore, always keep brake fluid containers tightly closed when not in use.

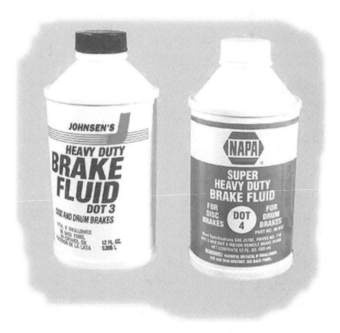

Figure 8. Commonly used types of polyglycol brake fluids.

Also, because brake fluids absorb moisture, they will absorb the oils from your skin and lead to dryness, irritation, and inflammation. Wear gloves when handling brake fluid and immediately wash off any brake fluid that contacts your skin.

BRAKE LUBRICANTS

Some special lubricants are used in brake service to aid assembly and to help prevent corrosion and mechanical seizure.

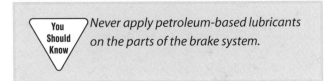

You Should Know *Never apply petroleum-based lubricants on the parts of the brake system.*

Assembly Fluid

If you overhaul hydraulic parts such as cylinders and calipers, assembly fluid will help to install pistons past seals or to install piston seals into bores. You can use brake fluid as an assembly lubricant, but special assembly fluid has a higher viscosity than brake fluid. The higher viscosity provides better lubrication and lets the fluid remain on the part until it is installed and put into service.

Brake Grease

Drum brake grease is applied to the shoe support pads on backing plates, and disc brake grease is applied to the moving caliper parts. Brake grease has a melting point above 500°F (260°C) and contains solid lubricants that will remain in place even if the temperature rises above the grease melting point. Wheel-bearing grease or chassis grease should not be used in place of brake grease because when the grease is heated by brake operation it can run onto the pads and linings and ruin them.

Rubber Lubricants

Some master cylinder overhaul kits contain a small package of nonpetroleum grease for application to the rubber boot of the cylinder where the pushrod enters. This special grease should be used only on rubber boots of cylinders. Do not use it in place of assembly fluid or brake grease.

Wheel-Bearing Grease

Brake service often includes repacking front or rear wheel bearings. Although some multipurpose greases can be used for both chassis lubrication and for wheel bearings, you must be sure that whatever grease you use is identified as suitable for wheel-bearing lubrication.

All greases are made from oils blended with thickening agents so that the grease will stick to the surfaces to be lubricated. Greases are identified by the National Lubricating Grease Institute (**NLGI**) number and by the kinds of thickeners and additives that the grease contains. The higher the NLGI number, the higher viscosity (thicker) the grease is. Almost all wheel-bearing and chassis greases are NLGI 2 greases. Most wheel-bearing grease uses a lithium-based thickener for temperature resistance. Molybdenum disulfide is another common additive that improves the antiseize properties of wheel-bearing grease.

There is no way to tell what kind of grease was used previously on a wheel bearing. Therefore, wipe away all old grease thoroughly when repacking wheel bearings and use only new grease that meets the vehicle maker's specifications for wheel-bearing service.

Summary

- The true measure of a good technician is an ability to find and correct the cause of problems.
- Diagnosis is not guessing and it is more than following a series of interrelated steps in order to find the solution to a specific problem.
- In the United States, brake systems are regulated by the Federal Motor Vehicle Safety Standards.
- FMVSS 105 defines the minimum performance requirements for the hydraulic brake system on any vehicle driven on the highway.
- Asbestos was used as the primary material for brake and clutch linings. Asbestos is a silicate compound that has excellent heat dissipation abilities and a high coefficient of friction. Even moderate quantities of inhaled asbestos fibers can lead to serious diseases, such as asbestosis and lung cancer.
- OSHA has set regulations to control asbestos exposure.
- The EPA regulations are concerned primarily with the handling and disposal of asbestos waste.
- Breathing protection can be provided by an approved and well-fitted respirator.
- Wetting the dirt and dust residue on a brake assembly with solvent keeps asbestos fibers and other toxic materials out of the air.

■   The various cleaning solvents used on brake systems, especially chlorinated hydrocarbon solvents, must be handled with specific precautions.

■   The most common brake fluids are polyglycol liquid compounds that act as solvents on most automotive paints and are hygroscopic.

■   Brake assembly fluid, brake grease, rubber lubricants, and wheel-bearing grease are some of the special lubricants used in brake service to aid assembly and to help prevent corrosion and mechanical seizure.

Review Questions

1. What must a technician do in order to become certified as an ASE Master Automobile Technician?

2. Explain the flat-rate pay system.

3. List the steps that should be followed while logically diagnosing a problem.

4. _____ is a progressive disease caused by asbestos fibers continually lodging in the lungs.

5. Technician A says that the U.S. Occupational Safety and Health Administration (OSHA) regulates asbestos exposure and the handling of materials that contain asbestos. Technician B says that the U.S. Environmental Protection Agency (EPA) regulates the handling and disposal of asbestos waste. Who is correct?
   A. Technician A only
   B. Technician B only
   C. Both Technician A and Technician B
   D. Neither Technician A nor Technician B

6. The most common source of hazardous asbestos fibers in a shop is:
   A. brake cleaning solvent
   B. brake dust
   C. outgassing of vapors from brake linings
   D. brake lining adhesive

7. Chemical poisoning can occur by any of the following ways *except*:
   A. inhalation
   B. refraction
   C. absorption
   D. ingestion

Section 2

Basic Theories and Services

SECTION OBJECTIVES

After you have read, studied, and practiced the contents of this section, you should be able to:

- List and describe the operation of the basic parts of a brake system.
- Explain how kinetic energy in a brake system is affected by mass, weight, momentum, and inertia.
- Explain the importance of kinetic and static friction in a brake system.
- Describe the effects of pressure, surface area, and friction material on the performance of a braking system.
- Explain how the coefficient of friction is determined for a material.
- Describe the three basic types of friction materials currently used in brake systems.
- Explain how hydraulics can be used to transmit force and motion.
- Explain the relationships of force, pressure, and motion in a hydraulic system.
- Explain how vacuum and atmospheric pressure can be used to increase force.
- Explain the basic principles of electricity.
- Explain how to use Ohm's Law to determine the amount of current, resistance, or voltage in a circuit.
- Name the various electrical components and their uses in electrical circuits.
- Diagnose electrical problems by logic and symptom description.
- Perform troubleshooting procedures using meters, test lights, and jumper wires.
- Inspect, test, and replace the brake warning lamp, parking brake indicator lamp, brake stop lamp switch, and master cylinder fluid level sensor.
- Repair electrical wiring and connectors.

Interesting Fact   *Hydraulic brakes were invented in 1918 in California in Malcolm Loughead's shop. The spelling of his name was later changed to Lockheed. The Lockheed brake system first appeared on the custom-made 1920 Dusenberg car. The first mass-produced car to use the Lockheed hydraulic brake system was the 1924 Chrysler.*

Chapter 6

Basic Brake System Operation

Introduction

The brake system is one of the most important systems on a vehicle. It must slow a moving vehicle, bring a vehicle to a stop, and hold a vehicle stationary when stopped. If the brake system does not operate properly, the driver and passengers could be injured or killed in an accident. Technicians that service the brake system must be highly skilled experts because the work they do can save lives. This chapter covers the principles of science that are at work in a brake system. Understanding the science of braking will help you troubleshoot and service late-model brake systems more accurately and quickly. Before going through the basic theories, let us first take a look at the types of brake systems and components used on vehicles.

BRAKE SYSTEM OVERVIEW

The complete brake system consists of the major components shown in **Figure 1**. A brake system can be divided into the **service brakes**, which slow and stop the moving vehicle, and the **parking brakes**, which hold the vehicle stationary. On most late-model vehicles, the antilock brake system (ABS) is a third major subsystem; and many vehicles now also include traction control that relies on the brake system for operation.

LEVERAGE AND BRAKE PEDAL DESIGN

Braking action begins when the driver applies force to the brake pedal, and the pedal transfers that force through a linkage to the pistons in the master cylinder. The brake pedal also multiplies the force of the driver's foot through leverage.

The brake pedal is mounted on a lever with its pivot near the top of the lever. The movement of the pedal causes a pushrod to move against a piston in the master cylinder. The master cylinder is mounted inside the engine compartment and acts as a hydraulic pump when activated.

Most brake pedal installations are an example of what is called a second-class lever. In the science of physics, a second-class lever has a pivot point (or **fulcrum**) at one end and a force applied to the other end. A second-class lever transfers the output force in the same direction as the input force and multiplies the input force depending on where the output load is placed. The brake pedal installation shown in **Figure 2** has a 10-inch lever, and the load (the master cylinder's pushrod) is 2 inches from the fulcrum (8 inches from the pedal). The pedal ratio, or the force multiplying factor, is the length of the lever divided by the distance of the load from the fulcrum. In this case, it is 10 inches divided by 2, or a ratio of 5:1.

If the driver applies 50 pounds of force to the pedal, the lever increases the force at the master cylinder to 250 pounds. When the driver applies 50 pounds of input force, the pedal may travel about 2.5 inches. When the lever applies 250 pounds of output force, the pushrod moves only 0.5 inch. Thus, as **leverage** in a second-class lever increases force, it reduces distance by the same factor: 2.5 inches divided by 5 equals 0.5 inch.

SERVICE BRAKE DESIGN

Modern vehicle brakes evolved from the relatively crude brakes of horse-drawn vehicles. The earliest vehicle

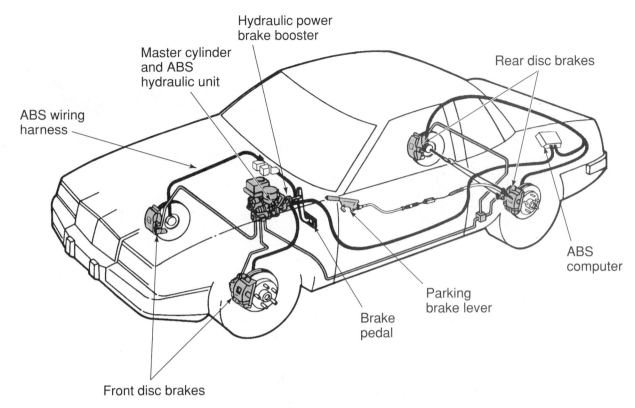

Figure 1. A typical automotive brake system is comprised of these major components and subsystems.

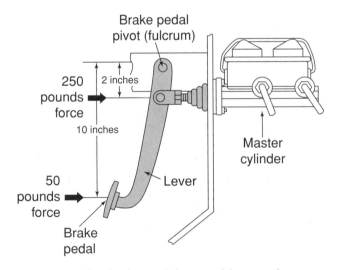

Figure 2. The brake pedal assembly uses leverage to increase force applied to the master cylinder.

brakes were pads or blocks applied by levers and linkages to the outside of a solid tire on a wooden-spoked wheel. The lever increased the force of the brake pad applied to the solid tire. These brakes worked fine with speeds of 10–20 mph and little traffic. As vehicle speeds and traffic increased, this type of brake system was no longer efficient.

By the end of the first decade of the twentieth century, vehicles were using either external-contracting band brakes or internal-expanding drum brakes. External-contracting brakes have a band lined with friction material wrapped around a drum located on the driveline or on the wheels. The band is anchored at one end or the center; levers and linkages tighten the band around the drum for braking force. The service brakes on Ford's Model T were a single contracting band applied to a drum inside the transmission.

Band brakes lose their effectiveness when higher braking force is needed. When you study drum brakes, you will learn about the mechanical servo action of brake shoes. It is very difficult to develop servo action with an internal band brake, thus higher brake force is needed. Servo action on an external band brake tends to make the brake grab at high brake forces and high drum speed. Other problems associated with band brakes included dirt and water damage and loss of friction with external bands and the tendency of these brakes to lock if the drum overheated and expanded too much. External band brakes also suffered from band and drum overheating and reduced braking force.

As drum brakes evolved, internal-expanding shoe-and-drum brakes became the standard. External-contracting band brakes were used as parking brakes until the late 1950s, but their days as service brakes were over by the late 1920s.

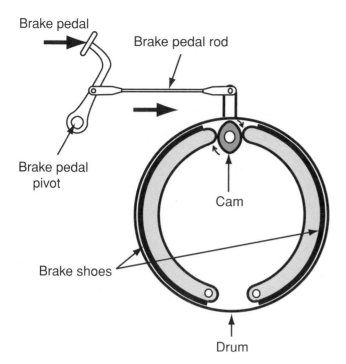

Figure 3.   A typical mechanical drum brake.

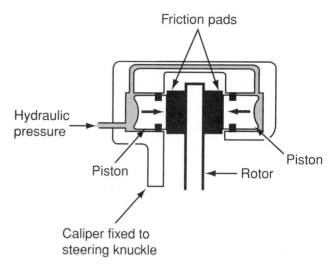

Figure 4.   Hydraulic pressure is applied equally to pistons on both sides of a fixed caliper.

Drum Brakes

By the mid-1920s, drum brakes with internal-expanding shoes were common. Early **drum brakes** were operated mechanically by levers and linkages (**Figure 3**). The 1921 Dusenberg Model A was one of the first vehicles to use hydraulic drum brakes. Hydraulic brakes started to become more common in the mid-1920s.

The rigid brake shoes used with drum brakes could be made stronger than the flexible bands of earlier brake designs. This eliminated breakage problems that occurred with the greater braking forces that were required as automobiles became more powerful and faster. With hydraulic actuation, four-wheel drum brakes remained the standard braking system for most vehicles into the middle and late 1960s. With the coming of FMVSS 105 in 1967, brake systems needed to pass specific performance tests, which led to the common use of disc brakes on the front wheels. Today, drum brakes are still used on the rear wheels of most cars and light trucks.

Disc Brakes

Disc brakes work by applying pressure to two brake pads on opposite sides of a spinning rotor attached to the hub of a wheel. Disc brake pads are mounted in a caliper that sits above the spinning rotor. The caliper is either fixed or movable on its mounting. With a fixed caliper, hydraulic pressure is applied to pistons, inside the caliper, on both sides to force the pads against the rotor

(**Figure 4**). With a movable caliper, pressure is applied only to the piston on the inboard side of the caliper. This forces the inboard pad against the rotor, and the reaction force moves the outboard side of the caliper inward so that both pads grip the rotor (**Figure 5**).

All the friction components of a disc brake are exposed to the air, which helps to cool the brake parts and maintain braking effectiveness during repeated hard stops from high speeds. A disc brake requires higher hydraulic pressure and greater force to achieve the same stopping power as a comparable drum brake. These pressure and force requirements for disc brakes are met easily, however, with large caliper pistons and power brake boosters. Because their advantages far outweigh any disadvantage, disc brakes have been standard equipment as the front

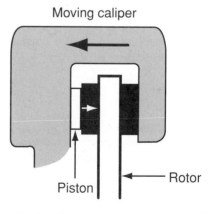

Figure 5.   Hydraulic pressure in a movable caliper forces the piston in one direction and the caliper body in the other. The resulting action and reaction force the pads against the caliper.

brakes on all cars and light trucks for many years. Additionally, four-wheel disc brakes are found on many vehicles and trucks.

BRAKE HYDRAULIC SYSTEMS

Common hydraulic brake systems consist of a master cylinder, steel lines, rubber hoses, various pressure-control valves, and wheel cylinders or caliper pistons at each wheel.

Master Cylinder

The master cylinder is sealed at one end, and the movable pushrod extends from the other **(Figure 6)**. The pushrod moves a pair of in-line pistons that produce the pumping action. When the brake pedal lever moves the pushrod, it moves the pistons to draw fluid from a reservoir on top of the master cylinder. Piston action then forces the fluid under pressure through outlet ports to the brake lines.

All master cylinders for vehicles built since 1967 have two pistons and pumping chambers. Motor vehicle safety standards require this in order to provide hydraulic system operation if one hose, line, or wheel brake assembly loses fluid. Because the brake hydraulic system is sealed, all the lines and cylinders are full of fluid at all times. When the master cylinder develops system pressure, the amount of fluid that is moved is only a few ounces.

Brake Lines and Hoses

The rigid lines, or pipes, of a brake hydraulic system are made of steel tubing. Flexible rubber hoses connect the

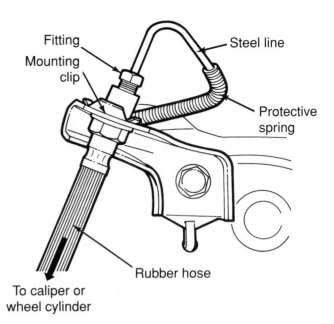

Figure 7. Rubber hoses and steel lines provide hydraulic pressure to the wheel brakes.

wheel brakes to the rigid lines mounted on the vehicle's body or frame **(Figure 7)**. The front brakes have a rubber hose at each wheel to allow for steering movement. Rear brakes can have separate hoses at each wheel brake or a single hose connected to a line on the body or frame if the vehicle has a rigid rear axle. Brake lines and hoses contain the high-pressure fluid, and the fluid acts as a solid rod to transmit force to the wheel cylinders and caliper pistons.

Pressure Control Valves

Nearly all brake hydraulic systems built since 1967 have one or more valves to control system pressure. Metering and proportioning valves modulate hydraulic pressure to front disc or rear drum brakes to provide smooth brake application and reduce the tendency of the brakes to lock. A valve, called the pressure differential switch, is used in most systems to illuminate the instrument-panel warning lamp if half of the hydraulic system loses pressure. The hydraulic system may have several individual valves or a single combination valve with multiple functions **(Figure 8)**.

The ABSs make some common valves obsolete. An ABS electronic control module can modulate hydraulic pressure for normal braking better than metering and proportioning valves can. As ABS installations become more widespread, some older hydraulic functions may be given over to electronic controls.

Wheel Cylinders and Caliper Pistons

Technically, the wheel cylinders in drum brakes and the caliper pistons in disc brakes are "slave" cylinders because they operate in response to the master cylinder.

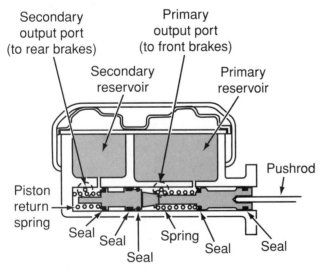

Figure 6. The master cylinder is a cylindrical pump with two pistons that develop pressure in the hydraulic lines to the front and rear brakes.

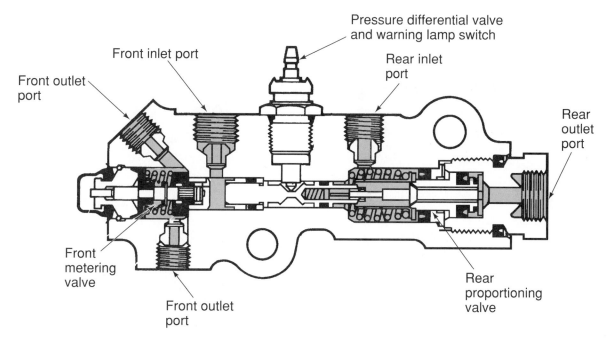

Figure 8.   This combination valve performs three hydraulic control functions in a single housing.

These hydraulic cylinders change hydraulic pressure into mechanical force to apply the brakes.

Most late-model systems have a single, 2-piston cylinder at each wheel with drum brakes **(Figure 9)**. Hydraulic pressure enters the cylinder between the two pistons and forces them outward to act on the brake shoes. As the shoes move outward, the lining contacts the drums to slow and stop the vehicle. Some drum brake installations have two cylinders at each wheel with one piston each. A few older systems had wheel cylinders with pistons of different diameters.

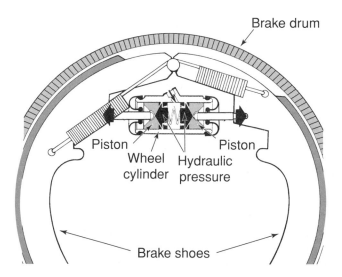

Figure 9.   Hydraulic pressure in the wheel cylinder moves the two pistons outward to force the shoes against the drum.

The caliper pistons for disc brakes also act in response to hydraulic pressure that enters a fluid chamber in the caliper. Hydraulic pressure in a stationary fixed caliper is applied to one or two pistons on each side of the caliper to force the pads against the rotor. Pressure is applied to a single piston in a movable caliper on the inboard side to force the inboard pad against the rotor. Because hydraulic pressure is equal in all directions in a sealed chamber, the equalized pressure creates a reaction force that moves the outboard side of the caliper inward so that both pads grip the rotor.

POWER BOOSTERS

Nearly all late-model brake systems have a power booster that increases the force of the driver's foot on the pedal **(Figure 10)**. Most cars and light trucks use a vacuum booster that uses the combined effects of engine vacuum and atmospheric pressure to increase pedal force. Some vehicles have a hydraulic power booster that may be separate from the brake system and supplied with fluid by the power steering system or a part of the brake system and driven by an electric motor.

PARKING BRAKES

A vehicle's parking brake is designed to hold the vehicle stationary after it has stopped. It is often mistakenly called an emergency brake, but its purpose is not to stop the vehicle in an emergency. The amount of potential stopping power available from a parking brake is much less

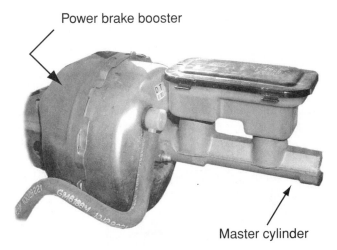

Power brake booster

Master cylinder

Figure 10. The power brake booster increases the brake pedal force applied to the master cylinder.

ANTILOCK BRAKE SYSTEMS

Whenever the brakes are applied with heavy pressure, a wheel can totally stop rotating. This is called wheel lockup. This does not help the car stop; rather, the tire loses some frictional contact with the road and slides or skids. As the tires slide, the car is no longer stopping under control and the driver is in a dangerous situation. Experienced drivers prevent wheel lockup by pumping the brake pedal up and down rapidly. This stops and starts hydraulic pressure to the brakes and gives the driver control during hard braking.

Many late-model cars have an ABS. ABS does the same thing as an experienced driver, only faster and more precisely. It senses when a wheel is about to lock up or skid. It then rapidly interrupts the braking pressure to the brakes at that wheel. Speed sensors, at the wheels, monitor the speed of the wheels and send this information to an onboard computer. The computer then directs the ABS unit to pulse the pressure going to the wheel that is starting to lock up.

than that from the service brakes. Because a parking brake works only on two wheels or on the driveline, it cannot adequately and safely stop a moving vehicle. In the rare case of total hydraulic failure, a parking brake can be used to stop a moving vehicle, but its application requires careful attention and skill to keep the vehicle from skidding or spinning.

The parking brake may be activated by a hand lever or a foot pedal. When the parking brake is being applied, a parking brake cable mechanically pulls on a lever that applies the brakes. Parking brakes are mechanically, not hydraulically, controlled.

The simplest parking brake system consists of cables and levers that apply the service brake shoes of rear drum brakes **(Figure 11)**. Vehicles with rear disc brakes can have a small drum formed on the inside of the rotor and small auxiliary brake shoes mounted on the hub or axle to act as parking brakes. Alternatively, some vehicles with rear disc brakes use a lever and a thrust screw at each rear wheel to apply the disc pads and serve as parking brakes.

THE PHYSICS OF A BRAKE SYSTEM

All brake systems work according to a few laws of physics and the concept of energy. Energy is the ability to do work and comes in many familiar forms. Chemical energy, mechanical energy, heat energy, and electrical energy are among the most obvious forms in all automotive systems.

A brake system converts one form of energy to another. To slow and stop a moving vehicle, the brakes change the kinetic energy of motion to heat energy through the application of friction devices. When the brakes change one form of energy to another, they are doing work. Work is the result of releasing, or using, energy.

Kinetic energy is the energy of mechanical work or motion. When an automobile starts, accelerates, decelerates, and stops, kinetic energy is at work. The amount of kinetic energy at work at any moment is determined by a vehicle's mass (weight) and speed and the rate at which the speed is changing.

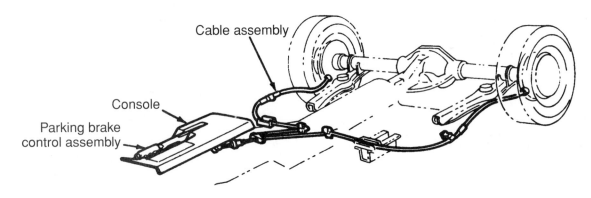

Cable assembly

Console

Parking brake
control assembly

Figure 11. Typical lever-operated parking brake installation.

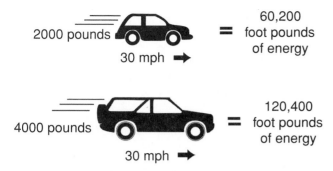

Figure 12. Kinetic energy increases proportionally with vehicle weight.

We can use the terms "mass" and "weight" interchangeably, but the two terms are not technically the same. **Mass** is a measurement of the number of molecules that make up an object. **Weight** is a measurement of the effect of gravity on that mass. Everything has mass, even air. The basic difference between mass and weight can be best explained by considering a space shuttle that weighs about one million pounds on earth. When the shuttle is in orbit, outside earth's gravity, it is weightless. However, its mass stays the same.

The combined effects of weight and speed make up kinetic energy, but speed has a much greater effect than weight. You can calculate the kinetic energy of any moving vehicle by multiplying the weight of the vehicle by the square of its speed. That product is then divided by a constant, 29.9. The result is the amount of kinetic energy expressed in foot-pounds.

Consider two cars, both traveling at 30 mph. One weighs 2000 pounds; the other weighs 4000 pounds **(Figure 12)**. The lighter car has 60,200 foot-pounds of kinetic energy; the heavier one has 120,400 foot-pounds. Doubling the car weight doubles the kinetic energy when the speeds are equal. Therefore, kinetic energy increases and decreases proportionally with weight. Now let us accelerate the lighter car to 60 mph **(Figure 13)**. It now has 240,802 foot-pounds of kinetic energy. Note that when we

doubled the speed, kinetic energy increased not two times, but four times. If we accelerate the same 2000-pound car to 120 mph, its kinetic energy is 16 times greater than it was at 30 mph. It is now almost 1,000,000 foot-pounds.

When we accelerate a car and then decelerate it and bring it to a stop, we are dealing with two forms of **inertia**. Inertia is simply the resistance to a change in motion. Static inertia is the inertia of an object at rest; dynamic inertia is the inertia of an object in motion. When the brake system slows and stops a vehicle, it overcomes dynamic inertia and imposes static inertia.

The brake system also must overcome the vehicle's **momentum**. Momentum is another way to view kinetic energy at work because it, too, is the mathematical product of an object's weight times its speed. Physical force starts an object in motion and gives it momentum. Another kind of force must overcome the momentum to bring the object to a stop. In a brake system, that force is friction.

FRICTION PRINCIPLES

Service brakes use friction to stop the vehicle. The parking brakes use friction to hold the vehicle stationary. Friction is used at the wheel brake units to stop the wheels. The friction between the tires and the road stops the vehicle. Because friction is so important to the brake system, you should understand some of its principles.

Types of Friction

Two basic types of friction are at work in the brake system. The first is called kinetic, or moving, friction. The second is called static, or stationary, friction.

When the brakes are applied on a moving vehicle, the frictional parts (brake shoes or pads) are forced against the rotating parts of the vehicle (brake drums or rotors). The friction causes the rotating parts to slow down and stop. This friction is called **kinetic friction**. Kinetic friction changes kinetic energy into **thermal energy** (heat). Brake parts and tires get very hot during braking because of kinetic friction. For example, the temperature of the brake friction material in a typical vehicle going 60 mph (95 kilometers per hour) can be more than 450°F (230°C) during an emergency stop.

Static friction holds the vehicle in place when it is stopped. The friction between the applied brake components and between the tire and the road resists movement. To move the vehicle, the brake components must be released. Then, the power of the engine must be great enough to overcome the static friction between the tires and the road.

Effects of Pressure on Friction

The amount of friction developed in a braking system is related to the amount of pressure used to force the fric-

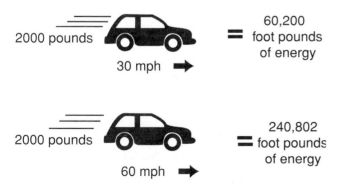

Figure 13. Kinetic energy increases exponentially with vehicle speed.

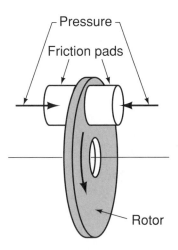

Figure 14. The greater the pressure against the friction pads, the greater the friction.

tion material against the rotating brake part. The more pressure that is applied to two frictional surfaces, the harder they grip each other and the harder they resist any movement between them.

Figure 14 shows the basic frictional parts of a disc brake system. The rotor is the part connected to, and rotating with, the vehicle's wheel. The friction pads are forced against the frictional surfaces or sides of the rotor. The friction causes the rotor to slow and stop, which stops the wheel from turning.

Braking pressure is created in the brake system by using mechanical leverage, hydraulic pressure, and different kinds of power-assist systems.

Surface Areas

The amount of friction produced is also affected by the area of the frictional surfaces in contact with each other. The larger of the two brake units shown in **Figure 15** has more frictional contact area than the smaller unit. The larger unit will stop a vehicle faster than the smaller unit. This is

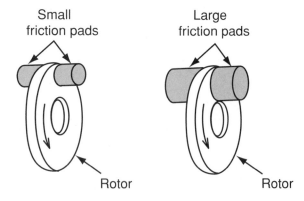

Figure 15. The larger the frictional contact area, the greater the braking force.

why large cars and trucks require larger brake components than smaller vehicles.

COEFFICIENT OF FRICTION

It takes more force to move some materials over a surface than others, even though the applied pressure and the amount of surface in contact are the same. Different materials have different frictional characteristics, or **coefficients of friction**. We can calculate the coefficient of friction by measuring the force required to slide an object over a surface and then dividing it by the weight of the object. The moving force that slides an object over a surface is **tensile force**. The weight of the object is the force pushing the object against the surface.

For example, it takes about 100 pounds of force to slide a 100-pound block of iron over a concrete floor, but only about 2 pounds of force to slide a 100-pound block of ice **(Figure 16)**. If we divide the amount of force by the weight of the object, we find that the coefficient of friction for the metal block is 1.0, whereas the ice block has a coefficient of friction of 0.02.

The coefficient of friction between brake friction components is always less than 1.0. A coefficient of friction greater than 1.0 means that some material actually is transferred from one surface to the other. Although brake friction surfaces wear, material is not transferred from pads to rotors or from shoe linings to drums. The coefficient of friction between a tire and the road can exceed 1.0. When material transfers from the tire to the road, we see a skid

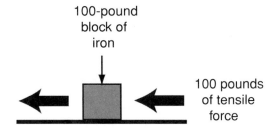

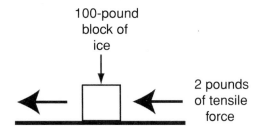

Figure 16. A high coefficient of friction requires greater force to move one surface against another.

mark. That means that the coefficient of friction momentarily exceeded 1.0.

If the coefficient of friction of a brake lining is too high, the brakes may grab and cause the wheels to lock up and slide. If the coefficient of friction is too low, excessive pressure on the brake pedal is required to stop the vehicle.

The friction materials selected for brake pads and rotors or brake shoes and drums are designed to give the best cold and hot temperature performance. The best materials have a coefficient that stays within narrow limits over a wide range of temperatures. If the selected material increases or decreases its coefficient of friction as temperature changes, then the brakes can fade, grab, or work erratically. This is one reason why only approved replacement brake parts should be installed in a vehicle.

Surface Finish

Let us go back to the 100-pound iron block that we discussed previously. If we polish the block's surface and wax the concrete floor, we might reduce the required tensile force to move the block from 100 pounds to 50 pounds. The metal block still weighs 100 pounds, but the coefficient of friction is now 0.5.

In a brake system, the surface finishes of pads, rotors, linings, and drums must be smooth for long life but must also be rough to provide good stopping power. Friction surfaces as rough as coarse sandpaper could provide excellent stopping action, but for just a few brake applications.

Materials

Iron and steel work best as rotors and drums because of their combined characteristics of strength, their ability to hold a surface finish for a long time, and their ability to transfer heat without being harmed by high temperatures.

Friction materials for pads and shoe linings are selected based on their coefficient of friction and their expected durability and useful life. If pads and linings are soft and wear fast, the vehicle may have very good braking ability, but their life will be short. On the other hand, if the friction material is hard and wears slowly, a low coefficient of friction and poor stopping ability will result.

Today, most friction materials are semimetallic or organic compounds. All have different coefficients of friction, and engineers select material that best matches the intended use of a vehicle and its braking requirements.

Temperature

Temperature also affects the coefficient of friction, but it affects different materials in different ways. A moderate amount of heat increases the coefficient of friction of most brakes. The semimetallic and carbon fiber materials of some racing brake linings must be heated quite a bit to work their best. Too much heat, however, reduces the coefficient

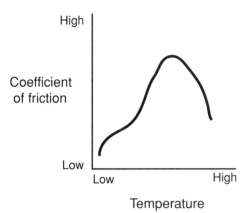

Figure 17.   Initially, the coefficient of friction increases with heat, but very high temperatures cause it to drop off.

of friction (**Figure 17**), and as heat continues to increase, the coefficient of friction continues to drop. This leads to **brake fade** and braking efficiency is reduced.

Heat Dissipation

The weight and speed of a vehicle during braking determines how much frictional energy is required to stop it. Because heat is a natural result of friction, the more frictional energy needed to stop the vehicle, the greater the amount of heat generated during braking. This heat must be removed or it will damage the brake system. Heat is removed from the friction surfaces as it passes through the friction material and metal of the brake components into the surrounding air. This is called heat dissipation.

BRAKE FADE

The buildup of heat from continuous braking can cause the friction material to become glazed or polished and the metal surface of the rotors or drums to become hardened. As a result, the driver will need to push harder on the brake pedal to stop the vehicle.

Brake drums and rotors absorb heat faster than they can dissipate it to the surrounding air. Drum or rotor temperature can rise 100°F (55°C) in just a couple of seconds during a hard stop. Dissipating that same heat to the air can take 30 seconds or more. Repeated hard stops can severely overheat brake components and reduce braking effectiveness. This is called brake fade.

A brake system is designed to provide the best possible heat dissipation. Parts are often vented to allow maximum airflow around the hot surfaces. The sizes of the friction surfaces are carefully designed with heat dissipation in mind. The larger the frictional contact area of the brake system, the better the heat dissipation and its resistance to brake fade.

BRAKING DYNAMICS

As you can see, much of a vehicle's ability to stop depends on friction. Other things, called dynamic factors, also affect a vehicle's braking ability. One important brake system dynamic is called weight transfer. During hard braking, the front of the vehicle lowers and the passengers feel they are being thrown forward. This is caused by the weight of the vehicle being transferred from the rear to the front during braking. This means that more of the braking must be done by the front wheels and less by the rear wheels.

The weight of a vehicle is not distributed evenly on all four wheels even when the vehicle is standing still. The location of the engine and powertrain components determines weight distribution. During braking, the weight of these components transfers forward. On a rear-wheel-drive vehicle, about 70 percent of the weight ends up on the front wheels during braking. On a front-wheel-drive vehicle, as much as 90 percent of the vehicle's weight ends up on the front wheels during braking. The vehicle's momentum and weight combine to cause the rear wheels to lift (have less downforce applied) and the front wheels to be forced down. This is why the front wheel brake units are larger than those used at the rear of the vehicle.

Summary

- The force applied to the brake pedal is multiplied by the brake pedal assembly that serves as a lever.
- Early brake systems relied on mechanical levers and linkages to apply the brakes.
- An automotive brake system consists of a master cylinder connected hydraulically through lines to disc and/or drum brake units that stop the wheels.
- A hydraulic or vacuum power assist is used on most vehicles to decrease the braking effort required from the driver.
- A mechanical brake, operated by levers and cables, is used for keeping the vehicle from moving after it has been stopped and is called the parking brake.
- Many vehicles have an ABS to improve brake operation and vehicle handling during emergency stopping.

- Kinetic energy is used both to accelerate and to brake a vehicle. Kinetic energy overcomes static inertia to accelerate a vehicle and overcomes dynamic inertia to stop it.
- The amount of friction in a brake system depends upon the pressure exerted on the friction surfaces and the frictional contact area.
- Different friction materials have different frictional characteristics, or coefficients of friction.
- Friction materials are made from asbestos or nonasbestos materials. The most common types of nonasbestos lining materials are nonmetallic materials, semimetallic materials, and fully metallic materials.
- A number of vehicle dynamics are related to braking. These include the shift of weight during braking and the tire contact area with the road.

Review Questions

1. When the brakes are applied on a moving vehicle, the frictional parts (brake _____ or _____) are forced against the rotating parts of the vehicle (brake _____ or _____). The friction causes the rotating parts to slow down and stop. Just as the energy of the rotating parts is called kinetic _____, the friction used to stop them is called kinetic _____.

2. The amount of kinetic energy at work at any moment is determined by a vehicle's _____ (_____) and _____ and the rate at which the _____ is changing.

3. The basic frictional parts of a brake system are being discussed. Technician A says that the harder the frictional parts are pushed together, the higher the friction. Technician B says that the harder the frictional parts are pushed together, the more heat is developed. Who is correct?
    A. Technician A only
    B. Technician B only
    C. Both Technician A and Technician B
    D. Neither Technician A nor Technician B

4. Brake friction used to hold a car in place after it is stopped is being discussed. Technician A says that this is kinetic friction. Technician B says that this is static friction. Who is correct?
    A. Technician A only
    B. Technician B only
    C. Both Technician A and Technician B
    D. Neither Technician A nor Technician B

5. Vehicle dynamics during braking are being discussed. Technician A says that the rear of the car rises during braking. Technician B says that the front of the car lowers during braking. Who is correct?
    A. Technician A only
    B. Technician B only
    C. Both Technician A and Technician B
    D. Neither Technician A nor Technician B

Chapter 7

Basic Hydraulic System Theory

Introduction

Automotive brake systems use the force of hydraulic pressure to apply the brakes. Because automotive brakes use hydraulic pressure, we need to study some basic hydraulic principles used in brake systems. These include the principles that fluids cannot be compressed, fluids can be used to transmit movement and force, and fluids can be used to increase force.

LAWS OF HYDRAULICS

Automotive brake systems are complex hydraulic circuits. To better understand how the systems work, a good understanding of how basic hydraulic circuits work is needed. A simple hydraulic system has liquid, a pump, lines to carry the liquid, control valves, and an output device. The liquid must be available from a continuous source, such as the brake fluid reservoir or a sump. In a hydraulic brake system, the master cylinder serves as the main fluid pump and moves the liquid through the system. The lines used to carry the liquid may be pipes, hoses, or a network of internal bores or passages in a single housing, such as those found in a master cylinder. Valves are used to regulate hydraulic pressure and direct the flow of the liquid. The output device is the unit that uses the pressurized liquid to do work. In the case of a brake system, the output devices are brake drum wheel cylinders (**Figure 1**) and disc brake calipers.

As can be seen, hydraulics involves the use of a liquid or fluid. Hydraulics is the study of liquids in motion. All matter, everything in the universe, exists in three basic forms: solids, liquids, and gases. A fluid is something that does not

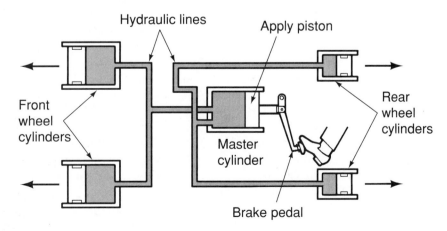

Figure 1. The master cylinder is an apply piston, working as a pump, to provide hydraulic pressure to the output pistons at the wheel brakes.

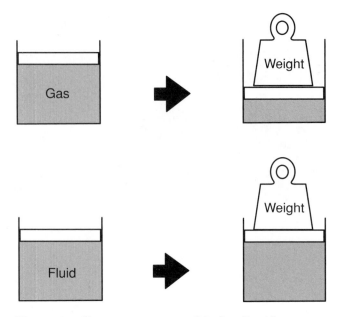

Figure 2. Gases are compressible, but liquids are not.

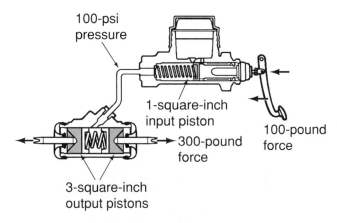

Figure 3. The mechanical force exerted on the brake pedal is transferred hydraulically to provide an increased mechanical force at the wheel brake unit.

have a definite shape; therefore, liquids and gases are fluids. A characteristic of all fluids is that they will conform to the shape of their container. A major difference between a gas and a liquid is that a gas will always fill a sealed container, whereas a liquid may not. A gas will also readily expand or compress according to the pressure exerted on it **(Figure 2)**. A liquid will typically not compress, regardless of the pressure on it. Therefore, liquids are considered noncompressible fluids. Liquids will, however, predictably respond to pressures exerted on them. Their reaction to pressure is the basis of all hydraulic applications. This fact allows hydraulics to do work.

Pascal's Law

More than 300 years ago a French scientist, Blaise Pascal, determined that if you had a liquid-filled container with only one opening and applied force to the liquid through that opening, the force would be evenly distributed throughout the liquid. This explains how pressurized liquid is used to operate and control the brakes on a vehicle. The action of the brake pedal on the pistons inside the master cylinder pressurizes the brake fluid and the fluid is delivered to the various wheel brake units **(Figure 3)**.

Pascal constructed the first known hydraulic device, which consisted of two sealed containers connected by a tube. The pistons inside the cylinders seal against the walls of each cylinder and prevent the liquid from leaking out of the cylinder and prevent air from entering into the cylinder. When the piston in the first cylinder has a force applied to it, the pressure moves everywhere within the system. The force is transmitted through the connecting

tube to the second cylinder. The pressurized fluid in the second cylinder exerts force on the bottom of the second piston, moving it upward and lifting the load on the top of it. By using this device, Pascal found he could increase the force available to do work, just as could be done with levers or gears.

Pascal determined that force applied to liquid creates pressure or the transmission of force through the liquid. These experiments revealed two important aspects of a liquid when it is confined and put under pressure. The pressure applied to it is transmitted equally in all directions and this pressure acts with equal force at every point in the container.

Interesting Fact *Pascal's work is known as Pascal's Law. Pascal's Law says that pressure at any one point in a confined liquid is the same in every direction and applies equal force on equal areas. One of the most important results of Pascal's work was the discovery that fluids may be used to increase force. Pascal was the person who first demonstrated the relationships of pressure, force, and motion and the inverse relationship of motion and force. In an automobile, Pascal's Laws are not applied just to the brake system. These same hydraulic principles are at work in the hydraulic system of an automatic transmission. Pascal's Laws are even at work in the movement of liquid fuel from a tank to the fuel injection system on the engine.*

Fluid Characteristics

If a liquid is confined and a force applied, pressure is produced. In order to pressurize a liquid, the liquid must be in a sealed container. Any leak in the container will decrease the pressure.

The basic principles of hydraulics are based on certain characteristics of liquids. Liquids have no shape of their own; they acquire the shape of the container they are put in. They also always seek a common level. Therefore, oil in a hydraulic system will flow in any direction and through any passage, regardless of size or shape. Liquids are basically incompressible, which gives them the ability to transmit force. The pressure applied to a liquid in a sealed container is transmitted equally in all directions and to all areas of the system and acts with equal force on all areas. As a result, liquids can provide great increases in the force available to do work. A liquid under pressure may also change from a liquid to a gas in response to temperature changes.

Fluids Can Transmit Movement

Liquids can be used to transmit movement. Two cylinders of the same diameter are filled with a liquid and connected by a pipe as shown in **Figure 4**. If you force piston A downward, the liquid will push piston B upward. Because piston A starts the movement, it is called the apply piston. Piston B is called the output piston. If the apply piston moves 10 inches, the output piston also will move 10 inches. This principle works not only for one output piston, but for any number of output pistons.

The principle that motion can be transmitted by a liquid is used in hydraulic brake systems. A master cylinder piston is pushed when the driver applies the brakes. The master cylinder piston is the apply piston. The brake fluid in the master cylinder is connected by pipes to pistons in each of the car's front and rear wheel brake units. Each of the wheel brake pistons is an output piston. They move whenever the master cylinder input piston moves.

Mechanical Advantage with Hydraulics

Hydraulics is used to do work in the same way as a lever or gear does work. All of these systems transmit energy or force. Because energy cannot be created or destroyed, these systems only redirect energy to perform work and do not create more energy. **Work** is the amount of force applied and the distance over which it is applied. **Force** is power working against resistance; it is the amount of push or pull exerted on an object needed to cause motion. We usually measure force in the same units that we use to measure weight: pounds or kilograms. **Pressure** is the amount of force exerted onto a given surface area. Therefore, pressure equals the applied force (measured in pounds or kilograms) divided by the surface area (measured in square inches or square centimeters) that is receiving the force. In customary English units, pressure is measured in pounds per square inch (psi). In the metric system it can be measured in kilograms per square centimeter, but the preferred metric pressure measurement unit is the pascal.

The pressure of a liquid in a closed system such as a brake hydraulic system is the force exerted against the inner surface of its container, which is the surface of all the lines, hoses, valves, and pistons in the system. Pressure applied to a liquid exerts force equally in all directions. If the hydraulic pump provides 100 psi, there will be 100 pounds of force on every square inch of the system **(Figure 5)**. When pressure is applied to a movable output piston, it creates output force.

If the system included a piston with an area of 30 square inches, each square inch would receive 100 pounds of force. This means there would be 3,000 pounds of force applied to that piston **(Figure 6)**. The use of the larger piston would give the system a mechanical advantage or increase

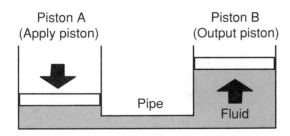

Figure 4.   Fluid can transmit motion through a closed system.

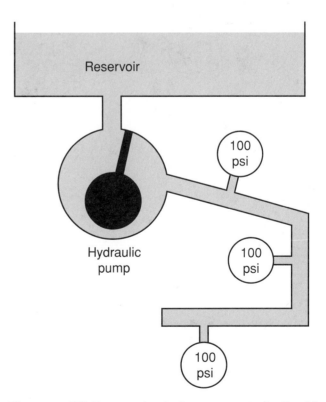

Figure 5.   While contained, the pressure of a liquid is the same throughout the container.

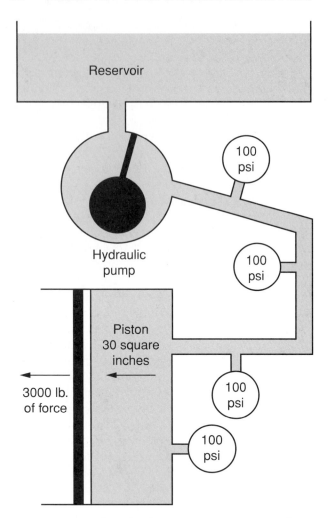

Figure 6. One hundred psi on a 30-square-inch piston generates 3000 pounds of force.

in the force available to do work. The multiplication of force through a hydraulic system is directly proportional to the difference in the piston sizes throughout the system. By changing the sizes of the pistons in a hydraulic system, force is multiplied, and as a result, low amounts of force are needed to move heavy objects. The mechanical advantage of a hydraulic system can be further increased by the use of levers to increase the force applied to a piston.

In **Figure 7**, input piston A is smaller than output piston B. Piston A has an area of 20 square inches; in the example, we are applying 200 pounds of force. Therefore,

$$\frac{200 \text{ pounds (F)}}{20 \text{ square inches (A)}} = 10 \text{ psi (P)}$$

where F is force, A is area, and P is pressure.

If that same 200 pounds of force is applied to a piston of 10 square inches, system pressure is 20 psi because

$$\frac{200 \text{ pounds (F)}}{10 \text{ square inches (A)}} = 20 \text{ psi (P)}$$

Therefore, pressure is inversely related to piston area. The smaller the piston, the greater the pressure that is developed.

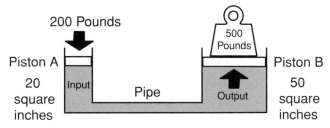

Figure 7. A hydraulic system also can increase force.

Let us apply the 10 psi of pressure in the first example to an output piston with an area of 50 square inches. In this case, output force equals pressure times the surface area:

$$P \times A = F$$

Therefore, 10 psi of pressure on a 50-square-inch piston develops 500 pounds of output force:

$$10 \times 50 = 500$$

Brake systems use hydraulics to increase force for brake application.

Figure 8 shows a hydraulic system with an input piston of 10 square inches. A force of 500 pounds is pushing on the piston. The pressure throughout the system is 50 psi:

$$500 \text{ (F)} \div 10 \text{ (A)} = 50 \text{ (P)}.$$

A pressure gauge in the system shows the 50-psi pressure.

There are two output pistons in the system. One has 100 square inches of area. The 50-psi pressure in the system is transmitted equally everywhere in the system. This means that the large output piston has 50 psi applied to 100 square inches to deliver an output force of 5000 pounds:

$$100 \text{ square inches} \times 50 \text{ psi} = 5000 \text{ pounds}$$

The other output piston in Figure 8 is smaller than the input piston with a 5-square-inch area. The 5-square-inch area of this piston has 50-psi pressure acting on it to develop an output force of 250 pounds:

$$5 \text{ square inches} \times 50 \text{ psi} = 250 \text{ pounds}$$

In a brake system, a small master cylinder piston is used to apply pressure to larger pistons at the wheel brake units to increase braking force. Importantly, the pistons in the front brakes have a larger surface area than the pistons in the rear brakes. This creates greater braking force at the

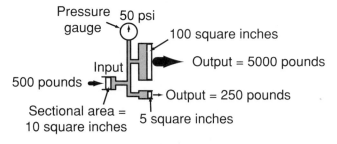

Figure 8. Different-sized output pistons produce different amounts of output force from the same hydraulic pressure.

front wheels to overcome the weight transfer created by momentum during braking.

Hydraulic Pressure, Force, and Motion

Although the force available to do work is increased by using a larger piston in one cylinder, the total movement of the larger piston is less than that of the smaller one. When output force increases, output motion decreases. If the 10-square-inch input piston moves 2 inches as it applies 50 psi to the 100-square-inch output piston, that output piston will move only 0.2 inch as it applies 5000 pounds of output force **(Figure 9)**. The ratio of input motion to output motion is the ratio of the input piston area to the output piston area, and you can use this simple equation to calculate it: The result from dividing the area of the input piston (A1) by the area of the output piston is multiplied by the stroke of the input piston or

(A1 ÷ A2) × S (the input stroke) = M (the output stroke)

or

$$\frac{10 \text{ square inches (input piston)}}{100 \text{ square inches (output piston)}} =$$

$$\frac{1}{10} \times 2 \text{ inches (input stroke)} =$$

0.2 inch output motion

If the output piston is larger than the input piston, it exerts more force but travels a shorter distance. The opposite also is true. If the output piston is smaller than the input piston, it exerts less force but travels a longer distance. Apply the equation to the 5-square-inch output piston in Figure 9:

$$\frac{10 \text{ square inches (input piston)}}{5 \text{ square inches (output piston)}} =$$

$$\frac{2}{1} \times 2 \text{ inches (input stroke)} =$$

4.0 inches output motion

In this case, the smaller output piston applies only half the force of the input piston, but its stroke (motion) is twice as long.

These relationships of force, pressure, and motion in a brake system are easily observed when you consider the

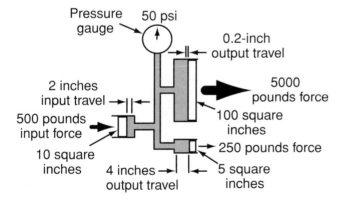

Figure 9. As output force increases, output travel (motion) decreases.

force applied to the master cylinder's pistons and the resulting brake force and piston movement at the wheels. Wheel cylinder pistons move only a fraction of an inch to apply hundreds of pounds of force to the brake shoes, but the wheel cylinder piston travel is quite a bit less than the movement of the master cylinder piston. Disc brake caliper pistons move only a few thousandths of an inch but apply great force to the brake rotors.

This demonstrates how the use of hydraulics provides a mechanical advantage similar to that provided by the use of levers or gears. Although hydraulic systems, gears, and levers can accomplish the same results, hydraulics is preferred when the size and shape of the system are of concern. In hydraulics, the force applied to one piston will transmit through the fluid, and the opposite piston will have the same force on it. The distance between the two pistons in a hydraulic system does not affect the force in a static system. Therefore, the force applied to one piston can be transmitted without change to another piston located somewhere else.

A hydraulic system responds to the pressure or force applied to it. The mere presence of different-sized pistons does not always result in fluid power. The force or pressure applied to the pistons must be different in order to cause fluid power. If an equal amount of pressure is exerted onto both pistons in a system and both pistons are the same size, neither piston will move; the system is balanced or is at equilibrium. The pressure inside the hydraulic system is called **static pressure** because there is no fluid motion.

When an unequal amount of pressure is exerted on the pistons, the piston receiving the least amount of pressure will move in response to the difference between the two pressures. Likewise, the fluid will move if the size of the two pistons is different and an equal amount of pressure is exerted on the pistons. The pressure of the fluid while it is in motion is called **dynamic pressure**.

HYDRAULIC BRAKE SYSTEMS

Engineers must consider these principles of force, pressure, and motion when designing a brake system for any vehicle. If an engineer chooses a master cylinder with relatively small piston areas, the brake system can develop very high hydraulic pressure, but the pedal travel will be extremely long. Moreover, if the master cylinder piston travel is not long enough, this high-pressure system will not move enough fluid to apply the large-area caliper pistons regardless of pressure. If, on the other hand, the engineer selects a large-area master cylinder piston, it can move a large volume of fluid but may not develop enough pressure to exert adequate braking force at the wheels.

The overall size relationships of master cylinder pistons, caliper pistons, and wheel cylinder pistons are balanced to achieve maximum braking force without grabbing or fading. Most brake systems with front discs and rear drums have large-diameter master cylinders (a large piston area) and a power booster to increase the input force.

HYDRAULIC BRAKE FLUID

The liquid used in a hydraulic brake system is brake fluid. The specifications for all automotive brake fluids are defined by Society of Automotive Engineers (SAE) Standard J1703 and Federal Motor Vehicle Safety Standard (FMVSS) 116. Fluids classified according to FMVSS 116 are assigned United States Department of Transportation (DOT) numbers: **DOT 3, 4, and 5**. Basically, the higher the DOT number **(Figure 10)**, the more rigorous the specifications for the fluid. These specifications list the qualities that brake fluid must have, such as:

- Free flow at low and high temperatures.
- A high boiling point (over 400°F or 204°C).
- A low freezing point.
- Ability to not deteriorate metal or rubber brake parts.
- Ability to lubricate metal and rubber parts.
- Ability to absorb moisture that enters the hydraulic system.

Choosing the right fluid for a specific vehicle is not based on the simple idea that if DOT 3 is good, DOT 4 must be better, and DOT 5 better still. The domestic carmakers all specify DOT 3 fluid for their vehicles, but Ford calls for a heavy-duty variation that meets the basic specifications for DOT 3 but has the higher boiling point of DOT 4. Import manufacturers are about equally divided between DOT 3 and DOT 4.

DOT 3 and DOT 4 fluids are polyalkylene-glycol-ether mixtures, called **polyglycol**. The color of both DOT 3 and DOT 4 fluid ranges from clear to light amber. DOT 5 fluids are all silicone based because only silicone fluid—so far—can meet the DOT 5 specifications. No vehicle manufacturer, however, recommends DOT 5 fluid for use in its brake systems. Although all three fluid grades are compatible they do not combine well if mixed together in a system. Therefore,

Figure 10. The DOT number is always clearly stated on a container of brake fluid.

the best rules are to use the fluid type recommended by the manufacturer and never mix fluid types in a system.

> **Interesting Fact** Some older European cars used a petroleum-based brake fluid. This fluid was colored green. Many technicians got into trouble by putting the wrong type of fluid in these cars or using this type of brake fluid in another system. The result was usually rapid and complete brake system failure.

Brake Fluid Boiling Point

The most apparent differences among the three fluid grades are the minimum boiling points as listed below:

	DOT 3	DOT 4	DOT 5
Dry boiling point	401°F (205°C)	446°F (230°C)	500°F (260°C)
Wet boiling point	284°F (140°C)	311°F (155°C)	356°F (180°C)

The boiling point is important because heat generated by braking can be transmitted into the hydraulic system. If the temperature rises too high, the fluid can boil and form a vapor in the brake lines. The stopping power of the system then will be reduced. As a result, the brake pedal can go to the floor and the vehicle will not stop.

The dry boiling point is the minimum boiling point of new, uncontaminated fluid. Because brake fluids are hygroscopic, their boiling points decrease due to water contamination after the fluid has been in service for some time. Brake systems are not completely sealed, and some exposure of the fluid to air is inevitable.

Other Brake Fluid Requirements

A high-temperature boiling point is not the only requirement brake fluid must meet. Brake fluid must remain stable throughout a broad range of temperatures, and it must retain a high boiling point after repeated exposure to high temperatures. Brake fluid must also resist freezing and evaporation and must pass specific viscosity tests at low temperatures. If the fluid thickens and flows poorly when cold, brake operation will suffer.

Besides temperature requirements, brake fluid must pass corrosion tests. It also must not contribute to deterioration of rubber parts and must pass oxidation-resistance tests. Finally, brake fluid must lubricate cylinder pistons and bores and other moving parts of the hydraulic system.

DOT 5 Silicone Fluid

Silicone DOT 5 fluid does not absorb water. This purple fluid has a very high boiling point, is noncorrosive to hydraulic system components, and does not damage paint like ordinary fluid does. DOT 5 fluid also has some other characteristics that are not so beneficial.

Silicone fluid compresses slightly under pressure, which can cause a slightly spongy brake pedal feel. Silicone fluid also attracts and retains air more than polyglycol fluid does, which makes brake bleeding harder; it tends to outgas slightly just below its boiling point, and it tends to aerate from prolonged vibration. DOT 5 fluid has other problems with seal wear and water accumulation and separation in the system. All of these factors mean that DOT 5 silicone fluid should *never be used* in an ABS.

Hydraulic System Mineral Oil (HSMO) Fluids

HSMO is the least common type of brake fluid, being used by only three carmakers: Citroen, Rolls Royce, and in some Audi models in their brake booster system. HSMO is not a polyglycol or silicone fluid, but rather is made from a mineral oil base. It has a very high boiling point, it is not hygroscopic, it is a very good lubricant, and it actively prevents rust and corrosion. HSMO fluid can be identified by its green color.

Because HSMO is petroleum based, systems designed for its use also require seals made of special rubber. If polyglycol or silicone fluid is used in a system designed for HSMO, these fluids will destroy the HSMO system seals.

> **You Should Know** *If HSMO is used in a system designed for polyglycol or silicone fluid,* it will destroy the seals of those systems. *HSMO is not covered by the DOT classifications of FMVSS 116 and is not compatible with DOT fluids.*

Fluid Compatibility

Although the performance requirements of DOT 3, 4, and 5 fluids are different, FMVSS 116 requires that DOT 3 and 4 fluids must be compatible with each other in a system. Mixing DOT 3 and DOT 4 in a system is not recommended, but it can be done without damaging the system or creating a damaging reaction. DOT 5 should never be mixed with DOT 3 or 4 fluids.

If DOT 3 and DOT 4 fluids are mixed in a system, the boiling point of the DOT 4 fluid will be reduced by the same percentage as the percentage of DOT 3 fluid in the mixture. Thus, overall system performance can be compromised by mixing fluids.

Reservoir

All hydraulic systems require a reservoir to store fluid and to provide a constant source of fluid for the system. In a brake system, the reservoir is attached to the top of the master cylinder, although some vehicles might use tubing to connect the reservoir to the master cylinder. Brake fluid is forced out of the pan by atmospheric pressure into the master cylinder and returned to it after the brake pedal has been let up.

Venting

In order to allow the fluid to flow into the master cylinder, the reservoir has an air vent that allows atmospheric pressure to force the fluid into the master cylinder when a low pressure is created by the movement of the pistons. The vent is positioned above the normal brake fluid level in the reservoir and keeps atmospheric pressure at the top of the fluid.

VACUUM AND AIR PRESSURE PRINCIPLES

A law of nature defines the role of **atmospheric pressure** on the operation of a brake system. The law simply states that whenever a high pressure is introduced to a lower pressure, it moves to equalize the pressures. In other words, something that has a high pressure will always move toward something that has a lower pressure. The force at which the higher pressure moves toward the lower pressure is determined by the difference in pressures. When the pressure is slightly lower than atmospheric, the force is low. When there is a large difference, the higher pressure will rush into the lower and the force will be great.

In the world of automotive technology, any pressure that is lower than atmospheric pressure is called a **vacuum**. Atmospheric pressure **(Figure 11)** is the pressure of the air

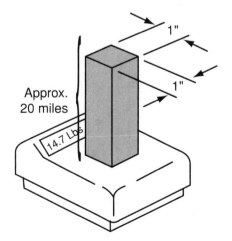

Figure 11.   A square-inch column of air the height of the earth's atmosphere exerts 14.7 pounds of pressure on the earth's surface at sea level.

around and on us and has a value of approximately 14.7 psi at sea level. When we are at higher elevations, there is less air above and on us and therefore the pressure of the atmosphere is less, but that air is still considered atmospheric pressure, and any pressure less than that is a vacuum.

When the pistons inside a master cylinder move, the volume of the piston's cylinder changes. When the volume decreases, the pressure increases. When the piston moves back to its original location, the pressure is lower and atmospheric pressure pushes fluid from the reservoir into the cylinder.

The relationship of vacuum and atmospheric pressure is used in most power brake systems to provide a power

assist for the driver **(Figure 12)**. The rush of high pressure toward an area of vacuum causes an increase in force, much like a lever.

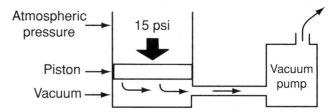

Figure 12. Vacuum (low pressure) works with atmospheric pressure to develop force.

Summary

- In a hydraulic brake system, the master cylinder moves brake fluid through the system. The lines used to carry the liquid may be pipes, hoses, or a network of internal bores or passages in a single housing, such as those found in a master cylinder. Valves are used to regulate hydraulic pressure and direct the flow of the liquid. The output devices are brake drum cylinders and disc brake calipers.
- Hydraulics is the study of liquids in motion.
- Liquids are considered noncompressible fluids.
- Pascal's Law says that pressure at any one point in a confined liquid is the same in every direction and applies equal force on equal areas.
- If a liquid is confined and a force applied, pressure is produced. If the pressure on the fluid is applied to a movable output piston, it creates output force.
- In a brake system, a small master cylinder piston is used to apply pressure to larger pistons at the wheel brake units to increase braking force.

- Most brake systems with front discs and rear drums have large-diameter master cylinders (large piston area) and a power booster to increase the input force.
- DOT 3 and DOT 4 fluids are polyglycol mixtures. The color of both DOT 3 and DOT 4 fluids ranges from clear to light amber.
- DOT 5 fluids are all silicone based, and no vehicle manufacturer recommends DOT 5 fluid for use in its brake systems.
- Although the performance requirements of DOT 3 and 4 fluids are different, FMVSS 116 requires that the grades of fluid be compatible with each other in a system; however, mixing different types of fluid in a system is not recommended.
- All hydraulic systems require a reservoir to store fluid and to provide a constant source of fluid for the system.
- The relationship of vacuum and atmospheric pressure is used in most power brake systems.

Review Questions

1. Explain how hydraulic fluid can be used to transmit motion.
2. Hydraulic systems work because fluids cannot be _____.
3. The pressure in a hydraulic system is the same in _____ directions.
4. A hydraulic system with a 1-square-inch input piston and a 3-square-inch output piston is being discussed. Technician A says that the output piston will have three times as much force as the input piston. Techni-

cian B says that the output piston will move one-third as far as the input piston. Who is correct?
   A. Technician A only
   B. Technician B only
   C. Both Technician A and Technician B
   D. Neither Technician A nor Technician B
5. Technician A says that DOT 3 brake fluid has a higher boiling point than DOT 5. Technician B says that DOT 4 brake fluid has a lower boiling point than DOT 5. Who is correct?
   A. Technician A only
   B. Technician B only
   C. Both Technician A and Technician B
   D. Neither Technician A nor Technician B

Chapter 8

Basic Electrical Theory

Introduction

Antilock brake systems rely on electronic actuation and control, but all modern brake systems also use simple electrical circuits for basic system status checks and safety functions. For example, an electrical circuit is energized if the parking brake is applied or if fluid level in the master cylinder reservoir falls below a certain point. An electrical circuit can also be energized if a pressure difference develops in the hydraulic brake lines. All of these circuits are tied into one or more brake system warning lamps on the instrument panel.

To understand the operation of these circuits and others, you must have a good understanding of electricity and electronics. Although the subject is normally covered in a separate course, a quick overview of electricity and its principles is presented here.

BASIC ELECTRICITY

All things are made up of atoms. Atoms are the smallest particles of something and are the building blocks of all materials. The basics of electricity focus on atoms. Understanding the structure of the atom is the first step in understanding how electricity works. In the center of every atom is a nucleus, which contains positively charged particles called protons and particles called neutrons that have no charge. Negatively charged particles called electrons orbit around every nucleus. Every type of atom has a different number of protons and electrons, but each atom has an equal number of protons and electrons. Therefore, the total electrical charge of an atom is zero, or neutral.

The release of energy as one electron leaves the orbit of one atom and jumps into the orbit of another is **electricity**. There is a natural attraction of electrons to protons. Electrons have a negative charge and are attracted to something with a positive charge. An electron moves from one atom to another because the atom next to it appears to be more positive than the one it is orbiting around. In order to have a continuous flow of electricity, three things must be present: an excess of electrons in one place, a lack of electrons in another place, and a path between the two places.

Two power or energy sources are used in an automobile's electrical system—the vehicle's battery **(Figure 1)**

Figure 1. A typical automotive battery.

Figure 2.   An AC generator (alternator).

and AC generator **(Figure 2)**. A chemical reaction in the battery provides for an excess of electrons and a lack of electrons in another place. Batteries have two terminals, a positive and a negative. The chemical reaction in a battery causes a lack of electrons at the positive (+) terminal and an excess at the negative (−) terminal. This creates an electrical imbalance, causing the electrons to flow through the path provided by a wire. A simple example of this process is shown in the battery and light arrangement in **Figure 3**.

The chemical process in the battery continues to provide electrons until the chemicals become weak. Fortunately, the vehicle's charging system restores the battery's supply of electrons. This allows the chemical reaction in the battery to continue indefinitely.

Moving a wire through an already existing magnetic field (such as a permanent magnet) can produce electricity. This process of producing electricity through magnetism is called **induction**. In a generator, a magnetic field is moved through a coil of wire, and electricity is induced. The amount

of induced electricity depends on a number of factors: the strength of the magnetic field, the number of wires that pass through the field, and the speed at which the wire moves through the magnetic field. A common component of ABSs, a wheel speed sensor, operates by the same principles.

ELECTRICAL TERMS

Electrical **current** is a term used to describe the movement or flow of electricity. The greater the number of electrons flowing past a given point in a given amount of time, the more current there is in the circuit. **Voltage** is electrical pressure. Voltage is the force developed by the attraction of the electrons to the protons. The more positive one side of the circuit is, the more voltage is available to the circuit. Voltage does not flow; rather, it is the pressure that causes current flow. When any substance flows, it meets resistance. The resistance to electrical flow can be measured.

Electrical Current

The unit for measuring electrical current is the **ampere**, usually called an amp. The instrument used to measure electrical current flow in a circuit is called an **ammeter**.

When electricity flows, millions of electrons are moving past any given point at the speed of light. The electrical charge from any one moving electron is extremely small. It takes millions of electrons to make a charge that can be measured.

There are two types of current: **direct current (DC)** and **alternating current (AC)**. In direct current, the electrons flow in one direction only. In alternating current, the electrons change direction at a fixed rate. Most automobile circuits operate on DC current, whereas the current in homes and buildings is AC.

Resistance

The resistance to current flow produces heat. This heat can be measured to determine the amount of resistance. A unit of measured resistance is called an **ohm**. Resistance can be measured by an instrument called an **ohmmeter**.

Voltage

In electrical flow, some force is needed to move the electrons from one atom to another. This force is the pressure that exists between a positive point and a negative point within an electrical circuit. This force, also called **electromotive force (EMF)**, is measured in units called **volts**. One volt is the amount of pressure (force) required to move 1 ampere of current through a resistance of 1 ohm. Voltage is measured by an instrument called a **voltmeter**.

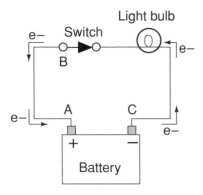

Figure 3.   A simple light circuit.

ELECTRICAL CIRCUITS

When electrons are able to flow along a path (wire) between two points, an electrical circuit is formed. An electrical circuit is considered a **complete circuit** when there is a path that connects the positive and negative terminals of the electrical power source. In a complete circuit, resistance must be low enough to allow the available voltage to push electrons between the two points. Most automotive circuits contain four basic parts:

- A power source, such as a battery or generator that provides the energy needed to create electron flow.
- Conductors, such as copper wires that provide a path for the electrons.
- **Loads** or devices, such as lights and motors, that use electricity to perform work.
- **Controllers**, such as switches that control or direct the flow of electrons.

A functioning electrical circuit must have a complete path from the power source to the load and back to the source. With the many circuits in an automobile, this would require hundreds of wires connected to both sides of the battery. To avoid this, vehicles are equipped with power distribution centers or fuse blocks that distribute battery voltage to the various circuits. The positive side of the battery is connected to the fuse block and power is distributed from there.

As a common return circuit, manufacturers use the vehicle's metal frame as part of the return circuit. The load is often grounded directly to the metal frame. Current passes from the battery, through the load, and into the frame. The frame is connected to the negative terminal of the battery through the battery's ground cable. This completes the circuit.

Electrical components are often mounted directly to the engine block, transmission case, or frame. This direct mounting effectively grounds the component without the use of a separate ground wire. In other cases, however, a separate ground wire must be run from the component to the frame or another metal part to ensure a sound return path. The increased use of plastics and other nonmetallic materials in body panels and engine parts has made electrical grounding more difficult. To ensure good grounding back to the battery, some manufacturers now use a network of common grounding terminals and wires.

CONDUCTORS AND INSULATORS

Controlling and routing electricity requires the use of materials known as conductors and insulators. **Conductors** are materials with a low resistance to the flow of current. Most metals, such as copper, silver, and aluminum are excellent conductors.

Copper wire is by far the most popular conductor used in automotive electrical systems. Wire wound inside of electrical units, such as ignition coils and generators, usually has a very thin baked-on insulating coating. External wiring often is covered with a plastic-type insulating material that is highly resistant to environmental factors such as heat, vibration, and moisture. Where flexibility is required, the copper wire will be made of a large number of very small strands of wire woven together.

Insulators resist the flow of current. Thermal plastics are the most common electrical insulators used today. They can resist heat, moisture, and corrosion without breaking down.

OHM'S LAW

To understand the relationship between current, voltage, and resistance in a circuit, you should become familiar with **Ohm's Law**. This law states that it takes 1 volt of electrical pressure to push 1 ampere of electrical current through 1 ohm of resistance. As such, the law provides a mathematical formula for determining the amount of current, voltage, or resistance in a circuit when two of these are known. The basic formula is: voltage equals current multiplied by resistance.

Although the basic premise of this formula is calculating unknown values in an electrical circuit **(Figure 4)**, it also helps to define the behaviors of electrical circuits. If voltage does not change, but there is a change in the resistance of the circuit, the current will change. If resistance increases, current decreases. If resistance decreases, current will increase. If voltage changes, so must the current or resistance. If the resistance stays the same and the current decreases, so will voltage. Likewise, if current increases, so will the voltage.

In a complete circuit, the flow of electricity is controlled and applied to do useful work, such as light a headlamp or turn over a starter motor. Components that use electrical power put a load on the circuit and consume electrical energy. The energy used by a load is measured in volts and is called **voltage drop**.

Voltage (E) = Current (I) times Resistance (R); therefore

$$E = I \times R.$$

Current (I) = Voltage (E) divided by Resistance (R); therefore

$$I = E/R.$$

Resistance (R) = Voltage (E) divided by Current (I); therefore

$$R = E/I.$$

Figure 4. Ohm's Law.

CIRCUIT COMPONENTS

Automotive electrical circuits contain a number of different types of electrical devices.

Resistors

Resistors are used to limit current flow (and thereby voltage) in circuits in which full current flow and voltage are not needed. Resistors are devices specially constructed to introduce a measured amount of electrical resistance into a circuit. Fixed value resistors are designed to have only one rating, which should not change. Some electrical loads use resistance to produce heat, such as lamps that get so hot they produce light.

Tapped or stepped resistors are designed to have two or more fixed values, available by connecting wires to the several taps on the resistor. Heater motor resistor packs, which provide for different heater fan speeds, are an example of this type of resistor.

Variable resistors are designed to have a range of resistances available through two or more taps and a control. Two examples of this type of resistor are rheostats and potentiometers. **Rheostats** have two connections (**Figure 5**), one to the fixed end of a resistor and one to a sliding contact with the resistor. Turning the control moves the sliding contact away from or toward the fixed-end tap, increasing or decreasing the resistance. **Potentiometers** have three connections (**Figure 6**), one at each end of the resistance and one connected to a sliding contact with the resistor. Turning the control moves the sliding contact away from one end of the resistance, but toward the other end.

Circuit Protective Devices

When overloads or shorts in a circuit cause excessive current flow, the wiring in the circuit heats up, the insulation melts, and a fire can result, unless the circuit has some kind of protective device. **Fuses**, fuse links, **maxi-fuses**, and **circuit breakers** are designed to prevent circuit damage caused by high current. Protection devices open the circuit when high current is present. As a result, the circuit no longer works but the wiring and the components are saved from damage.

Switches

Electrical circuits are usually controlled by some type of **switch**. Switches turn the circuit on or off or direct the flow of current in a circuit. Switches can be controlled by the driver or can be self-operating through a condition of the circuit, the vehicle, or the environment.

A hinged-pawl switch is the simplest type of switch. It either makes or breaks (completes or opens) a single conductor or circuit. It is a single-pole, single-throw switch (**SPST switch**). The throw refers to the number of output circuits, and the pole refers to the number of input circuits made by the switch. Switches can be designed with a great number of poles and throws and can be referred to as multiple-pole, multiple-throw switches (**MPMT switches**).

Relays

A **relay (Figure 7)** is an electrical switch that allows a small amount of current to control a circuit with high current. The low-current circuit is called the control circuit. When the control circuit switch is open, no current flows to the coil in the relay, so the coil windings are de-energized. When the switch is closed, the coil is energized, turning the soft iron core into an electromagnet and drawing the armature down. This closes the high-current circuit contacts and connects power to the load circuit. When the control switch is opened, the current stops slowing in the coil, the electromagnetic field disappears, and the armature is released, which opens the power circuit contacts.

Figure 5. A rheostat.

Figure 6. A potentiometer.

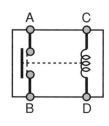

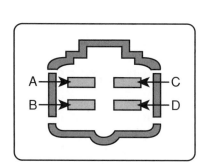
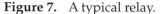

Figure 7. A typical relay.

Solenoids

Solenoids are electromagnetic devices with movable cores and are used to translate electrical current flow into mechanical movement. The movement of the core causes something else to move, such as a lever. Solenoids can also close electrical contacts, acting as a relay at the same time. Solenoids are commonly used to control fluid flow to the individual wheel service brakes in ABSs.

ELECTROMAGNETISM BASICS

Electricity and magnetism are related. One can be used to create the other. Current flowing through a wire creates a magnetic field around the wire. Moving a wire through a magnetic field creates current flow in the wire. Many automotive components, such as wheel speed sensors, operate using principles of electromagnetism.

A magnet has two points of maximum attraction, one at each end of the magnet. These points are called poles, with one being designated the north pole and the other the south pole. When two magnets are brought together, opposite poles attract, while similar poles repel each other. A magnetic field, called a **field of flux**, exists around every magnet **(Figure 8)**. The field consists of imaginary **flux lines** that show the attraction of the two poles.

Magnets can occur naturally or can be made by inserting a bar of magnetic material inside a coil of insulated wire and passing a heavy direct current through the coil. This principle is important in understanding certain automotive electrical components. Another way of creating a magnet is by stroking the magnetic material with a bar magnet.

Induced Voltage

Now that we have explained how current can be used to generate a magnetic field, it is time to examine the opposite effect of how magnetic fields can produce electricity. Consider a straight piece of conducting wire with the terminals of a voltmeter attached to both ends. If the wire is moved across a magnetic field, the voltmeter registers a small voltage reading **(Figure 9)**. A voltage has been induced in the wire.

> **You Should Know** *It is important to remember that the conducting wire must cut across the flux lines to induce a voltage. Moving the wire parallel to the lines of flux does not induce voltage. Holding the wire still and moving the magnetic field at right angles to it also induces voltage in the wire.*

BASICS OF ELECTRONICS

Antilock brake systems and other features of today's cars would not be possible if it were not for electronics. Electronics is best defined as the technology of controlling

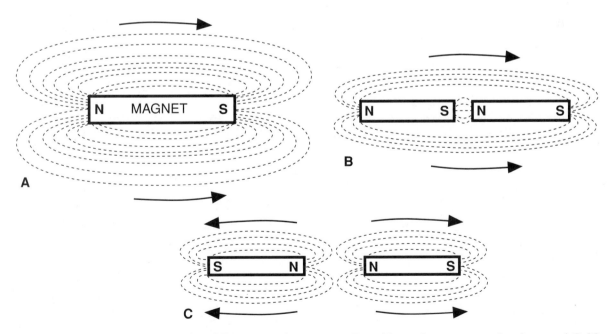

Figure 8.   Magnetic principles: A, field of flux around a magnet; B, unlike poles attract each other; and C, like poles repel.

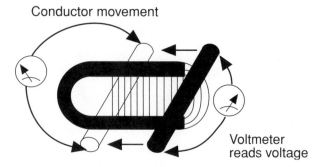

Conductor movement

Voltmeter
reads voltage

Figure 9. Moving a conductor through a magnetic field induces a voltage.

electricity. Transistors, diodes, semiconductors, integrated circuits, and solid-state devices are all considered to be part of electronics rather than just electrical devices. But keep in mind that all the basic laws of electricity apply to electronic controls.

Semiconductors

A **semiconductor** is a material or device that can function as either a conductor or an insulator, depending on its structure. Semiconductor materials have less resistance than an insulator but more resistance than a conductor. Some common semiconductor materials include silicon (Si) and germanium (Ge).

The **diode** is the simplest semiconductor device. A diode allows current to flow in only one direction. Therefore, it can function as a switch, acting as either a conductor or an insulator, depending on the direction of current flow.

A variation of the diode is the zener diode, which functions like a standard diode until a certain voltage is applied to the diode. When the voltage reaches this point, the zener diode allows current to flow in the reverse direction. Zener diodes are often used in electronic voltage regulators. Another semiconductor, the **transistor** is useful as a switching device, functioning as either a conductor or an insulator.

One transistor or diode is limited in its ability to do complex tasks. However, when many semiconductors are combined into a circuit, they can perform complex functions. An **integrated circuit** is simply a large number of diodes, transistors, and other electronic components, all mounted on a single piece of semiconductor material. This type of circuit is extremely small.

Many transistors, diodes, and other solid-state components are installed in a car to make logic decisions and issue commands to other areas of the engine. This is the foundation of computerized control systems. The computer has taken over many of the tasks in cars and trucks that were formerly performed by vacuum, electromechani-

cal, or mechanical devices. When properly programmed, computers can carry out explicit instructions with blinding speed and almost flawless consistency. A typical electronic control system is made up of sensors, actuators, and related wiring connected to a computer.

Sensors

All **sensors** perform the same basic function. They detect a mechanical condition (movement or position), chemical state, or temperature condition and change it into an electrical signal that can be used by the computer to make decisions. The computer makes decisions based on information it receives from sensors. Each sensor used in a particular system has a specific job to do. Although there are a variety of sensor designs, they all fall under one of two operating categories: reference voltage sensors or voltage-generating sensors.

Reference voltage (Vref) sensors provide input to the computer by modifying or controlling a constant, pre-determined voltage signal. This signal, which can have a reference value from 5 to 9 volts, is generated and sent out to each sensor by a reference voltage regulator located inside the **processor**. The term "processor" is used to describe the metal box that houses the computer and its related components. Because the computer knows that a certain voltage value has been sent out, it can indirectly interpret things such as motion, temperature, and component position, based on what comes back.

In addition to variable resistors, two other commonly used reference voltage sensors are switches and **thermistors**. Regardless of the type of sensors used in electronic control systems, the computer is incapable of functioning properly without input signal voltages from the sensors.

Voltage-generating sensors (Figure 10) are commonly **Hall effect switches**, oxygen sensors, and knock sensors, which are capable of producing their own input voltage signal. This varying voltage signal, when received by the computer, enables the computer to monitor and adjust for changes in the computerized control system.

Actuators

After the computer has processed the information, it sends output signals to control devices called **actuators**. These actuators—solenoids, switches, relays, or motors—physically act upon or carry out a decision made by the computer. Actuators are electromechanical devices that convert an electrical current into mechanical action. This mechanical action can then be used to open and close valves or open and close switches. When the computer receives an input signal indicating a change in one or more of the operating conditions, it determines the best

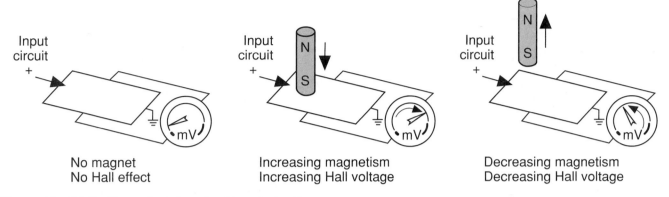

Input circuit +

No magnet
No Hall effect

Increasing magnetism
Increasing Hall voltage

Decreasing magnetism
Decreasing Hall voltage

Figure 10. Hall effect principles of voltage induction.

strategy for handling the conditions. The computer then controls a set of actuators to achieve the desired effect or strategy goal.

In order for the computer to control an actuator, it must rely on a component called an **output driver**. Out-put drivers are located in the processor and operate by the digital commands issued by the computer. Basically, the output driver is nothing more than an electronic on/off switch used to control the ground circuit of a specific actuator.

Summary

- There is a natural attraction of electrons to protons. Electrons have a negative charge and are attracted to something with a positive charge.
- In order to have a continuous flow of electricity, there must be an excess of electrons in one place, a lack of electrons in another place, and a path between the two places.
- The energy sources used in an automobile's electrical system are the vehicle's battery and its charging system.
- Electrical current is a term used to describe the movement or flow of electricity. Current is measured in amperes or amps with an ammeter.
- Voltage is the pressure that causes current flow and is measured with a voltmeter.
- The resistance to electrical flow is measured in ohms, and an ohmmeter is used to measure resistance.
- An electrical circuit is considered complete when there is a path that connects the positive and negative terminals of the electrical power source.
- Conductors are materials with a low resistance to the flow of current, and insulators resist the flow of current.
- Ohm's Law states that it takes 1 volt of electrical pressure to push 1 ampere of electrical current through 1 ohm of resistance.
- Resistors are used to limit current flow.

- Fuses, fuse links, maxi-fuses, and circuit breakers are designed to prevent circuit damage caused by high current.
- Switches turn the circuit on or off or direct the flow of current in a circuit.
- A relay is an electrical switch that allows a small amount of current to control a circuit with high current.
- Solenoids are electromagnetic devices with movable cores and are used to translate electrical current flow into mechanical movement.
- A substance is a magnet if it has the ability to attract such substances as iron, steel, nickel, or cobalt.
- A semiconductor is a material or device that can function as either a conductor or an insulator, depending on its structure.
- A typical electronic control system is made up of sensors, actuators, and related wiring connected to a computer.
- Sensors detect a mechanical condition (movement or position), chemical state, or temperature condition and change it into an electrical signal that can be used by the computer to make decisions.
- After the computer has processed input signals and its program, it sends output signals to control devices called actuators.

Review Questions

1. Name the two energy sources used in automobile electrical systems.
2. What is the difference between voltage and current?
3. What is the difference between a fixed resistor and a variable resistor?
4. What types of sensors are typically used in an automotive computer system?
5. Current is measured in _____, electrical voltage is measured in _____, and electrical resistance is measured in _____.
6. _____, _____ _____, _____ _____, and _____-_____ are used to protect circuits against current overloads.
7. A computerized circuit depends on two types of signals: _____ and _____.
8. While discussing the behavior of electricity Technician A says that if voltage does not change, but there is a change in the resistance of the circuit, the current will change; Technician B says that if resistance increases, current decreases. Who is correct?
   A. Technician A only
   B. Technician B only
   C. Both Technician A and Technician B
   D. Neither Technician A nor Technician B

9. Technician A says that rheostats are used to monitor wheel speed. Technician B says that potentiometers are used to monitor wheel speed. Who is correct?
   A. Technician A only
   B. Technician B only
   C. Both Technician A and Technician B
   D. Neither Technician A nor Technician B
10. Technician A says that electrical resistance is the pressure that causes current to flow in a circuit. Technician B says that if there is zero resistance in a circuit, a maximum amount of current will flow in the circuit. Who is correct?
   A. Technician A only
   B. Technician B only
   C. Both Technician A and Technician B
   D. Neither Technician A nor Technician B

Chapter 9

Common Brake System Electrical Components

Introduction

You can find a wide variety of electrical and electronic components in a vehicle's brake system, especially with an ABS. Electronic components are critical to the operation of an ABS. This chapter takes a look at some of the common electrical parts found in the brake systems of today's vehicles. These include the warning lamp switch operation of a pressure differential valve and the electrical switches to operate the failure warning lamp and the stop lamps, as well as sensors to indicate low brake fluid level and wheel speed.

FAILURE WARNING LAMP SWITCH

A **pressure differential valve** is a hydraulically operated switch **(Figure 1)** that controls the brake failure warning lamp on the instrument panel. Each side of the pressure differential valve is connected to half of the hydraulic system (one chamber of the master cylinder). Each master cylinder piston provides pressure to a separate front or rear hydraulic system **(Figure 2)**. On many vehicles, the brake system is split diagonally with each half operating one front brake unit and the rear brake unit on the other side of the vehicle **(Figure 3)**. Regardless of how the system is split, all modern brake systems are composed of two separate brake circuits. If one of the circuits fails, the brake pedal travel will increase and more brake pedal effort will be required to stop the car. The driver might not notice a problem, but the lamp on the instrument panel will provide a warning in case of hydraulic failure.

Failure in one half of the hydraulic system causes a pressure loss on one side of the pressure differential valve.

Pressure on the other side moves the valve's plunger into contact with the switch terminal. This closes the circuit and the warning lamp is illuminated. All pressure differential valves work in this basic way but differ in the details of the shape of the piston and the use of centering springs.

- If the center of the piston is higher than the adjacent sections, the switch plunger drops down to close the circuit.
- If the center of the piston is lower than the adjacent sections, the switch plunger moves up a ramp on either side of the piston center to close the circuit.
- If the center of the piston is open, the piston completes the circuit when it moves and contacts the switch plunger.
- If the piston *does not* have centering springs, the lamp will light the first time a pressure difference occurs and stay lit until the hydraulic problem is fixed. After the problem is fixed, the piston must be manually recentered. Additionally, a pressure differential valve without centering springs might need to be disabled when bleeding the hydraulic system.
- If the piston *has* centering springs, the lamp will light when a pressure difference occurs but will go out when the brake pedal is released. A valve with centering springs will automatically recenter the piston after the problem is repaired and after system bleeding.

The pressure differential valve on most late-model vehicles is part of a combination valve **(Figure 4)** or is built into the body of the master cylinder. Some vehicles have a switch in the float assembly of the fluid reservoir instead of a pressure differential valve. This float switch turns on the brake warning lamp when the fluid level changes to a dangerous point. This accomplishes the same thing as a pressure differential valve.

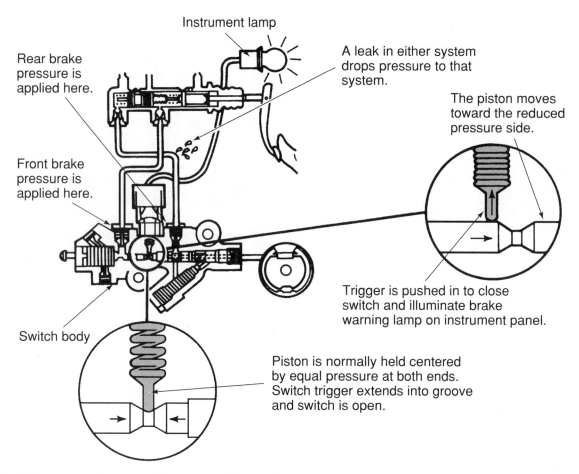

Instrument lamp

Rear brake pressure is applied here.

A leak in either system drops pressure to that system.

The piston moves toward the reduced pressure side.

Front brake pressure is applied here.

Trigger is pushed in to close switch and illuminate brake warning lamp on instrument panel.

Switch body

Piston is normally held centered by equal pressure at both ends. Switch trigger extends into groove and switch is open.

Figure 1.   This warning lamp switch is part of the combination valve, but its operation is the same whether it be an individual switch or part of the valve.

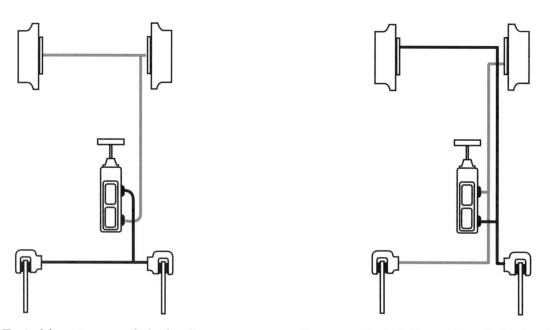

Figure 2.   Typical front-to-rear split hydraulic system.

Figure 3.   Typical diagonally split hydraulic system.

Figure 4. A three-function valve.

MASTER CYLINDER FLUID LEVEL SWITCH

Because brake fluid level is important to safe braking, many vehicles have a fluid level switch that causes illumination of the instrument panel's red brake warning lamp when the fluid level is too low **(Figure 5)**. This warning system is similar to the hydraulic failure warning provided by a pressure differential valve because fluid level in the reservoir will drop from a leak caused by hydraulic failure. Therefore, a fluid level switch has replaced the pressure differential valve on many vehicles. An added advantage of a fluid level switch is that it will alert the driver of a dangerous fluid level caused by inattention and poor maintenance practices.

Fluid level sensors are built into the reservoir body or cap. One type of switch has a float with a pair of switch con-

tacts on a rod that extends above the float. If the fluid level drops too low, the float will drop and cause the rod-mounted contacts to touch a set of fixed contacts and close the lamp circuit. Another type of switch uses a magnet in a movable float. If the float drops low enough, the magnet pulls a set of switch contacts together to close the lamp circuit. The contacts typically provide a ground path for the brake warning lamp.

Many European vehicles have a separate indicator lamp and circuit to indicate low brake fluid level. Manufacturers such as Mercedes-Benz, BMW, Jaguar, and Porsche use an electronic module to activate a special lamp on the dash.

> *Make sure the lamp works before assuming fluid levels are all right. Do this by turning the ignition switch on; the lamp should light for a few seconds.*

STOP LAMPS

Stop lamps are included in the right and left tail lamp assemblies. Vehicles built since 1986 also have a **center high-mounted stop lamp (CHMSL)** located on the vehicle centerline no lower than 3 inches below the rear window (6 inches on convertibles).

A three-bulb tail lamp contains bulbs for three separate vehicle functions: the tail lamps, the turn signals, and the stop lamps. In a three-bulb tail lamp system, the stop lamps are controlled directly by the stop lamp switch. The stop lamp switch receives direct battery voltage through a fuse. Therefore, the lamps will operate even when the ignition is off. When the normally open switch is closed, voltage is applied to the stop lamps. The lamps on both sides of the vehicle and in the CHMSL are wired in parallel. The bulbs are grounded to complete the circuit.

Many tail lamp assemblies are two-bulb assemblies with dual-filament bulbs that perform two functions. The stop lamp circuit and the turn signal and hazard lamp circuit usually share a single dual-filament bulb, with the stop lamp circuit connected to the high-intensity filament of the bulb. The tail lamps are the separate, single-filament bulbs in a two-bulb assembly.

In a two-bulb circuit, the CHMSL can be wired in one of two ways. The first way is to connect the stop lamp circuit between the stop lamp switch and the turn signal switch. However, this method increases the number of conductors needed in the wiring harness. Therefore, most manufacturers prefer to install diodes in the wires that are connected between the left and right side bulbs. If the brakes are applied when the turn signal switch is in its neutral position, the diodes allow voltage to flow to the CHMSL. When

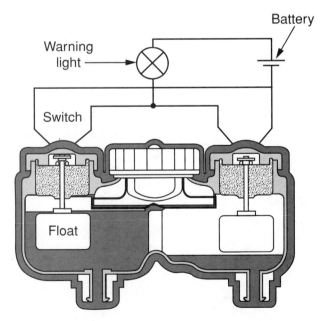

Figure 5. Float switches such as these are used in some master cylinders to warn of a low fluid level.

the turn signal switch is placed in the left turn position, the left lamp must receive a pulsating voltage from the flasher. However, the steady voltage being applied to the right stop lamp would cause the left lamp to light continuously if diodes were not used. One diode blocks the voltage from the right lamp, preventing it from reaching the left lamp. The other diode allows the voltage from the right stop lamp circuit to reach the CHMSL.

Stop Lamp Switch and Circuit

Brake stop lamp switches are operated hydraulically or mechanically. Hydraulic switches were used on older vehicles, were installed in the master cylinder's high-pressure chamber, and were activated by system pressure. A mechanical switch is mounted on the bracket for the brake pedal and activated by the movement of the pedal lever.

Mechanical switches are the typical type of switch found on today's vehicles because they can be adjusted to illuminate the stop lamps with the slightest pedal movement. A hydraulic switch, on the other hand, requires a certain amount of pressure before it will close the lamp circuit. Moreover, a dual-chamber master cylinder requires two hydraulic switches; therefore, a mechanical switch is simpler, more economical, and more reliable.

Stop lamp switches also can be single-function or multifunction units. Single-function switches have only one set of switch contacts that control electric current to the stop lamps at the rear of the vehicle **(Figure 6)**. Multifunction switches have one set of switch contacts for the stop lamps and at least one additional set of contacts for the torque converter clutch, the cruise control, or the ABS. Some multifunction switches have contacts for all of these functions.

The brake lamp contacts are often connected to the brake lamps through the turn signal and hazard flasher switch. If a vehicle has antilock brakes, there is a connection or a separate switch for the ABS control unit to sense when the brakes are being applied.

The switch contacts for the stop lamps are always normally open contacts. Contacts for other functions can be either normally open or normally closed. Additionally, some vehicles have separate brake pedal switches for the torque converter clutch, the cruise control, or ABS. Wiring diagrams are essential for accurate identification of brake pedal switches and their functions.

PARKING BRAKE SWITCH

The parking brake should be applied only when the vehicle is parked to hold it stationary. If the parking brake is even partially applied while the vehicle is moving, it will produce enough heat to glaze friction materials, expand drum dimensions, and increase pedal travel. On rear disc brake systems with integral parking brake actuators, driving with the parking brake applied will distort the brake rotors and reduce pad life.

A normally closed SPST switch is used to ground the circuit of the red brake warning lamp in the instrument cluster. This switch is located within the parking brake handle **(Figure 7)** or pedal assembly and is designed to turn on the light whenever the parking brake is applied. Vehicles with daytime running lights (DRLs) use the parking brake switch to complete a circuit that prevents the headlights from coming on if the parking brake is applied when the engine is started. When the parking brake is released, the DRLs operate normally.

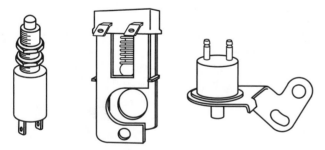

Figure 6. These different mechanical stop lamp switches are all activated by movement of the brake pedal.

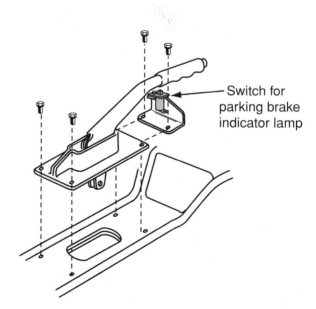

Switch for parking brake indicator lamp

Figure 7. A typical hand-operated parking brake control unit with a parking brake switch.

BRAKE PAD INDICATORS

Some models of vehicles have systems with an electronic wear indicator in the disc brake pads. As a pad wears to a predetermined point, a warning light in the instrument panel is illuminated by the wear sensors. In some systems, a small pellet is contained in the brake pad's friction material. The pellets are wired in series or in parallel to the red brake warning lamp circuit **(Figure 8)** and complete the lamp circuit when the pellets contact the rotor.

ABS SPEED SENSOR CIRCUITS

Wheel speed sensors are used in ABSs to measure the speed of the wheels. The tip of the sensor is located near a toothed ring or rotor. The toothed ring is typically attached to the drive axle or steering knuckle and rotates at the same speed as the wheel. As the ring spins, a voltage is induced in the sensor. The strength and frequency of this voltage varies with the wheel's speed. In some ABSs, the wheel speed sensor is mounted at each wheel. The toothed ring is part of the outer constant velocity joint (CV joint) or axle assembly **(Figure 9)**. In other systems, the toothed ring is mounted next to the differential ring gear **(Figure 10)**. In these cases, one sensor monitors the speed of the entire axle assembly instead of the individual wheels. Some transmissions are fitted with a vehicle speed sensor. This sensor is typically used to monitor vehicle speed, not just wheel speed.

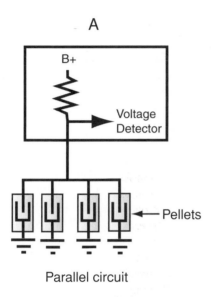

A

Parallel circuit

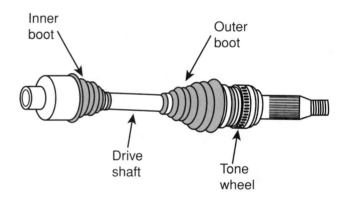

Figure 9.   Location of speed sensor toothed ring (tone wheel) on a half shaft.

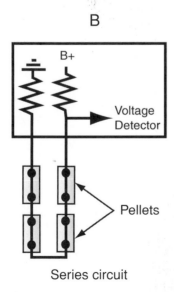

B

Series circuit

Figure 8.   Electronic wear sensors are used in both parallel (A) and series (B) circuits.

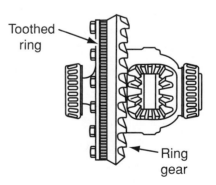

Figure 10.   Toothed ring (exciter ring) mounted to a differential's ring gear.

Summary

- A pressure differential valve is a hydraulically operated switch that controls the brake failure warning lamp on the instrument panel.
- Each side of the pressure differential valve, typically part of the combination valve or built into the body of the master cylinder, is connected to one chamber of the master cylinder. Failure in one half of the hydraulic system causes a pressure loss on one side of the pressure differential valve.
- A switch or sensor in the master cylinder causes the instrument panel's red brake warning lamp to light when the fluid level is too low.

- Stop lamps are controlled by the stop lamp switch and are included in the right and left tail lamp assemblies. Vehicles built since 1986 also have a center high-mounted stop lamp (CHMSL).
- Brake stop lamp switches are operated hydraulically or mechanically. On late-model vehicles, the mechanical switch is mounted on the bracket for the brake pedal and activated by the movement of the pedal lever.
- Wheel speed sensors are used in ABSs to measure the speed of the wheels. The tip of the sensor is located near a toothed ring or rotor. The toothed ring is typically attached to the drive axle or steering knuckle and rotates at the same speed as the wheel.

Review Questions

1. Technician A says that the pressure differential valve on most late-model vehicles is part of a combination valve or is built into the body of the master cylinder. Technician B says that some vehicles have a switch in the float assembly of the fluid reservoir that accomplishes the same thing as a pressure differential valve. Who is correct?
   A. Technician A only
   B. Technician B only
   C. Both Technician A and Technician B
   D. Neither Technician A nor Technician B
2. Describe how the brake system failure switch operates when there is a hydraulic failure to one side of the split system.
3. Describe the difference between a single-function and a multifunction stop lamp switch.

4. Wheel speed sensors are used in ABSs to measure the speed of the wheels. The toothed ring of the sensor is typically attached to the _____ _____ or is pressed onto the _____ _____ of the differential.
5. The operation of a brake system failure switch is being discussed. Technician A says that the lamp comes on when there is a pressure difference on one side of the switch. Technician B says that the lamp comes on when pressure is the same on both sides of the switch. Who is correct?
   A. Technician A only
   B. Technician B only
   C. Both Technician A and Technician B
   D. Neither Technician A nor Technician B

Chapter 10

Basic Electrical Testing and Service

Introduction

Certain diagnostic basics apply to all electrical systems. Many who have a difficult time diagnosing electrical problems do not understand what electricity is, what electrical meters can and do tell them, and how the different electrical problems affect a circuit. The purpose of this chapter is to help you understand these things. Also covered in this chapter are the testing and repair of basic electrical systems and the common electrical components found in brake systems.

A QUICK OVERVIEW OF ELECTRICITY

Resistance within an electrical circuit controls the flow of current within that circuit. The amount of voltage and resistance in a circuit determines how much current will flow in the circuit. Voltage pushes current through the circuit's resistors or loads. The amount of voltage within a load decreases as the current passes through the load. By the time the current passes through the last bit of resistance in a circuit, all of the voltage is lost.

Any device that uses electricity for operation has some resistance. This resistance converts the electrical energy into another form of energy. For example, a lamp illuminates because the filament inside the bulb resists the flow of electricity. The bulb shines from the heat given off, as the circuit's voltage pushes the current through the resistance of the filament. The amount of electrical energy converted to heat can be measured as voltage drop.

In a 12-volt light circuit with a bulb resistance of 3 ohms, 4 amps of current will flow through the circuit **(Figure 1)**. However, if there is a bad connection at the bulb

that creates an additional ohm of resistance, the current will decrease to 3 amps. As a result of the added resistance and decreased current, the bulb will lose about half of its brightness. This lack of brightness is also caused by the amount of

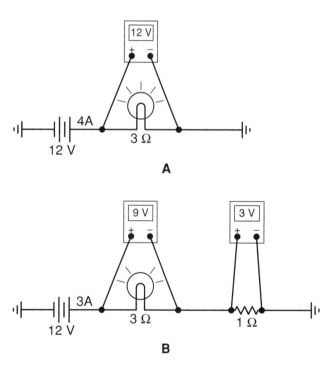

Figure 1. *A* represents a normal 3-ohm light circuit. *B* represents the same circuit but with a high resistance ground. Notice that the voltage drop across the lamp decreased, as did the circuit current. The bulb will burn dimmer.

voltage dropped by the bulb. In the normal circuit, the bulb drops 12 volts. However, in the circuit with the added resistance, 3 volts will be dropped by the bad connection, and the bulb will drop the remaining 9 volts.

The way resistance affects a circuit depends on the placement of the resistance in the circuit. If the resistances (wanted or unwanted) are placed directly into the circuit, the resistances are said to be in series. If a resistor is placed so that it allows an alternative path for current, it is a parallel resistor.

In a **series circuit (Figure 2)**, current follows only one path, and the amount of current flow through the circuit depends on the total resistance of the circuit. To calculate the total resistance in a series circuit, all resistance values are added together. At each resistor, voltage is dropped, and the total amount of voltage dropped in a series circuit is equal to the voltage of the source (battery). Regardless of the possible differences and resistance values of the loads, current in a series circuit is always constant and is determined by the total resistance in the circuit.

Parallel circuits (Figure 3) are designed to allow current to flow in more than one path. This allows one power source to power more than one circuit or load. A car's accessories and other electrical devices can be individually controlled through the use of parallel circuits. Within a parallel circuit, there is a common path to and from the power source. Each branch or leg of a parallel circuit behaves as if it were an individual circuit. Current flows only through the individual circuits when each is closed or completed. All legs of the circuit do not need to be complete in order for current to flow through one of them.

In parallel circuits, the total amperage of the circuit is equal to the sum of the amperages in all of the legs of the circuit. No voltage is dropped when the circuit splits into its branches; therefore, equal amounts of voltage are applied to each branch of the circuit. The total resistance of a parallel circuit is always less than the resistance of the leg with the smallest amount of resistance. The total resistance of two resistors in parallel can be calculated by dividing the product of the two by their sum.

The legs of a parallel circuit may contain a series circuit. To determine the resistance of that leg, the resistance values are added together. The resistance values of each leg are used to calculate the total resistance of a parallel circuit. Total circuit current flows only through the common power and ground paths; therefore, a change in a branch's resistance will not only affect the current in the branch but will also affect total circuit current.

ELECTRICAL PROBLEMS

All electrical problems can be classified into one of three categories: opens, shorts, or high resistance. Identifying the type of problem helps identify the correct tests to conduct when diagnosing an electrical problem.

An **open** occurs when a circuit has a break in the wire **(Figure 4)**. Without a completed path, current cannot flow, and the load or component cannot work. A disconnected or broken wire or a switch in the off position can cause an open circuit. Although voltage will be present up to the open point, there is no current flow. Without current flow, there are no voltage drops across the various loads. If there is an open in one leg of a parallel circuit, the remaining part of that parallel circuit will operate normally.

A **short** results from an unwanted path for current. Shorts cause an increase in current flow, which can burn

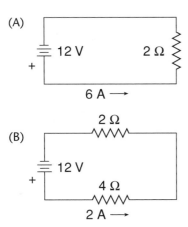

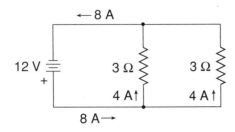

Figure 2.   In a series circuit, the same amount of current flows through the entire circuit.

Figure 3.   A simple parallel circuit.

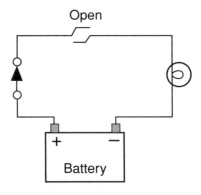

Figure 4.   An open prevents current flow.

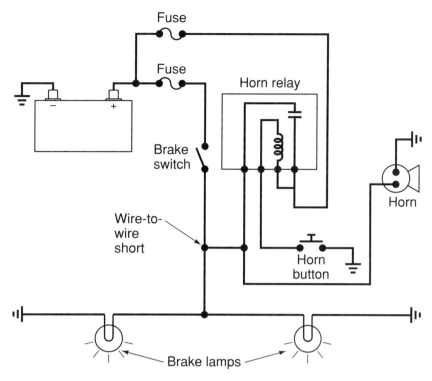

Figure 5. The horn circuit is shorted to the brake light circuit.

wires or components. Sometimes two circuits become shorted together, resulting in one circuit powering another. This can result in strange happenings, such as the horn sounding every time the brake pedal is depressed. In this case, the brake light circuit is shorted to the horn circuit **(Figure 5)**. A short to ground can be present before the load in the circuit or internally within the load. This problem provides a low resistance path to ground **(Figure 6)**. Improper wiring and damaged insulation are the two major causes of short circuits.

High-resistance problems occur when there is unwanted resistance in the circuit. The higher-than-normal resistance causes current to be lower than normal, and the

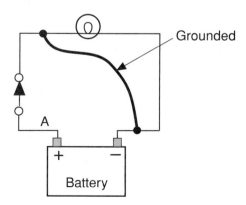

Figure 6. A short to ground causes a large increase in circuit current.

components in the circuit are unable to operate properly, if at all. A common cause for this problem is corrosion at a connector. The corrosion becomes an additional resistance in the circuit. This load not only decreases the circuit's current but also uses some of the circuit's voltage, which prevents full voltage to the normal loads in the circuit.

BASIC ELECTRICAL TESTING

To troubleshoot a problem, always begin by verifying the customer's complaint. Then operate the system and others to get a complete understanding of the problem. Often, there are other problems that are not as evident or bothersome to the customer that will provide helpful information for diagnostics. Refer to the correct wiring diagram and study the circuit that is affected. From the diagram you should be able to identify testing points and probable problem areas. Then test and use logic to identify the cause of the problem. A multimeter or several different meters are used to test and diagnose electrical systems. These can be used along with test lights and jumper wires.

Ammeters

An ammeter must be placed into or in series with the circuit being tested **(Figure 7)**. Normally, this requires disconnecting a wire or connector from a component and connecting the ammeter between the wire or connector and the component. The red lead of the ammeter should

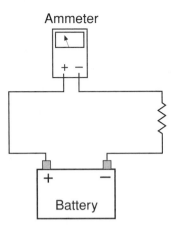

Figure 7.   An ammeter should be connected in series with the circuit.

always be connected to the side of the connector closest to the positive side of the battery, and the black lead should be connected to the other side.

> **You Should Know** *Never place the leads of an ammeter across the battery or a load. This puts the meter in parallel with the circuit and will blow the fuse in the ammeter or possibly destroy the meter.*

It is much easier to test with an ammeter if it has an inductive pickup. The pickup clamps around the wire or cable being tested and measures amperage by the strength of the magnetic field created by the current flowing through the wire. This type of pickup eliminates the need to separate the circuit to insert the meter. Inductive ammeters normally have three leads: positive, negative, and inductive. The positive and negative leads are connected to their

appropriate posts of the battery. When the circuit is activated, current flow will be read on the meter.

Ammeters can be used to diagnose many different circuit problems. For example, assume that a 6-amp fuse protects a circuit with four identical lights wired in parallel. If the circuit constantly blows the fuse, a short exists somewhere in the circuit. To find the short, disconnect all lights by removing them from their sockets. Then, close the switch and read the ammeter. With the load disconnected, the meter should read 0 amperes. If there is any reading, the wire between the fuse block and the socket is shorted to ground.

If zero amps were measured, reconnect each light in sequence; the reading should increase with each bulb. If, when making any connection, the reading is higher than expected, the problem is in that part of the light circuit.

Voltmeters

Electrical circuits and components cannot operate properly if the proper amount of voltage is not available. Not only must the battery be able to deliver the proper amount of voltage, the proper amount of voltage must also be available to the intended component. Because of this, voltage is measured at the source, at the electrical loads, and across the loads and circuits.

A voltmeter has two leads, a positive and a negative lead. Connecting the positive lead to any point within a circuit and connecting the negative lead to a ground will measure the voltage at the point where the positive lead was connected.

To measure the amount of available voltage, the negative lead can be connected to any point in the common ground circuit, and the positive lead can be connected to any point you wish to measure voltage at within the circuit. To measure the voltage drop of a light bulb in a simple circuit, connect the positive lead to the battery or power lead at the bulb and the other lead to the ground side of the bulb **(Figure 8)**. The meter will read the amount of voltage drop across the bulb. If no other resistances are present in

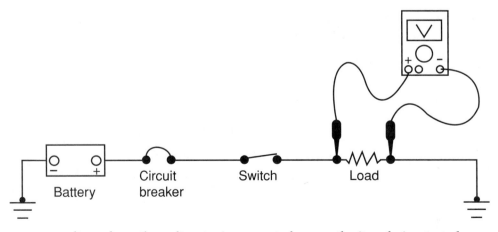

Figure 8.   To measure voltage drop, the voltmeter is connected across the item being tested.

the circuit, the amount of voltage drop will equal source voltage. However, if there is a resistance present in the ground connection, the bulb will have a voltage drop that is less than battery voltage. If the voltage drop across the ground is measured, it will equal the source voltage minus the amount of voltage dropped by the light. In order to measure voltage drop and available voltage at various points within a circuit, the circuit must be activated.

All wiring must have resistance values low enough to allow enough voltage to allow the load to operate properly. The maximum allowable voltage drop across wires, connectors, and other conductors in an automotive circuit is 10 percent of the system voltage. Therefore, in a 12-volt automotive electrical system, this maximum loss is 1.2 volts.

Ohmmeters

Ohmmeters use their own power to measure the resistance between two points. They should never be connected into an activated circuit. External power sources will damage the meter.

The amount of resistance, measured by the meter, is based on the voltage dropped across the component being tested. The scale of an ohmmeter reads from 0 to infinity (∞). A 0 reading means there is no resistance in the circuit and can indicate a short in a component that should show some resistance. An infinite reading indicates a number higher than the meter can measure. Most DMMs display an infinite reading as "O.L" **(Figure 9)**. This usually is an indication of an open circuit.

To measure the resistance of a circuit or component, connect one of the meter's leads to the power side of the component and the other lead to the ground side. If the resistance is greater than the selected range, the meter will show a reading of infinity. When this happens, the next highest range should be selected and the component retested. Whenever the range has been changed, the meter must be zeroed for that range before measuring the resis-

tance. If subsequent range changes result in continued readings of infinity, it can be assumed that there is no continuity between the two measured points.

Ohmmeters can also be used to compare the resistance of a component to the value it should have. Many electrical components have a specified resistance value that is listed in the service manuals. This resistance value is important because it controls the amount of voltage dropped and the amount of current that will flow in the circuit. If a component does not have the proper amount of resistance, the circuit will not operate properly.

> **You Should Know** *Before testing a component or circuit with an ohmmeter, the service manual should be checked for precautions regarding the impedance of the meter.*

Multimeters

It is not necessary to own separate voltmeters, ohmmeters, and ammeters. These meters are combined in a single tool called a multimeter **(Figure 10)**. The most commonly used multimeter is the digital DVOM or the DMM. These meters display the measurements digitally on the meter.

Most digital meters have high input **impedance**, usually at least 10 megaohms (10 million ohms). The high impedance reduces the risk of damaging sensitive components and delicate computer circuits. In addition to the basic electrical tests, some multimeters also measure engine revolutions per minute (rpm), ignition dwell, diode condition, distributor conditions, frequency, and even temperature.

A DMM has either an auto-range feature, in which the meter automatically selects the appropriate scale, or it must be set to a particular range. In either case, you should be familiar with the ranges and the different settings available

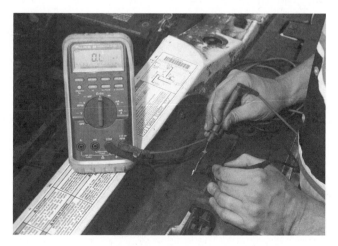

Figure 9. Most DMMs display an infinite reading as "O.L."

Figure 10. A DMM.

on the meter you are using. To designate particular ranges and readings, meters display a prefix before the reading or range. For example, if the meter has a setting for "mAmps," this means the readings will be given in milliamps or $1/1000$ of an amp.

Test Lights

There are two types of test lights commonly used in diagnosing electrical problems: nonpowered and self-powered. A **test light** looks like a stubby ice pick. Its handle is transparent and contains a light bulb. A probe extends from one end of the handle and a ground clip and wire from the other end. With the wire lead connected to ground and the tester's probe at a point of voltage, the light turns on with the presence of voltage. The brightness of the bulb is an indication of the amount of voltage present at the test point.

A self-powered test light is often used to test for continuity instead of an ohmmeter. A self-powered test light does not rely on the power of the circuit to light its bulb. Rather, it contains a battery for a power supply, and when connected across a completed circuit, the bulb will light. This type of test light is connected to the circuit in the same way as an ohmmeter.

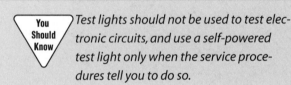

You Should Know *Test lights should not be used to test electronic circuits, and use a self-powered test light only when the service procedures tell you to do so.*

Jumper Wires

Jumper wires are used to bypass individual wires, connectors, components, or switches. Bypassing a component or wire helps to determine if that part is faulty. If the problem is no longer evident after the jumper wire is installed, the part bypassed is probably faulty. Technicians typically have jumper wires of various lengths that contain a fuse or circuit breaker in them to protect the circuits being tested.

ELECTRICAL WIRING

Two types of wire are used in automobiles: solid and stranded. Solid wires are single-strand conductors. Stranded wires are the most commonly used wire type and are made up of a number of small solid wires twisted together to form a single conductor. Computers and ABS components use specially shielded, twisted cable for protection from unwanted induced voltages that can interfere with computer functions. In addition, some use printed circuits.

The current-carrying capacity of a wire is determined by its length and gauge (size). Based on the **AWG (American Wire Gauge System)** system, wire size is identified by a numbering system ranging from 0 to 20 gauge, with 0-gauge wire having the largest cross-sectional area and 20-gauge wire the smallest. Most automotive wiring is 10- to 18-gauge wire; battery cables are 4-gauge wire.

ELECTRICAL WIRING DIAGRAMS

Wiring diagrams, sometimes called **schematics**, show how circuits are wired and how the components are connected. A wiring diagram does not show the actual position of the parts on the vehicle or their appearance, nor does it indicate the length of the wire that runs between components. It usually indicates the color of the wire's insulation and sometimes the wire gauge size. The first letter of the color coding is a combination of letters usually indicating the base color. The second letter usually refers to the strip color (if any). Tracing a circuit through a vehicle is basically a matter of following the colored wires.

Many different symbols are also used to represent components such as motors, batteries, switches, transistors, and diodes. Common symbols are shown in **Figure 11**. Wiring diagrams can become quite complex. To avoid this, the vehicle's electrical system can be divided into many diagrams, each illustrating only one system, such as the backup light circuit, oil pressure indicator light circuit, or wiper motor circuit.

BRAKE WARNING LAMP CIRCUIT TROUBLESHOOTING

The brake warning lamp on most vehicles performs multiple warning functions. Several circuits are connected to the same lamp on the instrument panel **(Figure 12)**. Typically, this lamp lights when:

- Pressure is lost in half of the split hydraulic system, and the pressure differential valve closes a switch. This warns the driver of a hydraulic system problem.
- The parking brake is applied.
- When the brake fluid level is low.
- As a circuit test while the engine is cranking.

Additionally, on most vehicles with an ABS, the control module will turn on the brake warning lamp if a problem causes the ABS to revert to the base brake system without ABS control.

Battery voltage reaches the brake warning lamp when the ignition switch is in the run, bulb test, and start positions. When any of the other switches connected to the brake warning lamp close, the circuit is grounded and the lamp is illuminated. The ignition switch completes the circuit to ground when it is in the bulb test and start positions.

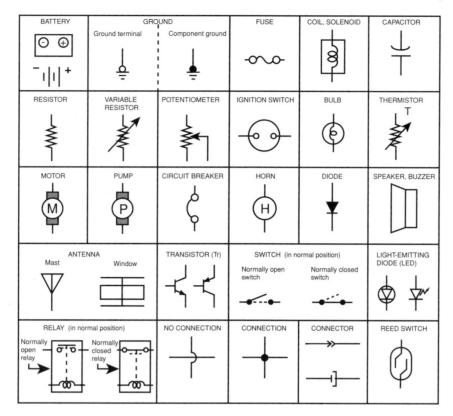

Figure 11. Common symbols used in electrical wiring diagrams.

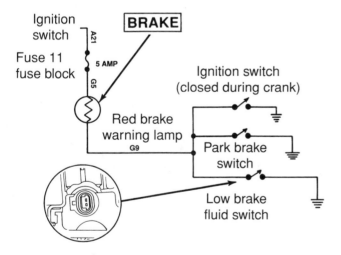

Figure 12. The brake warning lamp can be controlled by three different switches.

The parking brake switch provides a ground when the parking brake is applied. On vehicles with DRLs, the parking brake switch completes ground to the stop lamp through a diode in the DRL module.

The brake fluid level switch closes to light the brake warning lamp when the brake fluid in one of the two fluid reservoirs falls below switch level. This can be caused by a leak in one of the brake lines or by simply neglecting the fluid level.

Operational Check

Check the basic operation of the brake warning lamp as follows:

1. Turn the ignition switch to the start position or point halfway between on and start. The warning lamp should light.
2. Release the ignition switch to the run position. With the parking brake off, the warning lamp should turn off. On some vehicles, the lamp may light for a few seconds with the ignition on and then go off.
3. With the ignition on, apply the parking brake. The warning lamp should light.
4. Release the parking brake. The warning lamp should turn off.

If the warning lamp operates as described in these four steps, the system is working properly. If the warning lamp stays lit with the ignition on and the parking brake off, the following problems may be present:

- Hydraulic leak or failure in one half of the hydraulic system.
- Low fluid level in the master cylinder reservoir.

- Parking brake not fully released or parking brake switch shorted or grounded.
- ABS problem.

When checking the warning lamp bulb, it is often hard to tell if the filament is good. If you have any doubt about the bulb condition, replace it.

It is highly unlikely that all of the indicator lamps on the instrument panel would fail at the same time. If other indicator lamps are not operating properly, check the fuses first. Next, check for voltage at the last common connection. If no voltage is present here, trace the circuit back to the battery. If voltage is found at the common connection, test each branch of the circuit in the same manner.

To test the electrical circuit between the switch in a pressure differential valve and the warning lamp, disconnect the wire from the pressure switch terminal. Connect a jumper wire from the wire to a ground on the engine or chassis. Turn the ignition key on. The brake warning lamp should light. If the lamp does not light, inspect and service the bulb, wiring, and connectors as required. If the warning lamp lights with the jumper wire connected, turn the ignition switch off and reconnect the wire to the pressure differential switch terminal.

TESTING CIRCUIT PROTECTION DEVICES

Circuit protection devices are found in nearly all automotive electrical circuits. They are used to protect the circuit's components and wires. The common types of protection devices are fuses, fusible links, circuit breakers, and maxi-fuses.

Fuses

There are three basic types of fuses found in automobiles: cartridge, blade, and ceramic. The cartridge fuse is found in older domestic cars and a few imports, whereas many European imports use a ceramic fuse. Late-model domestic vehicles and many imports use blade, or spade, fuses. All fuses are comprised of a strip of low-melting metal on the outside of, or are enclosed inside of, the fuse assembly. To visually check a fuse, look for a break in the internal or external metal strip. Discoloration of the glass or plastic cover around the metal end caps is an indication of overheating.

All fuses are more accurately checked with an ohmmeter or test light. If the fuse is good, there will be continuity through it **(Figure 13)**.

Fuses are rated by the current at which they are designed to blow. A three-letter code is used to indicate the type and size of fuses. The code and the current rating are usually stamped on the end cap. The current rating for blade fuses is indicated by the color of the plastic case. It

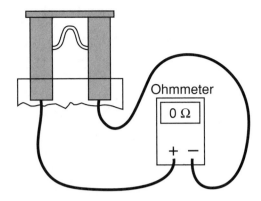

Figure 13.   An ohmmeter can be used to check the condition of a fuse.

may also be marked on the top. The insulator of ceramic fuses is color coded to indicate different current ratings.

Fusible Links

Fusible links are used in circuits in which limiting the maximum current is not extremely critical. They are often installed in the positive battery lead to the ignition switch and other circuits that have power with the key off. Fusible links are also used when it would be awkward to run wiring from the battery to the fuse panel and back to the load.

Because a fusible link is a lighter gauge wire than the main conductor, it melts and opens the circuit before damage can occur in the rest of the circuit. Fusible link wire is covered with a special insulation that bubbles when it overheats, indicating that the link has melted. If the insulation appears good, pull lightly on the wire. If the link stretches, the wire has melted. Of course, when it is hard to determine if the fuse link is burned out, check for continuity through the link with a test light or ohmmeter.

To replace a fuse link, cut the protected wire where it is connected to the fuse link. Then, tightly crimp or solder a new fusible link of the same rating as the original link. Never fabricate a fuse link from ordinary wire.

Maxi-Fuses

Many late-model vehicles use maxi-fuses instead of fusible links. Maxi-fuses look and operate like two-prong, blade or spade fuses, except they are much larger and can handle more current. Maxi-fuses are located in their own under-hood fuse block.

To check a maxi-fuse, look at the fuse element through the transparent plastic housing. If there is a break in the element, the maxi-fuse has blown. To replace it, pull it from its fuse box or panel. Always replace a blown maxi-fuse with a new one having the same ampere rating.

Circuit Breakers

Some circuits are protected by circuit breakers. Each circuit breaker conducts current through an arm made of two types of metal bonded together (bimetal arm). If the arm starts to carry too much current, it heats up. As one metal expands faster than the other, the arm bends, opening the contacts. Current flow is broken. A circuit breaker will either automatically reset or it must be manually reset by depressing a button.

TESTING SWITCHES

Switches can be tested with a voltmeter, test light, or ohmmeter. To check the operation of a switch with a voltmeter or a test light, connect the meter's positive lead to the battery side of the switch. With the negative lead attached to a good ground, voltage should be measured at this point. Without closing the switch, move the positive lead to the other side of the switch. If the switch is open, no voltage will be present at that point. The amount of voltage present at this side of the switch should equal the amount on the other side when the switch is closed. If the voltage decreases, the switch is causing a voltage drop due to excessive resistance. If no voltage is present on the ground side of the switch with it closed, the switch is not functioning properly and should be replaced.

If a switch has been removed from the circuit, it can be tested with an ohmmeter or a self-powered test light. By connecting the leads across the switch connections, the action of the switch should open and close the circuit.

PARKING BRAKE LAMP SWITCH

With the ignition on and the parking brake applied, the parking brake switch should close to light the red warning lamp on the instrument panel and remind the driver that the parking brake is applied. If the lamp does not light, the problem can be the switch, the circuit, or the lamp itself.

To test a typical parking brake switch, you must first gain access to the switch. Normally the switch is accessible, but the center console might need to be removed to gain access to the switch on a lever-operated parking brake.

If the switch connector has a single wire, connect a jumper from that wire to ground. If the switch connector has two wires, connect a jumper between the two wires in the connector. Turn the ignition on and check the brake warning lamp. If the lamp is lit, replace the parking brake switch. If the lamp is still off, find and repair the open circuit in the wiring harness between the lamp and the switch.

The switch can also be checked with an ohmmeter. The switch should be closed, and there should be continuity when the parking brake is applied. The switch should be open, with no continuity, with the pedal or lever released.

STOP LAMP TESTING AND SWITCH ADJUSTMENT

If only one stop lamp lights, you know you have electric power through the fuses and switches to the rear of the car. In this case, the problem is usually a burned-out bulb. Occasionally, an open circuit may exist in the branch of the circuit to one of the stop lamps, but this is much less common. If the bulb is not burned out, use a DMM or test light to pinpoint the open circuit.

If both stop lamps do not light, check the circuit's fuse. If the fuse is all right, check for the bulbs. Occasionally, both lamp bulbs may fail at the same time. If the fuse and bulbs are all right, continue testing. Locate the switch under the instrument panel or under the hood. Then disconnect the connector for the wiring harness and connect a jumper wire between the two terminals of the connector. Now check the stop lamps; they should be on. If they light, replace or adjust the switch. If the lamps do not light, continue testing for an open circuit condition between the switch and the lamps.

If the stop lamps are lit continuously, there is probably a short somewhere in the circuit. To locate the short, disconnect the connector at the switch. If the lamps are still lit, locate and repair the short circuit to battery voltage in the wiring harness. If the lamps turn off, adjust or replace the switch.

Stop Lamp Switch Adjustment

Since the introduction of split hydraulic systems in 1967, nearly all cars and trucks have been equipped with mechanical stop lamp switches operated by the brake pedal lever. Three basic adjustment methods exist for these mechanical switches:

1. If the switch has a threaded shank and locknut, disconnect the electrical connector and loosen the locknut. Then connect an ohmmeter to the switch. Screw the switch in or out of its mounting bracket until the ohmmeter indicates continuity while the brake pedal is depressed about $1/2$ inch. Tighten the locknut and reconnect the electrical harness.
2. If the switch is adjusted with a spacer or feeler gauge, loosen the switch mounting screw. Press the brake pedal and let it return freely. Then place a feeler gauge of the specified thickness between the pedal arm and the switch's plunger. Slide the switch toward the pedal arm until the plunger bottoms on the gauge. Tighten the mounting screw, remove the gauge, and check stop lamp operation.

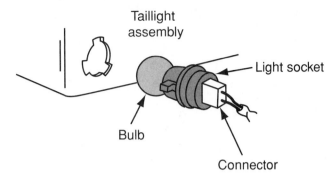

Figure 14. A tail/stop lamp with a removable socket.

3. If the switch has an automatic adjustment mechanism, insert the switch body into its mounting clip on the brake pedal bracket and press it in until it is fully seated. Then pull back on the brake pedal to adjust the switch position. You will hear a click as it ratchets back in its mount. Repeat this step until the switch no longer clicks in its mounting clip. Finally, check stop lamp operation to be sure it works correctly.

Replacing Stop Lamp Bulbs

Some tail lamp and stop lamp bulbs can be replaced without removing the lens assembly. Remove the bulb and socket by twisting the socket slightly and pulling it out of the lens assembly **(Figure 14)**. Push in on the bulb and turn it. When the lugs align with the channels of the socket, pull the bulb out to remove it.

On some vehicles, the complete tail lamp lens assembly must be removed for access to the bulbs. A typical lens assembly is held in position by several nuts or special screws and mounts to the vehicle.

BRAKE FLUID LEVEL SWITCH TEST

With the ignition on and the brake fluid level switch closed, the brake warning lamp alerts the driver of a low-fluid condition in the master cylinder. Some switches are built into the reservoir body; others are attached to the reservoir cap. Test procedures are similar for both types.

Begin by making sure the fluid level is at or near the full mark on the reservoir. Turn the ignition on and watch the warning lamp. If it is lit, disconnect the wiring connector at the switch. If the lamp then goes out, replace the switch. If the lamp does not go out, find and repair the short circuit between the switch and the lamp.

To verify that the warning lamp will light when the fluid level is low, manually depress the switch float or remove the cap with an integral switch and let the float drop. If the lamp does not light with the switch closed,

check for an open circuit between the switch and the lamp. If circuit continuity is good, replace the switch.

As a final check, disconnect the wiring harness from the switch and connect a jumper wire between the two terminals in the harness connector. The warning lamp should light. If it does not, find and repair the open circuit between the switch and the lamp.

BASIC ELECTRICAL REPAIRS

Faulty wiring can cause many electrical problems. Loose or corroded terminals; frayed, broken, or oil-soaked wires; and faulty insulation are the most common causes. Wires, fuses, and connections should be checked carefully during troubleshooting.

Wire end terminals are connecting devices. They are generally made of tin-plated copper and come in many shapes and sizes. They may be either soldered or crimped in place.

When working with wiring and connectors, never pull on the wires to separate the connectors. This can loosen the connector and cause an intermittent problem that can be very difficult to find later. Always follow the correct procedures and use the tools designed for separating connectors.

Nearly all connectors have push-down release locks. Make sure these are not damaged when disconnecting the connectors. Many connectors have covers over them to protect them from dirt and moisture. Make sure these are properly installed to provide for that protection.

Never reroute wires when making repairs. Rerouting wires can result in induced voltages from nearby components. These stray voltages can interfere with the function of electronic circuits.

Dielectric grease should be used to moisture proof and to protect connections from corrosion. If the manufacturer specifies that a connector be filled with grease, make sure it is. If the old grease is contaminated, replace it. Some car manufacturers suggest using petroleum jelly to protect connection points.

 Never crimp a terminal with the cutting edge of pliers. Although this method may crimp the terminal, it also weakens it.

When installing a terminal, select the appropriate size and type of terminal. Be sure it fits the unit's connecting post and has enough current-carrying capacity for the circuit. Also, make sure it is heavy enough to endure normal wire flexing and vibration.

To crimp a connector to a wire, make sure you use the correct-size stripping opening on the crimping tool. Remove enough insulation to allow the wire(s) to completely penetrate the connector. Place the wire(s) into the connector and crimp the connector. To get a proper crimp, place the open area of the connector facing toward the anvil. Make sure the wire is compressed under the crimp. Use electrical tape or heat-shrink tubing to tightly seal the connection. This will provide good protection for the wire and connector.

The preferred way to connect wires or install a connector is by soldering **(Figure 15)**. Some car manufacturers use aluminum in their wiring. Aluminum cannot be soldered. Follow the manufacturer's guidelines and use the proper repair kits when repairing aluminum wiring.

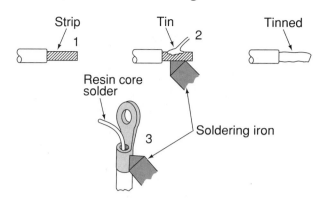

Figure 15.   The proper procedure for soldering a terminal to a wire.

Summary

- The amount of voltage within a load decreases as the current passes through the load. By the time current passes through the last resistance in a circuit, all of the voltage is lost.
- In a series circuit, current follows one path, and the amount of current through the circuit depends on the total resistance of the circuit. The total resistance in a series circuit is equal to the sum of the resistances, and current in a series circuit is always constant.
- Parallel circuits are designed to allow current to flow in more than one path. The total amperage of the circuit is equal to the sum of the amperages in all of the legs of the circuit. Equal amounts of voltage are applied to each branch of the circuit.
- The total resistance of two resistors in parallel can be calculated by dividing the product of the two by their sum.
- An open occurs when a circuit is not complete and therefore there is no current flow through the circuit.
- A short results from an unwanted path for current and causes an increase in current flow.
- High-resistance problems occur when there is unwanted resistance in the circuit, which causes lower than normal current flow and unwanted voltage drops.
- An ammeter must be placed into or in series with the circuit being tested.
- To measure available voltage, connect the negative lead to any point in the common ground circuit and the positive lead to the point where a measurement is sought.

- To measure the voltage drop of a component, connect the positive lead to the power lead of the component and the other lead to the ground side.
- Ohmmeters should never be connected into an activated circuit.
- On an ohmmeter, a 0 reading means there is no resistance in the circuit, and an infinite reading indicates a number higher than the meter can measure. Usually this indicates an open circuit.
- A DMM is the preferred electrical tester because it can measure volts, amps, and ohms, as well as conduct many other tests. A DMM displays the measurements digitally and has high input impedance, which reduces the risk of damaging sensitive components and delicate computer circuits.
- Wiring diagrams show how circuits are wired and how the components are connected.
- Faulty wiring can cause many electrical problems. Loose or corroded terminals; frayed, broken, or oil-soaked wires; and faulty insulation are the most common causes.
- The preferred way to connect wires or to install a connector is by soldering.
- Troubleshooting electrical problems involves the use of meters, test lights, jumper wires, circuit breakers, and short-detection devices.
- Fuses can be checked with an ohmmeter or a voltmeter.
- Switches can be tested with a voltmeter, test light, or ohmmeter.

Review Questions

1. If you were probing along with a voltmeter on a circuit that has a bad ground, how many volts would you read before and after the load? Why?

2. While discussing how to test a switch Technician A says that the action of the switch can be monitored by a voltmeter; Technician B says that continuity across

the switch can be checked by measuring the resistance across the switch's terminals when the switch is in its different positions. Who is correct?

A. Technician A only
B. Technician B only
C. Both Technician A and Technician B
D. Neither Technician A nor Technician B

3. Name the three types of electrical problems and their effects on current flow.

4. Describe what happens to the circuit when a bulb in a parallel circuit burns out.

5. While discussing electricity Technician A says that an open causes unwanted voltage drops; Technician B says that high resistance problems cause increased current flow. Who is correct?

A. Technician A only
B. Technician B only
C. Both Technician A and Technician B
D. Neither Technician A nor Technician B

6. Technician A says that a broken or frayed wire can cause an unintentional grounded circuit. Technician B says that dirt and grease buildup at terminals and connections can cause the same problem. Who is correct?

A. Technician A only
B. Technician B only
C. Both Technician A and Technician B
D. Neither Technician A nor Technician B

7. Technician A says that dual-filament lamp bulbs provide a backup filament in the bulb to double the life expectancy of the bulb. Technician B says that a dual-filament bulb serves two distinct functions, such as using one filament for the stop lamps and the other for turn signals. Who is correct?

A. Technician A only
B. Technician B only
C. Both Technician A and Technician B
D. Neither Technician A nor Technician B

8. Technician A states that the stop lamps in each tail lamp assembly and the CHMSL are wired as a series circuit. Technician B says that these components are wired in parallel. Who is correct?

A. Technician A only
B. Technician B only
C. Both Technician A and Technician B
D. Neither Technician A nor Technician B

9. Technician A says that a ground between the parking brake switch and the brake system warning lamp will cause the lamp to light continuously. Technician B says that if the brake system warning lamp does not light, replace the bulb with a new bulb before troubleshooting the entire circuit. Who is correct?

A. Technician A only
B. Technician B only
C. Both Technician A and Technician B
D. Neither Technician A nor Technician B

10. Technician A says that stop lamps on all vehicles receive fused voltage directly from the battery and will light with the ignition off if the brake pedal is depressed. Technician B says that the stop lamps are controlled by a normally open switch on the brake pedal. Who is correct?

A. Technician A only
B. Technician B only
C. Both Technician A and Technician B
D. Neither Technician A nor Technician B

Section 3

Hydraulic Systems

SECTION OBJECTIVES

After you have read, studied, and practiced the contents of this section, you should be able to:

- Check and adjust master cylinder fluid level.
- Bleed and flush the hydraulic brake system.
- Measure and adjust pedal height and brake pedal pushrod length.
- Diagnose poor stopping, pulling, or dragging, or poor brake pedal feel concerns caused by problems in the hydraulic system, master cylinder, and hydraulic valves; determine necessary action.
- Identify the main parts of a master cylinder.
- Explain the operation and purpose of a master cylinder.
- Check master cylinder for internal and external leaks and proper operation; determine necessary action.
- Remove, overhaul, bench bleed, and reinstall master cylinder.
- Describe the purpose and types of hydraulic brake lines, hoses, and fittings.
- Inspect flexible brake hoses, lines, flexible hoses, and fittings for leaks, kinks, cracks, bulging, or wear; make necessary repairs.
- Fabricate and install brake lines (double flare and ISO types); replace hoses, fittings, and supports as needed.
- Explain the purpose, parts, and operation of the various hydraulic valves used in a brake system.
- Test and service hydraulic valves as needed.
- Diagnose poor stopping and brake pull or grab conditions; perform needed repairs.

Interesting Fact

Single-piston master cylinders used before 1967 had one piston and one hydraulic circuit for all four wheel brake units. The problem with this system was that a hydraulic failure in the master cylinder could cause a complete loss of brakes.

Single-piston master cylinders were typically made from cast iron with the reservoir and cylinder body as one piece. One piston assembly had a primary cup in front of the piston head. The rear of the piston was often sealed with a secondary cup. A coil return spring and residual pressure check valve were located in front of the piston. A snapring at the rear of the cylinder held the components in the cylinder. An adjustable pushrod and boot were located at the rear of the cylinder. A single outlet at the front of the cylinder provided pressure to all the wheel brake units. A replenishing vent port was located in the bottom of the reservoir. A single-piston master cylinder operates exactly like one piston in a dual master cylinder.

A few General Motors cars of the mid-1980s had a single-piston master cylinder when the vehicle had an ABS. These systems rely on a proportioning valve that has a safety by-pass feature. If there is a pressure failure in the rear brake system, the proportioning valve closes. This stops pressure from being sent to the rear brakes and allows safe stopping with the front brakes.

Chapter 11

Basic Brake System Checks

Introduction

The action of the hydraulic system depends on the movement of the brake pedal. Not only does the pedal start the braking process, it also sets up the force on the hydraulic system that applies the brakes. The pedal assembly multiplies the force of the driver's foot and sends that modified pressure to the master cylinder. Because this is an important event, the brake pedal and its action are keys to diagnosing brake problems.

BRAKE PEDAL AND PUSHROD

Braking action begins when the driver pushes on the brake pedal. The brake pedal (**Figure 1**) is a lever that is piv-

oted at one end with the master cylinder pushrod attached to the pedal lever near the pivot point. One end of the pushrod engages the master cylinder's piston, and the other end is connected to the pedal linkage. The pushrod often is adjustable, with many made in two parts so they can be lengthened or shortened.

The brake pedal and master cylinder must be mounted close to each other so the pedal's pushrod can operate the master cylinder pistons. Older cars and some late-model trucks have a frame-mounted pedal (**Figure 2**). The pivot for the brake pedal is attached to the frame, and

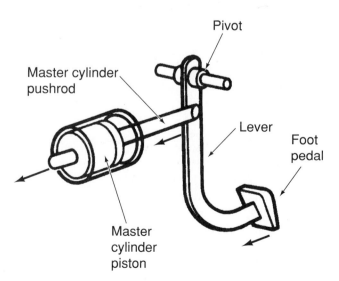

Figure 1. A basic brake pedal and linkage.

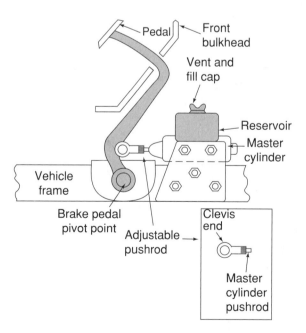

Figure 2. Brake pedals and master cylinders mounted between the vehicle's floorboards are found on older vehicles.

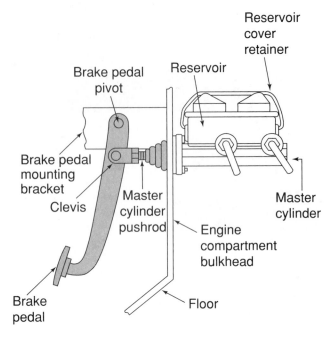

Figure 3. Nearly all brake pedal and master cylinder installations for the past 40 years have been a suspended pedal with the master cylinder mounted on the bulkhead in the engine compartment.

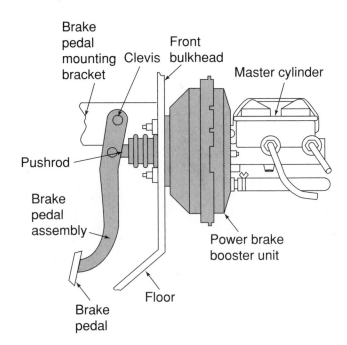

Figure 4. A suspended pedal installation with a vacuum power brake booster.

the pedal sticks up through the floor. The master cylinder also is mounted to the frame, using a threaded pushrod with an adjustable end called a clevis.

Most late-model vehicles use a suspended pedal assembly **(Figure 3)**. The pedal assembly is mounted to a support bracket that is attached to the inside of the engine compartment cowl or bulkhead. The pushrod that connects the pedal linkage to the master cylinder goes through a hole in the bulkhead. The master cylinder is mounted on the opposite side of the firewall. If the vehicle has manual brakes, the cylinder is mounted directly to the firewall. If a vacuum power brake booster is used, the booster is mounted to the firewall and the cylinder is mounted to the booster **(Figure 4)**.

BRAKE SYSTEM ROAD TEST

Leaks in the master cylinder or brake lines can rob the system of pressure and cause dangerous operating conditions. This is why the master cylinder and hydraulic system must be inspected whenever the brake pads or linings are changed or when a customer complains of poor braking. Any problems must be corrected immediately.

Check for the following nonbrake system problems that can cause poor brake performance:
- *Tire problems*—Worn, mismatched, underinflated, or overinflated tires will cause unequal braking.
- *Unequal vehicle loading*—A heavily loaded vehicle requires more braking power. If the load is unequal

from front to back or side to side, the brakes may grab or pull to one side.
- *Wheel misalignment*—Wheels that are out of alignment can cause problems that appear to be related to the brakes. For example, tires with excessively unequal camber or caster settings will pull to one side.

If the tires are in good shape and the wheel alignment and vehicle loading do not appear to have a problem, proceed with a road test of the brake system. Follow these guidelines when road testing a vehicle for brake problems:
- Test drive the vehicle on a dry, clean, relatively smooth roadway or parking lot. Roads that are wet or slick or have loose gravel surfaces will not allow all wheels to grip the road equally. Rough roads can cause the wheels to bounce and lose contact with the road surface.
- Avoid crowned roadways. They can throw the weight of the vehicle to one side, which will give an inaccurate indication of brake performance.
- First test the vehicle at low speeds. Use both light and fairly heavy pedal pressure. If the system can safely handle it, test the vehicle at higher speeds. Avoid locking the brakes and skidding the tires.
- Check the brake warning lamp on the instrument panel. It should light when the ignition switch is in the start position and go off when the ignition returns to the run position **(Figure 5)**.

If the brake warning lamp stays on when the ignition is on, verify that the parking brake is fully released. If it is, the problem may be a low brake fluid level in the master cylinder. Some vehicles have a separate master cylinder fluid level warning lamp. If either warning lamp remains on, check the fluid level in the master cylinder reservoir.

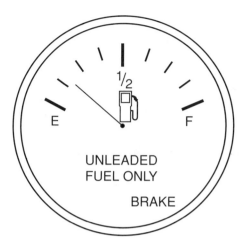

Figure 5. The brake warning light should light when the ignition is turned on and go off after the engine has started.

Listen for unusual brake noise during the test drive. Do you hear squeals or grinding? Do the brakes grab or pull to one side? Does the brake pedal feel spongy or hard when applied? Do the brakes release promptly when you take your foot off the brake pedal?

BRAKE PEDAL MECHANICAL CHECK

Checking the mechanical operation of the brake pedal is an important part of brake troubleshooting. Whether you do it as part of the brake system road test or during a system leak test, check the following points of pedal operation:

1. Check for friction and noise by pressing and releasing the brake pedal several times (with the engine running for power brakes). Be sure the pedal moves smoothly and returns with no lag or noise.
2. Move the brake pedal from side to side. Excessive side movement indicates worn pedal mounting parts.
3. Check stop lamp operation by depressing and releasing the brake pedal several times. Have a coworker check that the lamps light each time the pedal is depressed and go off each time it is released **(Figure 6)**.

Figure 6. Make sure to check for proper stop lamp operation before continuing your diagnostics of a brake system.

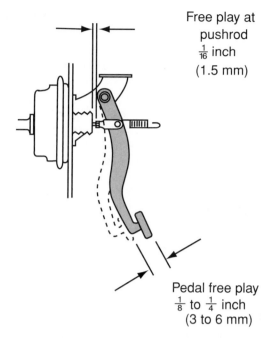

Free play at pushrod $\frac{1}{16}$ inch (1.5 mm)

Pedal free play $\frac{1}{8}$ to $\frac{1}{4}$ inch (3 to 6 mm)

Figure 7. The brake pedal multiplies the pedal's free play at the master cylinder.

LINKAGE FREE PLAY

All brake pedal linkage must provide some amount of free play between the master cylinder pistons and the pushrod. This free play is necessary to let the master cylinder's pistons completely retract in their bores. Free play at the master cylinder is usually very slight: about $\frac{1}{16}$ inch (1.5–2.0 millimeters). At the brake pedal, the free play is multiplied by the ratio of the pedal assembly **(Figure 7)**. Thus, if free play at the master cylinder is $\frac{1}{16}$ inch and the pedal ratio is 4:1, free play at the pedal will be $\frac{1}{4}$ inch.

The amount of desired free play measured at the pedal is usually specified by the vehicle manufacturer, at least as an inspection point. Free play is adjustable on some installations and not adjustable on others. Most adjustments are made by loosening a locknut at the pushrod clevis and turning the pushrod to lengthen or shorten it.

Vacuum power brake boosters have a second pushrod that transmits motion from the booster to the master cylinder. Booster pushrods can require adjustment separately from the brake pedal pushrod.

PEDAL FREE PLAY INSPECTION AND ADJUSTMENT

Brake **pedal free play** is the clearance between the brake pedal or booster pushrod and the primary piston in the master cylinder. A specific amount of free play must exist so that the primary piston is not partially applied when the pedal is released and, on the other hand, so that pedal travel is not excessive. Free play at the primary piston is usually only a small fraction of an inch or a few millimeters. The pedal ratio multiplies this to about 1/8 to 1/2 inch at the pedal.

Too much free play causes the pedal to travel too far before moving the pistons far enough to develop full pressure in the master cylinder. Excessive free play can severely reduce braking performance and create an unsafe condition. Too little free play causes the pedal to maintain contact with the primary piston. This can cause the piston cup to block the vent port and maintain pressure in the lines when the pedal is released. Unreleased pressure can cause the brakes to drag, overheat, fade, and wear prematurely. Additionally, with the piston cup blocking the vent port, each stroke of the pedal draws fluid from the low-pressure area behind the piston into the high-pressure area in front of the piston as the pedal is released. Eventually, enough fluid pressure can accumulate ahead of the piston to lock the brakes.

Check pedal free play by pumping the brake pedal with the engine off to exhaust vacuum in the booster. Place a ruler against the car floor, in line with the arc of pedal travel. Then depress the pedal by hand and measure the amount of travel before looseness in the linkage—the free play—is taken up. Measure at the top or bottom of the pedal, whichever provides the best view. Refer to vehicle specifications for the exact amount of free play to be provided.

Interesting Fact — American Motors Corporation introduced the dual hydraulic system on its 1960 cars. In 1963, several manufacturers added dual tandem in-line master cylinders to their cars. By 1967, the Department of Transportation (DOT) established FMVSS 105, which requires all cars sold in the United States to have split brake hydraulic systems. In 1978, Chrysler Corporation introduced the first mass-produced cars with a diagonally split hydraulic system.

Adjust the free play by lengthening or shortening the pushrod. On most vehicles, loosen the locknut on the pushrod at the pedal and rotate the pushrod while rechecking free play measurement. Tighten the locknut when adjustment is correct. If the car has a mechanical stop lamp switch on the brake pedal linkage, check switch operation and adjust it if necessary after adjusting pedal free play.

PEDAL TRAVEL AND FORCE TEST

Air in the hydraulic system causes most low-pedal problems, and bleeding the system usually solves the problems. Low pedal also can be caused by a leak in the hydraulic system, incorrect pushrod length adjustment, a parking brake that is out of adjustment, worn brake shoes, or a brake shoe adjuster that is not working.

When a given amount of force is applied to the pedal, brake pedal travel must not exceed a specified maximum distance. This maximum travel specification is normally approximately 2.5 inches (64 millimeters) when 100 pounds (445 newtons) of force is applied. The exact specifications can be found in the vehicle's service manual.

Use a brake effort gauge to measure force applied to the pedal as follows:

1. Turn off the engine. On vehicles with vacuum assist, pump the pedal until all reserve vacuum is exhausted from the booster.
2. Install the brake pedal effort gauge on the brake pedal.
3. Hook the lip of the tape measure over the top edge of the brake pedal and measure the distance from the pedal to the steering wheel rim.
4. Apply the brake pedal until the specified test force registers on the brake effort gauge.
5. Note the change in pedal position on the tape measure. The increased distance should not exceed the maximum specification listed in the vehicle service manual. If it does, look for a leak in the hydraulic system and check pushrod adjustment. Worn shoes, bad shoe adjusters, or a poorly adjusted parking brake also can cause excessive pedal travel.

Summary

- The brake pedal is a lever that is pivoted at one end, with the master cylinder pushrod attached to the pedal lever near its pivot point.
- Leaks in the master cylinder or brake lines can rob the system of pressure and cause dangerous operating conditions.
- The following should be checked before looking deeply into the brake system for braking problems: tires, unequal vehicle loading, and wheel misalignment.

- Test drive the vehicle on a dry, clean, relatively smooth roadway or parking lot. Avoid crowned roadways. First test the vehicle at low speeds. Check the brake warning lamp on the instrument panel.
- Check the pedal's action for friction and noise.
- Make sure the pedal has no side-to-side movement.
- Check stop lamp operation several times.
- All brake pedal linkage must provide some amount of free play between the master cylinder pistons and the pushrod.

- The amount of desired free play is measured at the pedal and is adjustable on some installations and not adjustable on others.
- Brake pedal free play is the clearance between the brake pedal or booster pushrod and the primary piston in the master cylinder.
- Too much free play causes the pedal to travel too far before moving the pistons far enough to develop full pressure in the master cylinder and can severely reduce braking performance and create an unsafe condition.
- Too little free play causes the pedal to maintain contact with the primary piston and can cause the brakes to drag, overheat, fade, and wear prematurely.

- Adjust the free play by lengthening or shortening the pushrod.
- Air in the hydraulic system causes most low-pedal problems, and bleeding the system usually solves the problems. Low pedal also can be caused by a leak in the hydraulic system, incorrect pushrod length adjustment, a parking brake that is out of adjustment, worn brake shoes, or a brake shoe adjuster that is not working.
- When a given amount of force is applied to the pedal, brake pedal travel must not exceed a specified maximum distance.

Review Questions

1. The leverage gained through the brake linkage is being discussed. Technician A says that the leverage comes from the lever length of the brake pedal. Technician B says that the leverage comes from the length of the pushrod. Who is correct?
   A. Technician A only
   B. Technician B only
   C. Both Technician A and Technician B
   D. Neither Technician A nor Technician B

2. Technician A says that too much free play allows the pedal to travel too far, preventing the system from developing full pressure and can severely reduce braking performance and create an unsafe condition. Technician B says that too little free play can cause the brakes to drag, overheat, fade, and wear prematurely. Who is correct?
   A. Technician A only
   B. Technician B only
   C. Both Technician A and Technician B
   D. Neither Technician A nor Technician B

3. As part of the brake system inspection and diagnosis, Technician A checks for correct wheel alignment, inspects the tires, and notes any unbalanced loading of the vehicle. Technician B performs a test drive on a smooth, level road, testing the brakes at various speeds. Who is correct?
   A. Technician A only
   B. Technician B only
   C. Both Technician A and Technician B
   D. Neither Technician A nor Technician B

4. Technician A says that checking brake pedal travel is best learned through experience, and good technicians develop a "feel" over time for a good travel range. Technician B says that brake pedal travel is a set specification that is found in the service manual and measured using a special gauge and/or tape measure. Who is correct?
   A. Technician A only
   B. Technician B only
   C. Both Technician A and Technician B
   D. Neither Technician A nor Technician B

5. What should be checked if the brake warning light stays on when the engine is running?

Chapter 12

Brake Fluid Service

Introduction

The fluid used in the brake hydraulic system is extremely important; in fact, without it there would be no brakes. Brake fluid service is more than simply adding fluid when the level is low in the reservoir. You already know the basics of the common types of brake fluid used today, and in this chapter you will learn more. This chapter also covers specific precautions that should be followed when working with, handling, and storing brake fluid. The chapter also covers the proper procedure for replacing the fluid, as well as checking the level and condition of the brake fluid in common brake systems.

The DOT rating is found on the container of brake fluid **(Figure 1)**. The vehicle's service manual and owner's manual clearly state the type and rating of the fluid that should be used in the car. Never use a brake fluid with a lower DOT rating than specified by the manufacturer. The lower-rated fluid could boil and cause a loss of brake effectiveness.

BRAKE FLUID PRECAUTIONS

When handling brake fluid, keep all of the following precautions in mind:

- Brake fluid is considered a toxic and hazardous material. Used brake fluid must be disposed of in accordance with local regulations and EPA guidelines. Do not pour used brake fluid down a wastewater drain or mix it with other chemicals awaiting disposal.
- Never mix polyglycol and silicone fluids because the mixture could cause a loss of brake efficiency and possible injury.

- Brake fluid can cause permanent eye damage. Always wear eye protection when handling brake fluid.
- If brake fluid gets on your skin, wash the area thoroughly with soap and water.
- Brake fluid is hygroscopic; it will attract moisture and must be stored in a tightly sealed container. When water enters brake fluid, it lowers the boiling point.
- Protect brake fluid from contamination by oil, grease, or other petroleum product. Never reuse brake fluid.

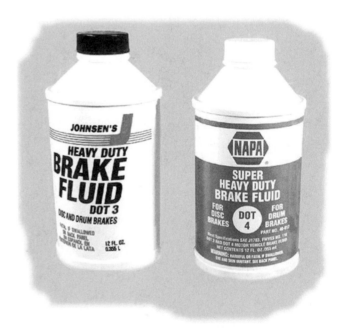

Figure 1. The DOT rating is found on every container of brake fluid.

- Polyglycol fluid will damage a painted surface. Always flush any spilled fluid immediately with cold water.

BRAKE FLUID STORAGE

Polyglycol fluids have a very short storage life. As soon as a container of DOT 3 or DOT 4 polyglycol fluid is opened, it should be used completely because it immediately starts to absorb moisture from the air. DOT 5 silicone fluids and HSMO fluids are not hygroscopic and can be stored for long periods of time. All brake fluids should be stored in the original container with all labeling intact. Containers must be kept tightly closed when not in use.

CONTAMINATED FLUID PROBLEMS

Always use the brake fluid type recommended for the vehicle. Using the wrong kind of fluid or using fluid that is contaminated can cause the brake fluid to boil or rubber parts in the hydraulic system to deteriorate.

Swollen master cylinder piston seals are the best indicator of contaminated brake fluid. Contamination also may be indicated by swollen or deteriorated wheel cylinder boots, caliper boots, or a damaged master cylinder cover diaphragm. If you find water or other contaminants in the brake system and the master cylinder piston seals have been damaged, replace all rubber parts in the system, including the brake hoses. Also check for brake fluid on the brake linings. If you find any, replace the linings.

If water or contaminants are in the system but the master cylinder seals appear undamaged, check for leakage throughout the system or signs of heat damage to hoses or components. Replace all damaged components. After repairs are made, or if no leaks or heat damage is found, drain the brake fluid from the system, flush the system with new brake fluid, and refill and bleed the system.

BRAKE FLUID SERVICE

Brake fluid service procedures are the most basic—but among the most important—brake system services. The following paragraphs provide instructions for checking the master cylinder fluid level and adding fluid to the system.

Checking the Fluid in the Master Cylinder

Master cylinder fluid level and fluid condition should be checked at least twice a year as part of a vehicle's preventive maintenance schedule. If the car has a translucent fluid reservoir, general fluid level can be checked every time the motor oil is checked or changed. Although normal brake lining wear will cause a slight drop in fluid level, an abnormally low

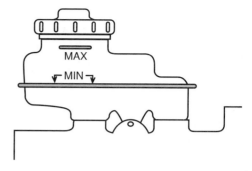

Figure 2. Most translucent reservoirs have markings for maximum and minimum fluid levels.

level in either chamber—especially an empty reservoir—usually means that there is a leak in the system.

When you check the fluid in the master cylinder, you are checking two things. First, be sure that the reservoir is filled to the correct level. A two-piece master cylinder with a plastic reservoir usually has graduated markings to indicate the correct fluid level. The markings may be on the outside of the reservoir if the reservoir is translucent **(Figure 2)** or they may be inside if the reservoir is opaque. Fill the reservoir to the "full" mark, or its equivalent.

If the master cylinder has a one-piece cast body with an integral reservoir, fluid level might not be marked in the reservoir. In this case, fill the reservoir to within 1/4 inch (6 millimeters) from the top. If the master cylinder is mounted on an angle, check the fluid level closest to the rim of the reservoir. Typically, the recommended fluid level is higher for a titled reservoir.

The second thing you look for when you check brake fluid is contamination. Most DOT 3 and DOT 4 fluids are clear or light amber when fresh, and, ideally, fluid in service should retain most of its original appearance. Fluid in good condition should be clear and transparent, although some darkening is allowable. Any of the following conditions may indicate the need for flushing and refilling the system or repair or replacement of system parts:

- *Cloudy fluid*—Cloudiness usually indicates moisture contamination.
- *Dark brown or murky, not transparent*, fluid—Very dark fluid usually indicates excessive contamination by rust and dirt.
- *Layering or separation*—These conditions indicate a mixture of two fluids that have not blended together. The contaminating fluid can be oil or some other petroleum-based product or it can be DOT 5 silicone fluid.
- *Layering or separation accompanied by cloudy or murky color and deteriorated rubber parts*—This condition almost always indicates oil contamination. In this case, the system must be flushed thoroughly and all seals and other rubber parts replaced.

> **You Should Know** *The separation of fluids occurs when a light-weight fluid separates from the fluid mixture and floats to the top of the mixture. A good example of this is the separation of water and oil. Oil will rise to the top, and water will sink to the bottom.*

Typical Procedure

Thoroughly clean the reservoir cover to prevent dirt from entering into the fluid. Remove the reservoir cover and diaphragm. Check that the vent hole in the reservoir cover is open. Then carefully inspect the diaphragm for holes, tears, and other damage **(Figure 3)**. Replace the diaphragm if it is damaged.

Check the level and condition of the brake fluid. The fluid level should be within ¼ inch of the top of the reservoir or at the reservoir's full or max level. The fluid should be clean, with no signs of discoloration or moisture. Note any differences in the level of the fluid in either reservoir. If there is a large difference between the two levels, a problem may be indicated. After checking the fluid, fill the reservoir with the recommended fluid and to the specified level.

Other Visual Checks

You might also see the following conditions when checking fluid in the master cylinder. These conditions by themselves do not always indicate a need to service the master cylinder or other parts of the brake system:

- Unequal fluid levels in the master cylinder reservoir chambers on front disc and rear drum systems can

Figure 3. The diaphragm for the reservoir cover or cap should be carefully inspected and replaced if it is damaged.

result as fluid moves from the reservoir into the calipers to compensate for normal lining wear. Fill both chambers to the full marks.

- A slight squirt of brake fluid from one or both master cylinder reservoir chambers when the brake pedal is applied is normal. It is caused by fluid moving through the reservoir vent ports as the master cylinder pistons move forward in the bore.
- Light fluid turbulence in the reservoir when the brake pedal is released is the result of brake fluid returning to the master cylinder after the brakes have been released.

Checking Boiling Point and Water Content

To check the water content and boiling point of brake fluid, use a **refractometer**. A refractometer works on the principle that different liquids have different **specific gravities**, and that liquids of different specific gravities will refract, or bend, light waves at different angles. As brake fluid absorbs water, its specific gravity changes. A brake fluid refractometer is graduated so that the specific gravity is correlated to the water content of the fluid and the resulting change in boiling point.

To use the refractometer, place a couple drops of fluid on the sample window, close the cover, point the instrument toward a light source, and focus the eyepiece. Light shining through the fluid sample will refract, or bend, and indicate a line across the measuring scale. SAE tests have shown that the brake fluid in the average year-old car is 2 percent water. This occurs just through the normal hygroscopic attraction of polyglycol fluids for water. As little as 4 percent water content can cut the boiling point of DOT 3 fluid in half.

Checking brake fluid with a refractometer can often be the deciding point on whether complete flushing and refilling of the hydraulic system should be recommended. A set of test strips is also available for checking the water content of brake fluid. The most common product is called "Wet Check" and works in less than one minute. Dip the end of the strip into the fluid and watch the color. The shade of the color indicates the amount of water or moisture in the brake fluid.

Checking Teves ABS Fluid Level

The Teves ABSs used by several carmakers employ an electrohydraulic booster that shares a fluid reservoir with the master cylinder. The reservoir is at atmospheric pressure, but design differences among Teves systems require that fluid level be checked for some systems with the accumulator charged and for others with the accumulator discharged. Checking the fluid level correctly on these systems is an example of why you should always refer to manufacturer's procedures for special instructions.

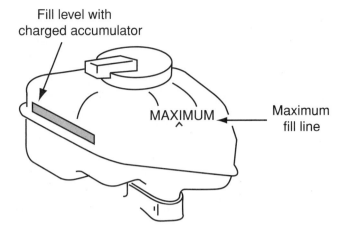

Figure 4. The fluid reservoir for a Teves ABS.

Systems that need the accumulator *charged* to check the fluid are typical of most Ford Teves ABS installations **(Figure 4)**. Generally, to check these systems, turn the ignition switch on and pump the brake pedal until the brake pump motor runs. When the pump stops, the accumulator is charged. Then verify that the fluid level is even with the max mark on the reservoir body.

Systems that need the accumulator *discharged* to check the fluid are typical of most GM Teves ABS installations. Generally, to check these systems, turn the ignition switch off and pump the brake pedal at least thirty times until pedal effort is noticeably harder. This discharges the accumulator. Then verify that the fluid level is at the full mark on the reservoir body.

Checking General Motors PowerMaster Fluid Level

The master cylinder reservoir of a General Motors PowerMaster system has three or four chambers **(Figure 5)**.

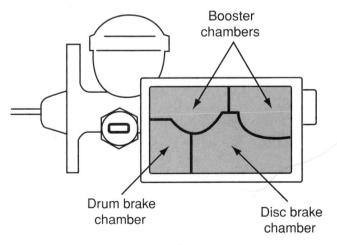

Figure 5. The fluid reservoir chambers for a Power-master system.

One side of the reservoir has two chambers that serve the master cylinder as in any other system. These chambers have conventional minimum and maximum fluid level marks. The other side of the reservoir serves the power booster accumulator.

Although the accumulator of the PowerMaster system operates at high pressure, the fluid reservoir is at atmospheric pressure. This allows you to remove the cover to check the fluid without depressurizing the system. Normally, when the accumulator is fully charged, its reservoir chamber appears almost empty. Fluid level barely covers the ports at the bottom of the chamber. This could mislead someone into adding too much fluid, which would overflow the next time the accumulator discharged.

To check PowerMaster fluid levels, leave the reservoir cover installed and the ignition switch off. Pump the brake pedal at least ten times with about 50 pounds of force to discharge the accumulator. Pedal effort will increase as the accumulator discharges. Then remove the reservoir cover. Accumulator fluid level with the accumulator discharged should be between the maximum and minimum markings on the reservoir.

Brake Fluid Replacement

Currently, more than a dozen manufacturers specify periodic brake fluid changes for some, or all, of their models built during the past 12 years. Change intervals vary from as often as every 12 months or 15,000 miles to as infrequently as every 60,000 miles. All of the manufacturers who call for brake fluid changes are import carmakers. None of the domestic manufacturers calls for periodic brake fluid changes.

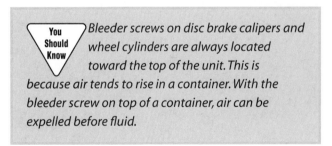

Bleeder screws on disc brake calipers and wheel cylinders are always located toward the top of the unit. This is because air tends to rise in a container. With the bleeder screw on top of a container, air can be expelled before fluid.

All brake systems accumulate sludge over some period of time. Flushing the system can remove this sludge, but once you have disturbed it, you want to be sure you get it *all* out of the system. Stirring up sludge from the master cylinder reservoir can cause it to get into ABS valves and pumps if you do not get it all out of the system. The control valves for some rear-wheel ABS installations on some trucks may be particularly susceptible to sludge and dirt contamination.

Brake hoses for disc brakes usually enter the caliper near the top of the caliper body. The bleeder valve is also located at the top of the caliper bore **(Figure 6)**. If sludge accumulates in the caliper bore, it collects at the bottom. A quick, superficial bleeding of the caliper will not flush out the sludge and all of the old fluid. To flush a caliper thoroughly, pump several ounces of fluid through it. During brake pad replacement on some vehicles, you may want to remove the caliper from its mounts and retract the piston to force out all the old fluid. Then reinstall it and thoroughly flush it with fresh fluid.

Flushing should be done at each bleeder screw in the same manner as bleeding. Open the bleeder screw approximately one and a half turns, and force fluid through the system until the fluid emerges clear and uncontaminated. Do this at each bleeder screw in the system. After all lines have been flushed, bleed the system using one of the common bleeding procedures.

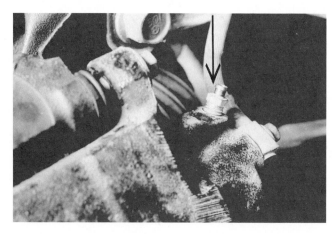

Figure 6. Bleeder screws are located at the top of wheel calipers and cylinders.

Summary

- Brake fluid is considered a toxic and hazardous material. Used brake fluid must be disposed of in accordance with local regulations and EPA guidelines.
- Never mix polyglycol and silicone fluids. The mixture could cause a loss of brake efficiency and possible injury.
- Always wear eye protection when handling brake fluid.
- Brake fluid is hygroscopic; when water enters brake fluid, it lowers the boiling point.
- Protect brake fluid from contamination. Contaminated fluid may cause system failure.
- Using the wrong kind of fluid or using fluid that is contaminated can cause the brake fluid to boil or rubber parts in the hydraulic system to deteriorate.
- Swollen master cylinder piston seals are the best indicator of contaminated brake fluid.
- Master cylinder fluid level should be checked at least twice a year as part of a vehicle's preventive maintenance schedule. An abnormally low level in either chamber—especially an empty reservoir—usually means that there is a leak in the system.
- Fluid in good condition should be clear and transparent, although some darkening is allowable. If the fluid is dark, cloudy, or murky, or if it shows signs of layering, a problem is indicated.
- A refractor is used to check the water content of brake fluid.
- Before checking fluid levels on some vehicles with ABSs, the hydraulic system might need to be depressurized.
- Some manufacturers recommend that the brake fluid be changed periodically.
- When the system is contaminated, it is wise to flush the entire hydraulic system and refill it with clean fluid.

Review Questions

1. Which of the following statements about brake fluid is *not* true?
    A. Clean DOT 3 and DOT 4 fluids are clear or light amber.
    B. Cloudy brake fluid usually indicates moisture contamination.
    C. Very dark fluid usually indicates excessive contamination by rust and dirt.
    D. Layering or separation of the fluid indicates excessive moisture contamination.

2. Which of the following conditions is not normal and requires additional diagnostics or service?
    A. Unequal fluid levels in the master cylinder reservoir chambers on front disc and rear drum systems.
    B. A slight squirt of brake fluid from one or both master cylinder reservoir chambers when the brake pedal is applied.
    C. Fluid turbulence in the reservoir when the brake pedal is applied.
    D. A slight squirt of brake fluid from a master cylinder reservoir chamber when the brake pedal is released.

3. As little as _____ percent moisture in the DOT 3 brake fluid can cut the fluid boiling point in half.

4. Brake fluid contamination is being discussed. Technician A says that swollen master cylinder seals are a sign that water is in the system. Technician B says that the system requires flushing if this condition exists. Who is correct?
   A. Technician A only
   B. Technician B only
   C. Both Technician A and Technician B
   D. Neither Technician A nor Technician B

5. While flushing a brake system Technician A forces fluid through the system and then bleeds the brakes. Technician B fills the system with brake flush before flushing the system. Who is correct?
   A. Technician A only
   B. Technician B only
   C. Both Technician A and Technician B
   D. Neither Technician A nor Technician B

Chapter 13

Hydraulic System Bleeding

Introduction

Bleeding air from the hydraulic system is probably the most often performed brake service procedure. The hydraulic system must be free of air to work properly. Air can enter the system when hydraulic parts are disconnected or if the system is operated with a low fluid level. Air in the system is compressed when the brake pedal is applied. This results in greater pedal travel and a spongy brake pedal.

The brake pedal symptoms of a system that contains air can be similar to the pedal symptoms caused by drum brakes that need adjustment. A low pedal caused by misadjusted drum brakes, however, will become firm after two or three applications and will not have the spongy feeling caused by air in the system. If air is trapped in the system, the brake pedal will be low and feel spongy when first applied. Rapidly pumping the pedal several times will compress much of the air and cause the pedal to rise and become more firm. As soon as pressure is released from the pedal, the air will expand again and the pedal will return to its spongy and low condition.

Whenever the hydraulic system is opened, either for a complete brake job or to replace just one part, the job must be completed by bleeding air from the system.

BRAKE BLEEDING

Brake bleeding is the process of removing air from the hydraulic brake system by opening a bleeder port at each wheel and sometimes elsewhere in the system. These ports are sealed with **bleeder screws** that are opened to allow fluid and air to escape the system. Bleeder screws are located at high points throughout the brake system **(Figure 1)**. A

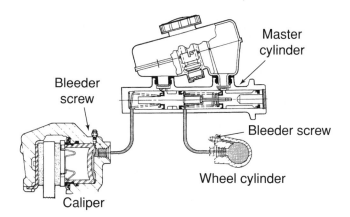

Figure 1. Bleeder screws are located at the highest points on calipers and wheel cylinders.

bleeder screw is normally installed in each drum brake wheel cylinder, in each disc brake caliper, next to the outlet ports of some master cylinders, and on some combination valves. Some bleeder screws have a threaded passage and a protective dust cap screw that must be removed before a drain hose can be installed on the bleeder screw.

Upward bends in the brake tubing as it is routed through the vehicle's chassis can trap air in the system **(Figure 2)**. This condition makes bleeding the system more important and difficult. All of the air must be removed to ensure proper brake operation and pedal feel. You can use the following air-entrapment test to help determine if air is in the system.

AIR-ENTRAPMENT TEST

Begin this test by removing the cover of the master cylinder and making sure the reservoirs are filled to the

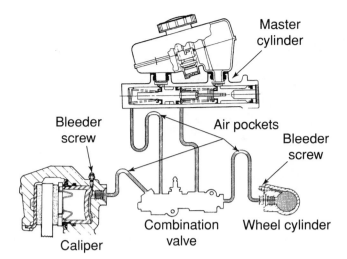

Figure 2. Bends in brake lines and hoses can trap air and make bleeding more difficult.

proper level. Hold the cover and gasket against the reservoir top but do not secure it with its clamp or screws. Then have an assistant pump the brake pedal ten to twenty times rapidly and maintain pressure after the last pedal application.

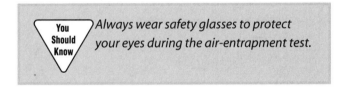

You Should Know ⟩ *Always wear safety glasses to protect your eyes during the air-entrapment test.*

Remove the reservoir cover and have the assistant quickly release pedal pressure. Watch for a squirt of brake fluid from the reservoirs. If air is compressed in the system, it will force fluid back through the compensating ports faster than normal and cause fluid to squirt in the reservoir. If a fluid squirt appears in one side of the reservoir but not the other, that side of the split hydraulic system contains the trapped air.

OVERALL BRAKE BLEEDING SEQUENCES

Bleeding may be performed on all or just part of the hydraulic system. When a master cylinder is replaced, it is normally bled on a bench before it is installed in the vehicle. Then, final bleeding can be performed at the master cylinder fittings or bleeder screws to remove any air trapped in the connections. If brake pedal and master cylinder operation is normal after these operations, it might not be necessary to bleed the wheel brakes.

Depending on where the hydraulic system was opened to air, bleeding might be needed at all four wheels and the master cylinder. If all fittings are disconnected from

the master cylinder, or if air was introduced into the system through low fluid level in both reservoirs, bleeding is required at the master cylinder and all four wheels.

If the tubes for one set of wheels are disconnected at the master cylinder, or if air entered through a low fluid level only in the reservoir for that set of wheels, only those wheels and lines need be bled. If there is any doubt about air in the system, the entire system should be bled.

Whether you are bleeding the entire system or just part of it, the first step is to fill the master cylinder with fresh fluid. Do not let dirt or other contamination enter the reservoir as you remove the cap or cover and fill the reservoir to the specified level with fresh fluid. If the fluid in the reservoir falls below the level of the compensating and replenishing ports, air will enter the system. If this has happened, bleeding steps must be repeated to ensure that all air is removed.

You Should Know ⟩ *On some vehicles, it is possible to install a right-hand caliper on the left-side wheel and vice versa. If you inadvertently do this, the bleeder screw will be located at the bottom of the caliper bore. This low position makes it impossible to bleed all air out of the system and is an indication that you did something wrong.*

When bleeding any part of the system, slip a hose over the end of the bleeder screw. Place the free end of the hose into a clear jar half filled with brake fluid. Always keep the end of the hose submerged in brake fluid. This prevents air from being drawn back into the system and lets you observe when air bubbles stop flowing from the bleeder to indicate that air has been removed from that portion of the system.

FREEING A FROZEN BLEEDER SCREW

When opening or removing a bleeder screw, use a bleeder screw box wrench. The shoulders of a bleeder screw are easily rounded if the wrong tool is used. Do not exert too much force on the screw; it is possible to break off the screw in the housing. Fixing this problem requires drilling and tapping, so avoid it by working carefully. If drilling and tapping are not possible, the entire cylinder or caliper must be replaced.

Before performing cylinder or caliper service, always check the bleeder screw to see if it can be loosened. If the caliper or cylinder is serviceable but the bleeder screw is frozen, apply a few drops of penetrating oil around the screw threads. Let the oil soak in for a few minutes and try again to turn the screw. Use a special wrench that allows you to exert pressure on the screw by striking it with a hammer. The shock to the screw from the blows of the hammer, and the tension against it, can free the screw.

> **You Should Know** *Never apply heat to a closed brake system. The fluid could boil and generate enough pressure to burst out of the weakest point in the system. If heat must be used, remove the piston and brake lines so the heat can dissipate through those openings.*

If this does not free the screw, use a torch to heat the housing around the screw to a dull red. When the caliper or cylinder body is hot, apply pressure to the screw with a wrench. When using this method, be sure to remove the rubber and plastic parts in the caliper or cylinder to prevent heat damage. This method is best done with the caliper or cylinder removed from the vehicle and clamped in a bench vise. Place the caliper or cylinder so the bleeder screw is downward. This will allow any residual fluid in the bore to run into the threads for the bleeder screw and help free the screw. If heat must be applied to brake parts while they are on the vehicle, remove any adjacent brake hoses or cover them with wet cloths to protect them from the heat.

WHEEL BRAKE BLEEDING SEQUENCES

All vehicle manufacturers recommend a specific sequence to follow when bleeding a vehicle's brakes. These recommendations can be found in service manuals **(Figure 3)**. Before bleeding the wheel brakes on any vehicle, refer to these recommendations and follow them during the bleeding procedure.

If the manufacturer's recommendations are not available, the following sequence will work on most vehicles:
1. Master cylinder
2. Combination valve or proportioning valve (if fitted with bleeder screws)
3. Right rear
4. Left rear
5. Right front
6. Left front
7. Height-sensing proportioning valve (if there is a bleeder screw)

This sequence is based on the principles of starting at the highest point in the system and working downward,

IMPORT CAR AND LIGHT TRUCK BRAKE BLEEDING SEQUENCE

Application	Year	Sequence	Remarks
Nissan			
Altima with ABS	1993–96	RR, LF, LR, RF	Diconnect actuator electrical connectors; bleed front and rear ABS actuators after bleeding brake lines.
Altima and Quest (without ABS)	1993–96	RR, LF, LR, RF	On models equipped with load-sensing proportioning valve (LSPV); bleed LSPV before bleeding brake lines.
Axxess	1990	LR, RF, RR, LF	Check fluid level frequently during bleeding sequence.
Maxima	1995–96	RR, LF, LR, RF	Turn ignition off; disconnect negative battery cable. Bleed wheel in sequence.
Maxima	1986–94	LR, RF, RR, LF	On models with ABS, turn ignition off and disconnect ABS actuator electrical connectors before bleeding.
Maxima	1984–85	RR, LR, RF, LF	Check fluid level frequently during bleeding sequence.
Pathfinder and Pickup	1980–95	LR, RR, LF, RF	Bleed Load Sensing Proportioning Valve (LSPV) before bleeding wheel cylinders. On models with rear antilock brakes, bleed ABS actuator after bleeding brake lines.
Pathfinder	1996	RR, LF, LR, RF	On four-wheel ABS models, disconnect negative battery cable before bleeding brakes in sequence.
Pickup	1978–79	RF, LF, LR, RR	Bleed master cylinder lines and front combination valve first. Bleed center and rear combination valve last.

Figure 3. The remarks section of this brake bleeding instruction table makes it clear why you should check the manufacturer's recommendations before bleeding any brake system. Also, note that the specified bleeding sequence varies with the different model vehicles from the same manufacturer.

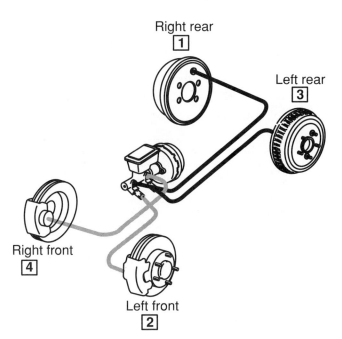

Right rear
1

Left rear
3

Right front
4

Left front
2

Figure 4. The sequence that is typically recommended for bleeding a diagonally split brake system.

and then starting at the wheel farthest from the master cylinder and working to the closest. A couple more general rules also are worth remembering. If the brake system is split between the front and rear wheels, the rear wheels (which are farthest from the master cylinder) usually are bled first. If the brake system is split diagonally, the most common sequence is: RR-LF-LR-RF **(Figure 4)**. This sequence also applies to most systems with a quick-takeup master cylinder. If you bleed a quick-takeup system in any other sequence, you may chase air throughout the system.

Exceptions to the general rules exist, however. Daimler-Chrysler, for example, recommends bleeding both rear brakes before the front brakes, regardless of how the hydraulic system is split.

Remember to press or pull the stem of a metering valve to lock it open when using a pressure bleeder. Many valves require a special clip to hold the metering valve stem open while bleeding the front brakes **(Figure 5)**. Also, the pistons of many pressure differential valves must be recentered after bleeding the system to be sure that the instrument panel warning lamp is off.

Methods for Bleeding

Five methods are commonly used for brake bleeding:
- Manual bleeding
- Pressure bleeding
- Vacuum bleeding
- Gravity bleeding
- Surge bleeding

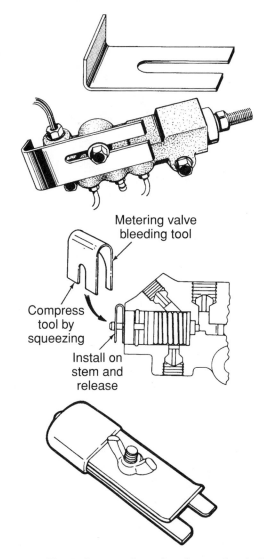

Metering valve bleeding tool

Compress tool by squeezing

Install on stem and release

Figure 5. Typical examples of tools used to hold the metering valve open during bleeding.

> **You Should Know** *Vehicles with ABSs often require special bleeding procedures or additional steps for the following general procedures. Failure to follow these special instructions may result in damage to ABS components or incomplete bleeding of the system.*

MANUAL BLEEDING

Manual bleeding uses the brake pedal and master cylinder as a hydraulic pump to expel air and brake fluid from the system when a bleeder screw is opened. Manual bleeding is a two-person operation needing one person to

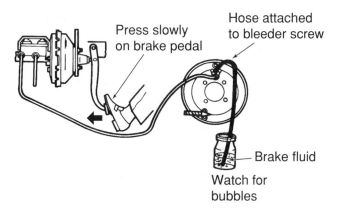

Figure 6. The basic setup for manual brake bleeding.

pump the brake pedal and another to open and close the bleeder screws.

Manual bleeding requires a bleeder screw wrench, a container partially full of fresh brake fluid, a length of clean plastic tubing that fits over the top of the bleeder screws, and several clean shop cloths. During manual bleeding, the brake pedal must be applied slowly and steadily **(Figure 6)**. Rapidly pumping the pedal will churn air in the system and make it harder to expel. Bleed the system as follows:

1. Check the fluid level in the master cylinder reservoir and be sure that both sections are full. Recheck the fluid level after bleeding each wheel brake and refill as necessary. Be sure that the reservoir cover or cap is installed securely during bleeding.

2. Discharge the vacuum or hydraulic pressure reserve in the booster by pumping the brake pedal with the ignition off until the pedal becomes hard.

3. Using a clean shop cloth, wipe dirt away from the bleeder screw on the first wheel in the recommended sequence.

4. Fit the plastic hose over the top of the bleeder screw and submerge the other end in the container of fresh brake fluid.

5. Loosen the bleeder screw one-half to one turn **(Figure 7)** and have your assistant depress the brake pedal slowly and steadily and hold it to the floor. Observe air bubbles flowing from the hose into the fluid container.

6. Tighten the bleeder screw and have your assistant slowly release the brake pedal.

7. Repeat steps 5 and 6 until no more air flows from the tubing into the brake fluid container.

8. Check the fluid level in the master cylinder and add fluid if necessary. Then proceed to the next wheel in the bleeding sequence. Repeat the bleeding sequence as necessary until the brake pedal is consistently firm. Check the fluid level a final time and install the reservoir cover or cap.

If you have trouble getting all air out of a disc brake caliper, tap the caliper lightly but firmly with a hammer to loosen trapped air bubbles and let them rise to the bleeder screw **(Figure 8)**.

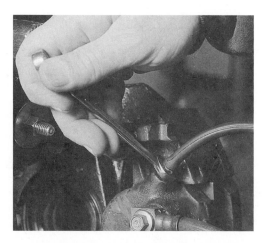

Figure 7. Loosen the bleeder screw and observe the fluid moving into the container on the other end of the hose.

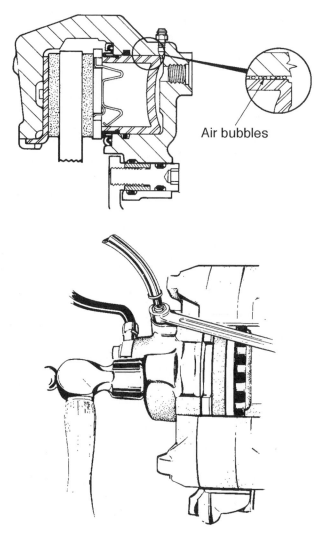

Figure 8. You can tap a caliper with a hammer to loosen air bubbles and let them rise to the bleeder screw.

PRESSURE BLEEDING

Pressure bleeding requires special equipment to force brake fluid through the system. Pressure bleeding **(Figure 9)** is a method commonly used because it has two advantages over manual bleeding. It is faster because the master cylinder does not need to be refilled several times, and it can be done by one person.

A pressure bleeder is a small tank that contains brake fluid pressurized by compressed air **(Figure 10)**. A hose from the pressure bleeder is connected to the master cylinder by an adapter that fits over the reservoir in place of the reservoir cap. One-piece cylinders with integral reservoirs generally use a flat plate adapter. Some plastic reservoirs require adapters that seal around the ports in the bottom of the reservoir **(Figure 11)**.

The hydraulic pressure generated by manual bleeding is enough to open the metering or combination valve and let fluid flow to the front disc brakes. When pressure bleeding, the valve must be held open manually because the pressure bleeder works with low pressure in the range where a metering valve is normally closed. To hold the

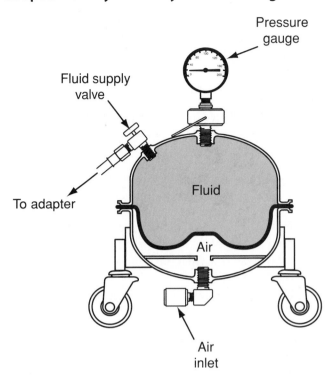

Figure 10. One chamber of the pressure bleeder holds the brake fluid. The other chamber contains compressed air. The two chambers are separated by a flexible, air-tight diaphragm.

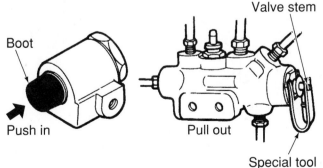

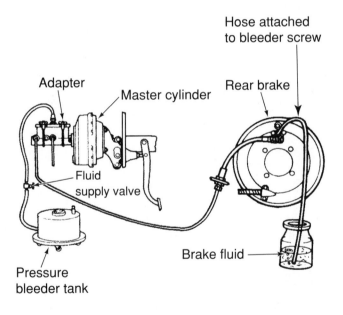

Holding metering valve open

Figure 9. The basic setup for pressure brake bleeding.

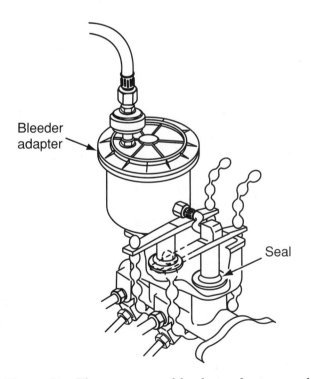

Figure 11. The pressure bleeder adapter seals directly against the ports in the bottom of the reservoir.

metering valve open, either push the valve stem in or pull it out, depending on the valve type. On some combination valves, such as those used by General Motors, loosen a valve mounting bolt and slide the valve tool under the bolt head. Move the end of the tool toward the valve body until it depresses the metering valve stem. Then tighten the mounting bolt to hold the tool in place. Other valves have a stem that must be held outward by a spring clip tool. On some valves, the stem must be held in by hand while the front brakes are bled.

Bleed the system as follows:

1. Fill the pressure bleeder with the specified type of fluid. Then charge the bleeder with 25–30 psi of compressed air according to the equipment instructions.
2. Clean the top of the master cylinder, remove the reservoir cover, and fill the reservoir about half full with fresh brake fluid.
3. Install the adapter on the reservoir and connect the fluid supply hose from the pressure bleeder to the adapter.
4. If required, install the correct override tool on the metering valve.
5. Open the fluid supply valve on the pressure bleeder to let pressurized fluid flow to the reservoir. Check the adapter and all hose connections for leaks and tighten if necessary.
6. Using a clean shop cloth, wipe dirt away from the bleeder screw on the first wheel brake to be bled.
7. Fit the plastic hose over the top of the bleeder screw and submerge the other end in the container of fresh brake fluid.
8. Loosen the bleeder screw one-half to one turn and observe air bubbles flowing from the hose into the fluid container.
9. Tighten the bleeder screw when clean fluid without any air bubbles flows into the container.
10. Repeat steps 6 through 9 at the next wheel in the bleeding sequence. Continue until the last brake in the sequence is bled. Repeat the bleeding sequence as necessary until the brake pedal is consistently firm.
11. Remove the metering valve override tool.
12. Close the fluid supply valve on the pressure bleeder.
13. Wrap the end of the fluid hose at the master cylinder adapter with a clean cloth and disconnect the hose from the adapter.
14. Remove the adapter from the master cylinder and be sure the reservoir is filled to the correct level. Install the reservoir cover or cap.

VACUUM BLEEDING

Vacuum bleeding is an alternative to pressure bleeding and is preferred by some technicians. As is pressure bleeding, vacuum bleeding is a one-person operation. Depending on the type of equipment, however, the master cylinder might require refilling during the bleeding operation. Two basic types of vacuum bleeding equipment are available: the hand-operated vacuum pump and the system operated by compressed air.

A hand-operated vacuum pump holds a small cup that contains fresh brake fluid, and a length of plastic tubing connects the pump to the bleeder screw **(Figure 12)**. Vacuum bleed a brake system as follows:

1. Check the fluid level in the master cylinder reservoir. Recheck the fluid level after bleeding each wheel brake and refill as necessary. Make sure that the reservoir cover or cap is installed securely during bleeding.
2. Install the small fluid container on the vacuum pump according to the equipment instructions. Be sure all connections are tight so that the pump cannot draw air past the fluid container.
3. Fill the small container about half full of fresh brake fluid. Be sure the short hose inside the container is submerged in fluid so that air cannot flow back into the brake system.
4. Using a clean shop cloth, wipe dirt away from the bleeder screw on the first wheel in the recommended sequence.
5. Fit the plastic hose from the vacuum pump over the top of the bleeder screw and operate the pump handle ten to fifteen times to create a vacuum in the container.
6. Using a bleeder screw wrench, loosen the bleeder screw one-half to three-quarters turn. Observe fluid with air bubbles flowing into the fluid container on the pump.
7. After evacuating about 1 inch of fluid into the container, tighten the bleeder screw.
8. Repeat steps 5 through 7 until no more air flows into the brake fluid container. Remove old fluid from the pump container as necessary during the bleeding procedure.

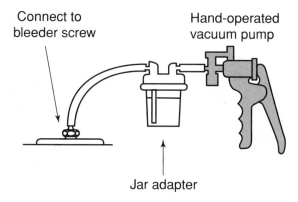

Connect to bleeder screw

Hand-operated vacuum pump

Jar adapter

Figure 12. With the correct adapters, a hand-operated vacuum pump can be used to bleed a brake system.

9. Check the fluid level in the master cylinder and add fluid if necessary. Then proceed to the next wheel in the bleeding sequence. Repeat the bleeding sequence as necessary until the brake pedal is consistently firm. Check the fluid level a final time and install the reservoir cover or cap.

Some vacuum bleeding equipment that uses compressed air includes a fresh fluid container that attaches to the master cylinder reservoir with an adapter similar to those used with a pressure bleeder. This eliminates the need to refill the reservoir repeatedly during bleeding. To use this kind of equipment **(Figure 13)**, connect the fresh fluid container to the master cylinder according to the equipment instructions. Then connect the vacuum pump to the first wheel to be bled. Open the wheel bleeder screw and the vacuum valve and let fluid flow from the wheel brake into the pump container until it is free of air bubbles. Close the bleeder screw and move the vacuum pump to the next wheel in sequence. Repeat the bleeding sequence as necessary until the brake pedal is consistently firm and no more air flows from any wheel brake. Remove old fluid from the pump container as necessary during the bleeding procedure. Check the fluid level in the master cylinder a final time and install the reservoir cover or cap.

When you use this type of vacuum bleeding equipment, it is common to see bubbles or foam in the fluid drawn from the brake system. Air is drawn into the evacuated fluid, past the threads of the bleeder screw. The air

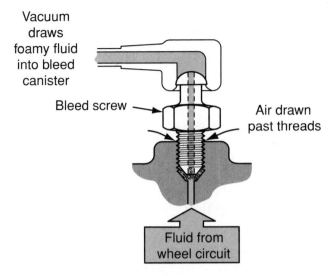

Figure 14. Air drawn past the threads for the bleeder screw may foam in the fluid, but this does not hurt the bleeding process.

mixes with the fluid being drawn out of the system and flows to the vacuum bleeder **(Figure 14)**. This does not affect the effectiveness of the bleeding process nor does it indicate the presence of air in the system.

Vacuum bleeding with either a hand-operated pump or a compressed air system generally does not require overriding the metering valve.

GRAVITY BLEEDING

Gravity bleeding is letting atmospheric pressure on the surface of the fluid in the master cylinder's reservoir eventually force the fluid through the hydraulic system and out the open bleeder screws. Gravity bleeding is the simplest and slowest way to bleed a brake system. It also can be the most effective on some systems. Gravity bleeding works best on a system that does not have a combination valve or a proportioning valve that requires bleeding. Gravity bleeding cannot be used on a system that contains a residual pressure check valve or any other valve that isolates any part of the system at low pressure. Before using the gravity bleeding method, the master cylinder should be thoroughly bench bled and then bled again after it is installed on the vehicle.

Gravity bleeding requires that all bleeder screws on the wheel brakes be open at the same time. Install lengths of clean plastic tubing on all bleeder screws and immerse the other end of each hose in a container of clean fluid before opening the bleeder screws. If all bleeder screws were opened to the air, air might be drawn back into the system because the total area of the bleeder openings can be greater than the area of the compensating ports through which fluid must flow to the master cylinder.

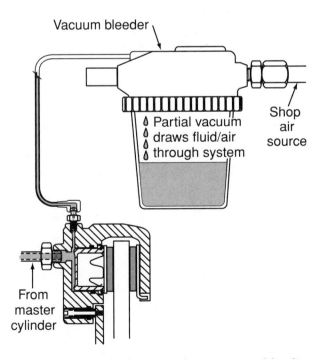

Figure 13. A typical setup for vacuum bleeding using compressed air.

Fill the master cylinder reservoir with fresh fluid before starting the bleeding process and check it periodically during gravity bleeding. Open each bleeder screw approximately one turn and verify that fluid starts to flow from each tube. Check the fluid level in the master cylinder periodically during bleeding. Gravity bleeding can take an hour or more to completely purge the system of air. When fluid flowing from the bleeder screws is clear and free of air bubbles, close the bleeder screws, disconnect the tubing, and check the master cylinder fluid level. Add fluid as required.

SURGE BLEEDING

Surge bleeding is a supplementary procedure that can be used to remove air pockets that resist other bleeding methods. Surge bleeding is a variation of manual bleeding in which an assistant pumps the brake rapidly to create turbulence in the system. Surge bleeding should not be used as the only bleeding procedure for a brake system. The agitation will often dislodge air trapped in pockets within the system.

The steps for surge bleeding are basically the same as for manual bleeding with one exception. After connecting plastic tubing to a wheel bleeder screw and immersing the tubing in a container of fresh fluid, open the bleeder screw about one full turn. Then, with the bleeder open, have an assistant pump the brake pedal rapidly several times. Watch for surges of air bubbles to flow from the tubing into the fluid container. Finally, have your assistant hold the pedal to the floor while you close the bleeder screw. Repeat the surge bleeding procedure at each wheel several times and then let the system stabilize for five to ten minutes. Follow the surge bleeding with any of the other bleeding procedures.

> **You Should Know**
>
> To simplify brake bleeding, a product called the "Speed Bleeder" is available from Russell Performance Parts. This tool has a built-in check valve and is installed in place of the normal bleeder screw at each wheel unit. To bleed the brakes, the Speed Bleeder is loosened one-quarter turn, and the brake pedal is pumped. The check valve allows air to leave the unit but prevents air from entering into it. Once the air is bled, the bleeder is closed. The Speed Bleeders remain on the wheel brake units and can be used again in the future. If you remove the Speed Bleeder, you will need to bleed the brakes all over again.

CHECKING THE PRESSURE AT THE WHEEL UNITS

When diagnosing brake drag, brake overheating, or premature wear of brake components, the pressure at the wheel units should be checked. To do this, connect a pressure gauge to the bleeder port on each, one at a time, caliper and wheel cylinder. When the brakes are applied, the pressure should rise quickly and drop just as quickly when the brake pedal is released. Any residual pressure at one brake usually indicates a restricted line or hose. However, some drum brake systems will retain about 10 psi to hold the wheel cylinder cup seals against the cylinder walls. If signs of a restriction are found, the problem should be corrected and the brakes bled before the vehicle is driven.

Summary

- Brake bleeding is the process of removing air from the hydraulic brake system by opening a bleeder port, sealed with a bleeder screw, at each wheel and sometimes elsewhere in the system.
- If air is trapped in the system, the brake pedal will be low and feel spongy when first applied.
- Trapped air is indicated when there is a squirt of fluid from the master cylinder's reservoir after the brake pedal has been pumped and is released.
- The bleeding process should follow this sequence: master cylinder, combination valve, wheel cylinders and brake calipers, and then the height-sensing proportioning valve.
- Before and during bleeding, make sure the reservoir is full of fresh fluid.
- When bleeding the system, slip a hose over the end of the bleeder screw and place the free end of the hose into a clear jar half filled with brake fluid. Always keep the end of the hose submerged in brake fluid to prevent air from being drawn back into the system.
- Five methods are commonly used for brake bleeding: manual bleeding, pressure bleeding, vacuum bleeding, gravity bleeding, and surge bleeding.
- Manual bleeding requires two people and uses pedal action to move the fluid and air out of the system.
- Pressure bleeding requires special equipment to force brake fluid through the system and can be completed quickly by one person.
- Two basic types of vacuum bleeding equipment are available: the hand-operated vacuum pump and the system operated by compressed air.
- Gravity bleeding is letting nature take its course during brake bleeding, and it takes a long time to complete.

- Surge bleeding is a supplementary procedure that can be used to remove air pockets that resist other bleeding methods.

- When diagnosing brake drag, brake overheating, or premature wear of brake components, the pressure at the wheel units should be checked.

Review Questions

1. Air in the hydraulic system is being discussed. Technician A says that air can enter the system whenever a line or component has been disconnected. Technician B says that air can enter the system when the fluid level in the master cylinder reservoir is low. Who is correct?
   A. Technician A only
   B. Technician B only
   C. Both Technician A and Technician B
   D. Neither Technician A nor Technician B

2. Bleeding is being discussed. Technician A says that bleeder screws are located at low points in the hydraulic system. Technician B says that lightly tapping a caliper assembly with a hammer can regroup tiny air bubbles into larger ones that can be expelled through bleeding. Who is correct?
   A. Technician A only
   B. Technician B only
   C. Both Technician A and Technician B
   D. Neither Technician A nor Technician B

3. Pressure bleeding is being discussed. Technician A says that metering and combination valves must be held open using a special tool during the bleeding operation to ensure good results. Technician B says that the pressure bleeder requires special adapters to connect it to the master cylinder reservoir. Who is correct?
   A. Technician A only
   B. Technician B only
   C. Both Technician A and Technician B
   D. Neither Technician A nor Technician B

4. Technician A says that the master cylinder should be bled before any individual wheel assembly. Technician B says that the bleeder screw should be closed before the brake pedal is released during manual bleeding of the system. Who is correct?
   A. Technician A only
   B. Technician B only
   C. Both Technician A and Technician B
   D. Neither Technician A nor Technician B

5. Technician A says that if air is trapped in the system, the brake pedal will become more firm. Technician B says that upward bends in the brake tubing as it is routed through the vehicle's chassis can also trap air in the system. Who is correct?
   A. Technician A only
   B. Technician B only
   C. Both Technician A and Technician B
   D. Neither Technician A nor Technician B

Chapter 14

Master Cylinders

Introduction

The master cylinder transmits the pressure on the brake pedal to each of the four wheel brakes to stop the vehicle. The master cylinder changes the driver's mechanical pressure on the pedal to hydraulic force, which is changed back to mechanical force at the wheel brake units. The master cylinder uses the fact that fluids are not compressible to transmit the pedal movement to the wheel brake units.

The master cylinder also uses the principles of hydraulics to increase the pedal force applied by the driver. A 100-pound force on the brake pedal can be used to push on a 1-square-inch master cylinder piston to create a 100-psi pressure in the hydraulic system. This 100 psi can be used to push on 4-square-inch output pistons at a wheel brake. The result is a 400-pound force at the 4-square-inch output pistons. The driver's 100-pound force has been multiplied to a force of 400 pounds.

DUAL-PISTON MASTER CYLINDERS

Figure 1 is a simplified illustration of how the master cylinder and the hydraulic system work. The master cylinder pushrod is connected to a piston inside the cylinder, and hydraulic fluid is in front of the piston. When the pedal is pressed, the master cylinder piston is pushed forward. The fluid transmits the force of the master cylinder piston to all the inner surfaces of the system. Only the pistons in the drum brake wheel cylinders and/or disc brake calipers can move, and they move outward to force the brake shoes or pads against the rotating brake drums and/or rotors. The master cylinder has two main parts: a reservoir and a master cylinder body **(Figure 2)**.

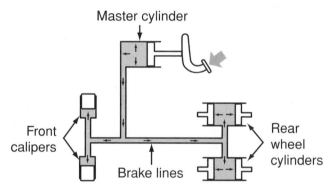

Figure 1. A simplified diagram of the operation of a master cylinder and a hydraulic circuit.

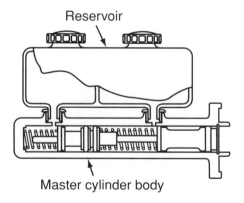

Figure 2. The two major parts of a master cylinder are the main body and the reservoir.

Master Cylinder Reservoir

The reservoir may be cast as one piece with the cylinder body or it may be a separate molded nylon or plastic container **(Figure 3)**. The one-piece body and reservoir casting is usually made from cast iron. The cylinder is directly under the reservoir. All reservoirs have a removable cover so that brake fluid can be added to the system. One-piece reservoirs typically have a single cover that is held on the reservoir with a retainer bail. Nylon or plastic reservoirs typically have two screw caps on top of the reservoir. Separate reservoirs can be clamped or bolted to the cylinder body or they can be pressed into holes in the top of the body and sealed with grommets or **O-rings**.

All master cylinder caps or covers are vented to prevent a vacuum lock as the fluid level drops in the reservoir. A flexible rubber **diaphragm** at the top of the master cylinder reservoir is incorporated in the screw caps or covers. The reservoir diaphragm separates the brake fluid from the air above it, while remaining free to move up and down with changes in fluid level. The diaphragm keeps the moisture and air from entering the brake fluid in the reservoir. Moisture in the brake fluid will lower the fluid boiling point.

If a vehicle with front disc and rear drum brakes has a hydraulic system split front-to-rear, the reservoir chamber for the disc brakes is larger than the chamber for the drum brakes. As disc pads wear, the caliper pistons move out farther in their bores. More fluid is then required to keep the system full from the master cylinder to the calipers. Drum brake wheel cylinder pistons always retract fully into the cylinders regardless of brake lining wear so the volume of fluid does not increase with lining wear. Vehicles with four-wheel disc brakes or diagonally split hydraulic systems usually have master cylinders with equally sized reservoirs because each circuit of the hydraulic system requires the same volume of fluid.

Plastic reservoirs often are translucent so that fluid level can be seen without removing the cover. Although this feature allows a quick check of fluid level without opening the system to the air, you should not rely on it for thorough brake fluid inspection. Stains inside the reservoir can give a false indication of fluid level, and contamination cannot be seen without removing the reservoir caps or cover.

Master Cylinder Ports

Different names have been used for the ports in the master cylinder. For the sake of uniformity, this text refers to

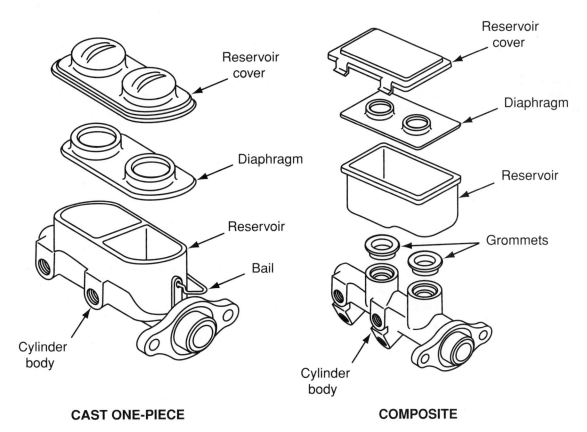

CAST ONE-PIECE   **COMPOSITE**

Figure 3. The master cylinder may be cast as one piece with an integral reservoir or it may be a composite assembly with a separate reservoir.

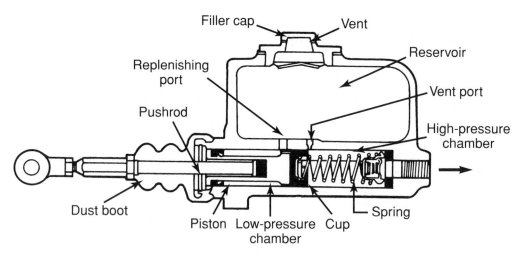

Figure 4. The vent port and the replenishing port have been called many different names, but the terms used here are established by an SAE standard for master cylinder terminology.

the forward port as the "vent" port and the rearward port as the "replenishing" port, which are the names established by SAE Standard J1153.

The **vent port** has been called the compensating port and the replenishing port by some manufacturers. To further confuse the nomenclature, the **replenishing port** has been called the compensating port by some manufacturers, as well as the vent port, the bypass port or hole, the filler port, or the intake port by many other manufacturers. The vent ports and replenishing ports let fluid pass between each pressure chamber and its fluid reservoir during operation. The names of these ports are not important as long as you understand their purposes and operations. **Figure 4** shows the names for the different parts of a typical master cylinder.

MASTER CYLINDER CONSTRUCTION

The location of the ports in relation to a master cylinder's pistons is shown in **Figure 5**. **Figure 6** is an exploded view of basically the same master cylinder. A single cylinder bore contains two piston assemblies. The piston assembly at the rear of the cylinder is the primary piston, and the one at the front of the cylinder is the secondary piston. Each piston has a return spring in front of it. There is a **cup seal** in front of each piston and a cup or seal at the rear of each piston. The seals retain fluid in the cylinder chambers and prevent seepage between the cylinders.

Inside the cylinder are two spool-shaped pistons **(Figure 7)**. The piston has a head on one end and a groove for an O-ring seal on the other end. The seal seats against the cylinder wall and keeps fluid from leaking past the pis-

ton. The smaller diameter center of the piston is the valley or spool area, which lets fluid get behind the head of the piston.

Each master cylinder piston works with a rubber cup seal, which fits in front of the piston head **(Figure 8)** . The cup has flexible lips that fit against the cylinder walls to seal fluid pressure ahead of the piston head. The cup lip also can bend to let fluid get around the cup from behind. When the brakes are applied, pressure in front of the cup forces the lip tightly against the cylinder wall and lets the seal hold very high hydraulic pressure. The lip of a cup seal is always installed toward the pressure to be contained, or away from the body of the piston. The cup seals only in one direction.

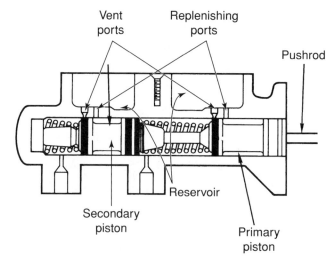

Figure 5. A cross sectional view of a dual master cylinder.

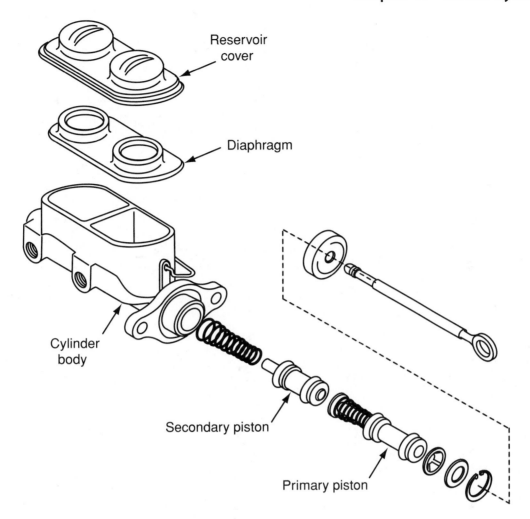

Figure 6. A disassembled view of a dual master cylinder.

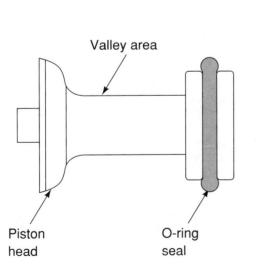

Figure 7. A simplified drawing of a spool-shaped piston for a master cylinder.

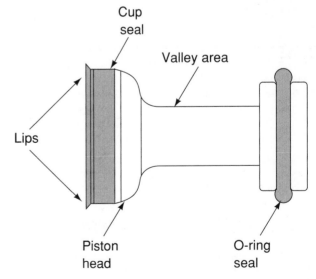

Figure 8. A cup seal and an O-ring seal the piston in the bore of the cylinder.

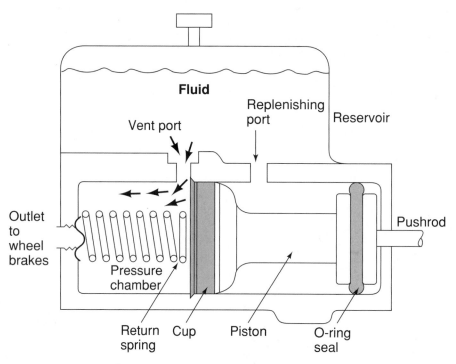

Figure 9.   The basic operating parts of a master cylinder.

If pressure behind the lip exceeds the pressure in front of it, the higher pressure will force the lip away from the cylinder wall and let fluid bypass the cup.

Pistons have small coil springs that return the pistons to the proper position when the brake pedal is released. Sometimes the springs are attached to the pistons; sometimes they are separate parts. A snapring holds the components inside the cylinder, and a rubber boot fits around the rear of the cylinder and pushrod to keep dirt from entering the cylinder.

A two-piece master cylinder has an aluminum body to lower the weight of the assembly. Because aluminum can be nicked or gouged easily, the bore of the aluminum cylinder is anodized to protect it from wear and damage. The removable nylon or plastic reservoir is also much lighter than a cast-iron unit. Because the master cylinder is made from two materials, it is often called a composite master cylinder. The pistons, cups, and springs used in a composite master cylinder are essentially the same and work the same way as those in a one-piece master cylinder.

MASTER CYLINDER OPERATION

The primary and secondary pistons of the master cylinder operate in the same way during normal braking. **Figure 9** shows a piston assembly and reservoir with the piston in the released position. The vent port in the bottom

of the reservoir is located just ahead of the piston cup. Fluid flows from the reservoir into the pressure chamber in front of the piston cup. The replenishing port is located above the valley area of the piston, behind the piston head. The O-ring seal on the piston keeps the fluid from leaking out the rear of the cylinder. The return spring in front of the piston and cup returns the piston when the brakes are released.

When the driver depresses the brake pedal, the pushrod pushes the master cylinder's piston forward **(Figure 10)**. As it moves forward, the piston pushes the cup past the

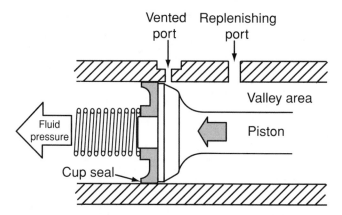

Figure 10.   As the piston moves forward, the cup seal closes the vent port and lets the pressure develop ahead of the piston.

vent port. As soon as the vent port is covered, fluid is trapped ahead of the cup. The fluid, under pressure, goes through the outlet lines to the wheel brake units to apply the brakes.

When the driver releases the brake pedal, the return spring forces the piston back to its released position. As the piston moves back, it pulls away from the fluid faster than the fluid can flow back from the brake lines to the pressure chamber. When this happens, low pressure is created ahead of the piston.

The piston must move back to the released position rapidly so it can be ready for another forward stroke, if necessary. The low-pressure area must be filled with fluid as the piston moves back. A path for fluid flow is provided by the valley area **(Figure 11)**, past the primary cup protector washer and through several small holes in the head of the piston, or by having enough clearance between the piston head and the cylinder bore. Fluid flows through the piston or around the lip of the cup and into the chamber ahead of the piston. This flow quickly relieves the low-pressure condition. The fluid that flows from the valley area to the pressure chamber must be replaced. **Figure 12** shows how the replenishing port lets fluid from the reservoir flow into this area.

When the piston is fully returned to its released position, the space in front of it is full of fluid. The piston cup again seals off the head of the piston. In the meantime, the fluid from the rest of the system has begun to flow back to the high-pressure chamber. If this pressure were not released, the brakes would not release. The returning fluid flows back to the reservoir through the vent port **(Figure 13)**. The vent port is covered by the piston cup at all times, except when the piston is released.

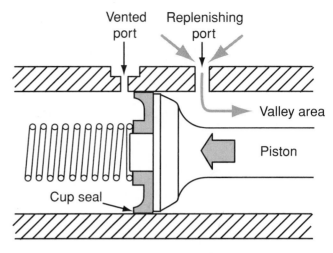

Figure 12. The low-pressure valley area is refilled through the replenishing port.

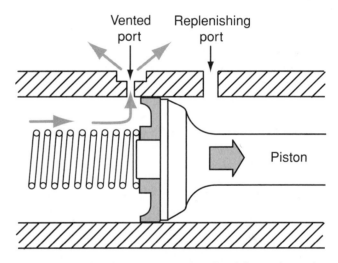

Figure 13. On the return stroke, fluid flows from the high-pressure chamber back to the reservoir through the vent port.

The replenishing port also has another important job. There are times when the amount of fluid in the wheel brake units and lines must be increased, such as when disc brake pads wear and when drum brake lining wears, before the automatic adjusters work.

We get more fluid ahead of the piston the same way we relieve low pressure ahead of the piston. On the return stroke, fluid flows through the head of the piston and around the lip of the cup **(Figure 14)**. When the piston returns, the vent port is open. There is not as much fluid returning from the wheel brake units and lines, so less fluid flows through the vent port and back into the reservoir. If the drum brakes are adjusted or new pads are installed on disc brakes, the system will automatically compensate for the amount of fluid needed on the next piston cycle.

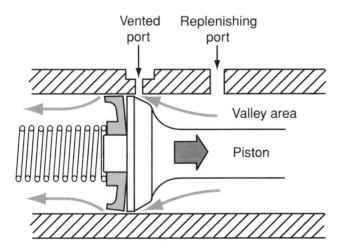

Figure 11. As the piston moves backward on the return stroke, fluid flows around the piston head and cup from the low-pressure area in the piston valley.

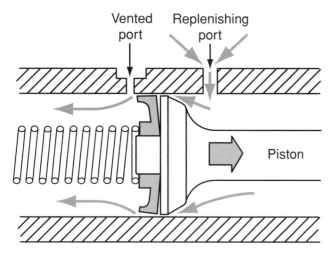

Figure 14. Fluid can flow around the cup seal when more fluid is required.

RESIDUAL PRESSURE CHECK VALVE

The pressure chamber in a master cylinder for some drum brake systems can have an additional part called a **residual pressure check valve**. This valve **(Figure 15)** can be installed in the pressure chamber or the outlet line of the master cylinder. A residual pressure check valve is the oldest type of pressure control valve used in a brake system. It can be found in older four-wheel drum brake systems, as well as in some late-model systems.

This valve usually is installed in the master cylinder outlet port to drum brakes. It maintains a residual pressure of 6–25 psi in the brake lines when the pedal is released. This residual line pressure maintains slight pressure on the wheel cylinder pistons to keep the sealing lips of the piston cups forced outward against the cylinder walls. When

the pedal is released, the retracting master cylinder piston creates a pressure drop in the lines. If pressure were to drop low enough, the piston cups could be pulled away from the wheel cylinder walls and draw air into the system. The slight residual pressure prevents this but is not high enough to overcome brake shoe spring tension. Maintaining pressure on the piston cups also maintains their fluid-sealing integrity and helps to prevent fluid leakage. Disc brake systems do not use this valve because residual pressure would cause the pads to drag on the rotor when the brakes are released.

The residual valve may be installed under the tubing seat in the master cylinder outlet port or inside the cylinder bore. When the brakes are applied, master cylinder pressure opens the valve and allows fluid flow to the wheel cylinders. When the brakes are released, pressure in the lines unseats the valve in the reverse direction to allow fluid return flow to the master cylinder. When line pressure drops below the pressure of the check valve spring, the valve closes to hold residual pressure in the lines.

With the redesign of wheel cylinders, many late-model master cylinders do not have a residual pressure check valve in their drum brake lines. Piston cup expanders were developed to hold the cups against the wheel cylinder walls **(Figure 16)**. Cup expanders are simpler, cheaper, and more reliable than check valves.

Diagonally split brake systems are another reason for the elimination of residual pressure check valves. A diagonally split system usually pairs one disc brake with one drum brake for half of the hydraulic system. Disc brakes rely on the action of a square-cut piston seal to retract the caliper piston and remove pressure from the brake pad. Any residual pressure at all would cause brake drag.

Although residual pressure check valves have been eliminated in many systems, they are still used in others.

Figure 15. A residual pressure check valve maintains slight fluid pressure on the wheel cylinder pistons in some drum brake systems.

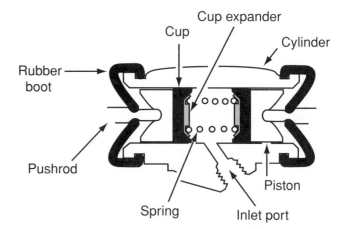

Figure 16. Cup expanders in wheel cylinders keep air from being drawn into the cylinder and eliminate the need for residual pressure check valves.

When you replace a master cylinder, it is very important to verify whether or not the vehicle requires a residual pressure check valve. Installing the wrong cylinder will cause improper brake operation and possible system failure.

SPLIT HYDRAULIC SYSTEMS

Most late-model cars have a diagonally split hydraulic system. If there is a hydraulic failure in the brake lines served by the secondary piston of the master cylinder, both pistons will move forward when the brakes are applied, as under normal conditions, but there is nothing to resist piston travel except the secondary piston spring. This lets the primary piston build up only a small amount of pressure until the secondary piston bottoms in the cylinder bore. Then, the primary piston will build enough hydraulic pressure to operate the brakes served by this half of the system.

In case of a hydraulic failure in the brake lines served by the primary piston, the primary piston will move forward when the brakes are applied but will not build up hydraulic pressure. Very little force is transferred to the secondary piston through the primary piston spring until the piston extension screw comes in contact with the secondary piston. Then, pushrod force is transmitted directly to the secondary piston and enough pressure is built up to operate its brakes.

FAST-FILL AND QUICK-TAKEUP MASTER CYLINDERS

Several manufacturers use fast-fill or **quick-takeup master cylinders**. These cylinders fill the hydraulic system quickly to take up the slack in the caliper pistons of low-drag disc brakes. Low-drag calipers retract the pistons and pads farther from the rotor than traditional calipers do. This reduces friction and brake drag and improves fuel mileage.

If a conventional dual master cylinder were used with low-drag calipers, excessive pedal travel would be needed on the first stroke to fill the lines and calipers with fluid and take up the slack in the pads. To overcome this problem, fast-fill and quick-takeup master cylinders provide a large volume of fluid on the first stroke of the brake pedal.

You can recognize a fast-fill or quick-takeup master cylinder by the bulge, or larger diameter, on the outside of the casting **(Figure 17)**. The cylinder has a larger diameter bore for the rear of the primary piston than for the front of the primary piston. Inside the cylinder, a fast-fill or **quick-takeup valve** replaces the conventional vent and replenishing ports for the primary piston. Some master cylinders for four-wheel disc brakes also have a quick-takeup valve for the secondary piston.

The quick-takeup valve contains a spring-loaded check ball that has a small bypass groove cut in the edge

Standard master cylinder

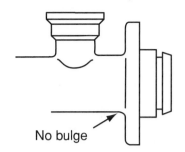

No bulge

Quick-takeup master cylinder

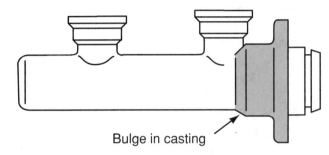

Bulge in casting

Figure 17. You can recognize a quick take-up master cylinder by the bulge or step in the casting.

of its seat. The outer circumference of the quick-takeup valve is sealed to the cylinder body with a lip seal. Several holes around the edge of the hole let fluid bypass the lip seal under certain conditions. Some valves (those more often called "fast-fill" valves) are pressed into the cylinder body and sealed tightly by an O-ring. A rubber flapper-type check valve under the fast-fill valve performs the same bypass functions as the lip seal of a quick-takeup valve.

Brakes Not Applied

When the brakes are off, both master cylinder pistons are retracted, and all vent and replenishing ports are open. Fluid to both ports of the primary piston must flow through the groove in the check ball seat, however.

Brakes Applied

As the brakes are applied, the primary piston moves forward in its bore. Remember that the diameter of the primary low-pressure chamber is larger than the diameter of the rest of the cylinder. As the primary piston moves forward into the smaller diameter, the volume of the low-pressure chamber is reduced. This causes hydraulic pressure to rise instantly in the low-pressure chamber. The higher pressure forces the large volume of fluid in the low-pressure chamber past the cup seal of the primary piston

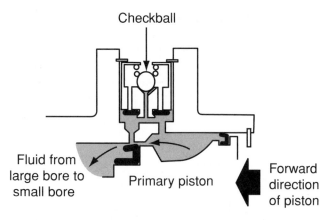

Figure 18. As the brakes are applied, fluid flows from the large bore to the small bore for the primary piston.

(Figure 18). This provides the extra volume of fluid to take up the slack in the caliper pistons.

The lip seal of the quick-takeup valve keeps fluid from flowing from the low-pressure chamber back to the reservoir. Initially, a small amount of fluid bypasses the check ball through the bypass groove, but this is not enough to affect quick-takeup operation.

As brake application continues, pressure in the low-pressure chamber rises to approximately 70–100 psi. The check ball in the quick-takeup valve then opens to let excess fluid return to the reservoir **(Figure 19).** Pressures in both chambers of the primary piston equalize, and the piston moves forward to actuate the secondary piston.

All of the actions described above apply to the primary piston if it is serving front disc brakes and the secondary piston is serving drum brakes. If the hydraulic system is split diagonally, or if the car has four-wheel, low-drag discs, the quick-takeup fluid volume must be available to both pistons. Some master cylinders have a second quick-takeup valve for the secondary piston. Others pro-

vide the needed fluid volume through the design of the cylinder itself. As long as the primary quick-takeup valve stays closed, the fluid bypassing the primary piston cup causes the secondary piston to move farther. This provides equal fluid displacement from both pistons and maintains equal pressure in the system. When the quick-takeup valve opens, both pistons move together just as in any other master cylinder.

Brakes Released

When the driver releases the brake pedal, the return springs force the primary and secondary pistons to move back. Pressure drops in the high-pressure chambers, and fluid bypasses the piston cup seals from the low-pressure chambers. Low pressure is created in the low-pressure chamber, which lets atmospheric pressure in the reservoir force fluid past the lip seal of the quick-takeup valve **(Figure 20).** Fluid from the reservoir then flows through both the vent and replenishing ports to equalize pressure in the pressure chambers and valley areas.

On the return stroke, fluid flow to the secondary piston is through a normal replenishing port unless the secondary piston also has a quick-takeup valve. If a secondary quick-takeup valve is installed, it works as described for a primary quick-takeup valve.

CENTRAL-VALVE MASTER CYLINDERS

Some ABSs use master cylinders that have central check valves in the heads of the pistons. If the master cylinder provides pressure during antilock operation (a so-called "open" system) and the system also has a motor-driven pump, the master cylinder's pistons can shift back and forth rapidly during antilock operation. This could cause excessive pedal vibration and—more importantly—wear on the piston cups where they pass over the vent ports.

To prevent seal damage and pedal vibration, spring-loaded check valves are installed in the piston heads. When the brakes are released, fluid flows from the replenishing ports to the low-pressure chambers, through the open cen-

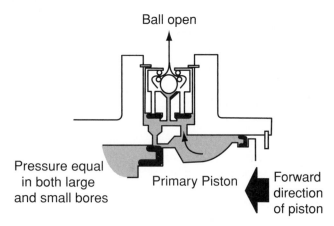

Figure 19. When pressures equalize in both chambers, fluid returns to the reservoir.

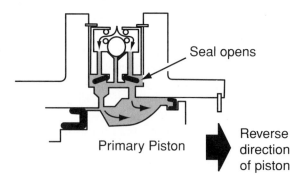

Figure 20. When the piston returns, low pressure draws fluid into the valley area.

tral check valves, and into the high-pressure chambers. As the brakes are applied, the central valves close to hold fluid in the high-pressure chambers. When the brakes are released again, the central check valves open to let fluid flow back through the pistons to the low-pressure chambers and the reservoir.

The central check valves in this kind of master cylinder provide supplementary fluid passages to let fluid move rapidly back and forth between the high- and low-pressure chambers during antilock operation. This is not much different in principle from non-ABS fluid flow, but the extra passages reduce piston and pedal vibration and cup seal wear.

Summary

- When the pedal is depressed, the master cylinder piston is pushed forward and the force is transmitted throughout the hydraulic system.
- The reservoir may be cast as one piece with the cylinder body or it may be a separate molded nylon or plastic container.
- All master cylinder caps or covers are vented to prevent a vacuum lock as the fluid level drops and have a diaphragm that separates the brake fluid from the air above it.
- The vent ports and replenishing ports let fluid pass between each pressure chamber and its fluid reservoir during operation.
- A master cylinder has two pistons: the piston assembly at the rear of the cylinder is the primary piston, and

the one at the front of the cylinder is the secondary piston.
- Each piston has a return spring and a cup seal in front of it and a cup or seal at the rear.
- The pressure chamber in a master cylinder for some drum brake systems might have a residual pressure check valve.
- Several manufacturers use fast-fill or quick-takeup master cylinders that fill the hydraulic system quickly to take up the slack in the caliper pistons of low-drag disc brakes.
- Some ABSs use master cylinders that have central check valves in the heads of the pistons to reduce the pedal and piston pulsations that are typically caused by ABS operation.

Review Questions

1. What is the purpose of the master cylinder vent port?
2. What is the purpose of the master cylinder replenishing port?
3. When the master cylinder piston is on a return stroke, fluid flow around the piston comes through the _____ port.
4. After the master cylinder piston returns, excess fluid in front of the piston returns to the reservoir through the _____ port.
5. The pushrod on a dual-piston master cylinder pushes on the _____ piston.
6. The _____ _____ in front of the piston traps fluid to build pressure in the pressure chamber.
7. Technician A says that the master cylinder primary cup seals pressure during brake application. Technician B says that the master cylinder primary cup allows fluid flow during brake release. Who is correct?
   A. Technician A only
   B. Technician B only
   C. Both Technician A and Technician B
   D. Neither Technician A nor Technician B
8. The operation of a master cylinder on the return stroke is being discussed. Technician A says that the replenishing port lets fluid flow from the reservoir into the low-pressure valley area. Technician B says that the

replenishing port allows fluid to return to the master cylinder reservoir. Who is correct?
   A. Technician A only
   B. Technician B only
   C. Both Technician A and Technician B
   D. Neither Technician A nor Technician B
9. The operation of a dual-piston master cylinder with a primary system leak is being discussed. Technician A says that the secondary piston is moved mechanically by the primary piston. Technician B says that the secondary piston is moved by hydraulic pressure from the primary piston. Who is correct?
   A. Technician A only
   B. Technician B only
   C. Both Technician A and Technician B
   D. Neither Technician A nor Technician B
10. The construction of a quick-takeup master cylinder is being discussed. Technician A says that the primary piston fits in two bore sizes. Technician B says that the primary and secondary pistons are interchangeable. Who is correct?
    A. Technician A only
    B. Technician B only
    C. Both Technician A and Technician B
    D. Neither Technician A nor Technician B

Chapter 15

Master Cylinder Service

Introduction

Master cylinder problems are common but not always readily evident. However, there are times when the master cylinder is suspect and the problem lies elsewhere. Accurate and logical troubleshooting is the only way to truly determine if the master cylinder is working properly.

Although brake pedal response and reservoir fluid levels are strong indicators of problems with the master cylinder or hydraulic system, other tests can be performed to help pinpoint the problem. Poor pedal feel or action may be caused by air trapped in the system. However, trapped air can be caused by a worn or defective master cylinder or some other part of the system. When the system acts as though the problem could be air, check it out before proceeding with your diagnosis. Air will also enter the system if there is a fluid leak. If there are no signs of leaks, check the system for trapped air **(Figure 1)**. If there is air trapped in the system, bleed the system and recheck.

> **You Should Know** The test described in Figure 1 can result in brake fluid bubbling or spraying out of the master cylinder reservoir. Wear safety goggles. Cover the master cylinder reservoirs with clear plastic wrap or another suitable cover to keep brake fluid off the vehicle's paint.

MASTER CYLINDER INSPECTION

If no air is trapped or you suspect a leak, check the master cylinder housing for cracks. Look for drops of brake fluid around the master cylinder. A slight dampness in the area surrounding the master cylinder is normal and is usually no reason for concern. However, if a reservoir chamber is cracked, it can be completely empty and the surrounding area might be dry. This is because the fluid drained very quickly and has had time to evaporate or wash away. But with only one-half of the brake system operational, the brake warning lamp should be lit and a test drive should reveal the loss of braking power. Refill the master cylinder reservoir section that is empty and apply the brakes several times. Wait five to ten minutes and check for leakage or fluid level drop in the reservoir.

If the master cylinder does not appear to be leaking, raise the vehicle on a lift and inspect all brake lines, hoses, and connections **(Figure 2)**. Look for brake fluid on the floor under the vehicle and at the wheels. Brake lines must not be kinked, dented, or otherwise damaged, and there should be no leakage. Brake hoses should be flexible and free of leaks, cuts, cracks, and bulges. Drum brake backing plates and disc brake calipers should be free of brake fluid and grease. Any parts attached to them should be tight.

LEAK TEST

Hydraulic brake system leaks can be internal or external. Most internal leaks are fluid bypassing the cups in the master cylinder. If the cups lose their ability to seal the pistons, brake fluid leaks past the cups and the pistons cannot develop system pressure.

Air may be indicated by spongy pedal, low pedal, or bottoming pedal.

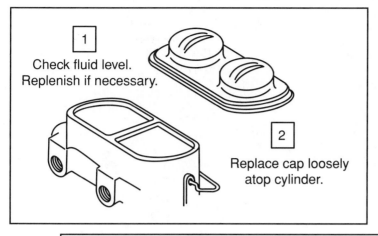

1 Check fluid level. Replenish if necessary.

2 Replace cap loosely atop cylinder.

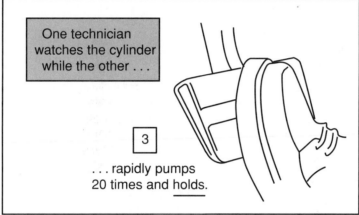

One technician watches the cylinder while the other . . .

3 . . . rapidly pumps 20 times and holds.

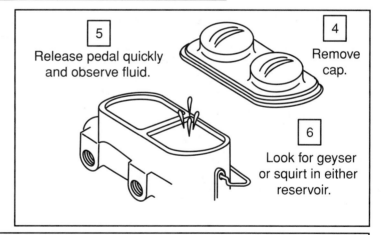

5 Release pedal quickly and observe fluid.

4 Remove cap.

6 Look for geyser or squirt in either reservoir.

RESULTS:

Geyser from reservoir:
Indicates air trapped in the system. It is compressed by pumping and causes a squirt when released. (If pedal is low, rear brake misadjustment can also cause a geyser.)

Action:
Bleed the affected system or systems. If one reservoir only, you need not bleed the other.

Figure 1. This quick test will help determine if the brake system has air trapped in it.

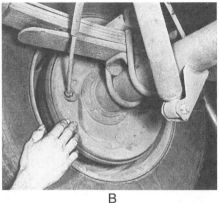

A   B   C

Figure 2.   Inspect the hydraulic system for leaks at these and other points: *A*, brake lines and hoses; *B*, drum brake backing plates; and *C*, disc brake calipers.

Internal and external rubber parts wear with use or can deteriorate with age or fluid contamination. Moisture or dirt in the hydraulic system can cause corrosion or deposits to form in the bore, resulting in the wear of the cylinder bore or its parts. Although internal leaks do not cause a loss of brake fluid, they can result in a loss of brake performance. This internal leakage, or fluid bypassing back to the reservoir, is the cause of many sinking pedal complaints and can be hard to pinpoint.

When external leaks occur, the system loses fluid. External leaks are caused by cracks or breaks in master cylinder reservoirs, loose system connections, damaged seals, or leaking brake lines or hoses. Check for a brake fluid leak as follows:

1. Run the engine at idle with the transmission in neutral.
2. Depress the brake pedal and hold it down with a constant foot pressure. The pedal should remain firm, and the foot pad should be at least 2 inches from the floor for manual brakes and 1 inch for power brakes (**Figure 3**) without ABS.

3. Hold the pedal depressed with medium foot pressure for about 15 seconds to make sure that the pedal does not drop under steady pressure. If the pedal drops under steady pressure, the master cylinder or a brake line or hose may be leaking.

If your inspection does not reveal any external leakage, but the pedal still drops under steady pressure, check for fluid bypassing the piston cups inside the master cylinder.

INTERNAL LEAK TEST (FLUID BYPASS TEST)

If the primary piston cup seal is leaking, the fluid will bypass the seal and move between the vent and replenishing ports for that reservoir or, in some cases, between reservoirs. If no sign of external leakage exists, but the brake warning lamp is lit, the master cylinder can be bypassing or losing pressure internally. Another sign of internal leakage, or bypassing, is a fluid level that rises slightly in one or both reservoirs when pressure is held on

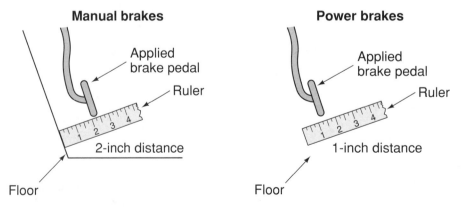

Figure 3.   With the brakes applied, the distance from the pedal to the floor should be about 2 inches for manual brakes and about 1 inch for power brakes.

the brake pedal. Test for fluid bypassing the piston cups as follows:

1. Remove the master cylinder cover and be sure the reservoirs are at least half full.
2. Watch the fluid levels in the reservoirs while a helper slowly depresses the brake pedal and then quickly releases it.
3. If fluid level rises slightly under steady pressure, the piston cups are probably leaking. Fluid level rising in one reservoir and falling in the other as the brake pedal is depressed and released also can indicate that fluid is bypassing the piston cups.
4. Replace or rebuild the master cylinder if it is bypassing fluid internally.

Another quick test for internal leakage is to hold pressure on the brake pedal for about one minute. If the pedal drops but no sign of external leakage exists, fluid is probably bypassing the piston cups.

TEST FOR OPEN VENT PORTS

Remove the cover from the master cylinder and observe the fluid in the reservoirs while a helper pumps the brake pedal. You should see a small ripple or geyser in the reservoirs as the brakes are applied. If you see no turbulence, loosen the bolts securing the master cylinder to the vacuum booster approximately $1/8$–$1/4$ inch and pull the cylinder away from the booster. Hold it in this position and have the helper pump the brakes again while you observe the fluid reservoirs. If turbulence (indicating compensation) now occurs, adjust the brake pedal pushrod length. If turbulence still does not occur, replace the master cylinder. Turbulence can only be seen in the front (secondary) reservoir of a quick-takeup master cylinder.

If turbulence does not occur in the master cylinder during this check, the pistons are probably restricting the vent ports. This means the pushrod is not allowing the pistons to return to the fully released positions. Adjusting the brake pedal pushrod at its connection to the pedal lever often fixes the problem on a nonpower-assist brake system. More often, however, the output pushrod of the power brake booster requires adjustment.

QUICK-TAKEUP VALVE TEST

The quick-takeup valve is used to provide a high volume of fluid on the first pedal stroke. This action takes up the slack in low-drag caliper pistons.

No direct test method exists for a quick-takeup valve, but excessive pedal travel on the first stroke may indicate that fluid is bypassing the valve. If this symptom exists, check for a damaged or unseated valve. If the pedal returns slowly when the brakes are released, the quick-takeup valve may be clogged so that fluid flow from the cylinder to the reservoir is delayed.

RESERVE STROKE TEST FOR VACUUM-BOOST SYSTEMS

When a low pedal or the feel of a bottomed-out pedal condition exists on a vehicle with vacuum power brakes, check the master cylinder reserve stroke as follows:

1. Put the transmission in either park or neutral and run the engine at idle.
2. Apply the brake pedal until it stops moving downward or there is increased resistance to the pedal travel.
3. While holding the pedal applied, raise the engine speed to approximately 2000 rpm.
4. Release the accelerator pedal and watch the brake pedal to see if it moves downward as the engine returns to idle speed. As the engine decelerates, the increased manifold vacuum exerts more force on the brake booster, causing the additional movement of the brake pedal.

If this test indicates that the master cylinder is not functioning properly, the unit should be serviced, rebuilt, or replaced with a new one.

REMOVING A MASTER CYLINDER

Often shops replace the unit to save time and reduce the chances of something going wrong during the rebuild. In either case, the master cylinder must be removed. On many vehicles with an ABS, the master cylinder is part of the ABS hydraulic modulator and master cylinder assembly. This ABS assembly is removed as a unit. The master cylinder is then separated from the modulator. To remove a master cylinder from a vehicle without an ABS:

1. Disconnect the battery ground (negative) cable.
2. Relieve any residual vacuum by pumping the pedal fifteen to twenty times until you feel a change in pedal effort.
3. Use a shop towel to remove any loose dirt or grease around the master cylinder.
4. Unplug the electrical connector from the fluid level sensor, if equipped.
5. Place a container under the master cylinder to catch brake fluid as the lines are disconnected.
6. Disconnect each brake line fitting from the master cylinder and plug the end of each line to keep dirt out of the lines and to prevent excessive brake fluid loss.
7. Remove the nuts attaching the master cylinder to the vacuum booster unit **(Figure 4)**.
8. Lift the master cylinder out of the vehicle. It may be necessary to insert a small pry bar between the booster and the master cylinder to free the master cylinder. Before removing the master cylinder in some vehicles, the proportioning valve assembly must be slid off the master cylinder mounting studs. On some vehicles, the vacuum valve from the booster must be removed, and the pressure warning switch connector must be

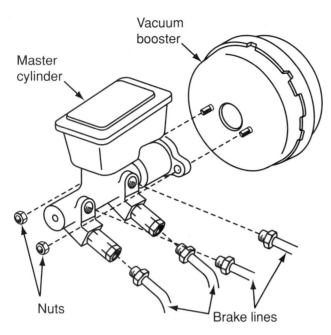

Figure 4.   Remove the attaching nuts and disconnect the brake lines and all electrical connectors to remove a master cylinder.

disconnected before removing the master cylinder. On vehicles with manual (nonpower) master cylinders, the pushrod must be disconnected from the brake pedal before the master cylinder can be removed.

9. Using a clean shop towel, clean the master cylinder and vacuum booster contact surfaces.

REBUILDING THE MASTER CYLINDER

Master cylinders should be totally rebuilt if they are not going to be replaced. To do this, install a master cylinder rebuild kit, which normally contains all replaceable seals, O-rings, and retainer clips **(Figure 5)**. Other components such

as the reservoir, external valves, and fluid sensors are replaced only if they are faulty.

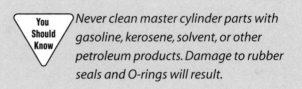

Never clean master cylinder parts with gasoline, kerosene, solvent, or other petroleum products. Damage to rubber seals and O-rings will result.

The following are the steps that should be followed when rebuilding a typical quick-takeup master cylinder:

1. Remove the reservoir cap and diaphragm and discard the old brake fluid.
2. Inspect the plastic fluid reservoir for cracks. If it is cracked or damaged, it must be replaced.
3. Remove the proportioning valves from the cylinder housing by unscrewing them **(Figure 6)**. Discard the O-rings for the valves.
4. Remove the fluid sensor switch by unlocking the switch tabs using snapring pliers. Discard the O-ring.
5. Depress the primary piston and remove the lock ring **(Figure 7)**.
6. Extract the primary and secondary piston assemblies from the bore **(Figure 8)**.
7. Clean the master cylinder and related parts with denatured alcohol or clean brake fluid. Blow dry with clean, unlubricated, compressed air.
8. Inspect the bore for pits, scoring, or corrosion. If any of these conditions is found, replace the master cylinder. Quick-takeup master cylinders should not be honed. Other types may be honed to remove minor imperfections.
9. To begin reassembly, install new primary and secondary cup seals **(Figure 9)**.
10. Install the return spring into the secondary bore.

Figure 5.   A typical master cylinder rebuild kit.

Figure 6.   Remove the proportioning valves from the cylinder housing by screwing them out.

Figure 7. Depress the primary piston and remove the lock ring.

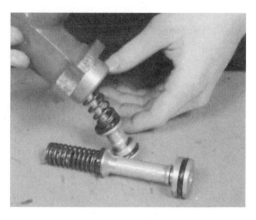

Figure 8. Extract the primary and secondary piston assemblies from the bore.

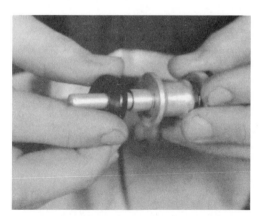

Figure 9. To begin reassembly, install new primary and secondary cup seals.

11. Lubricate the secondary piston with clean brake fluid.
12. Install the secondary piston into the cylinder bore **(Figure 10)**.

Figure 10. Lubricate the secondary piston with clean brake fluid and then install it into the cylinder bore.

13. Lubricate the primary piston with clean brake fluid and install it into the cylinder bore.
14. Depress the pistons into the bore and install the lock ring retainer.
15. Install the fluid level switch and the proportioning valves with new O-rings. Tighten the valves to specifications.

Master Cylinder Honing

Honing is a process whereby abrasive stones are rotated inside a cylinder to remove dirt, rust, or other slight corrosion and restore a uniform finish to the bore. Honing the bores of master cylinders, wheel cylinders, and calipers is similar to honing engine cylinders during overhaul or rebuilding.

An aluminum cylinder cannot be honed unless it has a steel insert or liner. The bore of an unlined aluminum cylinder is hardened by an anodizing process. This anodized finish would be removed by honing, and the cylinder would wear prematurely. Only a cast-iron cylinder bore or steel insert can be honed satisfactorily.

Brake cylinder honing is done less today than it was in the past for several reasons. Honing restores the finish of a cylinder but leaves it slightly rougher than the original finish. Therefore, cup seals can wear faster than in a new cylinder. Additionally, some manufacturers recommend against honing cylinders, preferring to replace them instead. Also, business economics and warranty issues often make it more practical to replace master cylinders and wheel cylinders than to hone and rebuild them. Nevertheless, honing may be necessary to save a master cylinder that is not easily replaced. Follow these guidelines when honing a master cylinder during rebuilding:

1. Disassemble the cylinder and clamp the cylinder in a vise. Clamp the cylinder by its mounting flange; do not clamp it on the bore section of the body. The vise can distort the cylinder bore and ruin the honing job.

Figure 11. A rigid hone has three or four abrasive stones to refinish a cylinder bore.

2. Select a suitably sized hone **(Figure 11)** and mount it in an electric drill motor or pneumatic die grinder.
3. Lubricate the cylinder bore thoroughly with fresh brake fluid and insert the hone into the bore.
4. Operate the hone at 300–500 rpm and move the hone back and forth in the bore with even strokes. Do not operate the hone while holding it stationary in the bore because an uneven diameter could result.
5. While the hone is rotating, let it extend partly out of the open end of the cylinder, but do not let the stones come completely out of the bore. This ensures an even finish on the bore without a ridge at the open end.

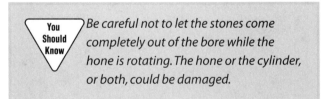

> **You Should Know**
>
> *Be careful not to let the stones come completely out of the bore while the hone is rotating. The hone or the cylinder, or both, could be damaged.*

6. Hone the bore for approximately 10–15 seconds while keeping the bore well lubricated with brake fluid.
7. Let the hone come to a complete stop and carefully remove it from the bore.
8. Wipe the bore with a clean rag and inspect its finish. The cylinder bore should be clean and free of rust and corrosion and have a uniform **crosshatch** finish **(Figure 12)**.

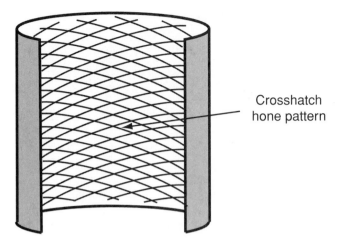

Crosshatch hone pattern

Figure 12. A crosshatch pattern is a uniform series of light grooves that cross each other at a 25- to 30-degree angle on the walls of the cylinder.

9. After inspecting the cylinder bore, clean it thoroughly with brake-cleaning solvent or alcohol. Do not use petroleum-based solvents.
10. Before assembling the cylinder, measure the piston-to-bore clearance. If honing has removed too much metal and clearance is excessive, the cylinder cannot be rebuilt.
11. Insert a narrow 0.006-inch (15-millimeter) feeler gauge into the cylinder bore and try to slide the piston into the bore, along with the gauge. If the piston fits, the bore is oversize. The cylinder must be replaced.

The 0.006-inch piston-to-bore clearance is a traditional guideline. Many late-model cylinders with small diameters require less clearance. Always check the vehicle service manual for exact specifications.

Brush Honing

Another kind of hone, called a flex hone or brush hone, also can be used to finish brake cylinder bores **(Figure 13)**. A brush hone is made of abrasive modules attached to flexible nylon cords. The flexible cords are, in turn, attached to the hone shaft. A brush hone removes peaks or high spots on the surface of a cylinder bore and produces a flattened finish, or **plateaued finish**.

A brush hone is driven by an electric drill motor, similarly to a rigid hone. Unlike a rigid hone, however, the brush hone should be rotating when it is inserted into and removed from the cylinder. To use a brush hone, lubricate the cylinder bore thoroughly with fresh brake fluid and insert the hone into the bore with the hone rotating at about 100 rpm. Operate the hone at 350–500 rpm and move it forward and backward in the cylinder 60–120 strokes to develop a suitable finish. Accelerate the final

Figure 13. A brush hone, or flex hone, is a series of abrasive balls on nylon cords that produce a flattened or plateaued finish on the cylinder walls.

stroking for ten to twenty seconds to develop a 45-degree crosshatch pattern. Brush hones of different diameters are available for honing wheel cylinders and caliper bores.

Reservoir Removal and Replacement

To remove a plastic reservoir without damaging it, secure the master cylinder in a vise. Clamp on the metal cylinder body flange to avoid damaging the cylinder body. Insert a small pry bar between the reservoir and cylinder body and push the reservoir body away from the cylinder. Once the reservoir is free, remove and discard the rubber grommets that seal the reservoir to the cylinder body. Make sure the reservoir is not cracked or deformed. Replace it if it is.

If the reservoir is serviceable, clean it with denatured alcohol and dry it with clean, unlubricated compressed air. Using clean brake fluid, lubricate the new grommets and the bayonets on the bottom of the reservoir. To reinstall the reservoir, place the reservoir top down on a hard, flat surface, such as a workbench. Start the cylinder body onto the reservoir at an angle, working the lip of the reservoir bayonets completely through the grommets until seated. Using a steady downward force and a smooth rocking motion, press the cylinder body onto the reservoir **(Figure 14)**.

BENCH BLEEDING MASTER CYLINDERS

To remove all air from a new or rebuilt master cylinder, bench bleed it before installing it on the vehicle. Bench bleeding reduces the possibility of air getting into the brake lines. Proper bench bleeding is particularly important with dual-piston cylinders, tandem chamber reservoirs, and master cylinders that mount on an angle other than horizontal.

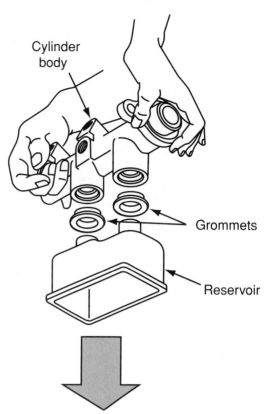

Place reservoir top down
on a hard, flat surface

Figure 14. Install new grommets on the cylinder and carefully push the reservoir onto the cylinder. Lubricate the grommets with brake fluid to aid in the assembly.

To bench bleed a master cylinder you need a bleeding kit. Bench bleeding kits are available that contain assorted fittings and tubing that will fit most vehicles. The tubing is installed in the outlet ports of the master cylinder and is routed back into the reservoir. You can also make your own bleeding kit using brake lines or hoses and fittings.

> **You Should Know** *Avoid spraying brake fluid or making it bubble violently during the bench bleeding procedure. Do not hold your face directly above the reservoirs. Wear safety goggles.*

A typical procedure for bench bleeding a master cylinder follows:

1. Mount the master cylinder in a bench vise so that the bore is horizontal. Make sure not to overtighten

Figure 15. Connect short lengths of tubing to the outlet ports.

Figure 17. Using a wooden dowel or the blunt end of a drift or punch, slowly push on the piston.

the vise because this will distort the bore of the cylinder.

2. Connect short lengths of tubing to the outlet ports; make sure the connections are tight **(Figure 15)**.
3. Bend the tubing so the ends are in each reservoir chamber **(Figure 16)**.
4. Fill the reservoirs with clean brake fluid, making sure the ends of the tubing are covered in fluid.
5. Using a wooden dowel or the blunt end of a drift or punch, slowly push on the piston at the rear of the master cylinder until you feel both master cylinders have bottomed out in their bore **(Figure 17)**.
6. Watch for air bubbles to appear at the ends of the tubing in the reservoir. Slowly release pressure on the piston and allow it to return to its rest position. On quick-takeup cylinders, wait fifteen seconds before pushing on the piston again. On other types of master cylinders, stroke the pistons again as soon as the piston returns to its rest position.
7. Repeat the process until no air bubbles appear in the fluid in the reservoir.

8. Remove the tubes from the outlet ports and plug the openings with your finger or temporary plugs. Keep the outlets plugged until the master cylinder is installed in the vehicle.
9. Install the master cylinder into the vehicle.
10. Attach the brake lines, but do not tighten them.
11. Slowly depress the brake pedal several times to force out any air that might be trapped at the connections. Before releasing the brake pedal, tighten the line nut slightly and loosen it before depressing the pedal again.
12. When there are no air bubbles in the fluid, tighten the line fittings and refill the reservoir.
13. Bleed the entire system.

Master Cylinder Bench Bleeding with a Syringe

Another bench bleeding technique for the master cylinder uses a special bleeding syringe to draw fluid out of the reservoir, remove air from it, and inject the fluid back into the unit **(Figure 18)** as follows:

1. Plug the outlet ports of the master cylinder. Carefully mount the master cylinder in a vise with the pushrod end slightly elevated **(Figure 18A)**. Do not clamp the cylinder by the bore or exert pressure on a plastic reservoir.
2. Pour brake fluid into the master cylinder until it is half full.
3. Remove a plug from one outlet port so you can use the syringe to draw fluid out of the cylinder.
4. Depress the syringe plunger completely and place its rubber tip firmly against the outlet port to seal it.
5. Slowly pull back on the plunger to draw fluid out of the cylinder. Fill the syringe body about one-half full **(Figure 18B)**.
6. Point the tip of the syringe upward. Slowly depress the plunger until all air is expelled **(Figure 18C)**.

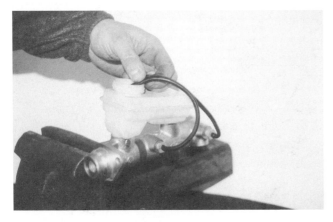

Figure 16. Bend the tubing so the ends are in each reservoir chamber.

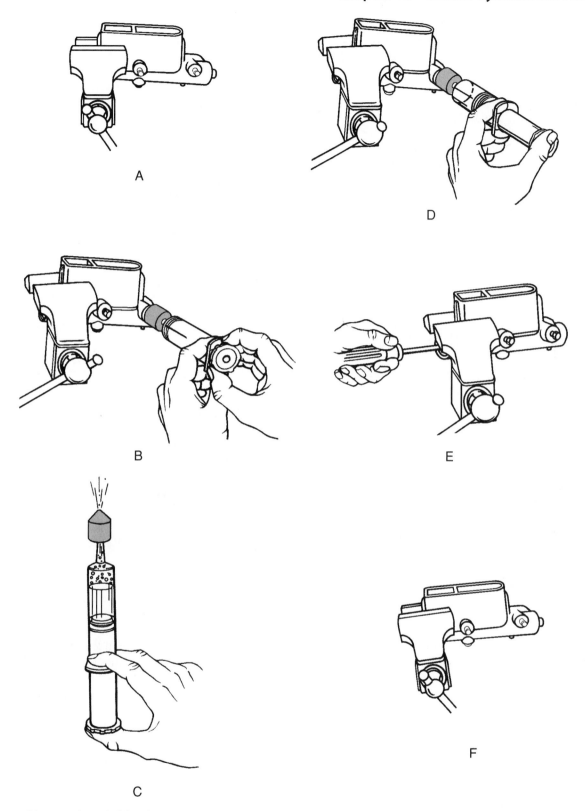

Figure 18.   You can bench bleed a master cylinder with a special syringe as shown in these six steps.

7. Place the tip of the syringe (with fluid) firmly against the same outlet and slowly depress the plunger to inject the fluid back into the cylinder **(Figure 18D)**. Air bubbles should appear in the reservoir.

8. When these bubbles stop, remove the syringe and plug the outlet. Repeat this procedure at the other outlet. Plug all outlets tightly.

9. With the pushrod end tilted downward slightly, reclamp the master cylinder in the vise.

10. Slowly slide the master cylinder pushrod back and forth about 1/8 inch until you see no air bubbles in the reservoirs **(Figure 18E)**.

11. Remount the master cylinder with the pushrod end up **(Figure 18F)**. Fill the syringe with brake fluid and expel the air as in step 6.

12. Remove one outlet plug at a time and repeat steps 6 and 7. The master cylinder is now completely bled.

INSTALLING A NON-ABS MASTER CYLINDER

After bench bleeding a master cylinder, install it as follows:

1. Install the master cylinder onto the vacuum booster studs.
2. Install the retaining nuts and torque to specifications.
3. Unplug each outlet port, then install and tighten the brake line fittings.
4. Reconnect the electrical connector to the brake fluid level sensor, if equipped.
5. Reconnect the battery ground (negative) cable.
6. Bleed the system.

PUSHROD ADJUSTMENT

Proper adjustment of the master cylinder pushrod is essential for safe and correct brake operation. If the pushrod is too long, the master cylinder piston will restrict the vent ports. This can prevent hydraulic pressure from being released and can result in brake drag. If the pushrod is too short, the brake pedal will be low and the pedal stroke length will be reduced, which can result in a loss of braking power. When the brakes are applied with a short pushrod, groaning noises might be heard from the vacuum booster.

If you suspect a problem with pushrod length or adjustment, ensure that the ports are open by observing how they affect the fluid in the reservoir. If fluid does not spurt from the vent ports when the brake pedal is released, the pushrod might be holding the master cylinder pistons in positions that partially restrict the ports.

In a vacuum power brake system, the pushrod is part of the booster and is matched to the booster during assembly. It is normally adjusted only when the vacuum booster or the master cylinder is serviced. Vacuum booster pushrod length is usually checked with a gauge.

Checking and adjusting brake pedal free play will help to ensure proper master cylinder piston travel.

MASTER CYLINDER BLEEDING ON THE VEHICLE

Whenever possible, bleed the master cylinder on the bench before installing it on the vehicle. Bleeding the cylinder after installing it on the car removes any final air bubbles in the outlet ports or air that may enter the fittings when they are connected.

Because the master cylinder is the highest point of the hydraulic system and because air rises, bleed the master cylinder before bleeding any of the wheel brakes. The following procedures explain how to bleed a master cylinder with and without bleeder screws.

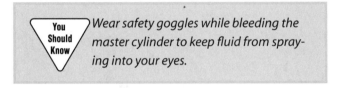

You Should Know ⟩ *Wear safety goggles while bleeding the master cylinder to keep fluid from spraying into your eyes.*

On-Vehicle Bleeding without Bleeder Screws

Most master cylinders do not have bleeder screws, but you can remove any trapped air at the outlet ports by loosening the line fittings and applying fluid pressure as follows:

1. Discharge the vacuum from the power booster, if equipped, by pumping the pedal with the engine off until it becomes hard.
2. Fill the master cylinder with fresh brake fluid and ensure that it stays at least half full during the bleeding procedure.
3. Loosen the forward, or highest, line fitting on the master cylinder outlet ports **(Figure 19)**.
4. Apply fluid pressure with a pressure bleeder or by having a helper press and hold the brake pedal.
5. Hold a clean rag or a container under the fitting to catch fluid that escapes.
6. While maintaining fluid pressure, tighten the fitting.
7. Repeat steps 3 through 6 until no air escapes from the fitting along with the fluid. Then repeat these steps at

each remaining fitting on the master cylinder, working from the highest to the lowest.

On-Vehicle Bleeding with Bleeder Screws

Some master cylinders have bleeder screws. These usually are cylinders that mount level on the vehicle. To bleed these master cylinders:

1. Discharge the vacuum from the power booster, if equipped, by pumping the pedal with the engine off until it becomes hard.
2. Fill the master cylinder with fresh brake fluid and ensure that it stays at least half full during the bleeding procedure.
3. Attach the plastic tubing to the end of the forward bleeder screw on the master cylinder and submerge the other end of the tubing in the container of fresh fluid.
4. Loosen the bleeder screw on the master cylinder and apply fluid pressure with a pressure bleeder or by having a helper depress and hold the brake pedal.
5. While maintaining fluid pressure, tighten the bleeder screw.
6. Repeat steps 3 through 5 until no air escapes from the bleeder screw along with the fluid. Then repeat these steps at each remaining bleeder screw on the master cylinder.

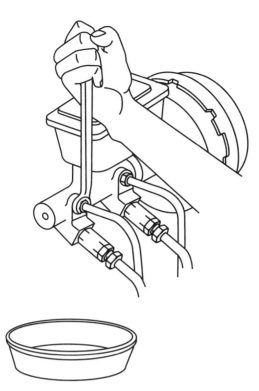

Figure 19. Loosen the fitting at the highest connection to the master cylinder and have an assistant press and hold the brake pedal. Make sure you have a catch basin or container below the fittings before you loosen them.

Summary

- It is possible for the master cylinder to have external leaks. If it does not appear to be leaking, raise the vehicle on a lift and inspect all brake lines, hoses, and connections.

- Most internal leaks are fluid bypassing the cups in the master cylinder. If the cups lose their ability to seal the pistons, brake fluid leaks past the cups, and the pistons cannot develop system pressure.

- Working and open vent ports are shown by a small ripple or geyser in the fluid reservoirs as the brakes are applied.

- When a low pedal or the feel of a bottomed-out pedal condition exists on a vehicle with vacuum power brakes, check the master cylinder reserve stroke.

- If the master cylinder is faulty, it must be replaced or rebuilt. In either case, the master cylinder must be removed. On many vehicles with an ABS, the master cylinder is part of the ABS hydraulic modulator and master cylinder assembly. This ABS assembly is removed as a unit. The master cylinder is then separated from the modulator.

- Rebuilding a master cylinder involves: draining the reservoirs; completely disassembling the unit; inspecting all parts, including the cylinder bore; replacing all rubber seals (cups) and O-rings; and reassembling the unit with new seals and O-rings.

- Honing is a process whereby abrasive stones are rotated inside a cylinder to remove dirt, rust, or other slight corrosion and restore a uniform finish to the bore.

- Two common types of hones used today are the rigid and the flex hone.

- To remove all air from a new or rebuilt master cylinder, it should be bench bled before installing it on the vehicle.

- Proper adjustment of the master cylinder pushrod is essential for safe and correct brake operation. If the pushrod is too long, the master cylinder piston will restrict the vent ports. If the pushrod is too short, the brake pedal will be low and the pedal stroke length will be reduced, which can result in a loss of braking power.

Review Questions

1. Technician A recommends cleaning a master cylinder body in the shop solvent tank cleaning system when rebuilding a master cylinder. Technician B says that the bore of aluminum master cylinder bodies should be honed when rebuilding a master cylinder to remove any small pits or burrs that have developed. Who is correct?
   A. Technician A only
   B. Technician B only
   C. Both Technician A and Technician B
   D. Neither Technician A nor Technician B

2. Technician A says that if the brake pedal drops under steady foot pressure, the master cylinder can have an internal leak or there can be an external leak in a brake line or hose. Technician B says that a slight trace of brake fluid on the booster shell below the master cylinder mounting flange indicates a master cylinder leak, and the unit should be replaced. Who is correct?
   A. Technician A only
   B. Technician B only
   C. Both Technician A and Technician B
   D. Neither Technician A nor Technician B

3. Testing for open replenishing ports on the master cylinder is being discussed. Technician A says that ripples or a small geyser should be visible in the master cylinder reservoir when the brake pedal is pumped. Technician B says that these ripples will be visible only in the front reservoir of a quick-takeup master cylinder. Who is correct?
   A. Technician A only
   B. Technician B only
   C. Both Technician A and Technician B
   D. Neither Technician A nor Technician B

4. Technician A says that if the master cylinder pushrod is too long, it causes the master cylinder piston to close off the vent port, resulting in brake drag. Technician B says that if the pushrod is too short, the brake pedal will be low and pedal stroke length will be reduced with a loss of braking power. Who is correct?
   A. Technician A only
   B. Technician B only
   C. Both Technician A and Technician B
   D. Neither Technician A nor Technician B

5. Technician A says that the master cylinder should be bench bled before being installed on the vehicle. Technician B says that after installing the master cylinder on the vehicle, the entire system should be bled. Who is correct?
   A. Technician A only
   B. Technician B only
   C. Both Technician A and Technician B
   D. Neither Technician A nor Technician B

6. Technician A says that unequal fluid levels in the master cylinder reservoir chambers can be caused by normal lining wear. Technician B says that a slight squirt of brake fluid from one or both master cylinder reservoir chambers when the brake pedal is applied is normal. It is caused by the fluid displacement through the reservoir replenishing ports. Who is correct?
   A. Technician A only
   B. Technician B only
   C. Both Technician A and Technician B
   D. Neither Technician A nor Technician B

Chapter 16

Brake Lines, Fittings, and Hoses

Introduction

Hydraulic lines made of tubes and hoses transmit fluid under pressure between the master cylinder and each of the wheel brake units **(Figure 1)**. Brake lines consist of steel tubes or pipes and flexible hoses connected with fittings. Rigid tubing is used everywhere except where the lines must flex. Flexing of the brake lines is necessary between the chassis and the front wheels and between the chassis and the rear axle or suspension. Brake tubing and hoses are manufactured to strict specifications developed by SAE and the International Standards Organization (ISO).

BRAKE TUBES OR PIPES

The hydraulic tubing used in the brake system is double-wall, welded steel tubing that is coated to resist rust and other corrosion. Double-wall tubing is made in two ways: seamless and multiple ply. Each must meet the specifications of SAE Standard J1047.

Seamless tubing is made by rolling steel sheet twice around a mandrel so the edges do not adjoin each other to form a seam **(Figure 2)**. The tubing is then run through a furnace where copper plating is applied and brazed to form the tubing into a single, seamless piece.

Multiple-ply tubing is formed as two single-wall tubes, one inside the other. The seams of each section must be at least 120 degrees apart. Then the two-ply tubing is furnace brazed just as seamless tubing is to form a single, seamless length. All brake tubing is plated with zinc or tin for corrosion protection.

Tubing Sizes

Brake tubing is made in different diameters and lengths **(Figure 3)**. The most common diameters for steel tubing or pipes in the inch system are $3/16$ inch, $1/4$ inch, and $5/16$ inch. Other tubing diameters from $1/8$ inch to $3/8$ inch also are available.

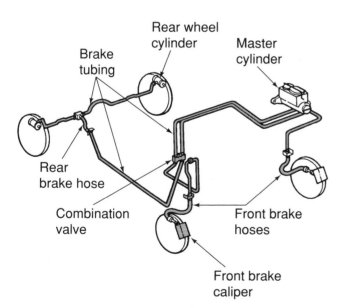

Figure 1. Brake lines consist of rigid tubing or pipes and flexible hoses that carry brake fluid from the master cylinder to the service brakes.

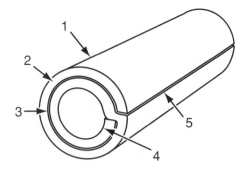

1. Tinned copper-steel alloy protects outer surface.
2. Long-wearing and vibration-resisting soft steel.
3. Fused copper-steel alloy unites two steel walls.
4. Copper-steel alloy lining protects inner surface.
5. Beveled edges and single close-tolerance
   strip result in no inside bead at joint. The
   tubing is uniformly smooth, inside and out.

Figure 2.   Seamless tubing is made by rolling steel sheets twice around a mandrel so the edges do not adjoin each other to form a seam. The tubing is then brazed with copper for corrosion protection.

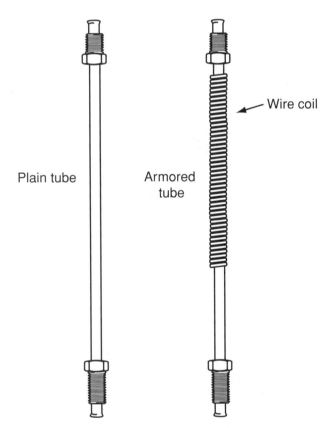

Figure 3.   Brake tubing is available in a variety of diameters and lengths. A wire coil is wound around armored tubing to protect it from abrasion.

Some vehicles use tubing sized in metric diameters. Metric diameters are specified in millimeters by SAE Standard J1290. Common metric diameters are 4.75, 6, 8, and 10 millimeters.

> **You Should Know** *Double-wall steel tubing is the only type of tubing approved for brake lines. Never use copper tubing as a replacement; it cannot withstand the high pressure or the vibrations to which a brake line is exposed. Fluid leakage and system failure can result.*

Tubing Shapes

The tubing installed on a new car is shaped properly to fit into the brake system. Like other parts made by the manufacturer, tubing is referred to as an OEM part. Aftermarket replacement tubes are most often available straight and in different lengths. If available, however, it is preferable to use an OEM-shaped, prefabricated tube as a replacement.

Each end of the tubing has a fitting, used to connect the tubing into the system. The fitting fits over the tube or pipe and seats against a specially formed end of the tubing or pipe. The formed end of the tubing is called a flare. There are two common types of flares formed on the end of brake tubing (**Figure 4**). The double flare has the end of the tubing flared out, and then it is formed back onto itself.

The **ISO flare** has a bubble-shaped end formed on the tubing. Each type of flare is used with a different type of fitting, and they are not interchangeable. Tools are available to form these flares when fabricating new tubing for a repair.

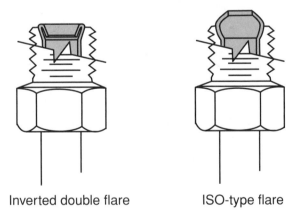

Figure 4.   Both the inverted double flare and the ISO flare are commonly used on brake lines. The double flare has the end of the tubing flared out, and then it is formed back on to itself.

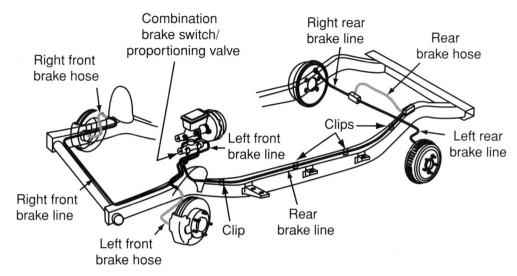

Combination
brake switch/
proportioning valve

Right front
brake hose

Right rear
brake line

Rear
brake hose

Left front
brake line

Clips

Left rear
brake line

Right front
brake line

Clip

Left front
brake hose

Rear
brake line

Figure 5. Typical installation of brake lines and hoses. The brake lines are held to the vehicle's frame by clips and brackets.

Flares and their fittings are discussed in more detail later in this chapter.

Brake lines are routed from the master cylinder along the car frame or body toward the wheel brake units **(Figure 5)**. Clips hold the tubes in position and keep the tubes from vibrating, which could cause metal fatigue and eventual rupture.

BRAKE HOSES

Brake hoses **(Figure 6)** are the flexible links between the wheels or axles and the frame or body. Hoses must withstand high fluid pressures without expansion and must be free to flex during steering and suspension movement.

Figure 7 shows the parts of a brake hose. The hose is made from materials that resist damage from both brake

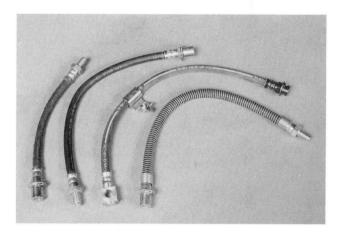

Figure 6. Brake hoses are the flexible sections of the brake lines.

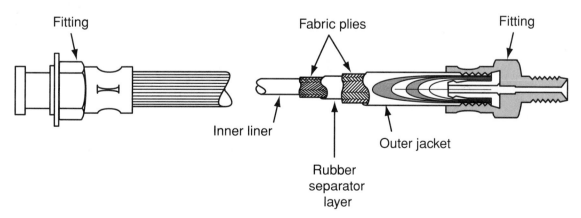

Fitting

Fabric plies

Fitting

Inner liner

Rubber
separator
layer

Outer jacket

Figure 7. Brake hoses are made with two fabric layers alternating with two rubber layers. The outer jacket is ribbed to indicate if the hose is twisted during installation.

fluid and petroleum-based chemicals. Brake hoses must withstand 4000 psi of pressure for two minutes without rupturing. The test pressure is then increased at a rate of 25,000 psi per minute until the hose bursts. These performance requirements make it clear why brake hoses are important safety devices.

Brake hoses must be marked with raised longitudinal ribs or two 1/16-inch colored stripes on opposite sides to indicate if the hose is twisted during installation. Twisting creates stress that could lead to rupture. Twisting also can cause the hose to kink and block fluid pressure. To further prevent twisting, at least one end fitting on a hose is usually a swivel fitting that will turn without twisting the hose.

Brake hoses are reinforced with metal or synthetic cords to withstand high pressures. Each end has a fitting so that it can be connected to other parts of the brake system. The brake hose length is specified from the end of one fitting to the end of the other fitting. Hose diameter is specified as the inside diameter of the hose. Hoses are available in both inch and metric diameters.

> ⚠ **You Should Know** *Never use low-pressure hydraulic hoses or oil hoses as replacements for brake hoses. These components cannot withstand the high pressure of the brake system. Fluid leakage, line rupture, and system failure can result.*

The fittings are crimped or swaged to the brake hose ends at very high pressures. Clamp-on or crimped fittings used with low-pressure hydraulic hoses or oil hoses cannot be used for brake hoses.

For all of their strength and durability, brake hoses are the weakest links in the brake hydraulic system. Atmospheric ozone attacks the rubber material and, over a long period, causes the hoses to deteriorate. Slight porosity of the hose material also lets air enter the system and contaminate the brake fluid—again, over a long period of time.

Hoses also are subject to wear, both externally and internally. Hoses must be installed so that they do not rub against vehicle parts. Some hoses have rubber ribs around their outer circumference to protect them from rubbing on suspension and chassis parts. A bulge in a hose usually indicates that the hose may fail, due to internal wear or damage.

Another internal hose problem can occur when the inside of the hose wears to the point that a flap of rubber loosens from the hose wall. If the flap stays secured to the wall at one end and its loose end faces the master cylinder, it can delay fluid pressure to the wheel brake and cause

uneven braking or pulling. If the loose end of the flap faces the wheel brake, it can delay pressure release from the wheel brake and cause brake drag. This kind of hose defect is impossible to see from the outside and is difficult to pinpoint with any kind of test. If the symptoms described above exist, hoses usually are replaced to try to eliminate the problem.

The point at which a hose connects to a rigid tube usually is secured to a bracket on the frame or body. A clip fits in a groove in the end of the hose fitting, and the end of the hose is inserted through a support bracket. The steel line is then threaded into the fitting on the end of the hose. The fitting on one side of the bracket and the clip on the other side hold the hose securely in position.

BRAKE FITTINGS

Fittings **(Figure 8)** are made from steel to withstand the brake system pressures and are threaded to allow connection to other brake parts. The ends of brake tubing are formed into either an inverted double flare or an ISO flare and fitted with a male flare nut. Brake hoses can have either male or female fittings. The threaded connections of master cylinders, wheel cylinders, calipers, and most valves are female. Fittings are made with both SAE inch-sized and ISO metric threads. Threads from one system do not fit threads of the other system. Adapters can be used to connect two different sizes of fittings.

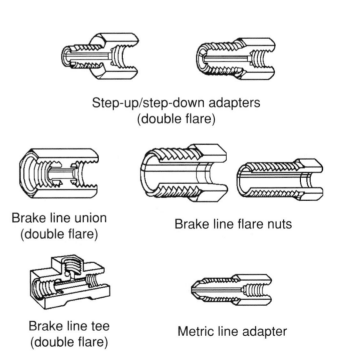

Step-up/step-down adapters
(double flare)

Brake line union
(double flare)

Brake line flare nuts

Brake line tee
(double flare)

Metric line adapter

Figure 8.   Assorted adapters and fittings that may be found in a hydraulic brake system.

Unions and tee fittings are used to connect two lines together.

SAE Flare Fittings

All SAE flare fittings used in brake systems have a 45-degree taper on the male nut and on the inside of the flare **(Figure 9)**. The tubing seat in the female fitting has a 42-degree taper. The 3-degree mismatch forms an interference fit that creates a leak-free, high-pressure seal.

A flare fitting with an external taper is called simply a standard flare. When the fitting is tightened, the tapered surfaces of the male and female fittings create the seal. A flare fitting with an internal taper is called an inverted flare or LAP flare. These are more common than standard flares on brake tubing. The male flare nut compresses the bell-mouthed inverted flare against the seat in the female fitting, and the tubing is sandwiched between the two halves of the fitting to form the seal.

The flared tubing end can be formed as either a single flare or a **double flare**. A single flare does not have the sealing power of a double flare and is subject to cracking. Preformed replacement brake tubing is sold with a double flare on each end and the flare nuts in place on the tubing.

ISO Flare Fittings

If you cut and form a brake pipe from bulk tubing, you will have to form double flares on both ends.

The ISO flare fitting is not folded back on itself like a double flare. The unique shape of an ISO flare causes it to be called a "bubble" flare **(Figure 10)**.

Like a standard or an inverted flare, an ISO flare uses interference angles between the flare and its seat to form a leak-free seal. The angle of the outer surface of an ISO flare

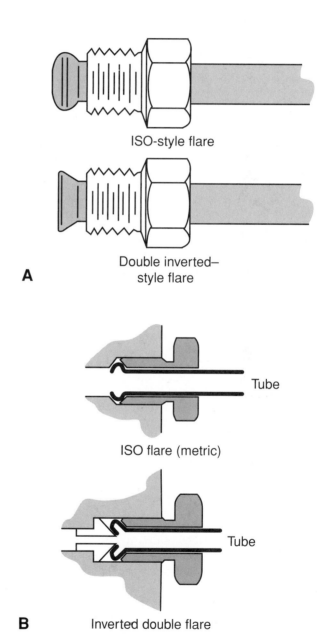

A

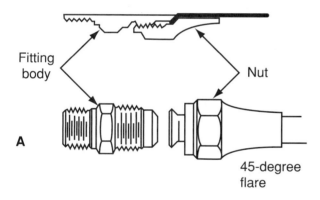

A

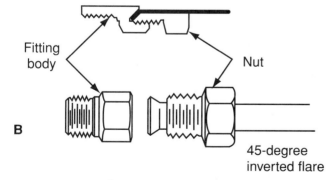

B

Figure 9. SAE 45-degree flare fittings commonly used for brake connections: *A*, standard flare with an external taper; and *B*, an inverted flare with an internal taper.

Figure 10. ISO and inverted flares are distinguishable by the shape of the flare. An ISO flare does not need a separate tubing seat as does an inverted flare.

is approximately 32.5 degrees. The flare seat is 30 degrees, and the angle at the end of the flare nut is 35 degrees. ISO flares have become popular because the outer surface of an ISO flare will form a leak-free seal against a mating surface in a cylinder or caliper body that is simply drilled and countersunk with the right taper.

Compression Fittings

Straight compression fittings are usually found on the ends of brake hoses that attach to calipers or wheel cylinders. As the rigid fitting on the end of the hose is tightened, it compresses a soft copper washer against a flat, machined surface on the cylinder or caliper **(Figure 11)**.

Repeated tightening and loosening of a compression fitting can permanently compress the copper washer. Whenever you disconnect a compression fitting, inspect the washer and replace it if it is significantly compressed, nicked, or deformed.

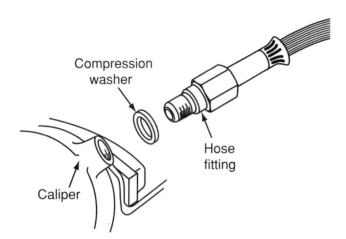

Figure 11. A brake line compression fitting uses a soft copper washer for a leak-free seal.

Banjo Fittings

A **banjo fitting** is a circular fitting that looks like the musical instrument of the same name. A banjo fitting is used to attach a hose or tube to a port on a cylinder or caliper at a close right angle **(Figure 12)**. Fluid passes from the brake line into the cutaway section inside the banjo fitting and then through the hollow bolt to the cylinder or caliper. A banjo fitting is a kind of compression fitting, and both flat surfaces of the fitting are sealed with soft copper washers.

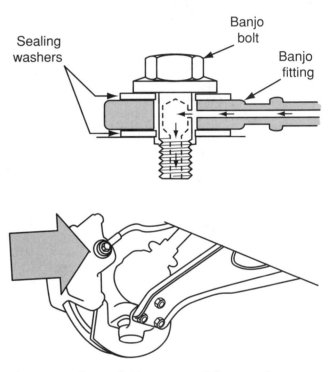

Figure 12. Banjo fittings are used for some hose connections, particularly those that are made at a right angle to a brake caliper.

Summary

- Hydraulic lines made of tubes and hoses transmit fluid under pressure between the master cylinder and each of the wheel brake units.
- Brake lines consist of steel tubes or pipes and flexible hoses connected with fittings. The hydraulic tubing used in the brake system is double-wall, welded steel tubing that is coated to resist rust and other corrosion.
- Each end of the tubing has a fitting, used to connect the tubing into the system.

- The double flare has the end of the tubing flared out, and then it is formed back onto itself.
- The ISO flare has a bubble-shaped end formed on the tubing.
- Brake hoses are the flexible links between the wheels or axles and the frame or body.
- Brake hoses are reinforced with metal or synthetic cords to withstand high pressures.
- The flared tubing end can be formed as either a single flare or a double flare.

- The ISO flare fitting uses interference angles between the flare and its seat to form a leak-free seal.
- Straight compression fittings are usually found on the ends of brake hoses that attach to calipers or wheel cylinders.

- A banjo fitting is commonly used to attach a hose or tube to a port on a cylinder or caliper at a close right angle.

Review Questions

1. Describe the difference between the double flare and the ISO-type brake line flare.
2. Brake lines are made up of solid brake _____ and flexible brake _____.
3. Which of the following statements about brake hoses is *not* true?
   A. The hose is made from materials that resist damage from both brake fluid and petroleum-based chemicals and must be able to withstand a pressure of 40,000 psi for twenty minutes without rupturing.
   B. Brake hoses are marked with raised longitudinal ribs or two $1/16$-inch colored stripes on opposite sides to indicate if the hose is twisted during installation.
   C. Brake hoses are reinforced with metal or synthetic cords to withstand high pressures.
   D. Each end has a fitting so that it can be connected to other parts of the brake system. The brake hose length is specified from the end of one fitting to the end of the other fitting.

4. Technician A says that copper is often used as a brake line material. Technician B says that brake lines are made of steel. Who is correct?
   A. Technician A only
   B. Technician B only
   C. Both Technician A and Technician B
   D. Neither Technician A nor Technician B
5. The flare on the end of a brake line has a bubble or chamfer shape. Technician A says that this is a double flare. Technician B says that this is an ISO flare. Who is correct?
   A. Technician A only
   B. Technician B only
   C. Both Technician A and Technician B
   D. Neither Technician A nor Technician B

Chapter 17

Brake Line Service

Introduction

All automobile manufacturers include brake line inspection in their vehicle maintenance schedules. Most recommend inspecting brake hoses twice a year. Brake hoses must be able to flex; therefore, they must not become hard and brittle. Steel brake lines and fittings should be checked for damage and leakage once a year or whenever the vehicle gets brake service. Exposure to the elements, road salts in winter, salted air, water, and contaminants in the system all contribute to the rusting and corrosion of brake fittings, lines, and hardware.

Brake line inspection is more than a quick glance to see if all the parts are in place. Physical damage may be apparent from the outside, but wear and deterioration can occur inside tubing and hoses, as well.

TUBING INSPECTION

Steel tubing is more durable than rubber hoses, but it can suffer rust, corrosion, impact damage, and cracking. Water trapped around brake tubes, fittings, and mounting clips can rust and corrode steel tubing. Corrosion can be particularly severe in areas that use salt to melt ice on the roads during the winter. Mounting clips are necessary to hold brake lines to the body or frame, but they can trap salt water and hide severe corrosion. Therefore, inspect all mounting points closely.

Missing mounting clips can cause other problems. If brake tubing is not mounted securely, vibration can cause the tubing to fracture and leak. Loose brake tubing hanging below the body or frame can be snagged and torn loose. Inspect all brake tubing for looseness and look for empty screw holes or scuff marks on body and frame parts that indicate missing clips.

Brake tubing can be damaged by objects thrown up by the tires, particularly if the vehicle is used off road. Road impact damage is far less common than impact damage caused by improperly installed towing chains or improper placement of a floor jack. Inspect brake tubing for dents, kinks, and cracks caused by careless service practices.

HOSE INSPECTION

Inspect brake hoses for abrasion caused by rubbing against chassis parts and for cracks at stress points, particularly near fittings. Look for fluid seepage indicated by softness in the hose accompanied by a dark stain on the outer surface. Look at each hose closely for general damage and deterioration such as cracks, a soft or spongy feel or appearance, stains, blisters, and abrasions.

To check for internal hose damage, have an assistant pump the brakes and feel the hose for swelling or bulging as pressure is applied internally **(Figure 1)**. Have your assistant apply and release the brakes, then quickly spin the wheel. If the brake at any wheel seems to drag after pressure is released, the brake hose may be restricted internally. No conclusive way exists to inspect or test for internal restriction, but replacing a hose if a brake has these symptoms can be good insurance.

BRAKE HOSE REMOVAL AND REPLACEMENT

A replacement brake hose should be the same length as the original one. A hose that is too long may rub on the

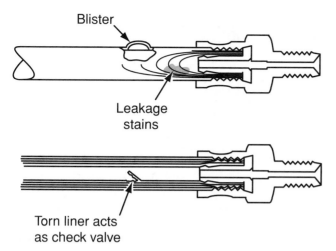

Figure 1. Possible internal defects in a brake hose.

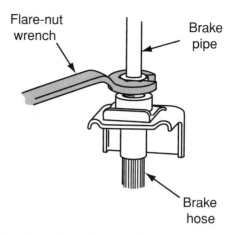

Figure 3. Use a line (flare-nut) wrench to disconnect the brake tubing from a hose or fitting.

chassis. One that is too short may break when the movable component reaches the limits of its travel.

> **You Should Know** *Always clean around any lines or fittings before loosening or removing them. Dirt and other contaminants can cause damage to the system.*

Hose Removal

Some brake hoses have a swivel fitting at one end and a fixed fitting that cannot be rotated at the other (**Figure 2**). Disconnect the swivel fitting first on this type of hose. Other hoses have a fixed male fitting on one end and a fixed female fitting on the other. The female end of such a hose is connected to a flare nut on a rigid brake tube; disconnect this end first. If the hose has a banjo fitting on one

end (usually for connection to a caliper) disconnect the banjo fitting first, then the other end of the hose. Occasionally, some caliper fittings and hoses have left-hand threads. Sometimes the left-hand fasteners are noted with a slash through the flat surfaces of the nut or bolt.

Follow these guidelines to remove a brake hose:
1. Clean dirt away from the fittings at each end of the hose to keep dirt from entering the system.
2. Disconnect the flare nut from the female end of the hose or disconnect the swivel end of the hose (**Figure 3**). When loosening one fitting at the end of a hose, hold the other half of the fitting with another flare-nut wrench. This will make it easier to remove the fitting and will prevent damage to mounting brackets and clips.
3. Remove the hose retaining clip from the mounting bracket with a pair of pliers.
4. Separate the hose from the mounting bracket and any clips used to hold it in place.
5. Disconnect the other end of the hose from the caliper or wheel cylinder. If the hose has a banjo fitting, hold the banjo fitting with a wrench while loosening the bolt.
6. If you are not going to immediately install a replacement hose, cap or plug open fittings to keep dirt out of the system.

Hose Installation

When installing a brake hose, be sure the new hose is the correct length. Route the new hose in the same location as the original and provide ¾–1 inch of clearance between the hose and suspension and wheel parts in all positions. If the original hose had special mounting clips or brackets, the replacement should have the same.

Follow these guidelines to install a brake hose:

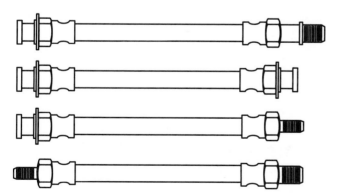

Figure 2. Typical brake hose end fittings.

1. If the hose has a fixed male end, install it into the wheel cylinder or caliper first. If the connection requires a copper gasket, install a new one.
2. If one end of the hose has a banjo fitting for attachment to a caliper, install the banjo bolt and a new copper gasket on each side of the fitting. Tighten the banjo bolt after connecting and securing the other end of the hose.
3. Route the hose through any support devices and install any required locating clips.
4. Insert the free end of the hose through the mounting bracket.
5. Depending on hose design, connect the flare nut on the steel brake line to the female end of the hose or connect the swivel end of the hose to the mating fitting.
6. Tighten the fitting and hold the hose with another wrench to keep it from twisting. Check the colored stripe or the raised rib on the outside of the hose to verify that the hose has not twisted during installation.
7. Install the retaining clip to hold the hose to its mounting bracket **(Figure 4)**. Install any other clips as required.

After installing the new brake hose, check the hose and line connections for leaks and tighten if needed. Check for clearance during suspension rebound and while turning the wheels. If any contact occurs, reposition the hose, adjusting only the female end or the swivel end.

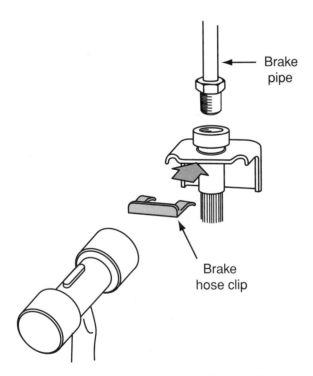

Brake pipe

Brake hose clip

Figure 4. Tap a new brake hose clip into the slot in the hose fitting.

BRAKE TUBING REMOVAL AND REPLACEMENT

To remove and replace a length of brake tubing:
1. Begin by cleaning all dirt away from the fittings at each end of the tubing.
2. Clean but do not remove the tubing mounting clamps. Leaving them in place will keep the tubing from moving around and make it easier to disconnect the fittings.
3. Disconnect the fittings at each end of the tubing. If the tubing is attached to a hose, use another flare-nut wrench to hold the hose fitting. If the tubing is attached to a rigidly mounted junction block or cylinder, a second wrench is not needed.
4. Cap or plug the open fittings to keep dirt out of the system.
5. Remove the mounting clips from the chassis and remove the brake tubing.
6. Inspect the clips and their screws to determine if they are reusable. If they are damaged, replace them.
7. If the brake tubing has any protective shields installed around it, save them also for installation with the new tubing.
8. If you must fabricate a new section of tubing, save the old section for a bending guide.
9. To install a length of brake tubing, position it on the chassis and install the mounting clips loosely. Leaving the new tubing slightly loose will help to align the tube's fittings.
10. Connect and tighten the fittings at both ends of the tubing.
11. Tighten the mounting clips and brackets.

FABRICATING BRAKE TUBING

Prefabricated brake tubing is available in various lengths with ends preflared and flare nuts installed **(Figure 5)**. Always use prefabricated tubing that is formed to the required bends for a specific vehicle, if possible; it can save time and money.

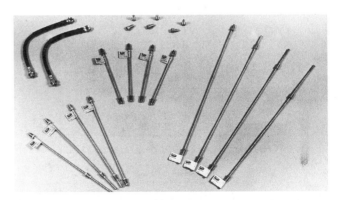

Figure 5. Prefabricated brake lines are available in a variety of lengths and with the fittings installed and the ends flared.

If a preformed brake line is not available, a straight section of brake tubing of the proper length is your next best choice. It may be necessary in many cases, however, to fabricate a replacement brake line from bulk tubing. The following sections explain how to cut and bend tubing and how to form the required flared ends.

Cutting Tubing

Bulk tubing comes in large rolls. To form a replacement length of tubing, you must cut the tubing to the desired length from a roll as follows:

1. Determine the exact length of replacement tubing that is needed; add 1/8 inch for each flare that is to be made. The use of a string can help you cut the tubing to about the right length **(Figure 6)**.
2. Hold the free outer end of the tubing against a flat surface with one hand, and unroll the roll in a straight line with the other hand **(Figure 7)**. Do not lay the roll flat

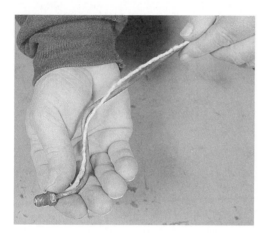

Figure 6. A string run along the shape of the old brake line can help you determine the correct length for the new brake line.

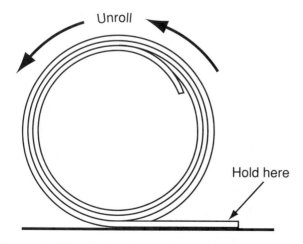

Figure 7. Unroll the desired length of tubing from the bulk roll in this manner.

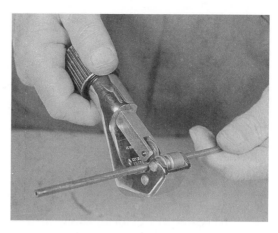

Figure 8. Cut the tubing at the required length with a tubing cutter.

and pull one end toward you; this will twist and kink the tubing.

3. Mark the tubing at the point to be cut and place a tubing cutter on the tubing. Tighten the cutter until the cutting wheel contacts the tubing at the marked point.
4. Turn the cutter around the tubing toward the open side of the cutter jaws **(Figure 8)**. After each revolution, tighten the cutter slightly until the cut is made.

> **You Should Know** *Do not use a hacksaw to cut tubing. The uneven pressure of the blade will distort the tubing end, and the teeth will leave a jagged edge that cannot be flared properly.*

5. **Ream** the cut end of the tubing with a reaming tool (usually attached to the cutter) to remove burrs and sharp edges. Hold the end downward so that metal chips fall out. Ream only enough to remove burrs; then use compressed air to be sure all chips are removed.

FLARE FITTINGS

After tubing is cut, flares must be formed on any unfinished ends. At times, the flares will be made after the tube has been bent to its required shape. Often it is easier to add the fittings and flares before the tube is bent to shape. This is a decision that should be made before doing either. If a bend is required close to an end, the flaring should take place before the bending. If there is doubt about the length, bending should take place before flaring.

Two kinds of flares are used on brake tubing: SAE 45-degree double flares and ISO flares. These flares and their

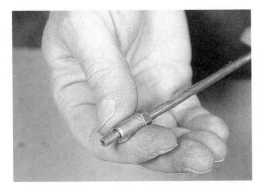

Figure 9.   Place the tube nut or fitting, facing the correct direction, before making the flare.

fittings are not interchangeable. Be sure to form the proper type of flare required for the vehicle's brake system. Special tools are required to flare tubing, and the tools for SAE and ISO flares are different. Always place the flare nut on the tubing with the threads facing the end of the tube *before* forming the flare **(Figure 9)**.

Flare nuts do not usually corrode or rust, but the tubing that passes through them may. If the line corrodes and freezes to the nut, the line will twist if you try to loosen the nut. To free a flare nut frozen to the line, apply penetrating oil to the connection. You also can heat the connection with a torch if all plastic and rubber parts are removed from the immediate area. Using a pencil-thin flame, apply heat to all sides of the flare nut, never to the line. When the steel nut begins to glow from the heat, try to loosen the nut. If the nut cannot be freed, cut the line. If heat is used to free a frozen flare nut, the entire length of tubing should be replaced.

> **You Should Know** *Never use spherical-sleeve compression fittings in brake lines. Spherical compression fittings are low-pressure fittings for applications such as fuel lines. They will fail and leak under the high pressures and vibrations of a brake hydraulic system.*

When connecting a tube to a hose, a tube connector, or a brake cylinder, use an inch-pound torque wrench to tighten the tube fitting nut to specifications. On fittings requiring gaskets, always install new copper gaskets. Used gaskets have taken a set and will not seal properly if reinstalled.

Forming an SAE 45-Degree Double Flare

A double flare is made in two stages using a special flaring tool. A typical flaring tool consists of a flaring bar and a

flaring clamp. The angle of the flare and the nut is 45 degrees, whereas the angle of the seat is 42 degrees. When the nut is tightened into the fitting, the difference in angles—called an interference angle—causes both the seat and the flared end of the tubing to wedge together. When correctly assembled, brake lines connected with flare fittings provide joints that can withstand high hydraulic pressure.

1. Select the forming die from the flaring kit that matches the inside diameter of the tubing.
2. Be sure the flare nut is installed on the tubing, then clamp the tubing in the correct opening in the flaring bar with the end of the tube extending from the tapered side of the bar **(Figure 10)** the same distance as the thickness of the ring on the forming die.
3. Place the pin of the forming die into the tube **(Figure 11)**, and place the flaring clamp over the die and around the flaring bar **(Figure 12)**.

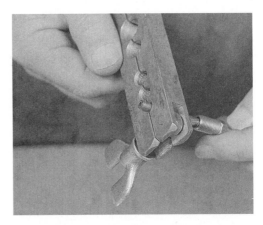

Figure 10.   Place the tubing in the flaring bar with the protruding end extended the same distance as the thickness of the ring on the forming die.

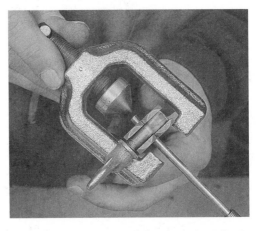

Figure 11.   Place the pin of the forming die into the tube, and place the flaring clamp over the die and around the flaring bar.

Figure 12. Place the flaring clamp over the die and around the flaring bar.

4. Tighten the flaring clamp until the cone-shaped anvil contacts the die. Continue to tighten the clamp until the forming die contacts the flaring bar.
5. Loosen the clamp and remove the forming die. The end of the tubing should be mushroomed.
6. Place the cone-shaped anvil of the clamp into the mushroomed end of the tubing. Be careful to center the tip of the cone and verify that it is touching the inside diameter of the tubing evenly. If the cone is not centered properly before tightening the clamp, the flare will be distorted.
7. Tighten the clamp steadily until the lip formed in the first step completely contacts the inner surface of the tubing.

8. Loosen and remove the clamp and remove the tubing from the flaring bar. Inspect the flare to be sure it has the correct shape **(Figure 13)**. If it is formed unevenly or is cracked, you must cut off the end of the tubing and start over again.

Forming an ISO Flare

A special tool kit is required to form an ISO flare **(Figure 14)**. To form an ISO flare:
1. Cut the tubing to length and install the fittings before forming the flare.
2. Clamp the ISO flaring tool in a bench vise. Select the proper size **collet** and forming **mandrel** for the diameter of steel tubing being used. Insert the mandrel into the body of the flaring tool. Hold the mandrel in place with your finger and thread in the forcing screw until it contacts and begins moving the mandrel **(Figure 15)**. After contact is felt, turn the forcing screw back one full turn.
3. Slide the clamping nut over the tubing and insert the tubing into the correct collet. Leave about 3/4 inch of tubing extending out of the collet **(Figure 16)**.
4. Insert the assembly into the tool body so that the end of the tubing contacts the forming mandrel. Tighten the clamping nut into the tool body very tightly to prevent the tubing from being pushed out during the forming process.
5. Using a wrench, turn in the forcing screw until it bottoms out. Do not overtighten the screw or the flare may be oversized.
6. Back the clamping nut out of the flaring tool body and disassemble the clamping nut and collet assembly.

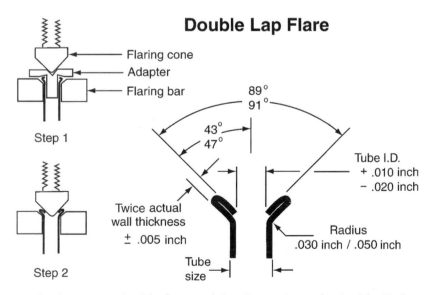

Figure 13. The two steps for forming a double flare and the dimensions of a double 45-degree flare.

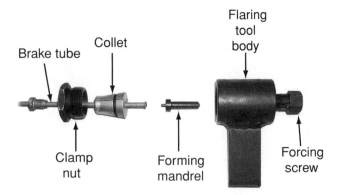

Figure 14. The components of an ISO flaring tool.

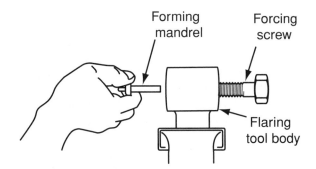

Figure 15. Insert the correct forming mandrel against the forcing screw in the flaring tool body.

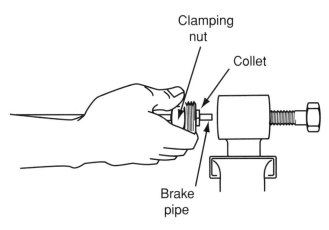

Figure 16. Slide the brake tube (pipe) through the clamping nut and collet.

7. Inspect the flare to be sure it has the correct shape **(Figure 17)**. If it is formed unevenly or is cracked, you must cut off the end of the tubing and start over again.

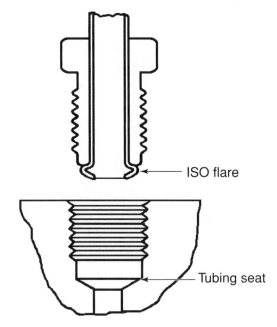

Figure 17. A completed ISO flare.

BENDING TUBING

Whether a replacement brake tube is a straight length of preflared tubing or is made from bulk material, it is usually necessary to bend the new line to match the old one. It is best to bend the tubing with bending tools.

> **You Should Know** *Steel tubing can be bent by hand to form gentle curves. However, never try to bend tubing into a tight curve by hand; you will usually kink the tubing. Because a kink in a brake tube weakens the line, never use a kinked tube. To avoid kinking, use a bending tool.*

On small-diameter tubing, you can use a bending spring. Slip the coil spring over the tubing and bend it slowly by hand **(Figure 18)**. Bend the tubing slightly further than required and back off to the desired angle. This releases spring tension in the bender so it can be easily removed. On larger diameter tubing, or where tighter bends are needed, use a lever-type or gear-type bender **(Figure 19)**. Slip the bender over the tubing at the exact point the bend is required **(Figure 20)**.

If you are bending the tubing near an end that is to be flared, leave about 1½ inches of straight tubing at the end. After the tubing is bent to the proper shape, assemble the flare nuts on the tubing before flaring the tube ends. Once

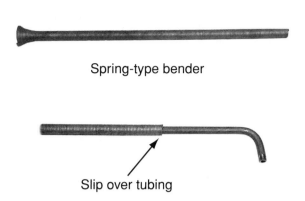

Spring-type bender

Slip over tubing

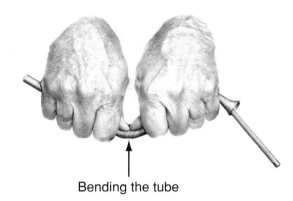

Bending the tube

Figure 18. Large-radius bends on small-diameter tubing can be made with a bending spring.

the ends are flared, the flare nuts will not fit over the end of the tubing.

Observe these additional guidelines when bending and fabricating tubing:

1. Avoid straight lengths, or runs, of tubing from fitting to fitting. They are hard to install and subject to vibration damage.
2. Ensure that the required clips and brackets will fit a replacement length of tubing, especially on long sections.

Figure 19. One type of tube bending tool that will form tight radius bends in steel tubing.

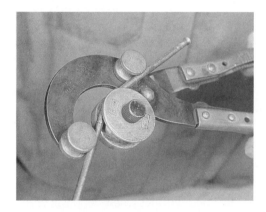

Figure 20. Slip the bending tool over the tube at exactly the point where the bend is needed.

3. Bend tubing to provide necessary clearance around exhaust components and suspension parts.
4. Be sure that tubing ends align with the fittings on mating components before mounting the tubing securely.
5. When installing tubing, connect the longest straight section first.

Summary

- Steel tubing can suffer rust, corrosion, impact damage, and cracking.
- Missing mounting clips can cause the tubing to fracture and leak or be snagged and torn loose.
- Inspect brake hoses for abrasion, fluid seepage, cracks, a soft or spongy feel or appearance, stains, blisters, and abrasions.
- Check hoses for internal hose damage.

- A replacement brake hose should be the same length as the original one.
- After installing the new brake hose, check the hose and line connections for leaks and check the clearance during suspension rebound and while turning the wheels.
- Always use prefabricated brake tubing, if available; it can save time and money.

- Bulk tubing comes in large rolls. To form a replacement length of tubing, you must cut it to the desired length from a large roll.
- After tubing is cut, flares must be formed on any unfinished ends.
- Two kinds of flares are used on brake tubing: SAE 45-degree double flares and ISO flares.

- A double flare is made in two stages using a special flaring tool.
- A special tool kit is required to form an ISO flare.
- To bend small-diameter tubing, use a bending spring.
- To bend larger diameter tubing, or where tighter bends are needed, use a lever-type or gear-type bender.

Review Questions

1. A _____ flare is made in two stages using a special flaring tool. A typical flaring tool consists of a flaring _____ and a flaring _____. The angle of the flare and the nut is 45 degrees, whereas the angle of the seat is 42 degrees. When the nut is tightened into the fitting, a(n) _____ angle causes both the seat and the flared end of the tubing to wedge together.

2. Brake line tubing is being discussed. Technician A says that double-wall steel tubing with single-flare fittings is acceptable. Technician B says that copper tubing with double-flare or ISO flare fittings is acceptable. Who is correct?
   A. Technician A only
   B. Technician B only
   C. Both Technician A and Technician B
   D. Neither Technician A nor Technician B

3. Replacing brake lines is being discussed. Technician A uses prefabricated lines whenever possible. Technician B always installs the flare fittings onto the tube before forming the flared ends. Who is correct?
   A. Technician A only
   B. Technician B only
   C. Both Technician A and Technician B
   D. Neither Technician A nor Technician B

4. When checking brake hoses Technician A has an assistant pump the brakes while feeling the hoses for swelling or bulging as the pressure is applied; Technician B spins the wheels after the brakes have been released to check for dragging. Who is correct?
   A. Technician A only
   B. Technician B only
   C. Both Technician A and Technician B
   D. Neither Technician A nor Technician B

5. Technician A says that no conclusive way exists to inspect or test for an internal restriction in a brake hose. Technician B says that a piece of rubber can partially break loose inside a brake hose and act like a check valve to trap pressure at the brake. Who is correct?
   A. Technician A only
   B. Technician B only
   C. Both Technician A and Technician B
   D. Neither Technician A nor Technician B

Chapter 18

Brake System Valves

Introduction

Before 1967, most brake systems had drum brakes at all four wheels, and master cylinders had only one chamber. Combining disc and drum brakes required that the timing of the application of required pressure be altered.

Metering valves, proportioning valves, and combination valves are used by auto manufacturers to regulate pressures within the system. Valves usually are mounted on or near the master cylinder, except for a height-sensing proportioning valve, which is mounted under the rear of the vehicle.

Different vehicles use different control valves or a combination of valves. All valves in the hydraulic system should be inspected whenever brake work is performed or a problem exists in the system. Uneven braking or premature wear of pad or shoe lining can indicate a faulty metering, proportioning, or combination valve.

RECENTERING A PRESSURE DIFFERENTIAL VALVE

After bleeding or flushing and refilling some brake systems, the pressure differential valve (or warning lamp switch) can cause the warning lamp to stay illuminated. Opening a bleeder screw creates a pressure differential between the two halves of the hydraulic system. This pressure differential, like a leak, causes the pressure differential valve's piston to move toward the low-pressure side to close the lamp circuit and turn on the warning lamp.

If the warning lamp stays lit after bleeding, the valve's piston might need to be recentered. Before doing this, make sure the parking brake is not applied and that its link-age is properly adjusted. Also check the fluid level in the master cylinder reservoir. Normally, both the parking brake warning circuit and the low-fluid warning circuit share the same warning lamp as the pressure differential valve. It is not uncommon to find the parking brake out of adjustment after relining the rear brakes or to find the fluid level low after bleeding the system.

Single-Piston Valve with Centering Springs

The most common pressure differential valve has a single piston and centering springs **(Figure 1)**. The warning lamp lights only when the brakes are applied and a pres-

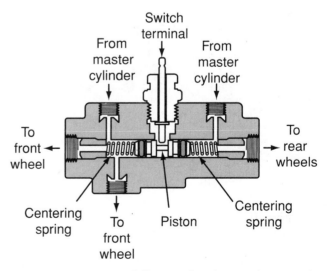

Figure 1. Pressure differential valve with a single piston and a centering spring.

sure difference exists between the two halves of the hydraulic system. When the brakes are released, pressure is low in both halves of the system. The springs then recenter the valve's piston and the lamp should turn off.

The piston in this type of valve should recenter automatically. It is possible, however, for the piston to stick at one side of its bore and leave the lamp lit. If the lamp is lit with the ignition on, apply the brakes rapidly with moderate to heavy force two or three times. Hydraulic pressure usually frees a stuck piston, and the springs can then recenter it. If the lamp stays lit and the parking brake or fluid level switch is not closed, try to recenter the piston. If the lamp is still lit, test the circuit for an accidental ground. If the circuit is not shorted to ground, replace the pressure differential valve.

Single-Piston Valve without Centering Springs

Some vehicles have a single-piston pressure differential valve without centering springs **(Figure 2)**. This type of valve often leaves the warning lamp lit after system bleeding. Recentering the piston is a two-person job.

Recentering of the piston should begin by verifying that the warning lamp is lit by turning on the ignition. Then determine if the hydraulic system is split front to rear or diagonally. Next, open a bleeder screw on the side opposite from the side that was last bled. Have your assistant slowly depress the brake pedal by hand until the warning lamp turns off. Tighten the bleeder screw.

When trying to recenter the piston in this type of valve, the piston often goes over center in the opposite direction. This causes the lamp to turn off momentarily and then light again. If this happens, open a bleeder screw on the opposite side of the hydraulic system and repeat the procedure. Two or three tries may be needed to get the piston properly centered.

Two-Piston Valve with Centering Springs

The least commonly used pressure differential valve is a two-piston, or double-acting piston, design. This type of valve has two small pistons head to head with recessed sections on both sides of a raised center **(Figure 3)**. When the pistons are centered, the switch plunger is held upward. When a pressure difference occurs, the piston moves and the switch's plunger drops into the recessed section on one side. This closes the circuit and lights the warning lamp. The double-acting piston design uses a single piston with recessed sections on both sides of a raised center.

Many technicians remove the switch's plunger before bleeding a hydraulic system with this type of valve. When pressure drops on one side of the system as one bleeder screw is opened, the switch's piston will move and light the lamp. When the opposite side of the system is bled, pressure will exist in the opposite direction within the valve. This can jam the piston against the switch's plunger and damage the piston or plunger, or both.

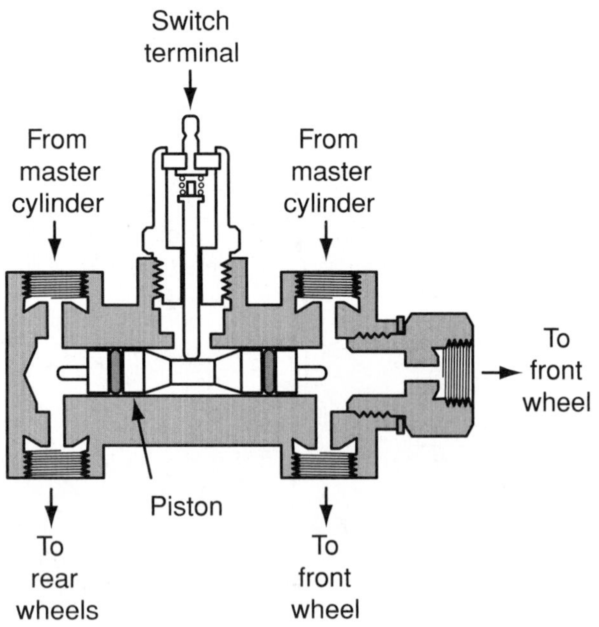

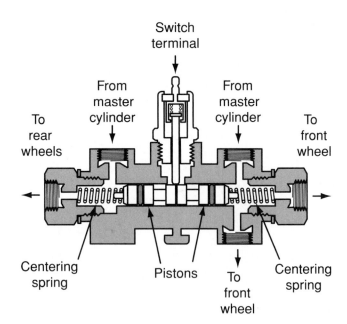

Figure 2. Pressure differential valve with a single piston and no centering spring.

Figure 3. Pressure differential valve with two pistons and centering springs.

METERING VALVE

Some vehicles with front disc and rear drum brakes have a **metering valve** in the hydraulic system to achieve balanced braking between the front and rear wheels. Metering valves are used primarily on rear-wheel-drive vehicles. A metering valve **(Figure 4)** is located in the line to the front brakes and keeps the front disc pads from operating until the rear drum brakes have started to work. The valve delays pressure application to the front disc brakes because disc brakes are fast acting, whereas drum brakes have spring tension and linkage clearance to overcome.

A typical metering valve **(Figure 5)** has an inlet connection from the master cylinder and two outlets, one for each front wheel. The metering valve works through the operation of fluid pressure against two springs of different tension. The valve piston is normally closed by a strong spring, and the valve stem is held in a normally open position by a weak spring to open by-pass passages between the stem and the inner bore of the piston.

When the brakes are applied, fluid pressure from approximately 3 to 30 psi immediately closes the head of the valve stem against the piston to block the lines to the front brakes. Pressure is applied to only the rear brakes until it increases enough to overcome spring tension and start to apply the brake shoes. The continuing pressure increase overcomes the metering valve spring pressure against the valve piston. This occurs from 75 to 300 psi, depending on system specifications. Fluid pressure then is applied to the caliper pistons to operate the front disc brake pads.

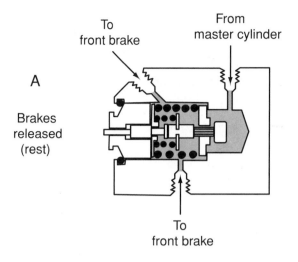

A

Brakes
released
(rest)

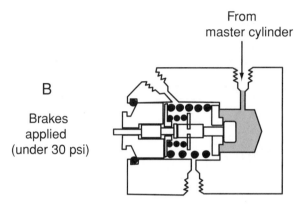

B

Brakes
applied
(under 30 psi)

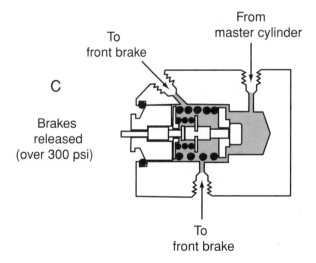

C

Brakes
released
(over 300 psi)

Figure 5. The basic operation of a metering valve.

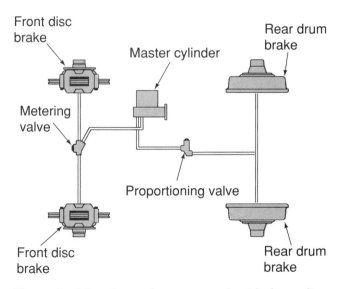

Figure 4. Metering valves are used with front disc brakes, and proportioning valves are used with rear drum brakes.

When the brakes are released, the metering valve piston closes, but the valve stem passages open and allow fluid to flow freely back to the master cylinder. These passages let fluid return to the master cylinder when the brake pedal is released and allow for fluid expansion and contraction due to temperature changes.

All of this valve action takes place in a fraction of a second. The effect of the metering valve is felt during the beginning stages of all brake applications and during light brake application.

Metering Valves Today

Metering valves appeared along with disc brake systems of the mid-1960s. Vehicles of that era were almost entirely rear-wheel-drive vehicles with front disc and rear drum brakes, and metering valves were needed. Today, with the increased use of front-wheel drive (FWD) and diagonally split hydraulic brake systems, metering valves have been eliminated from most vehicles.

On an FWD car, 80 percent of the braking is done by the front brakes, so it is desirable to apply them as quickly as possible. Until all of the clearance in the brake system is taken up, braking force is not great enough to overcome the torque of the front drive wheels. This driving torque and the forward weight bias of an FWD car eliminate the problem of front wheel lockup and any need for a metering valve. Furthermore, the diagonally split brake systems used on most FWD cars would require two metering valves, one for the front brake on each side of the hydraulic system. Avoiding the complication of extra parts is another good reason to eliminate the metering valve.

On a vehicle with four-wheel disc brakes, the application times for the brakes at all wheels are about equal. A metering valve, therefore, is unnecessary.

METERING VALVE SERVICE

Inspect the metering valve **(Figure 6)** whenever the brakes are serviced. Fluid leakage inside the boot on the end of the valve means the valve is defective and should be

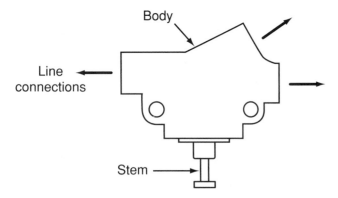

Figure 6. The metering valve should be inspected whenever the brakes are serviced. The valve stem or button must be pulled or pressed when the front brakes are bled with a pressure bleeder.

replaced. A small amount of moisture inside the boot does not necessarily indicate a bad valve.

Metering valves are not adjustable or repairable. If a valve is defective, replace it. Always make sure to mount the new valve in the same position as the old valve.

A faulty metering valve can allow the front brakes to apply prematurely and possibly lock, especially on wet pavement. Premature front pad wear or a tendency for front brakes to lock may indicate a bad metering valve.

If you suspect a metering valve problem, have an assistant apply the brakes gradually while you watch and feel the valve stem. It should move as pressure increases in the system. If it does not, replace the valve. You also can check metering valve operation with a pressure bleeder. Charge the bleeder tank with compressed air to about 40 psi and connect it to the master cylinder. Do not override the metering valve manually or with a special tool. Pressurize the hydraulic system with the pressure bleeder and open a front bleeder screw. If fluid flows from the bleeder, the metering valve is not closing at the right pressure and must be replaced.

Metering Valve Pressure Test

If a metering valve needs to be tested more precisely, two pressure gauges connected to the system can do this. The two pressure gauges must be able to measure from 0 to at least 500 psi and you will need an assistant.

1. Use a T-fitting to connect one gauge to the line from the master cylinder to the metering valve.
2. Use another T-fitting to connect the other gauge to the line from the metering valve to the front brakes.
3. Have an assistant apply the brakes gradually but firmly while you watch the gauges. The gauge readings should be as follows:
   a. As pressure is first applied, the readings of both gauges should increase together until the closing pressure of the valve is reached. This should be from 3 to 30 psi, depending on valve design.
   b. Above the closing pressure of the valve, the inlet pressure should continue to increase while the outlet pressure stays constant.
   c. As inlet pressure continues to increase, the valve will reopen. This should be from 75 to 300 psi, depending on valve design. At that point, the reading on the outlet gauge should rise to match the reading on the inlet gauge. Both gauges should then read the same as pressure continues to rise.
4. If the gauge readings do not follow the patterns described in step 3, replace the metering valve.

PROPORTIONING VALVE

The **proportioning valve** helps to balance front and rear pressures on cars with disc and drum brakes. A metering valve controls the *timing* of pressure application to front disc brakes. A proportioning valve controls the *actual pressure* applied to rear brakes.

Inertia and momentum cause weight to shift forward during braking. The weight shift is proportional to the braking force and the rate of deceleration. During hard braking, the weight shift unloads the rear axle and reduces traction between the tires and the road. With reduced traction, the rear brakes can lock, and the vehicle can spin. Rear brake lockup can be avoided, however, by modulating the hydraulic pressure applied to the rear brakes.

Drum brakes use mechanical **servo action** to increase force applied to the brake shoes. Because of this servo action, drum brakes require less hydraulic pressure to maintain braking force than to establish it. Disc brakes, on the other hand, require a constant hydraulic pressure for a given amount of braking force. Overall, disc brakes always require higher hydraulic pressure than do drum brakes. A proportioning valve does exactly what its name indicates—it proportions hydraulic pressure between disc and drum brakes to maintain equal braking force at the tires.

Although proportioning valves were originally designed for use with front disc rear drum combinations, some late-model vehicles with four-wheel disc brakes also have proportioning valves for the rear disc brakes.

A proportioning valve has an inlet passage from the master cylinder at one end and an outlet to the rear brakes at the other end. Inside, a spring-loaded piston slides in a stepped bore. One end of the piston has a larger area than the other. The actual proportioning is done by a spring-loaded, check-valve-type stem that moves in a smaller bore through the center of the piston.

When the brakes are first applied and under light braking, the proportioning valve does nothing. Fluid enters the valve at the end with the smaller piston area **(Figure 7)**, passes through the small bore around the stem, and exits to the rear brakes. The end of the valve piston at the outlet side of the valve is the end with the larger surface area. As outlet pressure rises in the valve, it exerts greater force on the piston than inlet pressure does and moves the piston toward the inlet, against spring pressure. This closes the center valve stem and blocks pressure to the rear brakes.

The pressure at which the proportioning valve closes is called the **split point** because the uniform system pressure splits at that point with greater pressure applied to the front disc brakes and lower pressure applied to the drum brakes. As pressure continues to increase from the master cylinder, inlet pressure at the proportioning valve overcomes the pressure at the large end of the piston and reopens the valve. Fluid again flows through the center of the valve, pressure at the large end of the piston rises, and the valve closes again. The opening and closing action repeats several times per second.

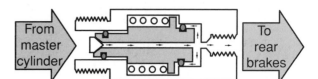

Approaching split point

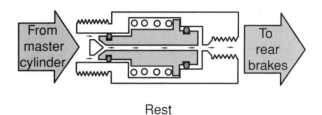

Rest

Figure 7. This proportioning valve serves a single rear brake. A valve that serves both rear brakes would have two outlet ports.

The piston cycles back and forth and lets pressure to the rear drum brakes increase but at a slower rate than pressure to the front disc brakes. The pressure increase to the drum brakes above the split point is called the **slope**. The slope is the numerical ratio—or proportion—of rear drum brake pressure to full system pressure **(Figure 8)**. If half of the system pressure is applied to the rear brakes, the slope is 1:2, or 50 percent. When the brakes are released, pressure drops and the spring moves the proportioning valve piston. This opens the valve for rapid fluid return to the master cylinder.

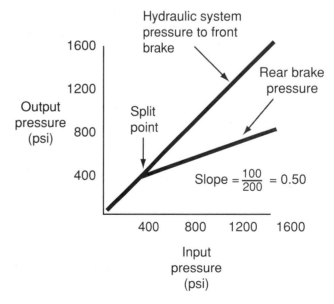

Figure 8. Split point and slope are the operating characteristics of a proportioning valve.

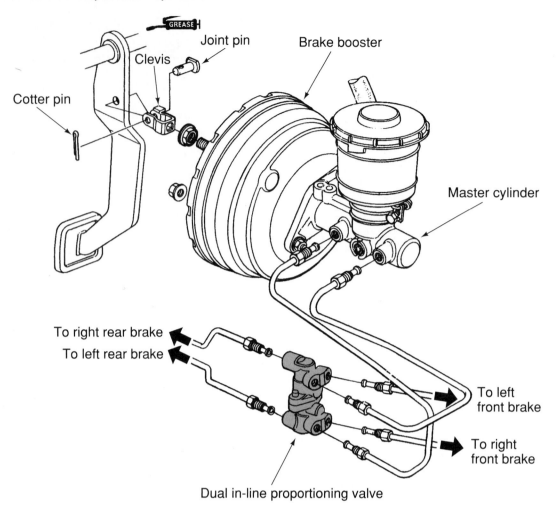

GREASE

Joint pin

Clevis

Brake booster

Cotter pin

Master cylinder

To right rear brake

To left rear brake

To left
front brake

To right
front brake

Dual in-line proportioning valve

Figure 9. This dual proportioning valve assembly is a separate housing and is mounted close to the master cylinder.

The first proportioning valves were installed in the single line to the rear brakes. One valve controlled pressure equally to both brakes. Diagonally split brake systems separated the rear brakes from each other; therefore, if a proportioning valve were required, two would be needed. These two valves may be housed in a single body mounted on or near the master cylinder **(Figure 9)** or the two valves may be built into the master cylinder body **(Figure 10)**. Height-sensing proportioning valves are installed under the rear of the vehicles and are connected to the rear axle or suspension. All proportioning valves work in the same way.

PROPORTIONING VALVE TESTING AND SERVICE

If a vehicle's rear drum brakes lock up during moderate-to-hard braking and all other possible causes of lockup

have been eliminated, the proportioning valve is the likely problem. If the valve is leaking, replace it.

Proportioning Valve Pressure Test

If a proportioning valve has fittings on the inlet and outlet lines that allow you to connect pressure gauges with T-fittings, you can test valve operation. If a proportioning valve is built into the master cylinder with no access for a gauge, it can still be tested by connecting pressure gauges to the bleeder ports on a front caliper and a rear wheel cylinder **(Figure 11)**. If the hydraulic system is split front to rear, only a single test is needed. If the hydraulic system is split diagonally, both gauges must be connected twice to individually test the left and right rear brakes. This test requires two pressure gauges that measure from 0 to 1000 psi, plus the help of an assistant. For accurate testing, you also should know the split point of the proportioning valve. This can be found in some service manuals.

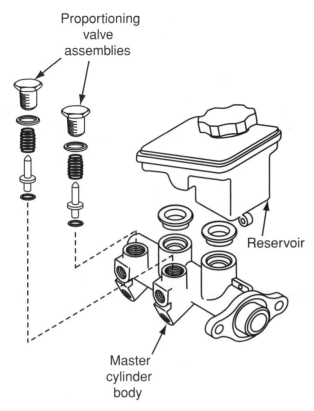

Figure 10.   These proportioning valves are built into the master cylinder.

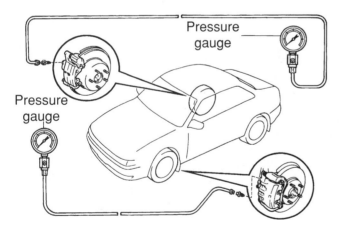

Figure 11.   Install the pressure gauges to the front and rear brake units in the same system.

1. Connect one gauge to the proportioning valve inlet pressure port by one of the following methods:
   a. Use a T-fitting to connect one gauge directly to the line from the master cylinder to the proportioning valve inlet port.
   b. If you cannot connect a T-fitting to the valve inlet line, remove a bleeder screw and connect the gauge to the bleeder port of one front caliper. Pressure to the front brakes should be the same as pressure at the proportioning valve inlet.

2. Connect the other gauge to the proportioning valve outlet pressure port by one of the following methods:
   a. Use another T-fitting to connect the other gauge directly to the line from the proportioning valve to the rear brakes.
   b. If you cannot connect a T-fitting to the valve outlet line, remove a bleeder screw and connect the gauge to the bleeder port of one rear wheel cylinder. If a diagonally split hydraulic system has two proportioning valves, the gauge must be connected twice: once to each wheel cylinder.

3. Have an assistant apply the brakes gradually but firmly while you watch the gauges. The gauge readings should be as follows:
   a. The readings of both gauges should increase together until the split point pressure of the valve is reached.
   b. Above the split point, the outlet pressure should increase more slowly than the inlet pressure.

4. If the gauge readings do not follow the patterns described in step 3, replace the proportioning valve.

Many manufacturers do not provide pressure test specifications for proportioning valves. As a general rule, however, the split point should occur at 300–500 psi. Maximum outlet pressure should be one-half to two-thirds of the maximum inlet pressure, or a slope of 50–67 percent.

Proportioning Valve Servicing

On some vehicles, proportioning valves are built into the master cylinder body. These valves often can be serviced with reconditioning kits.

HEIGHT-SENSING PROPORTIONING VALVE

Height-sensing proportioning valves were first used on pickup trucks. The weight on the rear axle of a truck can vary greatly from an unloaded to fully loaded condition. These weight extremes affect braking balance and the amount of hydraulic pressure needed at the rear drum brakes. Because the height of the vehicle also changes with the load at the rear of the truck, a proportioning valve that adjusts itself according to vehicle height is an effective way to provide variable pressure control to the rear brakes.

One common type of height-sensing proportioning valve has the valve mounted on a frame bracket with linkage that is attached to a rear spring eye **(Figure 12)**. During hard braking, the weight transfer will lift the chassis in relation to the axle. When this happens, the spring stretches and pulls on the linkage. A lever mechanically moves the

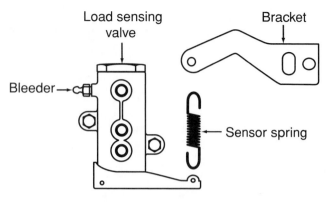

Figure 12. This height-sensing proportional valve is mounted on a bracket on the vehicle, and a linkage connects it to an eye on the rear spring.

proportioning valve to reduce pressure to the rear brakes. Hydraulic pressure is adjusted automatically according to the movement of the chassis in relation to the axle.

Another height-sensing proportioning valve uses a large round ball called a G-ball ("G" for gravity) in a valve assembly called a G-valve **(Figure 13)**. This assembly is mounted on the end of the proportioning valve. The ball moves back and forth in relation to the tilt of the rear of the vehicle. As the rear of the vehicle lifts during hard braking, the ball moves against the ball valve. The ball valve then controls the operation of the proportioning valve through a spring to lower pressure to the rear brakes. The advantage of this system is that there is no need for a mechanical linkage to be attached to the proportioning valve.

Height-sensing proportioning valves can also be found on many late-model passenger cars. The weight on the rear

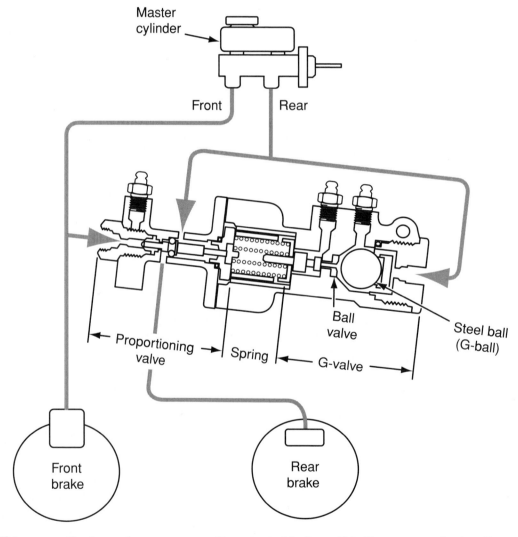

Figure 13. This proportioning valve uses a gravity-actuated ball, or G-ball, to correct hydraulic pressure during braking.

axle of an FWD four-door sedan can vary significantly with only the driver and an empty trunk to a full load of passengers and a trunk full of luggage. A height-sensing proportioning valve can modulate rear drum brake operation for the best braking action under variable load conditions.

HEIGHT-SENSING PROPORTIONING VALVE SERVICE

A height-sensing proportioning valve must be adjusted correctly if it is to balance rear braking force with the load. Height-sensing valves also are calibrated to work with the stock suspension. Any modification that improves load-carrying capability, such as helper springs or air-assist shocks, can adversely affect valve operation.

Modifications that make the suspension stiffer can prevent the suspension from deflecting the normal amount during hard braking or heavy load conditions. As a result, the proportioning valve may not increase rear brake effort enough, and stopping distance may increase dangerously.

Adjustment

There are as many ways to adjust a height-sensing proportioning valve as there are valve designs. All adjustments involve setting the operating rod to a specified stroke to ensure that the proportioning action takes place at the right pressure in relation to vehicle height and load.

The following example illustrates the adjustment principles for one design **(Figure 14)**, but the details will be different for other valve designs. Because it is difficult to accurately measure the operating rod in this example, a short length of ¼-inch plastic tubing is cut to size and used as a gauge to accurately set the rod length.

1. Raise the vehicle on a lift or an alignment rack so that its wheels are on a flat surface and the vehicle is at normal ride height.
2. Back off the valve adjuster setscrew but do not change the position of the upper nut.

3. Cut the length of ¼-inch tubing to the adjustment specification length (in this example 16.3 mm) and slit the tubing lengthwise so you can install it on the operating rod.
4. Slip the tubing onto the operating rod to set the proper operating length between the valve body and the upper nut.
5. With the adjuster sleeve resting on the lower mounting bracket, tighten the adjusting setscrew to lock the setting.
6. Remove the tubing from the operating rod. The operating rod adjustment is now set for normal driving conditions.

Test drive the vehicle. If you find that rear braking pressures are too little or too great, slight adjustments can be made. With the suspension at normal ride height, loosen the adjuster setscrew and move the adjuster sleeve toward or away from the brake pressure control valve. Each millimeter the adjuster sleeve is moved changes braking pressure by 60 psi. Move the adjuster sleeve down away from the valve body on the operating rod to increase braking pressure. Move the adjuster sleeve up toward the valve body to decrease braking pressure. When the setting is properly adjusted, tighten the setscrew.

Removal and Replacement

Remove and replace a typical height-sensing proportioning valve **(Figure 15)** as follows:

1. Raise the vehicle on a lift and disconnect the brake lines from the valve body. Tag the line positions to be certain they are reinstalled correctly.
2. Remove the fastener securing the height-sensing valve bracket to the rear suspension arm and bushing.

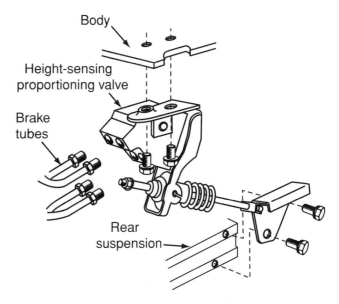

Figure 15.   The mounting of a typical height-sensing proportioning valve.

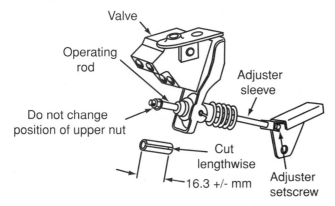

Figure 14.   The typical method of adjustment for a height-sensing proportioning valve.

3. Remove the screws that secure the valve bracket to the underbody and remove the assembly.
4. Before installing the new valve, make sure the rear suspension is in full rebound. If the valve has a red plastic gauge clip, make sure it is in position on the proportioning valve and that the operating rod lower adjustment screw is loose.
5. Position the valve and install the bolts that hold it to the underbody.
6. Secure the lower mounting bracket to the rear suspension arm and bushing using the retaining screw. Tighten all fasteners to specifications.
7. Ensure that the valve adjuster sleeve rests on the lower bracket and then tighten the lower adjuster setscrew.
8. Reconnect the brake lines to their original ports on the valve body and bleed the rear brakes. Remove the plastic gauge clip and lower the vehicle to the ground.

A new height-sensing proportioning valve will automatically become operational when the suspension is at the normal ride height. Remember that the addition of extra leaf springs to increase load capacity, spacers to raise the vehicle height, and air shocks to allow heavier loads without sagging should not be used on vehicles with height-sensing proportioning valves.

COMBINATION VALVE

As the name indicates, a **combination valve** has two or three valve functions in one valve body. The most common type is the three-function combination valve **(Figure 16)**.

Figure 16. A three-function valve.

A three-function valve has three separate sections: one end of the combination valve houses the metering valve, the center section houses the pressure differential valve (warning lamp switch), and the other end houses the proportioning valve. The operation of these three sections is the same as described for the separate units.

COMBINATION VALVE SERVICE

Inspect combination valve **(Figure 17)** parts whenever the brakes are serviced. If there is leakage around the large nut on the proportioning end, the valve is defective and must be replaced. A small amount of moisture inside the boot or a slight dampness around the large nut does not indicate a defective valve. Combination valves are nonadjustable and nonrepairable. If a valve is defective in any way, it must be replaced.

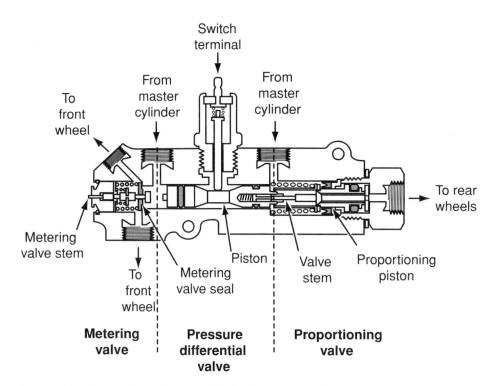

Figure 17. Inspect a combination valve whenever the brakes are serviced.

HYDRAULIC PRESSURE CONTROL WITHOUT VALVES

The installation of ABSs has increased rapidly in the past decade, and ABS is now standard equipment on many vehicles. The ABS has a single, very simple operating principle: to prevent wheel lockup by modulating hydraulic pressure to the brake at any wheel that is decelerating faster than the others and about to lock up. The ABS accomplishes this with speed sensors at the wheels or on the driveline and a computer that processes the speed information and controls hydraulic pressure with electrically operated valves or small high-speed pumps.

The pressure modulation provided by the ABS is really no different than the modulation provided by metering valves and proportioning valves. In fact, ABS pressure control is more precise than any control that could be provided by a metering valve or a proportioning valve. As the ABS becomes more common, engineers have the opportunity to eliminate these old, familiar hydraulic valves and replace them with a computer-controlled system.

Summary

- A pressure differential valve, a metering valve, and/or proportioning valve, or a combination valve can be used in late-model brake systems.
- After bleeding or flushing and refilling some brake systems, the pressure differential valve (or warning lamp switch) can be actuated and the valve's piston might need to be recentered.
- Some vehicles with front disc and rear drum brakes have a metering valve in the hydraulic system to achieve balanced braking between the front and rear wheels.
- Metering valves delay pressure application to the front disc brakes because disc brakes are fast acting, whereas drum brakes have spring tension and linkage clearance to overcome.
- Fluid leakage inside the boot on the end of a metering valve means the valve is defective and should be replaced.
- A faulty metering valve can allow the front brakes to apply prematurely and possibly lock, especially on wet pavement.

- The proportioning valve controls the actual pressure applied to rear brakes.
- The pressure at which the proportioning valve closes is called the split point because the uniform system pressure splits at that point with greater pressure applied to the front disc brakes and lower pressure applied to the drum brakes.
- The pressure increase to the drum brakes above the split point is called the slope, which is the numerical ratio of rear drum brake pressure to full system pressure.
- A height-sensing proportioning valve adjusts itself according to vehicle height and is an effective way to provide variable pressure control to the rear brakes.
- Adjusting a height-sensing proportioning valve involves setting the operating rod to a specified stroke to ensure that the proportioning action takes place at the right pressure in relation to vehicle height and load.
- A combination valve has two or three valve functions in one valve body: a metering valve, a pressure differential valve (warning lamp switch), and a proportioning valve.

Review Questions

1. Explain the purpose of a metering valve.
2. Explain the purpose of a proportioning valve.
3. Explain why a height-sensing proportioning valve is required on some vehicles.
4. Explain why dual proportioning valves are used on some brake systems.
5. A three-function combination valve has a brake system failure switch, a _____ valve, and a _____ valve.
6. Technician A says that slight moisture inside the protective boot of a combination valve or metering valve is normal. Technician B says that proportioning valves are never serviced and are always replaced as a unit. Who is correct?
    A. Technician A only
    B. Technician B only
    C. Both Technician A and Technician B
    D. Neither Technician A nor Technician B
7. Load- or height-sensing proportioning valves are being discussed. Technician A says that these valves are factory set and are nonadjustable. Technician B says that changing the suspension or load-carrying capacities of the vehicle can adversely affect valve operation. Who is correct?
    A. Technician A only
    B. Technician B only
    C. Both Technician A and Technician B
    D. Neither Technician A nor Technician B

Section 4

Power Brakes

SECTION OBJECTIVES

After you have read, studied, and practiced the contents of this section, you should be able to:

- Explain why power brake systems are desirable on today's vehicles.
- Name and explain the operational differences between the two common types of power brake systems used today.
- Test pedal free travel with and without engine running; check power assist operation.
- List the major types of vacuum booster assemblies and describe their differences.
- Explain the purpose and operation of an auxiliary vacuum pump.
- Diagnose vacuum power booster problems.
- Disassemble, repair, and adjust a vacuum booster as required to restore proper operation.
- Describe the parts and operation of a Hydro-Boost hydraulic power-assist system.
- Explain the parts and operation of the Power-Master power-assist system.
- Inspect and test a Hydro-Boost booster and accumulator.
- Repair, adjust, or replace components on a Hydro-Boost system as needed.
- Inspect and test a PowerMaster booster for leaks and proper operation.
- Repair, adjust, or replace components on a PowerMaster system as needed.

Interesting Fact

Today, power brakes are standard equipment on nearly all domestic and imported cars and light trucks. In fact, you will have to look long and hard to find a new vehicle without power brakes. But this was not always so.

Power brakes appeared as an extra-cost option on luxury cars in the early 1950s. The new, longer, lower, wider, heavier, and more powerful cars required more stopping power. Something needed to be done to safely stop these vehicles. Either the drivers needed to have very strong legs or the manufacturers needed to make something to increase the force on the brake pedal before the force reached the master cylinders. The obvious answer was the inclusion of a power brake system.

Originally, power brakes were a safety and convenience feature that provided braking power for drivers who might not have the strength (or weight) to stomp on the pedal for a hard stop. However, when vehicles became equipped with disc brakes, power boosters became a safety necessity.

The dual servo drum brakes of the 1950s and early 1960s used the self-energizing action of primary and secondary shoes to increase braking force when the driver pressed the pedal. Disc brakes, which began to appear in quantity by the mid-1960s, do not develop the self-energizing action of drum brakes. Although disc brakes develop greater braking energy at the wheels, they require a greater application force from the driver. Disc brakes made some kind of power assist a necessity for almost all drivers.

Chapter 19

Power Brake Systems

Introduction

A power brake system is used on most cars to reduce the braking effort required by the driver to stop a vehicle. Three things can be done to boost the force applied to the master cylinder or to reduce force that must be applied to the brake pedal by the driver:

1. **Mechanical advantage (leverage)**—The brake pedal ratio provides leverage to increase the force applied to the master cylinder **(Figure 1)**. Mechanical and space limitations allow the pedal ratio to be increased only to a given point. The pedal arm can be only so long and still fit into the car, and the longer the pedal arm, the greater the amount of pedal travel.

2. **Hydraulic advantage (force multiplication)**—Master cylinder piston size can be used to multiply the hydraulic pressure applied by the master cylinder to the wheel cylinders and caliper pistons. Hydraulic force multiplication has its limits too. If wheel cylinder and caliper pistons are made larger for greater force, master cylinder piston travel may increase along with brake pedal travel. All brake systems use hydraulic multiplication to increase braking force, but it has limitations just as pedal leverage does.

3. **Power boosters**—Installing a power booster will increase brake application force **(Figure 2)**. Power boosters are discussed in this chapter.

Power boosters do not alter basic brake system operation. They allow braking, even if the booster fails or loses its power supply. All boosters have a power reserve to provide at least one power-assisted stop after power is lost.

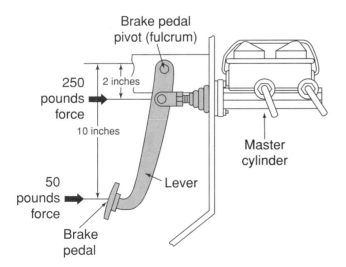

Figure 1. How the force applied to the brake pedal is affected by the pedal.

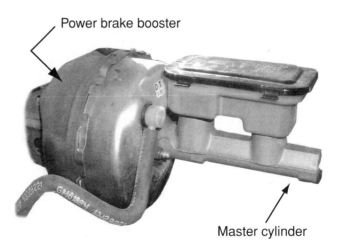

Figure 2. A typical power brake booster mounted to a master cylinder.

169

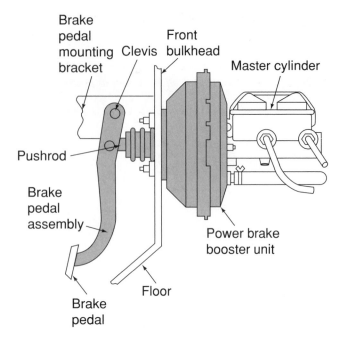

Figure 3. The power booster unit fits between the brake pedal assembly and the master cylinder.

TYPES OF POWER BRAKE SYSTEMS

Power brakes are nothing more than a standard hydraulic brake system with a booster unit located between the brake pedal and the master cylinder to help increase braking force **(Figure 3)**.

Two basic types of power-assist mechanisms are used. One type is vacuum assist. These systems use engine vacuum or an external vacuum pump to help apply the brakes. The second type of power assist is hydraulic assist. It is normally found on larger vehicles. This system uses hydraulic pressure developed by the power steering pump or other external pump to help apply the brakes.

Both vacuum and hydraulic assist act to multiply the force exerted on the master cylinder pistons by the driver. This increases stopping performance by increasing the hydraulic pressure delivered to the wheel cylinders or calipers while decreasing driver foot pressure.

VACUUM-ASSIST SYSTEMS

All vacuum-assisted power brake units **(Figure 4)** are similar in design. They generate their application energy

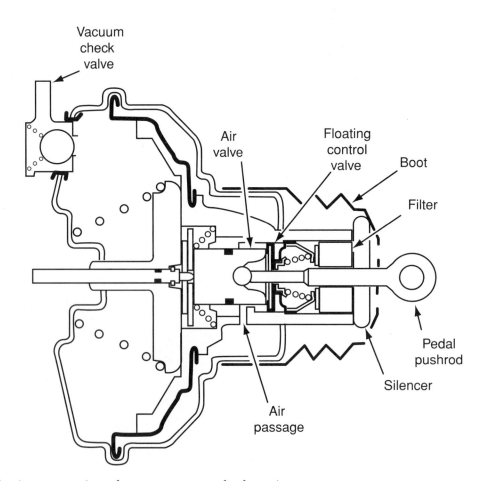

Figure 4. The basic construction of a vacuum power brake unit.

through the **pressure differential** between engine vacuum and atmospheric pressure. A flexible diaphragm and a power piston use this energy to provide brake assistance. Modern vacuum boosters are **vacuum-suspended** units. This means the booster diaphragm is suspended in a vacuum on both sides when the brakes are not applied. When the brake pedal is pressed, an air control valve attached to the brake pedal pushrod opens. This valve admits atmospheric pressure to the back of the diaphragm. Atmospheric pressure forces the diaphragm forward where it increases the amount of force applied to the pushrod and master cylinder piston.

Vacuum boosters can have one or two diaphragms, but most are single-diaphragm units. Single-diaphragm boosters are larger in diameter than dual-diaphragm vacuum boosters, or **tandem boosters**.

All vacuum boosters have vacuum check valves. The check valve is located between the engine manifold and the booster. Vacuum can reach the booster through the one-way check valve, but it cannot leak back past the valve. As a result, vacuum is maintained inside the booster even after the engine is turned off.

Hydraulic Power Brakes

Decreases in engine size, plus the continued use of engine vacuum to operate other engine-related systems, such as emission control devices, led to the development of hydraulic-assist power brakes. These systems use fluid pressure, not vacuum, to help apply the brakes. The two types of hydraulic boosters are:

1. A mechanical hydraulic power-assist system operated with pressure from the power steering pump. This unit is a Bendix design called **Hydro-Boost**.
2. An electrohydraulic power-assist system with an independent hydraulic power source driven by an electric motor. This unit is a General Motors design called **PowerMaster**.

Hydro-Boost Systems

Hydro-Boost systems use fluid pressure from the power steering pump to provide power assist to the brakes. The power brake booster is located between the cowl and the master cylinder. Hoses connect the pump to the booster assembly **(Figure 5)**.

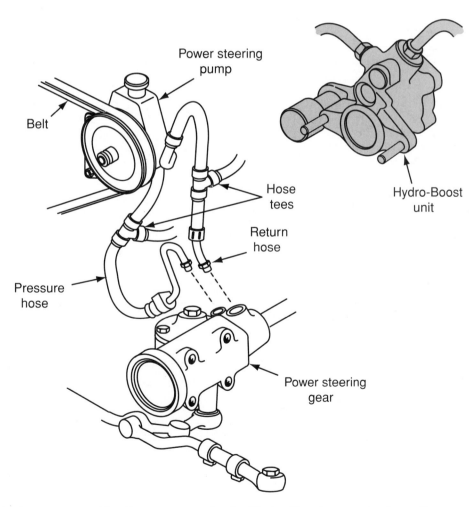

Figure 5.   A typical arrangement for the components of a Hydro-Boost power brake unit.

The power steering pump provides a continuous flow of fluid to the brake booster whenever the engine is running. Three flexible hoses route the power steering fluid to the booster. One hose supplies pressurized fluid from the pump to the booster. Another hose routes the pressurized fluid from the booster to the power steering gear assembly. The third hose returns fluid from the booster to the power steering pump.

Some systems have a nitrogen-charged pneumatic **accumulator** on the booster to provide reserve power-assist pressure. If power steering pump pressure is not available, due to belt failure or similar problems, the accumulator pressure is used to provide brake assist for one to three stops. A spring or compressed gas behind a sealed diaphragm provides the accumulator pressure.

The booster assembly consists of an open center spool valve and sleeve assembly, a lever assembly, an input rod assembly, a power piston, an output pushrod, and the accumulator. The booster assembly is mounted on the vehicle in much the same manner as a vacuum booster. The pedal rod is connected at the booster input rod end.

Power-steering fluid flow in the booster unit is controlled by a hollow center spool valve. The spool valve has **lands**, annular grooves, and drilled passages. These mate with grooves and lands in the valve bore. The flow pattern of the fluid depends upon the alignment of the valve in the bore.

PowerMaster Hydraulic Booster

The PowerMaster power brake system **(Figure 6)** is a self-contained hydraulic booster that is typically connected directly to the master cylinder. Instead of relying on the power steering pump for hydraulic pressure, as is done in the Hydro-Boost system, the PowerMaster has its own vane pump and electric motor to provide the hydraulic pressure required for booster operation.

The pump is driven by an electric motor that is mounted below the master cylinder. Brake fluid is drawn through a low-pressure hose from the reservoir to the pump. There it is pressurized and exits through a high-pressure hose to the accumulator. Pump motor operation is controlled through a dual-pressure switch and relay.

The accumulator stores brake fluid under pressure. It has a heavy, flexible diaphragm that separates brake fluid on one side from high-pressure nitrogen gas on the other.

The booster assembly consists of a power piston, inner and outer control valves, and the reaction components. The power piston uses the high-pressure fluid from the accumulator against its large diameter to produce the force needed to apply (push) the master cylinder pistons. The reaction components fit between the power piston and master cylinder primary piston. These components consist of a reaction body group, reaction disc, and reaction piston. In operation, the reaction components provide pedal feel. The apply (outer control) valve and discharge (inner control) valve regulate fluid movement through the booster according to brake pedal movement and brake pedal pressure (force).

Basic Diagnostics and Service

When troubleshooting and servicing power brake systems, keep in mind that the power system is a separate system. Check for faults in the master cylinder and hydraulic system first. As with conventional brakes, a spongy pedal in a power brake system is caused by air in the hydraulic lines. Brake grab may be caused by grease on the brake linings. Check out all basic brake components before moving on to the power-assist system.

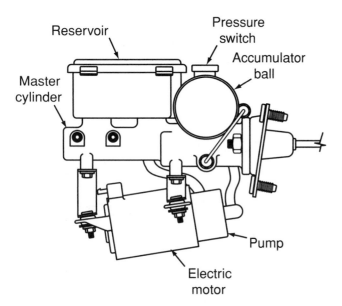

Figure 6. The main components of a PowerMaster power brake system.

> **You Should Know** *Always depressurize the accumulator of any hydraulic boost system before disconnecting any brake lines or hoses. This is usually done by turning the engine off and depressing and releasing the brake pedal more than ten times. Always follow the depressurizing procedures given in the service manual.*

Except for the master cylinder pushrod adjustment, vacuum and hydraulic power-assist units are not adjusted in normal service. If the booster is suspect, it is removed and replaced with a new or rebuilt unit or it can be rebuilt in the shop. Overhaul kits might be available.

Any investigation of a hydraulic boost compliant should begin with an inspection of the power-steering pump belt, power steering fluid level, and hose condition and connections.

Begin a basic operational test of these systems with the engine off. Pump the brake pedal numerous times to bleed off the residual hydraulic pressure that is stored in the accumulator. Hold firm pressure on the brake pedal and start the engine. The brake pedal should move downward and then push up against the foot.

BRAKE PEDAL CHECKS

The brake pedal must be adjusted properly for correct power-assist operation. Besides the mechanical checks and brake travel check, you should also check pedal free play and pedal height settings. Excessive play or low pedal height can limit the amount of power assist generated by the vacuum booster.

Pedal Free Play Inspection

Pedal free play is the first easy movement of the brake pedal before the braking action begins to engage. Pedal free play should be only about $1/16$–$3/16$ inch in most cases. To determine the amount of free play present, place your hand or foot on the pedal and gently push it down until you feel an increase in pushing effort (**Figure 7**). Hold a ruler alongside the pedal to measure the amount of free play.

Figure 7. Using your hand, rather than your foot, to check brake pedal free play will allow you to get an accurate measurement.

Pedal Height Adjustment

Brake pedal height specifications are listed in most vehicle service manuals. Typical heights range from 6 to 7 inches. Most pedal height measurements are taken from the floor mat to the base of the pedal, but in some cases the measurement is taken from a special point below the floor mat (**Figure 8**).

Before adjusting pedal height, loosen the brake switch locknut and back off the brake switch until it no longer touches the brake pedal. Then use pliers to screw the pushrod in or out as needed (**Figure 9**). When the proper pedal height is set, adjust the stop lamp switch as required. After adjusting pedal height, check for proper stop lamp and cruise control operation.

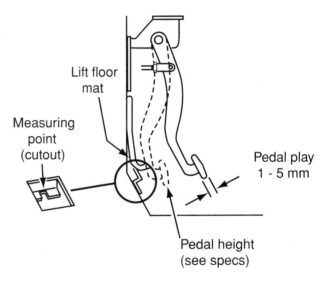

Figure 8. The pedal height on some vehicles is measured from a specific spot on the floor. Always check the service manual before taking a measurement.

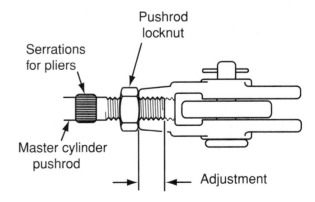

Figure 9. Pedal height on most vehicles is adjusted by shortening or lengthening the pushrod for the master cylinder.

Summary

- Three things can be done to boost the force applied to the master cylinder or to reduce force that must be applied to the brake pedal by the driver: increase the length of the brake pedal, increase the size of the pistons in the hydraulic system, and/or install a power booster.
- Power boosters do not alter the basic brake system operation.
- Two basic types of power-assist mechanisms are used: vacuum assist and hydraulic assist.
- All vacuum-assisted power brake units generate their application energy through the pressure differential between engine vacuum and atmospheric pressure.
- Modern vacuum boosters are vacuum-suspended units.
- There are two types of hydraulic boosters: Hydro-Boost and PowerMaster.
- Hydro-Boost systems use fluid pressure from the power steering pump to provide power assist to the brakes.
- The PowerMaster power brake system incorporates a self-contained hydraulic booster pump.
- The brake pedal travel, pedal free play, and pedal height settings must be correct in order to have efficient power-assist operation.

Review Questions

1. List and describe the different common types of brake vacuum boosters used on today's vehicles.
2. There is atmospheric pressure on one side of the booster diaphragm when the brakes are _____.
3. While checking power brake booster systems, Technician A pumps the brake pedal several times with the engine off and then holds firm pressure on the brake pedal while she starts the engine. If the hydraulic booster is working correctly, the brake pedal should move downward and then push up against her foot. Technician B says that excessive play or low pedal height may limit the amount of power assist generated by the vacuum booster. Who is correct?
   A. Technician A only
   B. Technician B only
   C. Both Technician A and Technician B
   D. Neither Technician A nor Technician B
4. The hydraulic brake systems of the Hydro-Boost and PowerMaster are being discussed. Technician A says that the Hydro-Boost uses power steering fluid. Technician B says that the PowerMaster uses brake fluid. Who is correct?
   A. Technician A only
   B. Technician B only
   C. Both Technician A and Technician B
   D. Neither Technician A nor Technician B
5. The hydraulic brake systems of the Hydro-Boost and PowerMaster are being discussed. Technician A says that the Hydro-Boost uses a power steering pump. Technician B says that the PowerMaster uses an electric motor-driven vane pump. Who is correct?
   A. Technician A only
   B. Technician B only
   C. Both Technician A and Technician B
   D. Neither Technician A nor Technician B

Chapter 20

Vacuum Power Brakes

Introduction

To understand how vacuum booster systems work, you must understand the relationship between atmospheric pressure and vacuum. Atmospheric pressure is caused by the weight of the gasses surrounding the earth, which varies with altitude and temperature. At sea level and at 68°F, atmospheric pressure is 14.7 psi. If you were to drive up in the mountains, you would find that the atmospheric pressure is lower.

For our purposes, vacuum is any pressure lower than atmospheric pressure. When an engine is running, the intake strokes in the cylinders create a low pressure. This low pressure draws in the mixture of air and fuel so the engine can run. We call this low pressure a vacuum. A true vacuum, however, is a complete absence of air and is found only in a laboratory or in outer space.

PRESSURE DIFFERENTIAL

Atmospheric pressure and vacuum can be used together as a force to make things move. In nature, a higher pressure will always move toward a lower pressure. The force and the speed at which the higher pressure moves depend on the difference between the high and the low pressures. This difference is called pressure differential. **Figure 1** shows a piston that is free to move up and down in a cylinder. One end of the cylinder is connected to a vacuum pump. The vacuum pump is used to lower the pressure at the bottom of the piston. When atmospheric pressure is applied on the other side of the cylinder, the piston will move down, toward the lower pressure.

If a perfect vacuum exists on one side of the piston and atmospheric pressure is applied to the other side, the pressure differential equals atmospheric pressure, or 14.7 psi. These values do not apply to a vacuum power booster because a total vacuum is not formed by the engine.

Air pressure (low or high) can be expressed in several different measurement units. To compare one pressure to another, both pressures must be expressed in the same units. Atmospheric pressure at sea level can be expressed as either 14.7 psi or 29.9 inches of mercury. Similarly, complete vacuum can be expressed as either 29.9 inches of mercury *below* atmospheric pressure or 14.7 psi *below* atmospheric pressure or zero psi. If we round off 14.7 to 15 and 29.9 to 30, we have a ratio of 2 between pounds per square inch and inches of mercury. This ratio of 2 can be used to determine the approximate values of pressures when expressed in inches of mercury or pounds per square inch:

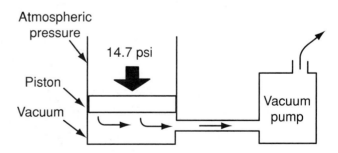

Figure 1. Because a higher pressure always moves toward a lower pressure, the presence of vacuum on one side of a piston and atmospheric pressure on the other side can be used to develop force.

- To convert pounds per square inch to inches of mercury, multiply by 2.
- To convert inches of mercury to pounds per square inch, divide by 2.

In a vacuum brake booster, the piston in the previous example is actually a flexible diaphragm that divides a metal chamber almost in half, with each half sealed from the other. Most vacuum brake boosters work with 15–20 inches of mercury on one side of the diaphragm. When a booster develops force to assist brake application, atmospheric pressure is admitted to the other side of the diaphragm as the brake pedal is depressed.

Therefore, if atmospheric pressure is approximately 15 psi and the vacuum on one side of the diaphragm is 15–20 inches of mercury (7.5–10 psi), the pressure differential is about 7.5 psi. This means the booster develops about 7.5 pounds of force for every square inch of the surface area of the diaphragm. Therefore, if the diaphragm has 50 square inches of area, the total booster force is approximately 375 pounds. This 375 pounds of force helps the driver apply the brakes.

VACUUM AND AIR FOR POWER BOOSTERS

To work properly, enough vacuum and air must be delivered to the power booster. Most power boosters receive air in a similar manner, and the vacuum is provided by either the intake manifold or an auxiliary vacuum pump.

Air Systems

Typically, air enters through passages in the pedal pushrod boot **(Figure 2)**. The air passes through a fine mesh material called a silencer, which slows down the air and reduces noise caused by the air moving in. The boot and air inlet are inside the car, so any noises could be heard by the driver. The air then passes through a filter to remove any dirt that could damage the valve. Air then flows into the power piston passages to the air valve.

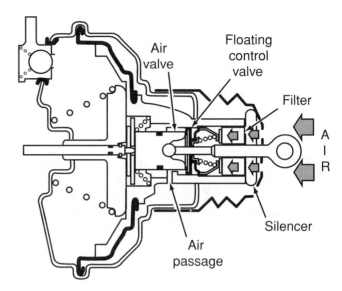

Figure 2. Air enters the brake booster through a filter around the pedal pushrod.

Vacuum Systems

The vacuum for most power brake systems is supplied by the engine's intake manifold. On older cars, a simple vacuum hose was attached from the intake manifold directly to the housing of the brake booster. A check valve was used to protect the booster against loss of vacuum. The problem with this system was that intake manifold vacuum decreases and increases as the engine is accelerated and decelerated. Engine vacuum also begins to drop as the engine wears.

A **check valve (Figure 3)** is installed between the manifold and the reservoir to prevent air from entering the reservoir during wide-open throttle or other operating conditions that cause low engine vacuum. The check valve is also a safety device, protecting the system from losing vacuum in case of a leaking supply line or other failure in the vacuum supply. The check valve is typically mounted on

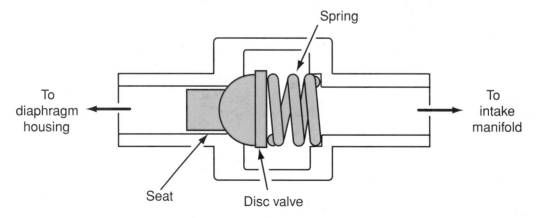

Figure 3. A cutaway view of a typical check valve.

the front of the booster at the connection for the vacuum hose.

A typical vacuum check valve has a small disc backed up by a spring. Manifold vacuum pulls and holds the disc off its seat and allows vacuum to enter into the booster. When vacuum drops below the calibration of the spring, the spring moves the disc against its seat. When the disc is seated, the valve closes and prevents vacuum from leaking out of the system.

When vacuum is supplied by an auxiliary vacuum pump, vacuum action on the booster diaphragm is the same as it is for manifold vacuum. Only the vacuum source is different in these systems.

Auxiliary Vacuum Pumps

Vacuum is used to power or control air-conditioning and heating control systems, cruise control, and several emission control systems on late-model vehicles. The more vacuum-operated devices a vehicle has, the less vacuum that is available to power the brakes. Additionally, diesel engines do not develop much intake manifold vacuum; therefore, diesel-powered vehicles need something other than the engine to provide the vacuum to operate the power brakes. These conditions led to the introduction of auxiliary vacuum pumps.

An auxiliary vacuum pump can be driven by an engine's drive belt, directly by the engine with a cam or gear, or by an electric motor. A typical belt-driven pump is mounted on a bracket at the front of the engine, and a pulley is attached to the pump's drive shaft. The pulley is driven by a drive belt. Some systems do not use a belt to drive the pump directly but mount the pump on the back of the generator **(Figure 4)**. The generator is driven by the belt,

and the pump is driven by an extension of the generator shaft.

A vacuum pump can also be driven directly by the engine and is often installed in place of an ignition distributor on a diesel engine and driven by a gear on the camshaft. An electric vacuum pump can be located any place in the engine compartment. The pump controller is connected to the wiring harness and gets electrical inputs from a low-vacuum switch in the booster or from the vehicle onboard computer. The controller has an on/off switch and a timer relay that control the motor. The electric motor gets its power from the controller. The electric motor drives either a vane pump or a diaphragm pump.

Vacuum Check Valves

Regardless of the vacuum source, a power brake system must have one or more vacuum check valves. The check valve uses a spring-loaded ball or disc to close the vacuum line if pressure in the line gets higher than the vacuum in the booster or reservoir.

Most systems have a single check valve as part of the inlet fitting of the vacuum booster **(Figure 5)**. Some vehicles, however, have an in-line check valve in the vacuum line. If the power brake system has a vacuum reservoir as well as the booster, it will have two check valves. One valve is part of the booster inlet fitting; the other is between the intake manifold and the reservoir **(Figure 6)**.

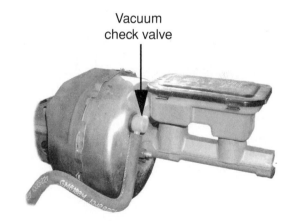

Vacuum
check valve

Figure 5. A vacuum check valve at the vacuum inlet of the brake booster unit.

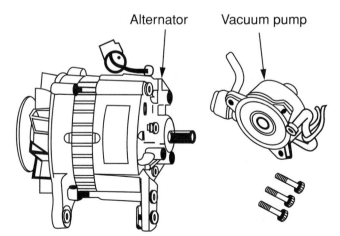

Alternator   Vacuum pump

Figure 4. An auxiliary vacuum pump that is mounted to the rear of a generator (alternator) and is driven by the shaft of the generator.

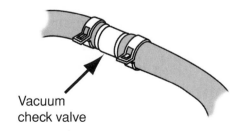

Vacuum
check valve

Figure 6. An in-line vacuum check valve.

The vacuum check valve has a secondary function of keeping fuel vapors out of the vacuum booster. Without a check valve, vacuum in the booster could draw part of the air-fuel mixture into the booster when the engine has low manifold vacuum. As an extra safety precaution, some systems have a charcoal filter between the manifold and the vacuum check valve to trap fuel vapors before they can get near the booster.

VACUUM POWER BOOSTERS

Most vacuum boosters have essentially the same parts **(Figure 7)**. These parts are contained in a steel housing or shell that is divided into front and rear halves held together with interlocking tabs. The rear housing has studs to mount the unit onto the firewall.

The diaphragm is a large, hemispherically shaped part made from rubber. The flexible diaphragm moves back and forth as it is acted upon by pressure and vacuum. The center of the diaphragm is supported by a metal or plastic diaphragm support.

The brake pedal is connected to the brake pedal pushrod, which is contained in the booster. One end of the pushrod sticks out of the center of the rear housing when the booster is assembled. A rubber boot seals the area between the pushrod and the housing. The other end of the pushrod is attached to the power piston.

The power piston is attached to the center of the diaphragm. The piston is often described as being suspended in the diaphragm. The power piston contains and operates the vacuum and air (atmospheric) valves that control the diaphragm. The power piston also transmits the force from the diaphragm through a piston rod to the master cylinder. Because the pushrod is mechanically connected through the booster, any vacuum failure will not cause loss of braking action.

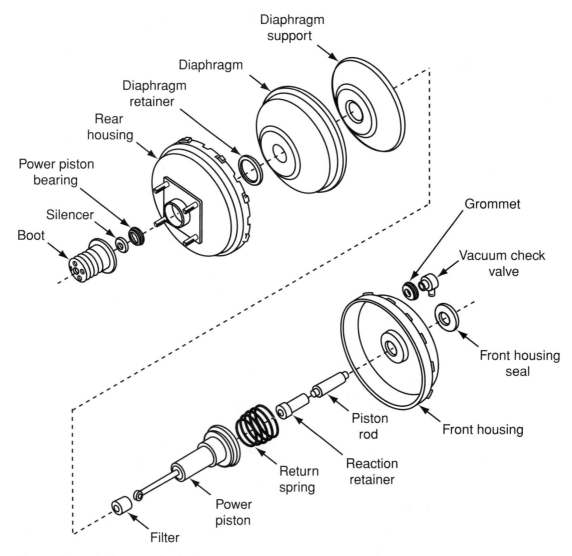

Figure 7.   Parts of a typical power brake booster.

The piston rod also may be attached to another rod called a reaction retainer. A large coil spring is mounted between the front housing and the power piston and diaphragm assembly. The spring returns the diaphragm and piston to the unapplied position when the driver releases the brake pedal.

Two valves are located on the pedal side of the diaphragm. One, called the air valve, controls the flow of air at atmospheric pressure into one side of the booster. The other, called the vacuum valve, controls the buildup of vacuum in both sides of the booster. Both valves are connected to, and operated by, the pushrod.

Operation and Construction

Vacuum boosters are often described as being vacuum suspended or **atmospheric suspended**. The brake boosters on nearly all vehicles built for the past thirty years have vacuum-suspended diaphragms. This means that when the brakes are released and the engine is running, vacuum is present on both sides of the diaphragm. When the pedal is pressed, atmospheric pressure is admitted to the rear of the diaphragm to develop booster force.

On some older vehicles, the vacuum boosters had atmospheric-suspended diaphragms. When the brakes are released, atmospheric pressure is present on both sides of the diaphragm. When the pedal is pressed, vacuum is admitted to the front of the diaphragm to develop booster force.

An atmospheric-suspended booster has no vacuum reserve for power braking with the engine off. A vacuum-suspended booster holds enough residual vacuum for two or three moderate power-assisted stops or at least one hard stop with the engine off. Also, if an atmospheric-suspended booster is applied with low vacuum (braking immediately after acceleration, for example), power assist may be reduced.

There are three basic types of vacuum suspended power brake boosters **(Figure 8)**:

1. Single diaphragm with a reaction disc.
2. Single diaphragm with a reaction lever.
3. Tandem booster (dual diaphragm) with a reaction disc.

The single diaphragm with a reaction disc provides force (or reacts back) to the brake pedal through a rubber reaction disc. The single diaphragm with a reaction lever reacts back to the pedal through a lever arrangement. The tandem diaphragm has two diaphragms that work together to provide force. The following paragraphs explain how a single diaphragm booster with a reaction disc works. The other two booster types operate similarly.

Brakes Not Applied (Released)

When the brake pedal is at rest, the return spring for the input pushrod holds the pushrod and the air control

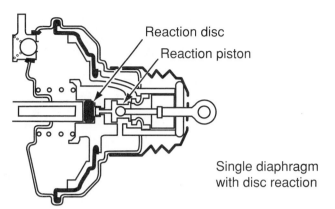

Single diaphragm
with disc reaction

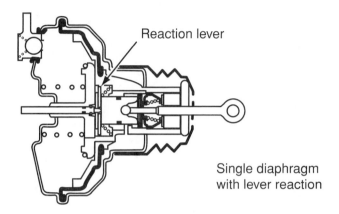

Single diaphragm
with lever reaction

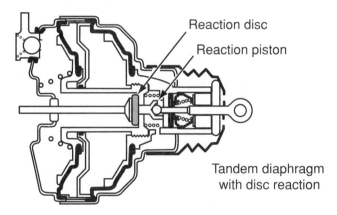

Tandem diaphragm
with disc reaction

Figure 8. The three basic types of vacuum brake boosters.

valve rearward in the power piston **(Figure 9)**. The rear of the vacuum valve plunger seats against the air control valve in the power piston to close the atmospheric port. The plunger also compresses the air valve against its spring to open the vacuum port. This valve action closes the booster to the atmosphere and opens a passage between the front and rear of the booster chamber. Equal vacuum is present on both sides of the diaphragm. The power piston return spring holds the diaphragm rearward so no force is applied to the output pushrod and master cylinder.

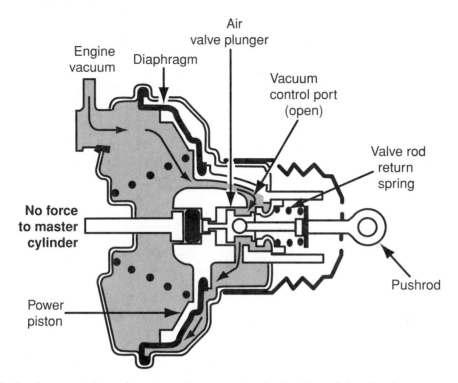

Figure 9. When the brakes are released, vacuum is present on both sides of the diaphragm.

Moderate Brake Application

When the driver applies the brakes, pedal pressure overcomes the input pushrod's return spring to move the pushrod and vacuum valve plunger forward **(Figure 10)**.

Spring pressure then moves the air valve to close the vacuum port to the rear chamber. As the plunger continues to move forward, it opens the atmospheric port to the rear chamber. As a result, vacuum (low pressure) is present in

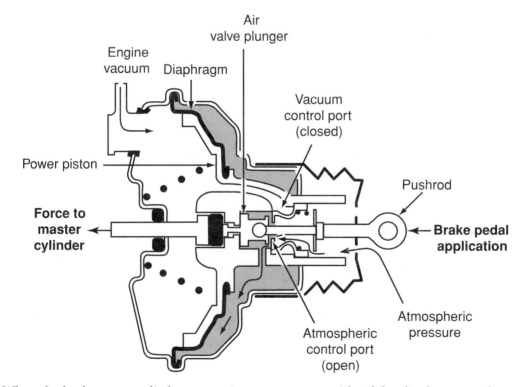

Figure 10. When the brakes are applied, vacuum is present on one side of the diaphragm and atmospheric pressure is on the other side.

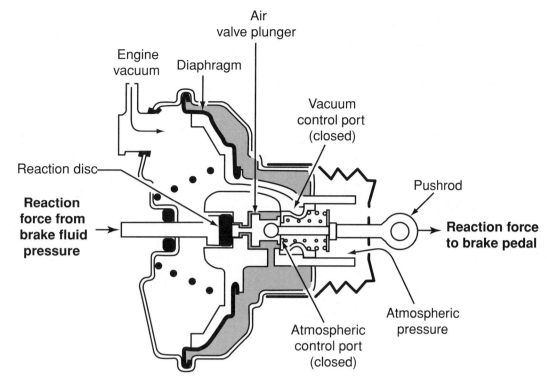

Figure 11. When the brake pedal is held in position, both valves are closed to trap vacuum and atmospheric pressure.

the front chamber of the booster and higher pressure (atmospheric) is in the rear. The resulting pressure differential moves the diaphragm and power piston forward against the return spring to apply force to the master cylinder pushrod.

Brakes Holding

When the driver maintains constant pressure on the brake pedal, the booster's input pushrod does not move. The diaphragm and power piston continue to move forward until the air valve on the piston seats against the rear of the vacuum valve's plunger to close the atmospheric port **(Figure 11)**. All of this happens very quickly, and as long as the atmospheric port is closed, the diaphragm does not move. It is suspended by a fixed pressure differential between the front and rear chambers of the booster. The booster always seeks the holding position when pedal force is constant, or unchanging.

Full Brake Application

If the driver increases pressure on the pedal and applies enough force to move the vacuum valve's plunger fully forward against the power piston, the vacuum port to the rear chamber closes and opens the atmospheric vent. This action closes the vacuum port to the rear chamber and fully opens the atmospheric port. Maximum atmospheric pressure then exists in the rear chamber, and the

booster supplies its maximum power assist. This condition sometimes is called the booster vacuum runout point. At this point, additional braking force can be applied to the master cylinder, but it must come entirely from foot pressure on the pedal. Because the booster is supplying its maximum force at this point, the driver will feel the pedal become harder to press.

Brakes Being Released

When the driver releases the pedal, the input pushrod's spring moves the pushrod and the vacuum valve's plunger rearward in the power piston. The rear of the valve's plunger closes the atmospheric port, and the plunger compresses the air control valve to open the vacuum port. Vacuum is then applied to the rear chamber, and pressure equalizes on both sides of the diaphragm. The large return spring moves the diaphragm, the power piston, and the output pushrod rearward. The booster returns to the released position.

BRAKE PEDAL FEEL

A power brake booster must provide some kind of physical feedback to the driver. This is called brake pedal feel. Pedal feel is provided in modern vacuum boosters by a **reaction disc** or a **reaction plate and levers**. Reaction simply means that as the driver's foot, the pedal, and the booster apply force to the master cylinder, an equal force

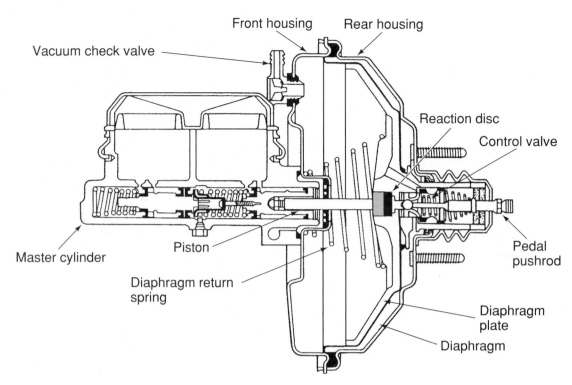

Figure 12. A reaction-disc type vacuum booster.

develops in the opposite direction. You normally feel the reaction force as the resistance to application force. Simply put, when you apply the brake pedal as hard as you can, it stops moving. This means that the reaction force has equaled the maximum force that you are capable of applying to the pedal.

Reaction-Disc Booster

In a brake booster with a reaction disc, the input pushrod and vacuum valve plunger bear on a rubber disc **(Figure 12)**. The reaction disc is located in the power piston and compresses under the force of the pedal. Its ability to compress lets it absorb reaction force back from the master cylinder when the brakes are applied. As the disc compresses and feeds back reaction force to the pedal pushrod, it also modulates the action of the vacuum and air control valves to adjust pressure on the diaphragm. The harder the brake pedal is pressed, the more the disc compresses and the greater the feedback feel applied to the pedal.

Plate-and-Lever Booster

A plate-and-lever booster **(Figure 13)** uses a lever mechanism to provide brake pedal feel. The connection between the pedal pushrod is through the reaction plate and levers in the power piston. When the brakes are first applied, the fixed ends of each lever are in contact with the power piston. The other ends are spring loaded and free to

move. The force back to the driver's foot on the brake pedal is kept low. As brake application continues, the springs deflect enough to allow contact between the movable ends of the levers and the vacuum and air valves. The operation of the air

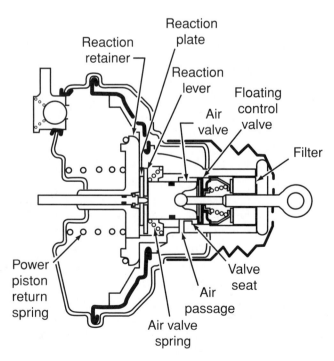

Figure 13. A reaction plate-and-lever vacuum booster.

valve and vacuum valve begins to bleed out vacuum and add atmospheric pressure to the brake pedal side of the diaphragm. The reaction force back to the brake pedal increases. The lever and reaction plate provide a resistance and feel to the pedal similar to a non–power brake system.

Tandem Boosters

Automotive engineers are constantly working to reduce vehicle weight. Lighter cars have resulted from making components out of lighter materials and from making components smaller. Master cylinders are one example. They are now much smaller than they were in the past and are made from lighter materials. Brake power boosters have not escaped this trend.

The amount of braking power from a vacuum booster is directly related to the area of the diaphragm. A booster with a smaller diaphragm would provide less power assist. Engineers solved this problem by designing a smaller diaphragm housing fitted with two smaller diaphragms in tandem **(Figure 14)**. The amount of force is proportional to the total area of both diaphragms. Two 10-square-inch diaphragms can provide the same power as one 20-square-inch diaphragm.

The front and rear housings are the same as those of a single-diaphragm booster. Some units have a housing divider between the diaphragms and others do not. Both diaphragms are connected to the power piston. A reaction disc or lever assembly is used, just as in a single-diaphragm booster.

The two diaphragms work the same way as a single diaphragm except that the air and vacuum valves must

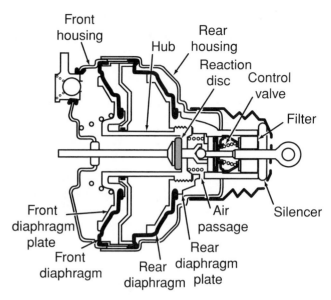

Figure 14. A cross-sectional view of a tandem vacuum brake booster.

control vacuum and atmospheric pressure on both diaphragms at the same time. When the brakes are released, vacuum is on both sides of each diaphragm. During brake application, the air valve and vacuum valve operate to admit air on the brake pedal side of both diaphragms. With a vacuum on the master cylinder side of both diaphragms and atmospheric pressure on the opposite sides, power assist is developed.

Summary

- In nature, a higher pressure will always move toward a lower pressure. The force and the speed at which the higher pressure moves depends on the pressure differential.
- To work properly, enough vacuum and air must be delivered to the power booster.
- Typically, air enters through passages in the pedal pushrod boot.
- The vacuum for most power brake systems is supplied by the engine, although a few use an auxiliary vacuum pump.
- A check valve is installed between the manifold and the reservoir to prevent air from entering the reservoir and to keep fuel vapors out of the vacuum booster.
- Most vacuum boosters have a steel housing or shell that is divided into front and rear halves with a diaphragm separating the two halves.

- The brake pedal is connected to the brake pedal pushrod, and the other end of the pushrod is attached to the power piston.
- The power piston is attached to the center of the diaphragm and contains and operates the vacuum and air valves that control the diaphragm.
- Vacuum boosters are either vacuum suspended or atmospheric suspended.
- There are three basic types of vacuum suspended power brake boosters: single diaphragm with a reaction disc, single diaphragm with a reaction lever, and tandem booster (dual diaphragm) with a reaction disc.
- A power brake booster must provide some kind of physical feedback or pedal feel to the driver.
- Pedal feel is provided in modern vacuum boosters by a reaction disc or a reaction plate and levers.
- A tandem brake booster is made lighter and smaller than a normal brake booster because it contains two smaller diaphragms connected in tandem.

Review Questions

1. Explain how vacuum is used to provide a power assist.
2. The _____ ____ _____ valve controls airflow to the pedal side of the vacuum booster diaphragm.
3. The _____ valve controls vacuum to the pedal side of the vacuum booster diaphragm.
4. The use of auxiliary vacuum pumps for power brakes is being discussed. Technician A says that the pump can be driven by a drive belt. Technician B says that the pump can be driven by an electric motor. Who is correct?

   A. Technician A only
   B. Technician B only
   C. Both Technician A and Technician B
   D. Neither Technician A nor Technician B

5. Tandem vacuum power boosters are being discussed. Technician A says that they contain two diaphragms instead of the usual one. Technician B says that tandem boosters are smaller in diameter than standard vacuum boosters. Who is correct?

   A. Technician A only
   B. Technician B only
   C. Both Technician A and Technician B
   D. Neither Technician A nor Technician B

Chapter 21

Vacuum Power Brake Service

Introduction

Safe and controllable braking on today's vehicles is possible only if the power brake unit is working properly. Although the power brake system is a separate system from the brake system and the normal brake system will still work if there is a malfunction in the power brakes, extreme amounts of pressure are required to stop the vehicle quickly. Fortunately, vacuum power boosters are generally trouble free, and many last the lifetime of a vehicle. Like anything, the units may break or begin to operate poorly. When the power brakes do not work as they should, the defective part(s) should be replaced or repaired. The power brake system should be inspected whenever a problem is suspected and whenever you perform other brake work. Along with an inspection, it is wise to conduct an operational check of the vacuum booster.

VACUUM BOOSTER TESTING AND DIAGNOSIS

Any condition that reduces the amount of vacuum formed by the intake strokes of the engine will affect power brake performance. These conditions include internal and external vacuum leaks, faulty valves, improper valve timing, and incorrect ignition timing. An engine with a nonstock high-performance camshaft can also produce lower vacuum. When investigating poor brake performance, check engine vacuum before inspecting or suspecting the vacuum booster system. In most systems, a vacuum of at least 14 inches of mercury is required for proper power brake operation.

To check engine vacuum, connect a vacuum gauge to the engine's intake manifold. Start the engine and observe the gauge **(Figure 1)**. Ideally, the reading should be 17+ inches of mercury, and the reading should be steady. If the reading is 14 inches of mercury or less, diagnose the engine to determine the cause of the low reading before testing the power brake system. On a vehicle with an engine-driven or electric vacuum pump, test it according to the manufacturer's recommendations.

A hard brake pedal is usually the first signal that the booster system is failing. Insufficient manifold vacuum, leaking or collapsed vacuum lines, punctured diaphragms, or leaky piston seals can cause weak booster operation. A steady hiss when the brake is held down indicates a leak that also can cause poor booster operation.

If the brakes do not release completely, there might be a misadjusted or misaligned connection between the booster's pushrod and the linkage for the brake pedal. If the

Figure 1. To check source vacuum, start the engine and observe the vacuum gauge.

pedal-to-booster linkage appears in good condition, loosen the connection between the master cylinder and the brake booster. If the brakes release, the trouble is in the power unit, and a piston, diaphragm, or bellows return spring may be broken. If the brakes do not release when the master cylinder is loosened from the booster, one or more brake lines might be restricted or a problem might exist in the brake system's hydraulic circuit.

> **You Should Know** *Do not overlook the brake vacuum booster as the cause of a drivability or comfort complaint. A leak in the booster vacuum hose or at the point that the hose connects to the manifold or the booster can cause a rough idle, misfire, hesitation, or surge. Depending on the size of the hole and the point where the hose connects to the manifold, the leak can affect the air-fuel mixture to one cylinder or several cylinders. Complaints about the air-conditioning system can also serve as a clue for leaks. If the air-conditioning plenum uses vacuum diaphragms to move the air delivery doors, the complaint may be an uncontrollable delivery of heat through the vents, most likely the defroster vents.*

If the brakes grab, look for common causes such as greasy linings or scored drums before checking the booster. If the problem is in the booster, it might be a damaged reaction control. The reaction control assembly is a diaphragm, spring, and valves that tend to resist pedal action.

Basic Vacuum-Boost Operational Test

To check the overall operation of a vacuum power booster, turn the engine off. Then repeatedly pump the brake pedal to remove all residual vacuum from the booster. Hold the brake pedal down firmly and start the engine. If the system is working correctly, the pedal should move downward slightly and then stop. Only a small amount of pressure should be needed to hold down the pedal. If this is not what happened, further testing is necessary.

Vacuum Supply Tests

If the booster is giving weak braking assistance or no assistance at all, a problem might exist with the vacuum supply to the unit. Vacuum boost efficiency is affected by loose or kinked vacuum lines **(Figure 2)** and clogged atmospheric air intake filters. Soft vacuum lines can also affect boost efficiency because they can collapse under a vacuum.

Figure 2. Carefully inspect all vacuum lines for cracks and other damage.

Another cause might be the check valve, which retains vacuum in the booster when engine vacuum is low. You can check this valve with a vacuum gauge to determine if it is restricted or stuck open or closed. The check valve can also be checked by removing it and blowing through the intake manifold end of the valve **(Figure 3)**. No air will pass through it. If no air passes through, connect a vacuum pump to the booster end of the valve **(Figure 4)**. A good check valve will hold the vacuum.

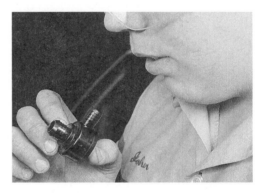

Figure 3. The vacuum check valve can be checked by attempting to blow through it.

Figure 4. With a hand-operated vacuum pump, apply a vacuum to the booster side of the check valve.

You can make some quick tests of a vacuum booster with a hand-held vacuum pump. Start by disconnecting the booster vacuum hose from the intake manifold and connect your vacuum pump to the hose. Apply 17–20 inches of mercury of vacuum. The gauge reading on the pump should hold steady. If it drops, the booster or the hose is leaking. Keep 17–20 inches of mercury of vacuum applied to the booster and have an assistant apply and hold the brake pedal for thirty seconds. Vacuum should drop to no less than 6 inches of mercury when the brakes are first applied and should leak down no more than another 2 inches of mercury during thirty seconds.

A vacuum supply hose that is restricted but not completely blocked will allow a normal reading on a vacuum gauge but will delay the buildup of full vacuum in the booster. That is, it can reduce the volume of vacuum for rapid, repeated brake applications. Check for a restricted vacuum hose by disconnecting it from the booster with the engine running. If the engine does not stumble suddenly and almost stall, the hose is probably restricted. Install a new hose and recheck booster operation. Also, if the vacuum hose contains a vapor separator that is clogged, the same kind of delayed vacuum application symptoms can occur.

If the vacuum booster seems to have a normal vacuum supply but the brakes take more effort than normal to apply, the booster might have an internal vacuum leak. Check for this by starting the engine and letting it idle to develop normal vacuum at the booster. Then roll up the windows to reduce outside noise and slowly but firmly apply the brakes. Listen to the engine. If it stumbles or runs roughly, and a hissing sound increases around the pedal pushrod, the booster has an internal leak in the diaphragm.

Do not mistake normal booster breathing for a vacuum leak. When the pedal is pressed, air rushing through the filter in the rubber boot of the booster input pushrod causes a slight breathing sound, which is normal. A diaphragm vacuum leak will cause a louder, continuous hiss.

Fluid Loss Test

If the fluid level in the master cylinder reservoir level is low, but there is no sign of an external leak, remove the vacuum hose from the intake manifold to the booster. Inspect it carefully for traces of brake fluid. If you find evidence of brake fluid, the master cylinder's front seal might be leaking. If so, the master cylinder requires rebuilding or replacement, and the booster should be replaced. If a trace of brake fluid is present on the front of the booster, suspect a bad seal at the rear of the master cylinder.

VACUUM BOOSTER REMOVAL AND INSTALLATION

The following procedure represents a typical one for removing and installing a vacuum power brake booster if it needs to be removed or replaced **(Figure 5)**. Detailed steps

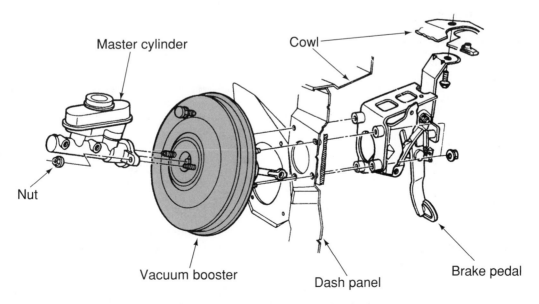

Figure 5. The mounting and placement of a typical vacuum power brake booster.

will vary from one vehicle to another, so always follow the procedures given in the service manual.

Booster Removal

Follow the following procedure for booster removal:

1. Set the parking brake and disconnect the battery ground (negative) cable.
2. On some vehicles, it is necessary to disconnect the brake lines at the outlet ports of the master cylinder. If you need to do this, plug both the cylinder ports and the lines to prevent fluid loss and to keep dirt out of the system.
3. Disconnect the vacuum hose at the booster check valve.
4. Remove all fasteners securing the master cylinder to the brake booster and carefully move the master cylinder away from the booster.
5. If the stop lamp switch is located on the brake pedal assembly, move inside the vehicle and disconnect the stop lamp switch wiring connector from the switch. Then remove the switch from its mounting pin.
6. Remove the nuts fastening the vacuum booster to the passenger compartment side of the bulkhead (firewall).
7. Slide the booster pushrod and bushing off the brake pedal pin.
8. Return to the engine compartment and clear the area around the booster. This may require removing a vacuum manifold, moving a wiring harness, or removing the transmission shift cable and bracket.
9. When the area is cleared of obstacles, move the booster forward so that its studs clear the engine compartment bulkhead.
10. Lift the booster out of the engine compartment.

Booster Installation

Before installing a new or rebuilt vacuum brake booster, check the booster pushrod length and then install the vacuum booster following these general steps:

1. Line up the brake support bracket from inside the vehicle.
2. Have a helper set the booster in position on the engine compartment bulkhead.
3. Thread the nuts onto the studs from inside the vehicle.
4. Reinstall the pushrod onto the brake pedal pin using a new pushrod bushing.
5. Tighten the booster retaining nuts to the specified torque.
6. If removed, reinstall the stop lamp switch on the pedal pin. Reconnect the electrical connector to the switch body.
7. Return to the engine compartment and reposition and install the wiring harness, transmission shift cable, and vacuum manifold.

8. Connect the manifold vacuum hose to the booster check valve.
9. Reinstall the master cylinder, connect all brake lines, and bleed the system of all air.
10. On a vehicle with a manual transmission or cruise control, adjust the manual shift linkage and cruise control dump valve.
11. Reconnect the battery and test drive the vehicle to be sure the booster is operating properly.

BOOSTER OVERHAUL

Many power brake booster assemblies are not designed to be rebuilt or repaired by service technicians. In fact, most repair shops replace the booster if it is faulty. Vacuum power boosters are typically rebuilt only by technicians at specialty shops. If it is necessary to rebuild a booster, make sure you have the correct overhaul kit before beginning. Always follow the instructions and precautions given in the service manual.

VACUUM BOOSTER PUSHROD LENGTH CHECK

The master cylinder's pushrod must be properly adjusted to allow for safe and efficient braking. This is especially true if the brake system is equipped with a vacuum brake booster. If the pushrod is too long, the master cylinder's piston will block the compensating port. This prevents the hydraulic pressure from being released, which causes brake drag. If the pushrod is too short, the brake pedal will be low, and the pedal stroke length will be reduced, resulting in a loss of braking power. When the brakes are applied with a short pushrod, groaning noises might be heard from the booster.

During assembly, the pushrod is matched to the booster. It is normally adjusted only when the vacuum booster or the master cylinder is serviced. You can check pushrod length by observing fluid action at the master cylinder's compensating ports when the brakes are applied. To do this, remove the master cylinder cover and have a helper apply pressure onto the brake pedal. Observe the fluid reservoirs. You should see a small ripple or geyser in the reservoirs as the brakes are applied. If you see no turbulence, loosen the bolts securing the master cylinder to the booster about $1/8$–$1/4$ inch and pull the cylinder forward, away from the booster. Hold it in this position and have your helper apply the brakes again. If turbulence (indicating compensation) now occurs, the brake pedal pushrod or the booster pushrod needs adjustment.

Booster pushrod adjustment is usually checked with a gauge that checks the distance from the end of the pushrod to the booster shell. Two basic gauge designs are used: Bendix and Delco-Moraine (Delphi Chassis). Make sure the pushrod is properly seated in the booster when checking the pushrod length with a gauge.

Bendix Pushrod Gauge Check

Bendix vacuum boosters are used on most Ford vehicles, as well as on vehicles from other manufacturers. Check and adjust pushrod length as follows with the booster installed on the vehicle and vacuum applied:

1. Disconnect the master cylinder from the vacuum booster housing, leaving the brake lines connected. Secure or tie up the master cylinder to prevent the lines from being damaged.
2. Start the engine and let it run at idle.
3. Place the gauge over the pushrod and apply a force of about 5 pounds to the pushrod **(Figure 6)**. The gauge should bottom against the booster housing.
4. If the required force is more or less than 5 pounds, hold the pushrod with a pair of pliers and turn the self-locking adjusting nut with a wrench until the proper 5 pounds of preload exists when the gauge contacts the pushrod.
5. Reinstall the master cylinder.
6. Remove the reservoir cover and observe the fluid while a helper applies and releases the brake pedal. If the fluid level does not change, the pushrod is too long. Disassemble and readjust the rod length.

Delphi Chassis Pushrod Gauge Check

On most Delphi Chassis brake systems, used primarily by General Motors, the master cylinder pushrod length is

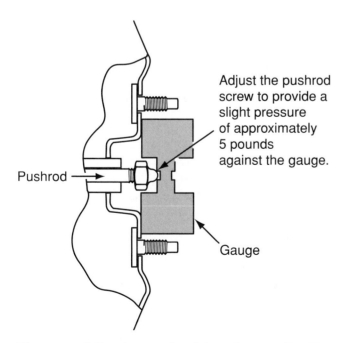

Figure 6.   Adjusting pushrod length on a Bendix-type booster with a gauge.

Adjust the pushrod screw to provide a slight pressure of approximately 5 pounds against the gauge.

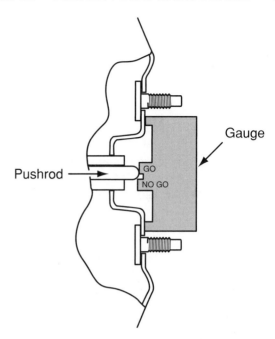

Figure 7.   The adjustment gauge for the pushrod in a Delphi Chassis–type vacuum booster.

fixed. If the pushrod length needs to be adjusted after master cylinder or booster service, an adjustable pushrod must be installed. Check the pushrod length with the booster on or off the vehicle as follows:

1. With the pushrod fully seated in the booster, place the go/no-go gauge over the pushrod **(Figure 7)**.
2. Slide the gauge from side to side to check the pushrod length. The pushrod should touch the longer "no-go" area and miss the shorter "go" area.
3. If the pushrod is not within the limits of the gauge, replace the original pushrod with an adjustable one. Adjust the new pushrod to the correct length.
4. Install the vacuum booster and check the adjustment. The master cylinder compensating port should be open with the engine running and the brake pedal released.

AUXILIARY VACUUM PUMPS

Some vehicles have an auxiliary pump to provide the vacuum necessary to operate the vacuum brake booster. These auxiliary vacuum pumps are usually the diaphragm or vane type. Auxiliary vacuum pumps are driven by an electric motor or by a belt, gear, or cam from the engine.

If you suspect that a belt-driven pump is performing poorly, check for worn or misadjusted belts or pulleys. Connect a vacuum gauge to the pump's vacuum port. The gauge should read approximately 15 inches of mercury after running the engine at idle for about thirty seconds. If the pump is overly noisy or does not have the correct or desired output, replace it.

Gear-driven vacuum pumps are diaphragm pumps that require no maintenance. Replace the entire pump if approximately 20 inches of mercury of vacuum cannot be attained at the pump with the engine running at idle.

Electric-Motor-Driven Pumps

A typical electric auxiliary vacuum pump consists of a small electric motor, a pump, and an electronic controller combined in a single assembly **(Figure 8)**. The controller has an integral on/off switch, a timer relay, and a piston and umbrella valve assembly. Vacuum hoses connect the vacuum pump to the vacuum-operated components. The vacuum pump outlet hose is usually connected to the intake manifold. A charcoal filter located in the outlet line prevents fuel vapors from entering the vacuum pump.

During low vacuum conditions, the vacuum pump motor is activated and generates auxiliary vacuum to help maintain the 14 inches of mercury of vacuum the brake booster requires. Once 14 inches of mercury of vacuum is present, the motor shuts off. A low-vacuum warning switch in the inlet hose line activates an instrument panel warning lamp when a low-vacuum condition exists.

Besides providing increased vacuum power for the brake system, an auxiliary vacuum pump provides extra vacuum for the heating and air-conditioning system, various emission controls, and cruise control.

Electric Pump Troubleshooting

Before troubleshooting the auxiliary vacuum pump, check the hydraulic system, vacuum hoses and connections, and the brake booster. If these appear to be fine, check for power at the pump motor's electrical connector to ensure the problem is in the pump motor, not the circuit.

The ports on a vacuum pump typically include the inlet port, a vacuum switch port, a pump housing port, and the pump outlet port. You will use a DMM and a vacuum gauge (or the gauge on a hand-operated vacuum pump) for most testing. Connect the gauge to the various ports on the auxiliary vacuum pump. Reading vacuum at these connections can pinpoint problems in the pump.

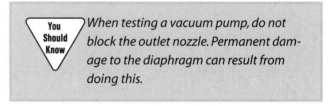

| You Should Know | *When testing a vacuum pump, do not block the outlet nozzle. Permanent damage to the diaphragm can result from doing this.* |

Typically, how you should check the pump depends on the suspected problem. Begin all testing by referring to the service manual's electrical wiring diagrams to locate

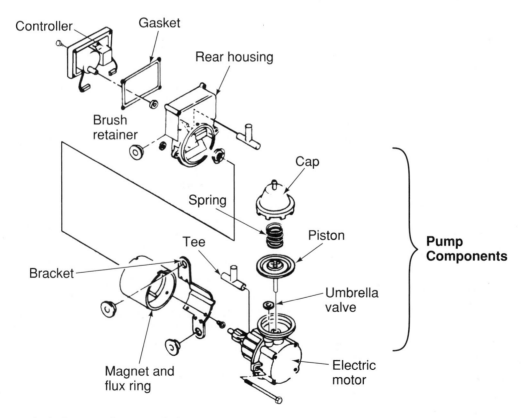

Figure 8.   An exploded view of a typical electric vacuum pump.

the proper power and ground terminals at the electric vacuum pump connector.

If excessive brake pedal effort is needed, the brake warning lamp is on, or the vacuum pump is inoperative:

1. Turn on the ignition. Disconnect the electrical connector from the vacuum pump. Use a digital voltmeter to check for battery voltage across the appropriate connector pins. Also check for proper ground. If battery voltage is present at the connector, proceed to step 2. If there is no power, check the fusible link, fuse, and wiring circuits and repair as needed.

2. Connect a 12-volt battery directly to the appropriate connector terminals and ground the circuit via the correct terminal pin. If the pump is still inoperative, proceed to step 3. If the pump now operates, check its output capacity.

3. Remove the vacuum pump. Disassemble and check for sticking or shorted brushes. Check the controller for damaged wires and repair or replace as needed.

If the vacuum pump operates but the brake warning lamp still lights and excessive brake pedal effort is needed:

1. Remove the pump inlet hose and connect a vacuum gauge (or pump) to the fitting **(Figure 9)**. Turn on the ignition.

2. If the gauge indicates a vacuum of 10–15 inches of mercury and the electric pump runs for five to ten seconds, check for vacuum leaks in other parts of the system. If the pump runs intermittently and fails to maintain steady vacuum, proceed to step 3.

3. Remove the auxiliary vacuum pump from the vehicle. Remove the controller and vacuum connector from the pump. Connect a hand-operated vacuum pump to the vacuum switch inlet port and apply approximately 20 inches of mercury of vacuum. If the switch leaks more than 2 inches of mercury per minute, replace the controller. If the switch holds vacuum, go to step 4.

4. Connect the hand-operated vacuum pump to the pump inlet. Apply 20 inches of mercury of vacuum. If the pump leaks more than 2 inches of mercury per minute, the pump umbrella valve is leaking and must be replaced. If the pump holds vacuum, proceed to step 5.

5. Connect the hand-operated vacuum pump to the pump outlet. Plug the pump housing inlet and apply 20 inches of mercury of vacuum. If the pump leaks more than 2 inches of mercury per minute, check for a loose bonnet. Repeat the test. If the pump continues to leak, replace the piston. If the pump holds vacuum, the system is operating properly.

If the vacuum pump runs continuously with normal or excessive brake pedal effort and the brake warning light is on or off:

1. Check all vacuum hoses for leaks and repair or replace as needed. If no leaks are found, go to step 2.

2. With the ignition off, disconnect the inlet hose from the vacuum connection. Attach the hand-operated vacuum pump and turn on the ignition. If the pump runs and then stops when the vacuum reaches 10–15 inches of mercury, the system is normal. Bleed off the vacuum to below 10 inches of mercury. At levels below 10 inches of mercury, the pump should start, run until the vacuum exceeds 10–15 inches of mercury, and automatically shut off. If the pump runs continuously and draws a vacuum of 10 inches of mercury or greater, the controller is defective and should be replaced.

Electric Pump Component Service

If the pump or related part is found to be defective, the pump or that particular part must be replaced. At times, it might be necessary to raise the vehicle on a lift or safety stands to gain access to the pump. If you need to do this, make sure the vehicle is resting securely on the stands or hoist before beginning to remove the pump. Remove any protective shields that might surround or be close to the pump. Then carefully disconnect the electrical connector and vacuum hoses from the pump. Remove all retaining nuts and carefully remove the pump. To install a new pump, follow the same steps but in reverse order.

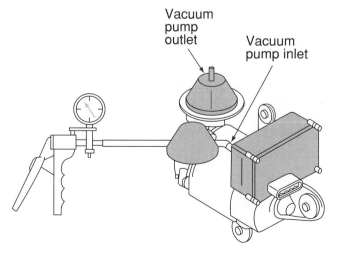

Figure 9. Checking the capacity of an electric vacuum pump with a vacuum gauge.

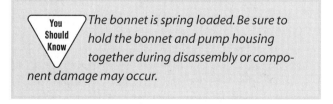

The bonnet is spring loaded. Be sure to hold the bonnet and pump housing together during disassembly or component damage may occur.

To remove and replace the pump's piston assembly, remove the pump's upper and lower shields. Carefully release the tabs of the cap and then remove the cap and spring. Remove the piston from the housing and the umbrella valve. Replace the umbrella valve, piston assembly, and spring. Place the bonnet over the spring and compress the spring and bonnet onto the pump housing. Bend the bonnet's tabs around the pump housing, making sure the bonnet is secure. Reinstall the shields and install the pump.

To remove the electronic controller, first remove the vacuum pump. Then remove all protective shields from around the controller and remove the fasteners that hold the rear housing and pump housing together. Carefully separate the rear housing from the pump housing and remove the fasteners securing the controller to the rear housing. Remove the brushes from the brush holders; then remove the controller, gasket, and washer.

Install the controller with a new gasket and washer. Place both brushes in the rear brush holder cavity. Preload the brush springs by locking them in the slots provided above the spring access slots. Position the brushes in channels. Install the brush retainer in the proper position and return the springs to the load position. Install the rear housing on the pump housing, replace all controller shields, and reinstall the vacuum pump.

Summary

- Any condition that reduces the amount of vacuum formed by the intake strokes of the engine will affect power brake performance.
- Insufficient manifold vacuum, leaking or collapsed vacuum lines, punctured diaphragms, or leaky piston seals can cause weak booster operation. A steady hiss when the brake is held down indicates a leak that also can cause poor booster operation. Hard brake pedal is usually the first signal that the booster is failing.
- If the brakes do not release completely, there might be a misadjusted or misaligned connection between the booster's pushrod and the linkage for the brake pedal.

- If the booster is giving weak braking assistance or no assistance at all, a problem might exist with the vacuum supply to the unit or the vacuum check valve might be faulty.
- If the master cylinder's pushrod is too long, the master cylinder's piston will block the compensating port. If the pushrod is too short, the brake pedal will be low and the pedal stroke length will be reduced, resulting in a loss of braking power.
- Pushrod adjustment is usually checked with a gauge that checks the distance from the end of the pushrod to the booster shell.
- Auxiliary pumps should be checked for proper operation and output capacity.

Review Questions

1. Which one of the following is not a common way to check a vacuum check valve?
   A. Filling the booster end of the valve with water and looking for leaks.
   B. Using a vacuum gauge to determine if it is restricted or stuck open or closed.
   C. Blowing through the intake manifold end of the valve to see if air passes through.
   D. Applying vacuum to the booster end of the valve with a vacuum pump to see if it can hold a vacuum.

2. While discussing the adjustment of a master cylinder's pushrod, Technician A says that if the pushrod is too short, the master cylinder's piston will block the compensating port. This prevents the hydraulic pressure from being released, which causes brake drag. Technician B says that if the pushrod is too long, the brake pedal will be low and the pedal stroke length will be reduced, resulting in a loss of braking power. When the brakes are applied with a short pushrod, groaning noises may be heard from the booster. Who is correct?
   A. Technician A only
   B. Technician B only
   C. Both Technician A and Technician B
   D. Neither Technician A nor Technician B

3. Brake fluid is found in the vacuum hose from the intake manifold to the booster. Technician A says that a faulty vacuum check valve is a likely cause. Technician B says that the master cylinder's secondary piston cup seal might be leaking. Who is correct?
   A. Technician A only
   B. Technician B only
   C. Both Technician A and Technician B
   D. Neither Technician A nor Technician B

4. Vacuum booster testing is being discussed. Technician A says that any condition that reduces engine vacuum will also reduce braking performance. Technician B

says that a blocked passage in the power piston, a sticking air valve, and a broken piston return spring are all causes of hard pedal. Who is correct?

A. Technician A only

B. Technician B only

C. Both Technician A and Technician B

D. Neither Technician A nor Technician B

5. While discussing the cause of weak braking assistance from a brake booster, Technician A says that a problem may be loose or kinked vacuum lines. Technician B says that the problem may be a ruptured diaphragm in the booster. Who is correct?

A. Technician A only

B. Technician B only

C. Both Technician A and Technician B

D. Neither Technician A nor Technician B

Chapter 22

Hydraulic Power Brakes

Introduction

Hydraulic power brake systems were developed as an alternative to using a vacuum pump for vacuum power brake systems. Hydraulic systems eliminated vacuum as a power source. Two basic types of hydraulic brake boosters are commonly used today: the Hydro-Boost and the Power-Master systems. The Hydro-Boost system operates with pressure from the power steering pump. The PowerMaster system has its own independent hydraulic power source driven by an electric motor.

HYDRO-BOOST SYSTEM

The Hydro-Boost power booster **(Figure 1)** fits in the same place as a vacuum booster, between the brake pedal and the master cylinder. Similar to a vacuum booster, the

Figure 1. A Hydro-Boost power brake unit.

Hydro-Boost unit multiplies the force of the driver's foot on the brake pedal.

The hydraulic pressure in the hydraulic booster should not be confused with the hydraulic pressure in the brake lines. Remember that they are two separate systems and require two different types of fluid: power steering fluid for the pump and brake fluid for the brake system. The brake hydraulic system uses DOT 3 or DOT 4 brake fluid as specified by the manufacturer. Never put power steering fluid in the brake reservoir, and always check the service manual for the proper type of fluid to put into the power steering unit.

> **You Should Know** *Do not mix brake fluid and power steering fluid. Power steering fluid is petroleum based and can damage the seals and cups in the brake hydraulic system. Brake system failure can result.*

The power steering pump develops pressure that is routed to the hydraulic booster assembly **(Figure 2)**. The hydraulic booster helps the driver apply force to the master cylinder pushrod. The booster has a large **spool valve** that controls fluid flow through the unit. When the brakes are applied, the spool valve directs pressure to a power chamber. The boost piston in the power chamber reacts to this pressure and moves forward to provide force to the master cylinder primary piston. The master cylinder operates in the same way as a conventional master cylinder. The boost pis-

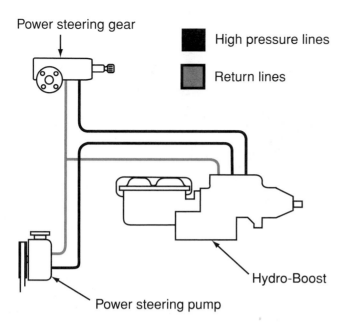

Figure 2. The Hydro-Boost system shares hydraulic fluid pressure with the power steering system.

ton in the Hydro-Boost does the same job as the vacuum diaphragm and power piston in a vacuum unit.

Spool Valves

Before examining the operation of the booster, you must understand how a spool valve works. Spool valves **(Figure 3)** are used in many hydraulic components, such as the valve body of an automatic transmission. A spool valve is a round bar that has been ground with highly polished surfaces called lands. These lands have specific widths and affect how the valve works. In between the lands are **annular** grooves, sometimes called **valleys**. The grooves (valleys) allow fluid flow around the spool valve. The valve lands fit in a bore with a very close clearance. This fit controls how the spool valve seals and moves in the bore. Only enough fluid is allowed between the bore and lands to provide lubrication.

HYDRO-BOOST CONSTRUCTION AND OPERATION

The Hydro-Boost booster is made up of a spool valve and sleeve assembly, a lever assembly, an input rod, a power piston, and an output pushrod. The input rod is connected to the brake pedal. The output pushrod goes into the master cylinder to push on its primary piston.

Three hydraulic lines are connected to the Hydro-Boost booster. The power steering pump provides hydraulic fluid under pressure to the booster through one of the lines. A second line routes pressurized fluid from the Hydro-Boost booster to the power steering gear. The third line is a low-pressure return hose from the booster to the steering pump reservoir.

Brakes Released

Figure 4 shows the booster in the unapplied position (brake pedal released). Fluid from the power steering pump

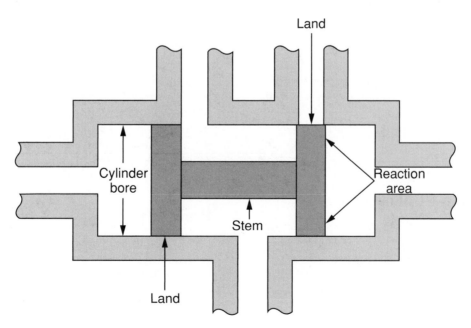

Figure 3. A typical spool valve.

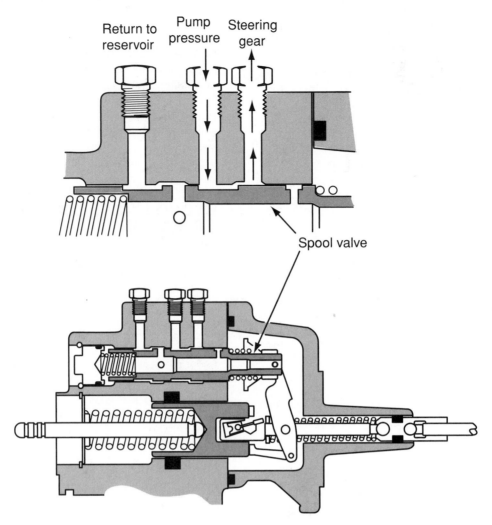

Figure 4. When the brake pedal is released or not applied, pressurized fluid flows through the Hydro-Boost valve to the steering gears.

enters the booster and is directed to the spool valve. The spool valve is held rearward by its spring so that the lands and grooves direct the fluid from the pump directly through the valve to the steering gear. The spool valve also opens a passage to vent the power cavity back to the reservoir. The power cavity has no pressure, so no force is applied to the output pushrod.

Moderate Brake Application

Moderate brake application causes the input pushrod to press on the reaction rod in the end of the power piston. The reaction rod moves and causes the lever to pivot on the power piston and move the spool valve forward in its bore **(Figure 5)**. This opens the pump inlet port to the power chamber and closes the vent port. Hydraulic pressure now increases in the power chamber and moves the power piston forward to operate the master cylinder.

As the spool valve moves forward, it also restricts fluid flow out to the steering gear. Closing the vent port and restricting flow to the steering gear causes pressure to rise in the brake booster. The farther the valve moves, the more the flow to the steering gear is restricted and the more the booster pressure increases. At maximum boost, hydraulic pressure in the brake booster can exceed 1400 psi.

Brakes Holding

As long as the driver maintains unchanging foot pressure on the pedal, the input pushrod does not move. As pressure rises to this holding point, the spool valve moves back toward the rear of its bore. This closes the fluid inlet port and reopens the bypass port to the steering gear. The vent port stays closed, however, so pressure in the power chamber reaches a steady state.

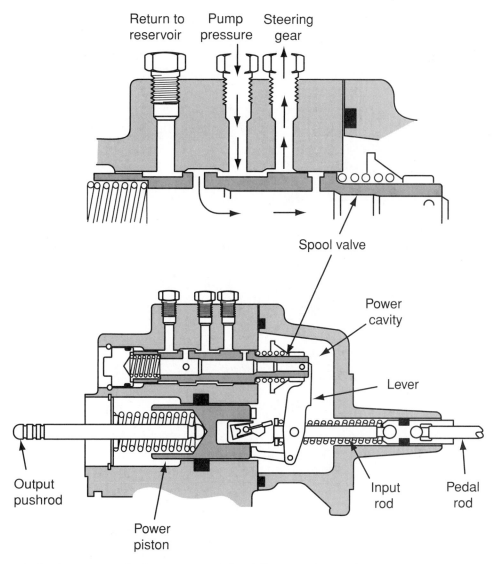

Return to reservoir   Pump pressure   Steering gear

Spool valve

Power cavity

Lever

Output pushrod

Power piston

Input rod

Pedal rod

Figure 5. When the brakes are applied, pressurized fluid flows to the Hydro-Boost power cavity and to the steering gears.

Brakes Being Released

When the brake pedal is released, the spool valve moves fully rearward to the unapplied position and vents pressure from the pressure chamber. Return springs on the power piston and lever quickly return the piston to the released position. Spool valve movement also blocks the fluid inlet port to the power chamber and lets fluid bypass the booster and flow to the steering gear.

Reserve Brake Application

A failure in the power steering system, such as a broken power steering hose, a broken power steering drive belt, or a pump failure could cause a loss of pressure to the Hydro-Boost system. The Hydro-Boost has a backup system powered by an accumulator that allows two or three power brake applications. When the pressure from the accumulator is needed, the gas or spring pressure inside the accumulator moves the pressurized hydraulic fluid to the brake booster.

Figure 6 shows a Hydro-Boost booster with an accumulator. During normal operation, the pump pressure fills the accumulator cavity in front of the piston. The piston compresses the accumulator spring. A check ball and plunger in the passage to the accumulator allow pressure in but prevent it from escaping. This keeps the spring compressed, or charged. A loss of pressure from the pump

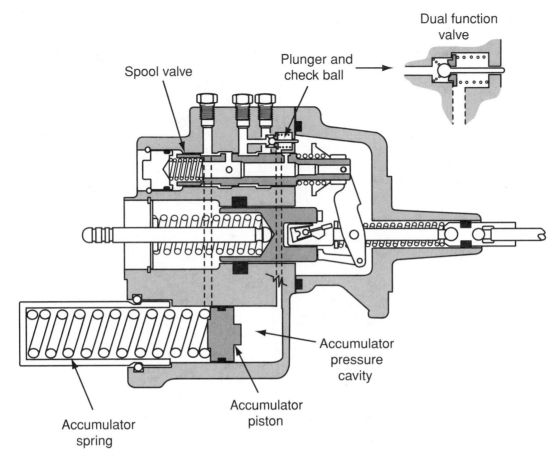

Figure 6. The accumulator provides reserve hydraulic pressure if the hydraulic supply fails.

opens a passage for the pressure stored in the accumulator to be routed to the power chamber to help the driver apply the brakes.

Loss of hydraulic power from the Hydro-Boost means only that the driver loses power assist. It does not mean that there is a loss of braking. A mechanical connection exists from the brake pedal through the pedal rod, through the input rod, through the power piston to the output pushrod. The driver's pedal effort will increase, but the brakes will still work.

> **You Should Know** *The parts and fluid inside an accumulator are under high pressure. If you attempt disassembly without discharging the pressure safely, parts and fluid could fly out and cause injury. Do not try to disassemble an accumulator without specific manufacturer's discharging instructions. Always wear eye protection when working around an accumulator.*

POWERMASTER CONSTRUCTION AND OPERATION

The potential problems of a power brake booster sharing a hydraulic system with the power steering led to the development of the PowerMaster system. These systems require DOT 3 brake fluid, which provides adequate lubrication for the pump. Fluid is supplied by the master cylinder reservoir both for normal operation and for the Power-Master booster.

The PowerMaster master cylinder has three or four fluid partitions instead of the two found in other dual master cylinders. Two reservoir chambers serve the wheel brakes as in a common dual-chamber master cylinder. The largest reservoir chamber contains fluid for the Power-Master booster. The bottom of this chamber has two ports: one supplies fluid to the pump, and the other is a return port from the booster. The fluid level in the PowerMaster booster chamber or the reservoir always looks as if it were low except when the accumulator is completely discharged. Do not add fluid to this chamber unless the accumulator is discharged or the fluid might overflow when the accumulator does discharge.

Although the PowerMaster booster uses the same DOT brake fluid as the wheel brakes, fluid should never be interchanged between the booster chamber of the reservoir and the chambers for the wheel brakes. The rubber diaphragm in the reservoir cover isolates only the wheel brake chambers from the atmosphere. Fluid flows to and from the booster chamber much faster and in greater volume than it does from the wheel brake reservoir chambers. Therefore, the booster reservoir chamber is vented to the atmosphere to prevent pressure or vacuum locking in the reservoir. Moisture absorbed from the air is not critical to booster operation because fluid for the booster is not subjected to high temperatures that could cause it to boil.

The PowerMaster system has an accumulator and a pressure switch mounted on the power booster section of the master cylinder. The accumulator stores fluid under pressure on one side of a diaphragm. Pressure is created by high-pressure nitrogen gas on the other side of the accumulator diaphragm. In the PowerMaster system, the pump charges the accumulator with fluid; and the accumulator provides boost pressure for *all* brake operation, not just for emergency operation. The accumulator pressure switch turns the pump on and off to maintain accumulator fluid pressure from 510 to 675 psi. A check valve retains pressure in the accumulator when the pump is off. The accumulator will provide boost for emergency braking until accumulator pressure is discharged if the pump motor fails. The pressure switch also is connected to a warning lamp on the instrument panel to warn the driver if there is a problem with the pressure in the accumulator. The assembly controlling the fluid is called the discharge or inner control valve. The discharge valve fits inside the apply valve. Chamfered surfaces on the two valves form the valve seats.

The PowerMaster system has four stages of operation: brakes not applied, moderate brake application, brakes holding, and brakes being released.

Brakes Released

When the brakes are not being applied, the assembly's outer apply valve is held closed by a spring. With this valve closed, fluid cannot get into the booster cavity and pressurize the power piston. A spring also holds the inner discharge valve open. If any fluid leaks past the closed apply valve, the open discharge valve directs it back to the reservoir **(Figure 7)**.

Brakes Applied

When the brake pedal is depressed, the input pushrod moves to close the discharge valve **(Figure 8)**, and fluid cannot return to the reservoir. The pedal pushrod continues to move forward and then opens the apply valve to allow fluid to flow to the booster cavity behind the power piston.

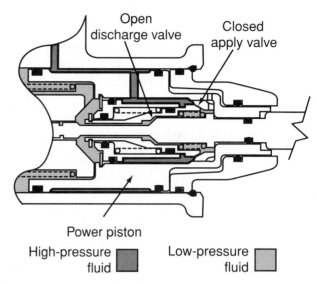

High-pressure fluid ▮ Low-pressure fluid ▯

Figure 7. PowerMaster valve positions when the brake pedal is not depressed and the system is at rest.

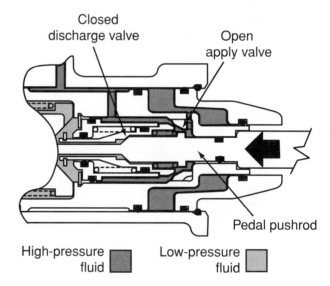

High-pressure fluid ▮ Low-pressure fluid ▯

Figure 8. The valve positions in a PowerMaster unit when the brake pedal is depressed.

The power piston provides a force to the master cylinder to help the driver apply the brakes.

Brakes Holding

As long as the driver maintains constant foot pressure on the pedal, the input pushrod does not move. As pressure rises to this holding point, the apply valve seats and prevents any more pressure from entering. The discharge port stays closed, and pressure in the booster reaches a steady state. The booster always seeks the holding position when pedal force is constant, or unchanging.

Pedal feedback is transmitted through the discharge valve and input pushrod to the brake pedal. If application force increases, the cycle repeats until the booster again reaches the holding position. During a panic stop, the valve opens fully and provides a high-pressure power assist for the brakes.

Brakes Being Released

When the driver releases the brake pedal, the pushrod force no longer holds the apply valve open. The apply valve spring closes the apply valve and prevents additional pressure from entering the booster cavity. Then the discharge valve spring opens the discharge valve, allowing booster cavity pressure to bleed off quickly to the fluid reservoir **(Figure 9)**. Because no pedal force or booster pressure is present, the master cylinder piston springs return the master cylinder pistons to the unapplied position.

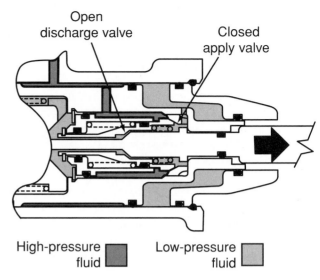

Figure 9.   PowerMaster valve positions and fluid flow right after the brake pedal is let up.

Summary

- The Hydro-Boost system operates with pressure from the power steering pump.
- The PowerMaster system has its own independent hydraulic power source driven by an electric motor.
- Pressure from the power steering pump is routed to the hydraulic booster assembly. The hydraulic booster has a large spool valve that directs pressure to a power

chamber when the brakes are applied. The boost piston in the power chamber reacts to this pressure and provides a force to the master cylinder primary piston.

- The PowerMaster system has apply and discharge valves that control a power piston. The power piston controls the hydraulic boost to the pistons of the master cylinder.

Review Questions

1. Which of the following statements about Hydro-Boost systems is *not* true?
   A. When the brake pedal is not depressed, pressurized fluid flows through the Hydro-Boost valve to the steering gears.
   B. When the brakes are applied, pressurized fluid flows to the Hydro-Boost power cavity and to the steering gears.
   C. When the brake pedal is released, the spool valve moves fully rearward to the unapplied position and vents pressure from the pressure chamber.
   D. A failure in the power steering system, such as a broken power steering hose, a broken power steering drive belt, or a pump failure could cause a loss of brakes.

2. Technician A says that application pressure can be stored by a spring in an accumulator. Technician B says

that application pressure can be stored by gas under pressure in an accumulator. Who is correct?
   A. Technician A only
   B. Technician B only
   C. Both Technician A and Technician B
   D. Neither Technician A nor Technician B

3. While discussing a Hydro-Boost power brake system, Technician A says that the brake hydraulic system uses DOT 3 or DOT 4 brake fluid as specified by the manufacturer. Technician B says that the power steering pump forces fluid into the master cylinder's reservoir. Who is correct?
   A. Technician A only
   B. Technician B only
   C. Both Technician A and Technician B
   D. Neither Technician A nor Technician B

4. While discussing a PowerMaster power brake system, Technician A says that the brake hydraulic system uses

DOT 3 brake fluid as specified by the manufacturer. Technician B says that the electric pump mounted on the master cylinder forces fluid into the master cylinder's reservoir. Who is correct?

A. Technician A only
B. Technician B only
C. Both Technician A and Technician B
D. Neither Technician A nor Technician B

5. The hydraulic system of the Hydro-Boost and Power-Master is being discussed. Technician A says that the Hydro-Boost uses a check valve to direct fluid. Technician B says that the PowerMaster uses a discharge and apply valve in the power piston to direct fluid. Who is correct?

A. Technician A only
B. Technician B only
C. Both Technician A and Technician B
D. Neither Technician A nor Technician B

Chapter 23

Hydraulic Power Brake Service

Introduction

It is important to remember that a hydraulic power brake system works with the normal hydraulic brake system and is not part of it. Therefore, all diagnosis and service to the wheel brake system is the same except for a few procedures. These are covered in this chapter. Also covered are specific service procedures for both Hydro-Boost and PowerMaster power brake systems.

HYDRO-BOOST SYSTEM INSPECTION

Before any detailed testing of Hydro-Boost components, check basic engine conditions and power steering operation that could affect Hydro-Boost performance. If both brake application and steering require more than normal effort, the cause is probably related to fluid pressure and delivery.

> ▽ **You Should Know** *Before loosening or disconnecting any system lines or covers, clean them and the surrounding area to prevent dirt from entering into the system. Dirt and other contaminants can damage the system.*

1. Inspect the fluid level in the power steering pump. Some pump reservoirs have dipsticks marked to check the fluid only when at normal operating temperature. Others have dipsticks with fluid level markings for both warm and cool fluid.

2. Also inspect the condition of the power steering fluid. If it is dirty, smells burned, or appears to be contaminated in any way, flush and refill the power steering system before proceeding further.
3. Inspect the condition of the power steering pump drive belt and replace it if it is cracked, glazed, grease soaked, or otherwise badly worn. Check the tension of a V-belt or be sure that a serpentine belt is installed properly and correctly positioned on the belt tensioner.
4. Inspect all hoses and lines in both the Hydro-Boost system and the power steering system for leakage.

> ▽ **You Should Know** *Do not hold the steering wheel at full lock for more than five seconds while testing system pressure. System damage can result from prolonged high-pressure operation.*

5. To verify a leak, have an assistant run the engine at fast idle and alternately apply the brakes and turn the steering from lock to lock. These actions develop high pressure and will force fluid from small leaks. Tighten connections or replace lines and hoses as required.
6. If you find leakage around the pump, clean and tighten all fittings and bolts. If the leak continues, rebuild or replace the pump. **Figure 1** shows common leak locations in a power steering pump and in a steering gear.

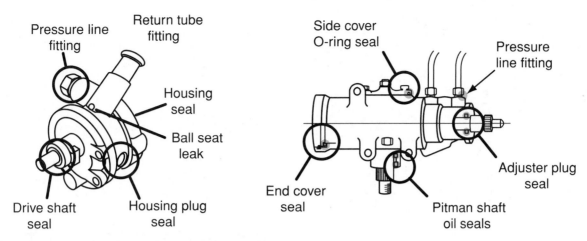

Figure 1. Possible fluid leakage points in a power steering system.

7. Check the brake fluid level in the master cylinder and add fluid if necessary. If there is air in the brake system and there is a spongy pedal, it will be difficult or impossible to accurately troubleshoot Hydro-Boost operation. Bleed the brake system if you suspect air in the system.

8. Check engine idle speed and adjust it if necessary. Also check for engine speed control by the powertrain control module (PCM) as you turn the steering to full lock.

HYDRO-BOOST SYSTEM DIAGNOSIS

A Hydro-Boost system (**Figure 2**) requires a continuous supply of power steering fluid, from the power steering pump, at the proper pressure and volume. Besides inspecting power steering components, you might need to measure the pressure developed by the pump. To do this, use

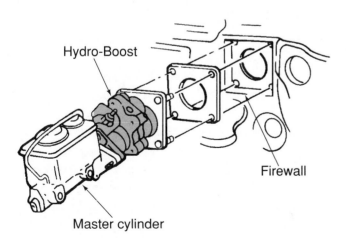

Figure 2. The Hydro-Boost booster is mounted between the master cylinder and the firewall in the engine compartment.

the instructions and specifications from the manufacturer of the pressure-test equipment and those found in the vehicle's service manual.

The power steering pump should not be operated without fluid in the reservoir, or damage to the pump's bearings and seals can result. If the pump has failed, check the fluid carefully for abrasive dirt and metal particles that can damage the Hydro-Boost unit. If you find such particles in the fluid from any source, flush the system thoroughly. Depending on the severity of the contamination, you might need to remove the booster and disassemble it to clean it completely. Also, flush the power steering and Hydro-Boost lines thoroughly to remove particulate contamination.

Troubleshooting Basics

If you suspect that the problem is in the steering system, refer to that section of your service manual and diagnose that system. Problems in the power steering system should be repaired before continuing your diagnosis of the power brake system.

Certain noises often occur with the Hydro-Boost system and can be cause for customer complaint. Some noises are normal and usually occur for a short time. Other noises are a sign of wear in the system or the presence of air in either the Hydro-Boost booster or the steering system.

Booster Fluid Leakage

Figure 3 shows possible leakage points at seals in the Hydro-Boost booster. If the booster is leaking at any of these points, it is often most practical to replace the booster with a new or rebuilt unit. All of these seals can be replaced individually, however, with seal repair kits. Most require that the booster be removed from the vehicle and disassembled. The spool valve plug seal and the accumulator seal can be replaced with the booster installed in the

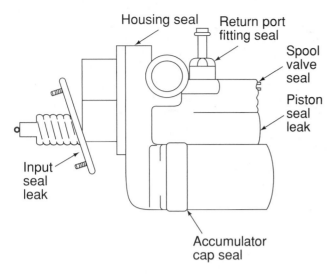

Figure 3. Possible leakage points in a Hydro-Boost brake booster.

vehicle if component access permits. If leakage appears around the return port fitting, torque the fitting to specifications. If leakage continues, replace the O-ring under the fitting.

Basic Operational Test

Check the overall operation of a Hydro-Boost system by turning the engine off and pumping the brake pedal several times to bleed off the residual hydraulic pressure stored in the accumulator. Then firmly hold the brake pedal down and start the engine. If the system is working correctly, the brake pedal should move downward and then push up against your foot.

> **You Should Know** *When fully charged, the accumulator holds more than 1000 psi of hydraulic pressure. Before removing the Hydro-Boost booster, discharge the accumulator by firmly applying the brake pedal at least six times with the engine off. Failure to discharge the accumulator could cause serious injury.*

Accumulator Test

To test the operation of the accumulator:
1. With the engine running, rotate the steering wheel until it stops and hold it in that position for no more than five seconds.
2. Return the steering wheel to the center position and shut off the engine.

3. Pump the brake pedal. You should feel two to three power-assisted strokes.
4. Now repeat steps 1 and 2. This will pressurize the accumulator.
5. Turn the engine off and wait one hour; then pump the brake pedal. There should be one or two power-assisted strokes.

Bad valves are the most common accumulator problem. If the valves are leaking, the accumulator might not hold a charge or might hold it for only a short period. In either case, the booster must be replaced or disassembled and the valves replaced.

HYDRO-BOOST BOOSTER SERVICE

A Hydro-Boost unit can be overhauled or rebuilt by installing the parts included in an overhaul kit. In many cases, however, it is more practical to replace the booster with a new or rebuilt unit.

The procedure for removing a typical Hydro-Boost booster follows. Always check the service manual for the exact procedure for doing this.
1. Disconnect the master cylinder from the booster, but leave the service brake hydraulic lines connected to the master cylinder.
2. Carefully lay the master cylinder aside, being careful not to kink or bend the steel tubing. Support the master cylinder from a secure point on the vehicle with wire or rope. Do not support the master cylinder on the brake lines.
3. Disconnect the hydraulic hoses from the booster ports. Plug all tubes and the booster ports to prevent fluid loss and system contamination.
4. Detach the pedal pushrod from the brake pedal. Remove the nuts and bolts from the booster support bracket and remove the booster from the vehicle.

> **You Should Know** *Do not try to disassemble or cut into the accumulator. The accumulator contains compressed gas or a spring compressed under high pressure. When removing the accumulator from the booster, always use the proper tools and follow shop manual procedures to avoid injury. If the accumulator is inoperative, damaged, or otherwise defective, replace it as an assembly. Do not apply heat to the accumulator or try to repair it. Dispose of an inoperative gas-charged accumulator by drilling a $1/16$-inch-diameter hole through the end of the accumulator opposite the O-ring.*

The installation procedure is reverse of removal. Make sure you keep all openings, fittings, and lines free of dirt. If the booster was disassembled, clean all parts with denatured alcohol or clean power steering fluid. Do not use transmission fluid, brake fluid, or petroleum-based solvent. If the parts are left exposed for eight hours or more, they must be recleaned at the time of assembly.

Flushing a Hydro-Boost System

If the hydraulic fluid in the Hydro-Boost system becomes contaminated, flush the entire system with clean power steering fluid; make sure the fluid is the type specified by the manufacturer.

1. Raise the front of the vehicle on a hoist or on safety stands so that the front wheels are off the ground.
2. Disconnect the low pressure hydraulic line of the system. This is usually the low-pressure return line at the reservoir. Route the disconnected hydraulic line into a large drain container.
3. Disable the ignition system or disconnect the power supply to the injection pump on a diesel engine. Connect a remote starter switch to the starter circuit.
4. Crank the engine for intervals no longer than thirty seconds while applying the brakes and turning the steering wheel. While cranking, have an assistant steadily add power steering fluid to the reservoir to keep the fluid level near the full mark.
5. Flush approximately three quarts of fluid through the system. Continue flushing until no dirt or contamination is seen in the draining fluid. If foreign material still shows in the fluid, disassemble and clean the booster, hydraulic pump, and power steering gear. If necessary, replace seals and hoses according to the manufacturer's instructions. After new parts are installed, repeat the flushing process.
6. After flushing, reconnect all lines and hoses and fill the pump reservoir with fluid.
7. Lower the front of the vehicle to the ground, start the engine, and turn the steering from lock to lock several times.
8. Stop the engine, check the fluid level, and add fluid if necessary. Proceed to the air-bleeding procedure.

Bleeding a Hydro-Boost System

Whenever the Hydro-Boost booster or power steering components are removed and reinstalled, the hydraulic system must be bled to rid the system of all air. Air can also enter the Hydro-Boost system if the fluid level drops below the minimum level in the pump reservoir. Air can also be drawn into the fluid through loose fittings in hydraulic lines and hoses.

Check the steering fluid for signs of air. Aerated fluid looks milky. The level in the steering fluid reservoir will also rise when the engine is turned off if air has been compressed in the fluid. If the fluid has air in it that cannot be purged using the following procedure, the problem may lie in the steering system pump.

1. Fill the power steering pump reservoir to the full mark and allow it to sit undisturbed for several minutes.
2. Start the engine and run it for approximately one minute.
3. Stop the engine and recheck the fluid level. Repeat steps 1 and 2 until the level stabilizes at the full mark.
4. Raise the front of the vehicle on a hoist or safety stands.
5. Turn the wheels from lock to lock. Check and add fluid if needed.
6. Lower the vehicle and start the engine.
7. Apply the brake pedal several times while turning the steering wheel from lock to lock.
8. Turn off the engine and pump the brake pedal five or six times.
9. Recheck the fluid level. If the steering fluid is extremely foamy, allow the vehicle to stand for at least ten minutes with the engine off. Then repeat steps 7 through 9 until fluid in the pump reservoir is clear and free of air bubbles.

POWERMASTER SYSTEM DIAGNOSIS

Before suspecting a PowerMaster problem or troubleshooting the system, verify that the service brakes are working properly. The PowerMaster system cannot cause a low or spongy brake pedal; it cannot cause brake pull or pulsation, and it cannot cause noises such as squealing or grinding sounds. Like any other power booster, however, PowerMaster problems can increase brake pedal effort.

Some PowerMaster tests require running the pump **(Figure 4)** to pressurize the system. To prevent possible

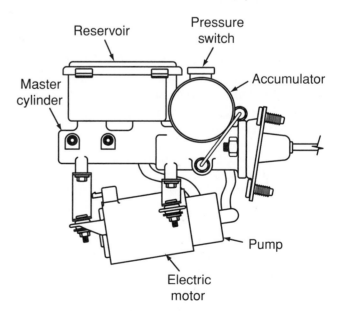

Figure 4.   The PowerMaster pump is built into the vehicle's master cylinder.

pump overheating and damage, do not let the pump run for more than twenty seconds at a time.

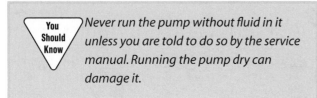

> You Should Know
>
> *Never run the pump without fluid in it unless you are told to do so by the service manual. Running the pump dry can damage it.*

Checking Fluid Level

If you suspect a PowerMaster problem, first check the fluid level in the reservoir and add fluid if necessary. The master cylinder reservoir has chambers for the service brakes as well as the PowerMaster system. When the accumulator is fully charged, it is normal for its reservoir chamber to be almost empty, with the fluid level barely covering the ports at the bottom of the chamber. Too much fluid in the reservoir would cause an overflow the next time the accumulator discharged.

To check the PowerMaster fluid level, leave the reservoir cover installed and the ignition switch off. Discharge the accumulator, then remove the reservoir cover. Accumulator fluid level with the accumulator discharged should be between the maximum and minimum markings on the reservoir.

Leakage Inspection

Leakage from a PowerMaster system usually is easy to see because the accumulator is constantly pressurized to 500 psi (3450 kilopascals). Even a small leak will cause a rapid fluid loss.

To pinpoint a leak, clean the PowerMaster assembly and all lines and hoses with a rag dampened in brake cleaner or alcohol. Discharge the accumulator and check the fluid level. Add fluid if needed. Turn the ignition on and let the pump run to charge the accumulator. Check for leaks at all fittings, at the accumulator, at the reservoir cover and grommets, at the pressure switch, and at the pedal pushrod under the instrument panel.

Tighten any fittings and components at leakage points and retest. If tightening the connections does not fix a leak, replace components. If you find leakage at the pedal pushrod, replace the PowerMaster booster and master cylinder assembly.

If you find a large amount of fluid spilled around the master cylinder reservoir, it might be due to overfilling the reservoir with the accumulator charged. In this case, discharge the accumulator and check and adjust the fluid level. Clean the outside of the reservoir and master cylinder thoroughly and recheck for leakage.

PowerMaster Operational Test

Test the general operation of the PowerMaster system as follows:

1. With the ignition off, apply and hold the brake pedal firmly. The pedal should move and then stay at a fixed height as it does for normal brake application. If it sinks to the floor under normal pressure, replace the PowerMaster booster and master cylinder assembly.
2. Discharge the accumulator. Pedal application should get firmer as the accumulator discharges.
3. Release the parking brake.
4. Turn the ignition on but do not start the engine. The brake warning lamp on the instrument panel should light and the PowerMaster pump should run for about twenty seconds to charge the accumulator. After about twenty seconds, the pump should stop and the lamp should turn off. If the pump does not run at all, check the electrical circuit. If the pump does not turn off, check the motor for a short circuit and retest the hydraulic system for leakage.
5. Leave the ignition on and the engine off for five minutes. Do not apply the brakes. If the pump motor runs without brake application, perform the internal leak test.

Pressure Tests

A complete pressure test of the PowerMaster system requires a special tool kit, which includes a 1000-psi pressure gauge, a gauge manifold with a bleeder valve, fittings, and a hose. To install the gauge, apply the brakes at least ten times with the ignition off to discharge the accumulator as explained previously. Then remove the pressure switch from the booster and install the gauge in the switch port. Finally, reinstall the pressure switch in the port provided in the gauge manifold.

To check the upper **limit pressure**, close the bleeder valve on the gauge manifold and turn on the ignition but do not start the engine. Let the pump run until it turns off; then check the gauge reading. The pump should turn off at a pressure from 635 to 735 psi. If the pressure exceeds 735 psi, replace the pressure switch. If the pump continues to run but never reaches the pressure range, replace the pump.

To check the lower limit pressure, close the bleeder valve on the gauge manifold and turn on the ignition but do not start the engine. Let the pump run until it turns off. Route the bleeder hose from the gauge manifold to the PowerMaster chamber of the master cylinder reservoir. Slowly open the bleeder valve and watch the gauge; note the pressure when the pump motor turns on. The pump should turn on at a pressure from 490 to 530 psi. If the pump turns on at a higher pressure or does not turn on at the lower pressure, replace the pressure switch.

To check the accumulator precharge pressure, discharge the accumulator. Note the pressure gauge reading just before the pedal becomes hard. This is the accumulator precharge pressure, which should be 200–300 psi.

Internal Leak Test

If the pump motor runs without brake application after a five-minute period with the ignition on and the engine off and no external leakage exists, fluid is bypassing valves inside the PowerMaster assembly. Check for this kind of internal leakage as follows:

1. Turn the ignition on and pump the brake pedal until the pump turns on.
2. Stop pumping the brakes and turn the ignition off after the pump stops.
3. Remove the reservoir cover and hold a clear plastic tube tightly over the fluid return port in the bottom of the PowerMaster fluid chamber. If fluid rises in the tube, the internal check valve is leaking; replace the Power-Master assembly. If fluid does not rise in the tube, go to step 4.
4. Hold the clear plastic tube tightly over the pump supply port in the bottom of the PowerMaster fluid chamber. If fluid rises in the tube, the internal control valves are leaking; replace the PowerMaster assembly.

POWERMASTER SERVICE

The PowerMaster **(Figure 5)** motor and pump, pressure switch, accumulator, pressure tube and hose, and master cylinder reservoir can be replaced individually with the assembly mounted to the vehicle.

 Before removing the PowerMaster assembly, discharge the accumulator. Failure to discharge the accumulator could cause serious injury.

PowerMaster Replacement

To remove and replace a typical PowerMaster assembly, follow these general steps:

1. Disconnect the brake lines from the master cylinder.
2. Disconnect the pushrod from the brake pedal arm.
3. Disconnect the wiring connector from the pressure switch and position any other wires out of the way for PowerMaster removal.
4. Remove the nuts that hold the assembly to the vehicle and remove the PowerMaster unit.

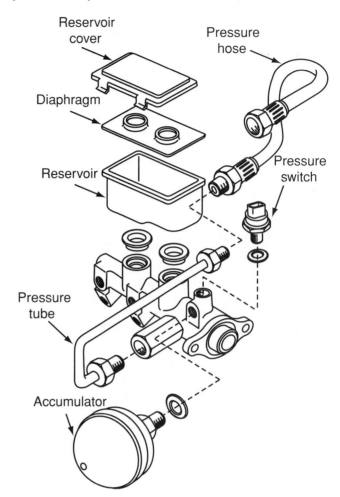

Figure 5. PowerMaster parts that can typically be replaced without removing the booster unit from the vehicle.

5. Bench bleed the master cylinder reservoirs of the replacement PowerMaster assembly.
6. Install the new assembly in the vehicle.
7. Connect the pushrod, the brake lines, and the electrical connector.
8. Fill and bleed the PowerMaster booster.

PowerMaster Filling and Bleeding

The service brakes on a vehicle with a PowerMaster system can be bled by common bleeding methods. After a PowerMaster assembly has been bench bled and installed, fill and bleed the booster section as follows:

1. Fill the booster chamber of the reservoir with fresh DOT 3 fluid.
2. Turn the ignition on and let the pump run for 20 seconds or until the fluid level drops to about ¼ inch above the bottom of the chamber.

3. Turn off the ignition. Add fluid, if required, and repeat step 2.
4. When the pump stops automatically, adjust the fluid level so that the ports in the bottom of the reservoir are just covered by fluid. Install the reservoir cover.
5. Turn the ignition off and pump the brake pedal at least ten times to exhaust accumulator pressure.
6. Remove the reservoir cover and adjust the fluid level to the maximum mark on the reservoir.

7. Reinstall the reservoir cover, turn the ignition on, and watch the fluid level in the reservoir. Add fluid if required to ensure that the ports remain covered and that air is not drawn into the system.
8. Repeat this procedure ten to fifteen times to ensure that all air is purged from the system. When the fluid level returns to the maximum mark on the reservoir consistently each time the accumulator is discharged, all air should be out of the system.

Summary

■ Before any detailed testing of Hydro-Boost components, check basic engine conditions and power steering operation that could affect Hydro-Boost performance.

■ Problems in the power steering system can affect the booster, and a problem in the booster can affect the steering.

■ Check the overall operation of a Hydro-Boost system by discharging accumulator pressure, then firmly hold the brake pedal down and start the engine. If the system is working correctly, the brake pedal should move downward and then push up against your foot.

■ If the hydraulic fluid in the Hydro-Boost system becomes contaminated, flush the entire system with clean power steering fluid.

■ Whenever the Hydro-Boost booster or power steering components are removed and reinstalled, the hydraulic system must be bled to rid the system of all air.

■ When you suspect a PowerMaster problem, first check the fluid level and condition in the reservoir and add fluid if necessary.

■ If the pump motor runs without brake application after a five-minute period with the ignition on and the engine off and no external leakage exists, fluid is bypassing valves inside the PowerMaster assembly.

■ A PowerMaster master cylinder and booster assembly should be bench bled before it is installed on the vehicle.

Review Questions

1. Which of the following statements about checking a PowerMaster system is *not* true?
   A. With the ignition off, pump the brake pedal at least ten times to discharge the accumulator. Pedal application should get firmer as the accumulator is discharged.
   B. With the ignition off, apply and hold the brake pedal firmly. The pedal should move and sink to the floor under normal pressure.
   C. Discharge the accumulator, then turn the ignition on but do not start the engine. The brake warning lamp on the instrument panel should light and the PowerMaster pump should run for about twenty seconds to charge the accumulator.
   D. With the ignition on and the engine off for five minutes, do not apply the brakes. If the pump motor runs without brake application, perform the internal leak test.

2. Hydro-Boost servicing is being discussed. Technician A says that whenever the Hydro-Boost booster is serviced the power steering system must be bled. Technician B says that power steering fluid and brake fluid are interchangeable in small amounts. Who is correct?
   A. Technician A only
   B. Technician B only
   C. Both Technician A and Technician B
   D. Neither Technician A nor Technician B

3. Hydro-Boost service is being discussed. Technician A cleans Hydro-Boost parts in clean transmission fluid or brake fluid. Technician B cleans the parts in a solvent cleaner. Who is correct?
   A. Technician A only
   B. Technician B only
   C. Both Technician A and Technician B
   D. Neither Technician A nor Technician B

4. While diagnosing a PowerMaster system, Technician A says that a faulty system can cause a low or spongy brake pedal. Technician B says that a faulty system can cause squealing or grinding sounds. Who is correct?
   A. Technician A only
   B. Technician B only
   C. Both Technician A and Technician B
   D. Neither Technician A nor Technician B
5. Hydro-Boost systems are being discussed. Technician A says that problems in the steering system can affect the booster. Technician B says that abnormal noises are a sign of wear in the system or the presence of air in either the Hydro-Boost booster or the steering system. Who is correct?
   A. Technician A only
   B. Technician B only
   C. Both Technician A and Technician B
   D. Neither Technician A nor Technician B

Section 5

Wheel Brake Units

SECTION OBJECTIVES

After you have read, studied, and practiced the contents of this section, you should be able to:

- Describe the basic parts of a drum brake assembly.
- Describe the types and functions of drum brake hubs.
- Describe the components that make up a wheel cylinder and the purpose of each one.
- Describe the two major drum brake designs and how they differ in operation.
- Describe the different types of self-adjusters used on duo-servo and leading-trailing shoe brake systems and how they function.
- Describe the basic components of a disc brake assembly.
- Describe the two different types of hub-and-rotor assemblies.
- Describe the two main categories of caliper designs and the variations in each category.
- Describe how a low-drag caliper operates compared with other types of calipers.
- Describe how rear disc brakes operate in relation to their emergency brake function.
- Name the types of brake pad wear indicators and how they operate.
- Name the types of friction material used on brake shoes and pads.

Interesting Fact

Automotive disc brakes were first developed for racing. The first practical disc brake system was used in 1953 on a C-type Jaguar that won the twenty-four-hour Le Mans endurance race. The efficiency and effectiveness of the brakes in this event led automotive engineers to develop disc brakes for passenger vehicles, and these became available in the 1960s.

Chapter 24

Drum Brake Construction

Introduction

For more than forty years, having drum brakes at all four wheels was the standard of the automobile industry. Although disc brakes are now found on the front wheels of nearly all cars and light trucks, drum brakes continue to be commonly used at the rear wheels. Drum brakes have certain features that have prevented them from becoming obsolete. They also have some characteristics that have made them less desirable than disc brakes.

Drum Brake Self-Energizing

One end of the lining of one shoe in a drum brake contacts the drum before the other end does and becomes a pivot point as friction quickly increases. The brake shoe becomes a **self-energizing** lever and adds its own mechanical leverage to hydraulic force to help the shoe move tightly against the surface of the brake drum **(Figure 1)**.

This self-energizing can occur in only one shoe of a drum brake. That shoe would be the leading shoe in relation to the direction of wheel rotation. When the self-energizing of one shoe applies mechanical force to the other shoe and helps it move against the wall of the brake drum, it is called servo action **(Figure 2)**. Self-energizing drum brakes that use servo action can always apply more stopping power for a given amount of pedal force than disc brakes can.

Lack of Noise

Drum brakes are almost noise free. Heavy return springs and holddown springs hold the brake shoes against the wheel cylinders, anchor pins, and backing plates to prevent rattles. Linings are securely bonded or riveted to the brake shoes, and the entire assembly is encased in the brake drum. About the only time that noise is a concern with drum brakes is when it is the sound of steel brake shoes grinding against the brake drum after the linings have worn away.

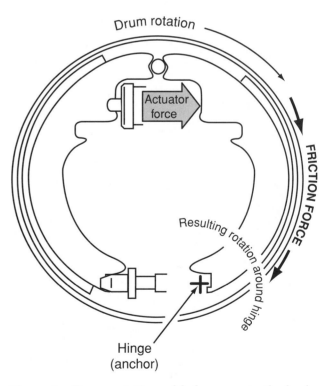

Figure 1. Drum rotation adds leverage to the brake shoe as it contacts the drum.

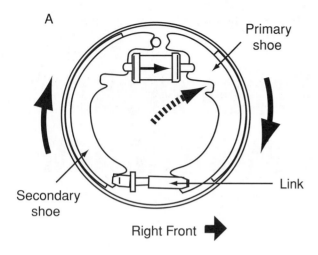

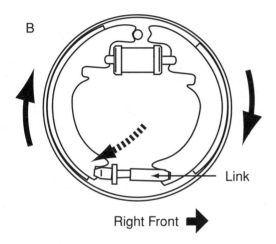

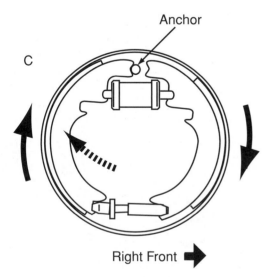

Figure 2. The self-energizing action of the primary shoe applies force onto the secondary shoe. This process is called servo action.

Parking Brake Operation

Drum brakes provide a better static coefficient of friction than disc brakes do. The brake linings grab and hold the drums more tightly than disc pads can hold a rotor. Additionally, the self-energizing and servo actions of drum brakes contribute to this feature, as does the larger surface area of brake shoe linings compared with the area of the linings on disc brake pads.

Compared with the complicated mechanisms required to mechanically apply parking brakes on rear disc brakes, parking brake linkage for drum brakes is quite simple. In almost all cases, the driver operates a pedal or a lever that pulls on cables attached to the rear brakes. The cables operate levers that mechanically apply the brake shoes. All mechanical motion is in a straight longitudinal line from the front to the rear of the vehicle.

Fade Resistance

Brake fade is the loss of braking power due to excessive heat that reduces friction between brake linings and the rotors or drums. One factor that contributes to heat dissipation and fade resistance is the swept area of the brake drum or rotor, which is the total area that contacts the friction surface of the brake lining. The greater the swept area, the greater the surface available to absorb heat. Although a brake drum can have a relatively large swept area, the entire area is on the inner drum surface. The swept area of a disc brake comprises both sides of the rotor. For any given wheel size, the swept area of a disc brake will always be larger than the swept area of a drum brake **(Figure 3)**. For example, a 10-inch-diameter rotor will have almost 50 percent more swept area than a 10-inch drum.

Mechanical fade is a problem that occurs with drum brakes when the drum becomes very hot and expands outward, away from the brake shoes. The shoes then must travel farther to contact the drum surface with normal braking force, and the pedal drops lower as the brakes are applied. The increased heat at the braking friction surfaces also reduces the coefficient of friction. The combined result is brake fade.

Lining fade occurs when the linings are overheated and the coefficient of friction drops off severely **(Figure 4)**. Some heat is needed to bring brake linings to their most efficient working temperature. The coefficient of friction rises, in fact, as brakes warm up. If temperature rises too high, however, the coefficient of friction decreases rapidly.

Water fade occurs when water is trapped between the brake linings and the drum or rotor and reduces the coefficient of friction. **Gas fade** occurs under hard braking when hot gases and dust particles are trapped between the brake linings and the drum or rotor and build up pressure that acts against brake force. These hot gases actually

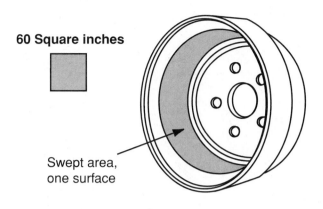

60 Square inches

Swept area,
one surface

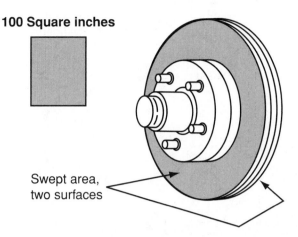

100 Square inches

Swept area,
two surfaces

Figure 3. For any given wheel size, a brake disc always has 35–50 percent more swept area than a brake drum to dissipate heat.

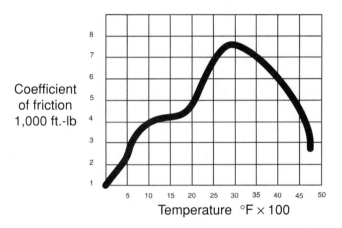

Coefficient
of friction
1,000 ft.-lb

Temperature °F × 100

Figure 4. Although the coefficient of friction of a lining material increases with heat, there is a temperature where it drops quickly and causes brake fade.

lubricate the friction surfaces and reduce the coefficient of friction.

Due to the design of the drum brake, it will trap heat, water, and gases inside the drum. This can cause the brake fluid to boil and can make the drum brake more susceptible to brake fade and pull.

Drum Brake Self-Adjustment

Brake adjustment compensates for lining wear and maintains the correct clearance between brake linings and the drum or rotor surfaces. Disc and drum brakes both require adjustment, but self-adjustment is a basic feature of the disc brake design. Drum brakes need extra cables, levers, screws, struts, and other mechanical linkage just to provide self-adjustment and proper lining-to-drum clearances.

Drum Brake Pulling and Grabbing

Servo action increases braking force at the wheels, but it must develop smoothly as the brakes are applied. If servo action develops too quickly or unevenly, brake grabbing or lockup will result. This is usually experienced as a severe pull to one side or the other or reduced vehicle control during braking.

DRUM BRAKE CONSTRUCTION

The basic parts of a drum brake are: a drum and hub assembly, brake shoes, a backing plate, a hydraulic wheel cylinder, shoe return springs, holddown springs, an adjusting mechanism, and a parking brake (**Figure 5**). The drum

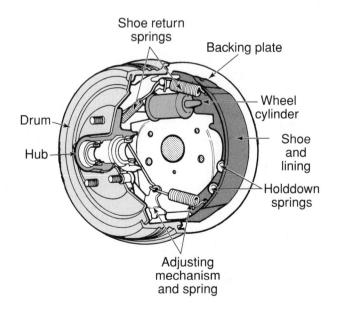

Shoe return
springs

Backing plate

Drum

Wheel
cylinder

Hub

Shoe
and
lining

Holddown
springs

Adjusting
mechanism
and spring

Figure 5. Basic parts of a drum brake assembly.

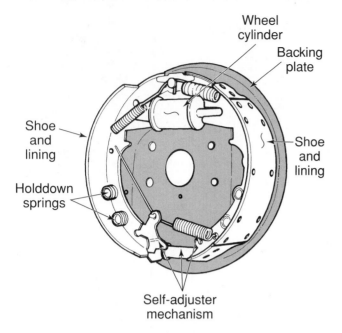

Figure 6. The hydraulic and friction components are attached to the backing plate.

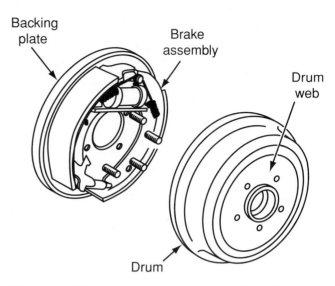

Figure 7. The major components of a drum brake.

and shoes provide the friction surfaces for stopping the wheel. The drum has a machined braking surface on its inside circumference. The wheel is mounted to the drum hub by nuts and studs. The hub and drum can be a one-piece casting, or the drum can be separate from the hub or axle flange and fit over the wheel studs for installation. The hub houses the wheel bearings that allow the wheel to rotate.

The hydraulic and friction components are attached to the **backing plate (Figure 6)**, which is mounted on the axle housing or suspension. The drum encloses and rotates over these components. When the brakes are applied, hydraulic pressure forces the pistons in the wheel cylinder outward. The pressure on the pistons is transmitted to the shoes as the shoes are forced against a pivot or anchor pin and into the rotating drum. Through this action, the shoes tightly wrap up against the drum to provide the stopping action.

The frictional energy on the drum surface creates heat. This heat is dissipated to the surrounding air as the drum rotates with the wheel. Some drums are finned to help them dissipate heat more easily.

BRAKE DRUMS AND HUBS

The brake drum mounts on the wheel hub or axle and encloses the rest of the brake assembly except the outside of the backing plate. Brake drums are made of cast iron, steel and cast iron, or aluminum with an iron liner. In all of these variations, iron provides the friction surface because

of its excellent combination of wear, friction, and heat-dissipation characteristics.

The drum is a bowl-shaped part with a rough cast or stamped exterior and a machined friction surface on the interior. The open side of the drum fits over the brake shoes and other parts mounted on the backing plate **(Figure 7)**. The closed side of the drum is called the **drum web** and contains the wheel hub and bearings or has mounting holes through which the drum is secured to the axle flange or separate hub.

Some iron and aluminum drums have fins cast into the outer circumference to aid in cooling. Some drums also may have a coil spring wrapped around the outer circumference. The spring is a vibration and noise damper that quiets high-pitched noise as the linings contact the drum. If the noise were not subdued, it would be heard as annoying brake squeal.

The braking surface area of a drum is determined by the diameter and the depth (width) of the drum. Large cars and trucks, which require more braking energy, have drums that measure 12 inches or more in diameter. Smaller vehicles use smaller drums. Generally, the manufacturers try to keep parts as small and light as possible, while still providing efficient braking. All brake drums have maximum diameter specifications stamped or cast onto their outer circumference. These specifications are used to determine if the drum has worn to an unsafe diameter and should be replaced.

Hubs and Bearings

Tapered roller bearings, installed in the hubs, are the most common bearings used on the rear wheels of front-wheel-drive vehicles. A tapered roller bearing has two main parts: the bearing cone and the outer cup. The bearing cone contains steel tapered rollers that ride on an inner cone and

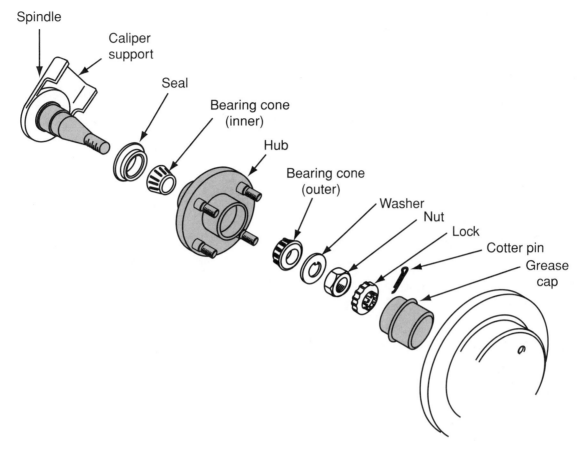

Figure 8. A typical arrangement of a wheel hub and tapered wheel bearings.

are held together by a cage. The bearing fits into the outer cup, which is pressed into the hub to provide two surfaces, an inner cone and outer cup, for the rollers to ride on.

A large tapered roller bearing is installed between the hub and the spindle on the inboard side, and a smaller tapered roller bearing is installed on the outboard side. The bearings are held in place with a thrust washer, nut, nut lock, and cotter pin. A dust cap fits over the assembly to keep dirt out and lubricant in. A seal on the inboard side prevents lubricant from escaping. **Figure 8** is a sectional view of a bearing assembly on a hub.

Solid Cast Iron Drums

A solid drum is a one-piece iron casting. Cast iron has excellent wear characteristics and a coefficient of friction that make it ideal as a braking friction surface. Iron also dissipates heat very well, and it is easy to machine when refinishing is necessary. However, a large iron casting can become brittle and can crack if overstressed or overheated. Small cracks can be almost invisible to the naked eye, but they can lead to drum failure or heat checking and glazing.

A one-piece cast iron drum is the heaviest of all drum types. The weight and mass of a one-piece drum make it

very good at absorbing and dissipating heat. The drum weight, however, adds a lot of weight to the vehicle, and it is all unsprung weight at the wheels.

Steel-and-Iron Drums

Steel-and-iron **composite drums** are made in two ways. The most common type has a stamped steel web mated with the edge of a cast iron drum. The other type of composite drum has a centrifugally cast iron liner inside a stamped steel drum. To make this kind of drum, a stamped steel drum is rotated at high speed while molten iron is poured into it. Centrifugal force causes the molten iron to flow outward and bond tightly to the inner circumference of the steel drum.

Steel-and-iron composite drums are lighter and cheaper to make than one-piece cast iron drums, but they are less able to absorb and dissipate heat and resist fade. They work well on the rear drum brakes of compact cars.

Bimetallic Aluminum Drums

Bimetallic drums are made with an aluminum outer drum cast around an iron liner. Bimetallic drums are almost three times lighter than a one-piece cast iron drum and they cool much better, but they are much more expensive. Aluminum-iron bimetallic drums were more common thirty or forty years ago.

BRAKE SHOES AND LININGS

Brake shoes are the components that mount the friction material or linings that contact the inside of the brake drum. Brake shoes are typically made from welded steel, although some are aluminum. The outer part is called the **table** and is curved to match the curvature of the drum. The brake lining is riveted or bonded to the brake shoe table. Many brake shoes have small notches, or nibs, along the edge of the table that bear against the backing plate and help to keep the shoe aligned in the drum. The **web** is the inner part of the shoe that is perpendicular to the table and to which all of the springs and other linkage parts attach.

All drum brakes have a pair of shoes at each wheel. In most drum brake designs, the size of the two shoes is the same. In other designs, the size of the shoes is different. Brake shoes also vary in the placement of the lining on the shoe and the lining's coefficient of friction. Some wheel drum brake assemblies have one lining centered from top to bottom on the table and the other offset toward either the top or the bottom. Other designs can have one shoe, the primary shoe, with a shorter lining made of material that has a different coefficient of friction than the other shoe, the secondary shoe.

BACKING PLATE

The backing plate is either bolted to the steering knuckle on the front suspension or to the axle flange or hub at the rear. The backing plate is the mounting surface for all other brake parts except the drum **(Figure 9)**. The cir-

cumference of the backing plate is curved to form a lip that fits inside the drum circumference and helps to keep dirt and water out of the brake assembly.

Brake **shoe anchors** are attached to the backing plate to support the shoes and keep them from rotating with the drum. Most brakes have a single anchor that is either a round post or a wedge-shaped block. Some older drum brakes and a few late-model versions have separate anchors for each shoe.

All backing plates have some form of shoe-support pads stamped into their surfaces. The edges of the shoes slide against these pads as the brakes are applied and released. The pads thus keep the shoes aligned with the drum and other parts of the assembly. A light coat of brake grease should be applied to the pads to aid shoe movement and reduce noise.

Many backing plates have **piston stops**, which are steel tabs at the ends of the wheel cylinder. These stops keep the pistons from coming out of the cylinder. The wheel cylinder must be removed from a backing plate with piston stops for disassembly.

WHEEL CYLINDERS

Wheel cylinders (Figure 10) convert hydraulic pressure from the master cylinder to a mechanical force to apply the brakes. Many types of wheel cylinders have been used in the different designs of drum brakes. Some designs used two single-piston cylinders or stepped-bore cylinders with pistons of different diameters. Some designs installed the wheel cylinders so they could slide on the backing plate as they applied the brake shoes. Regardless of their design, wheel cylinders have one purpose—to force a brake shoe against the brake drum.

In today's vehicles, nearly all wheel cylinders are two-piston, straight-bore cylinders mounted rigidly to the back-

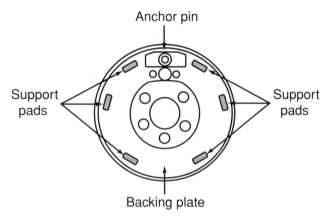

Figure 9. The entire brake assembly is mounted to the backing plate. The brake shoes slide across the support or contact pads as they are applied and released.

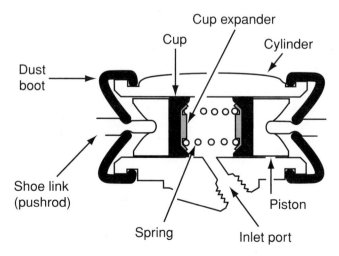

Figure 10. The parts of a typical wheel cylinder.

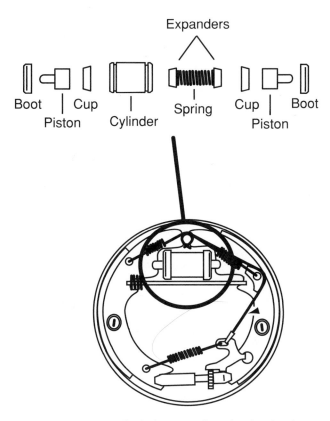

Figure 11. An exploded view of a wheel cylinder.

ing plate. The basic parts of a wheel cylinder are the body, the pistons, the seals (cups), the **cup expanders**, and two dust boots **(Figure 11)**. Some cylinders have two shoe links, or pushrods, to transfer piston movement to the shoes. In other designs, the pistons work directly on the web of the shoes.

The wheel cylinder body is usually cast iron, but a few aluminum cylinder bodies have been used. The cylinder bore is finished to provide a long-wearing, corrosion-resistant surface. A fitting for a hydraulic line is provided at the center of the cylinder, between the two pistons. The bleeder screw also is tapped into the cylinder body at its highest point.

Some imported vehicles have been built with rear wheel cylinders plumbed in series. In this kind of design, a brake line runs from the master cylinder to one rear wheel that does not have a bleeder screw. Instead, a second line runs from the first wheel cylinder to the other rear wheel cylinder that does have a bleeder. Both rear cylinders must be bled from the single bleeder in one cylinder.

A piston is installed in each end of the wheel cylinder, and the inside end of each piston is sealed with a cup seal. Hydraulic pressure against the seal forces the seal lip to expand against the cylinder bore and form a leak-free seal. Most wheel cylinders also have spring-loaded metal cup expanders that bear against the inner sides of the cups. When the brakes are released, the cup expanders hold the seal lips against the cylinder bore to keep air from entering the cylinder past the retracting pistons and seals. The spring also takes up any slack between the pistons and their pushrods and helps to center the pistons in the cylinder bore.

Each end of the wheel cylinder is sealed with a dust boot to keep dirt and moisture out of the cylinder. The dust boots also prevent minor fluid seepage from getting onto the brake linings.

RETURN AND HOLDDOWN SPRINGS

Strong **return springs** retract the shoes when the brakes are released and hold them against their anchors and the wheel cylinder's pushrods. Most return springs are tightly wound coil springs in which the coils touch each other when retracted. Return springs of the same size and shape are often color coded to indicate different tension values **(Figure 12)**. The type, location, and num-

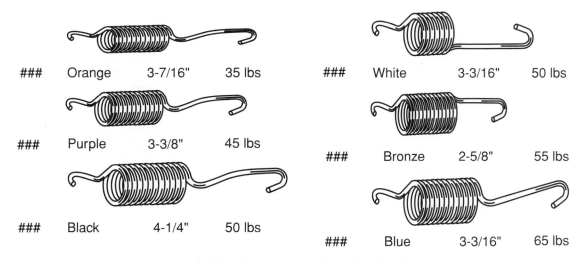

###	Orange	3-7/16"	35 lbs		###	White	3-3/16"	50 lbs
###	Purple	3-3/8"	45 lbs		###	Bronze	2-5/8"	55 lbs
###	Black	4-1/4"	50 lbs		###	Blue	3-3/16"	65 lbs

Figure 12. Return springs can be identified by their part number, length, shape, and tension.

ber of springs vary, but return springs are installed either from shoe to shoe or from each shoe to an anchor post **(Figure 13)**.

Holddown springs, clips, and pins also come in various shapes and sizes **(Figure 14)**, but all have the same purpose of holding the shoes in alignment with the backing plate. Holddowns must be designed to hold the shoes in position while providing flexibility for their application and release. Some late-model General Motors and imported vehicles use a single large horseshoe-shaped spring that acts as both a shoe return spring and a holddown spring.

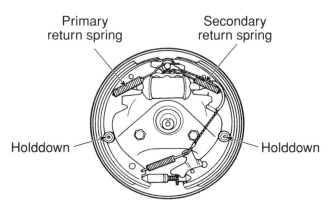

Figure 13. Typical return spring and holddown spring installation.

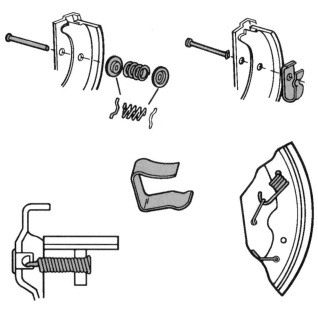

Figure 14. Examples of the various types of holddown springs used in drum brake assemblies.

SELF-ADJUSTERS

Each time the brakes are applied, the lining wears and becomes slightly thinner. As the distance between the shoe and the drum becomes greater, the shoes must move farther before the lining contacts the drum. Therefore, the pistons of the wheel cylinder also must move farther, and more brake fluid must come from the master cylinder to apply the brakes. To provide more brake fluid, brake pedal travel will increase. These changes are why the brake shoes require periodic adjustment.

Interesting Fact Self-adjusters on drum brakes became common in the late 1950s and early 1960s and quickly became standard equipment. Mercury and Edsel models from Ford are credited with leading the industry to self-adjusting brakes with their 1958 models. But Ford was not the first manufacturer to make self-adjusting brakes standard equipment. The 1947 Studebaker featured self-adjusting brakes, and they remained a Studebaker exclusive through 1954. Public disinterest (and technician distrust) caused Studebaker to drop this feature until the early 1960s, when the rest of the industry recognized the benefits.

Brake adjustment compensates for lining wear and maintains correct clearance between brake linings and the drum. Disc and drum brakes both require adjustment, but self-adjustment is a basic feature of disc brake design **(Figure 15)**. Drum brakes, however, need extra cables,

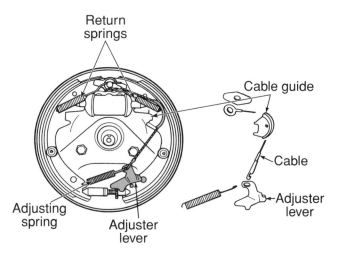

Figure 15. A typical drum brake self-adjuster setup.

levers, screws, struts, and other linkage to provide self-adjustment and proper lining-to-drum clearance. Many different kinds of manual and automatic adjustment devices have been used on drum brakes over the years, but today automatic **self-adjusters** that operate starwheel or ratchet adjustment mechanisms are the most common. Automatic adjusters use the movement of the brake shoes to maintain proper shoe-to-drum clearance.

PARKING BRAKE LINKAGE

Almost all rear drum brake installations include a mechanical parking brake linkage. The linkage basically consists of a cable, a lever, and a strut. The lever and strut spread the shoes against the drum when the cable pulls on the lever. The parking brake strut might contain part of the self-adjuster mechanism.

Summary

- When compared with disc brakes, drum brakes are self-energizing, make less noise, and provide for efficient parking brake operation without a complicated linkage, but drum brakes have poor heat dissipation, less fade resistance, do not self-adjust without a special linkage, and have a greater tendency to pull and grab.
- One end of one shoe contacts the drum before the other end does and becomes a pivot point as friction quickly increases. The brake shoe becomes a self-energizing lever and adds its own mechanical leverage to hydraulic force to help the shoes move tightly against the surface of the brake drum.
- When the self-energizing of one shoe applies mechanical force to the other shoe and helps it move against the wall of the brake drum, it is called servo action.
- The basic parts of a drum brake are: a drum and hub assembly, brake shoes, a backing plate, a hydraulic wheel cylinder, shoe return springs, holddown springs, an adjusting mechanism, and a parking brake.
- The drum has a machined braking surface on its inside circumference. The wheel is mounted to the drum hub by nuts and studs. The hub houses the wheel bearings that allow the wheel to rotate.

- The hydraulic and friction components are attached to the backing plate.
- Brake drums can be categorized as either solid or composite and by the different ways in which the drums are made.
- Brake shoe anchors are attached to the backing plate to support the shoes and keep them from rotating with the drum.
- All backing plates have some form of shoe-support pads stamped into their surfaces that serve as a sliding surface for the edges of shoes when they are applied and released.
- Wheel cylinders convert hydraulic pressure from the master cylinder to a mechanical force to apply the brakes.
- Strong return springs retract the shoes when the brakes are released and hold them against their anchors and the wheel cylinder's pushrods.
- Holddown springs, clips, and pins are used to hold the shoes in alignment with the backing plate.
- Cables, levers, struts, and other devices are used to provide for self-adjustment of brake shoes.

Review Questions

1. List the components of a typical drum brake assembly.
2. How are the brake shoes returned to their released position when the brake pedal is released?
3. Brake linings react against the brake _____ and create friction that is dissipated as _____.
4. The hydraulic component that is mounted to the backing plate and moves the brake shoes is called the _____ _____.

5. Technician A says that drum brakes typically make less noise than disc brakes. Technician B says that drum brakes have more fade resistance than disc brakes. Who is correct?
   A. Technician A only
   B. Technician B only
   C. Both Technician A and Technician B
   D. Neither Technician A nor Technician B

Chapter 25

Drum Brake Design Variations

Introduction

Through the years, there have been many design variations of drum brakes. For the most part, the variations have been dictated by the size, weight, and performance level of the vehicle. Full servo, partial servo, nonservo, two leading shoe, two trailing shoe, and center plane are just a few of the brake designs that are part of automobile history. Nearly all of the drum brakes installed on today's vehicles are one of two designs. These are:

- Nonservo brakes **(Figure 1)**, also called partial-servo or leading-trailing brakes.
- Duo-servo brakes **(Figure 2)**, also called dual-servo or full-servo brakes.

 Interesting Fact You can determine which shoe is the leading shoe in a nonservo brake assembly by putting your hand at the position of the wheel cylinder and then pointing in the direction of drum rotation: clockwise or counterclockwise. The first shoe that you point to is the leading shoe.

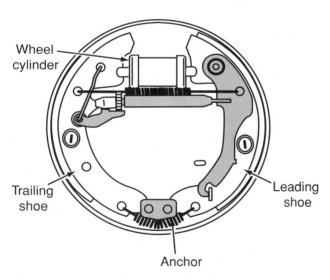

Figure 1. A typical nonservo brake assembly.

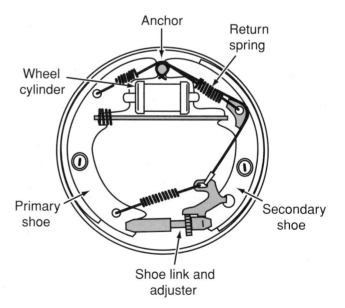

Figure 2. A typical duo-servo brake assembly.

SELF-ENERGIZING AND SERVO ACTIONS

Two terms commonly used to describe drum brake operation are self-energizing and servo. These refer to the leverage developed on the brake shoe as it contacts the drum and the action of one shoe to help apply the other.

In nonservo brake assemblies, the brake shoes are described as leading or trailing shoes. The **leading shoe** is not always the forward shoe in the brake assembly. It can be the front or the rear shoe, depending on whether the drum is rotating forward or in reverse and whether the wheel cylinder is at the top or the bottom of the backing plate.

When the brakes are applied, the wheel cylinder forces one end of the leading shoes outward against the drum. The other end of the shoe is forced back solidly against its anchor post or anchor block. The cylinder end of the leading shoe contacts the drum first and develops friction against the rotating drum. The drum friction pulls the shoe into tighter contact with the drum **(Figure 3)** and aids the hydraulic force of the cylinder to apply the brake shoe. As drum-to-lining contact increases, the rest of the lining is forced by the cylinder and pulled by friction against the drum to stop rotation. This is self-energizing action by the leading shoe **(Figure 4)**.

In a nonservo brake, the reaction of the **trailing shoe** to drum rotation is opposite to that of the leading shoe. As the wheel cylinder tries to force the trailing shoe outward against the drum, rotation tries to force the shoe back

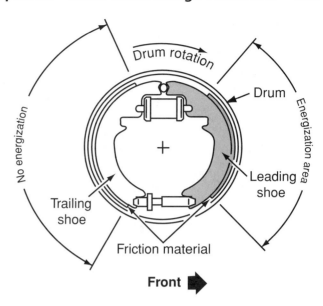

Figure 4.   Only the leading shoe is self-energizing in a nonservo brake assembly.

against the cylinder. Eventually, hydraulic force overcomes drum rotation as the wheel slows, and the trailing shoe contributes to braking action. The trailing shoe, however, is said to be non-self-energizing.

In another brake design, the self-energizing action of one brake shoe can be used to help apply the other shoe. In a duo-servo brake, the **primary shoe** has the same position as the leading shoe in a nonservo brake. The ends of both shoes are opposite the wheel cylinder and are not mounted on a rigid anchor attached to the backing plate as in a nonservo brake. The ends of the two shoes are linked to each other through the starwheel adjuster, and the ends of the shoes float in the drum.

As the primary shoe is applied by the wheel cylinder, it develops self-energizing action as in a nonservo brake. Instead of being forced against an anchor, however, the primary shoe is forced against the **secondary shoe** and applies leverage to force the secondary shoe against the drum. This is servo action **(Figure 5)**, or the action of mechanically multiplying force. The secondary shoe is then forced against the drum by the wheel cylinder force at one end and the servo action of the primary shoe at the other.

NONSERVO BRAKES

The **nonservo**, or leading-trailing, brake design was common in four-wheel drum brake systems fifty or sixty years ago, but was steadily replaced by duo-servo four-wheel drum brakes as cars became heavier and faster. Duo-servo brakes are a more powerful drum brake design, and nonservo brakes nearly vanished from the industry in the 1960s. However, as smaller and lighter cars were introduced, many of them were equipped with nonservo brake

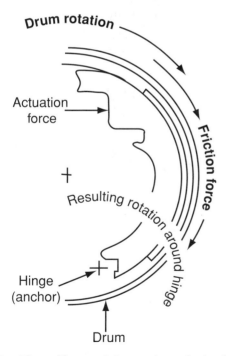

Figure 3.   The self-energizing action of a brake shoe develops as friction against the rotating drum and pulls the shoe into tighter contact with the drum.

A

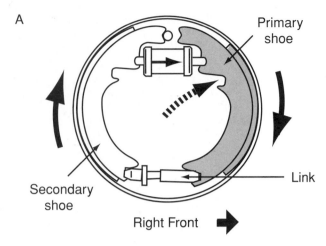

Primary shoe

Secondary shoe

Link

Right Front ➡

B

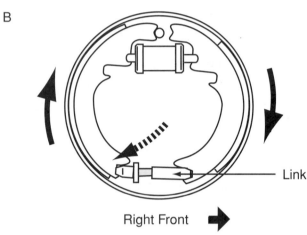

Link

Right Front ➡

C

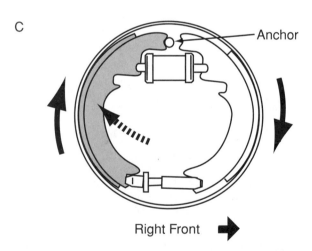

Anchor

Right Front ➡

Figure 5. The self-energizing action of the primary shoe begins when the wheel cylinder forces it against the rotating brake drum. Then the reaction force of the primary shoe is transferred through the link to the secondary shoe (A). This transfer of force, or servo action, causes the secondary shoe to become self-energizing (B). This causes the top of the secondary shoe to move against its anchor and the brakes are applied (C).

systems. Today, nonservo brakes are used as rear drum brakes on many front-wheel-drive (FWD) automobiles, as well as on some light trucks with front discs.

Nonservo brakes are used on these vehicles because FWD automobiles generate up to 80 percent of their braking force on the front wheels. Overly powerful duo-servo brakes at the rear could easily cause the rear wheels to lock as weight shifts forward and the rear of the vehicle becomes very light during braking. Equally important, antilock brakes make overall brake balance at all four wheels a critical factor. To make brake lockup easier to control while maintaining the best total braking efficiency, nonservo brakes are used at the rear wheels of many vehicles.

In a typical nonservo brake installation **(Figure 6)**, the wheel cylinder is mounted at the top of the backing plate, and the cylinder pushrods push against the upper ends of the shoe webs. The lower end of each shoe is held against an anchor block or anchor post toward the bottom of the backing plate. One or two strong return springs usually hold the lower ends of the shoes against the anchor, and another return spring usually is installed between the upper ends of the shoes to hold them together and hold them against the wheel cylinder pushrods.

The wheel cylinder of a nonservo brake acts equally on each brake shoe. The cylinder forces the top of each shoe outward toward the drum, and each shoe pivots on the anchor at the bottom. Each shoe thus operates separately and independently from the other. The leading shoe develops self-energizing action as described previously, however, and provides most of the braking force. The force of drum rotation works against the wheel cylinder force on the trailing shoe so that it is not self-

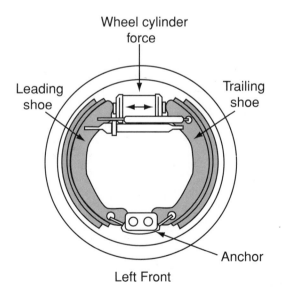

Wheel cylinder force

Leading shoe

Trailing shoe

Anchor

Left Front

Figure 6. The force exerted by the wheel cylinder pushes both the leading and trailing shoes against the fixed anchor in a nonservo brake assembly.

energizing. The trailing shoe often is said to be nonenergized even though it is technically energized by the wheel cylinder.

The front shoe is the leading shoe for forward braking in the nonservo brake as described here. For reverse braking, however, the roles are reversed, and the rear shoe becomes the self-energized leading shoe.

DUO-SERVO BRAKES

Duo-servo drum brakes develop braking force through combined hydraulic and mechanical action. From the 1950s through the mid-1960s, duo-servo brakes with large diameter drums and wide linings provided efficient braking for powerful and heavy vehicles. Duo-servo drum brakes are still used on the rear wheels of many vehicles, however, particularly rear-wheel-drive cars, trucks, and sport utility vehicles.

Like nonservo brakes, duo-servo brakes have a single, two-piston wheel cylinder mounted toward the top of the backing plate. A return spring holds each shoe against the cylinder's pushrods. A duo-servo brake has a single anchor post at the top of the backing plate. The top of each shoe web has a semicircular notch, and the return springs hold the shoes tightly against the anchor post. The cylinder acts on the shoes a couple of inches below the anchor. The bottoms of the shoes are not anchored to the backing plate but are linked to each other by the adjuster link. Another spring holds the shoes tightly together and against the adjuster link. You can think of the shoes of a duo-servo brake as hanging from the anchor post and floating within the drum at the bottom, and that is the key to their operation.

The forward shoe of a duo-servo brake is called the primary shoe, and its lining is shorter than the lining on the rear shoe, which is called the secondary shoe. As the wheel cylinder applies force against both brake shoes during forward braking, the cylinder force pushes the top of the primary shoe away from the anchor. The top edge, or leading edge, of the primary shoe contacts the rotating drum first and develops self-energizing action just as the leading shoe of a nonservo brake does. The self-energized primary shoe is both drawn and forced against the drum, but its lower end does not bear against a fixed anchor. Instead, the lower end of the primary shoe applies servo force to the bottom of the secondary shoe.

The secondary shoe then becomes self-energized, with the self-energizing action beginning at the bottom of the shoe. This combined servo and self-energized action actually works against the wheel cylinder, but servo force applied by the primary shoe to the secondary shoe is greater than the hydraulic force of the wheel cylinder alone. Servo force and the self-energizing action of the secondary shoe force its upper end against the anchor post and wrap the shoe tightly into the drum. The secondary shoe, with its longer and larger lining, therefore applies most of the braking force for forward braking.

During reverse braking, the roles of primary and secondary shoes are reversed. The forward shoe becomes the secondary shoe and receives servo action from the rear shoe. Although the forward shoe must apply most of the braking force, safe braking is available because vehicle speed is usually much lower and the vehicle travels a shorter distance in reverse.

Although it allows for great braking efficiency, excessive servo action can lead to brake grabbing and locking. This is the reason why duo-servo brake assemblies vary in design. Engineers must balance several factors to take full advantage of servo power but maintain smooth braking. These factors include lining size on both shoes and placement of the lining on the primary shoe, which can be centered or mounted high or low on the shoe. Primary and secondary linings also can have different coefficients of friction to achieve smooth brake application.

SELF-ADJUSTERS

Brake shoe adjustments are made automatically and by technicians servicing the brake system. The operation and construction of automatic or self-adjusters depend on the design of the drum brake and on the manufacturer.

Duo-Servo Starwheel Adjusters

The bottoms of the shoes in a duo-servo brake assembly are connected to each other by a link, and the shoes are held against the link by a strong spring **(Figure 7)**. The link keeps the shoes aligned with each other and transfers the servo action of one shoe to the other during braking. The link is not attached to the backing plate but moves back and forth with the shoes.

One half of the adjuster link is internally threaded, and the other is externally threaded. The externally threaded part has a **starwheel** near one end. When the two halves of the link are assembled, rotating the starwheel moves them together or apart. In this way, the adjustment link is lengthened or shortened to adjust the lining-to-drum clearance.

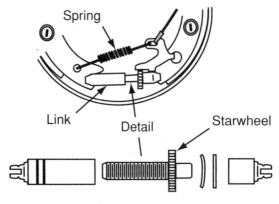

Figure 7. A strong spring holds the ends of the shoes tightly against the adjuster link.

On manually adjusted duo-servo brakes, an adjusting tool is inserted through a hole in the backing plate to engage the starwheel and rotate it to adjust the link. Even with self-adjusters, the initial adjustment of a duo-servo brake is made this way after new shoes are installed and the drum is resurfaced. Further adjustment is done automatically, however, by the self-adjustment mechanism for the life of the linings.

Duo-servo brakes commonly have self-adjusters operated by the secondary shoe **(Figure 8)**. Either a cable, a heavy wire link, or a lever is attached to the secondary shoe.

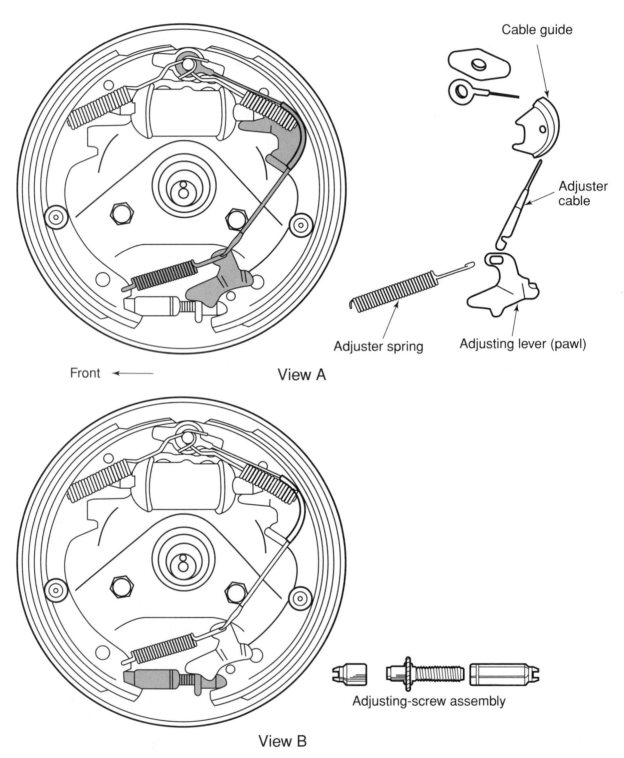

Front ◄——

View A

Cable guide

Adjuster cable

Adjuster spring

Adjusting lever (pawl)

Adjusting-screw assembly

View B

Figure 8. A cable-operated self-adjuster in a duo-servo brake moves the adjusting lever when the secondary shoe moves off its anchor during reverse braking.

The cable, link, or lever is attached to a smaller lever, or **pawl**, that engages the starwheel. During braking in reverse, the secondary shoe moves away from the anchor post. If the lining is worn far enough and the shoe can move far enough, it will pull the cable **(Figure 9)**, link, or lever **(Figure 10)** to move the pawl to the next notch on the

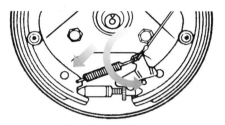

Figure 9. As the adjuster spring pulls the lever downward, it turns the starwheel to expand the adjuster link.

starwheel. Lever-operated self-adjusters usually move the pawl to rotate the starwheel as the brakes are applied. The rotation of the starwheel then expands the link to take up the clearance between the linings and the drum.

Some cable-operated self-adjusters move a pawl mounted under the starwheel to adjust the brakes during application, not release. These cable-operated adjusters usually have an **overload spring** in the end of the cable that lets the cable move without breaking if the pawl or starwheel is jammed.

The overload spring also prevents overadjustment during very hard braking when the drum might distort and let the shoes move farther out than normal. To prevent this, the overload spring stretches with the cable and does not let the pawl actuate. Overadjustment will not occur during hard braking when the brakes are in normal adjustment.

Left-hand and right-hand threaded starwheel adjusters are used on the opposite sides of the car. Therefore, parts

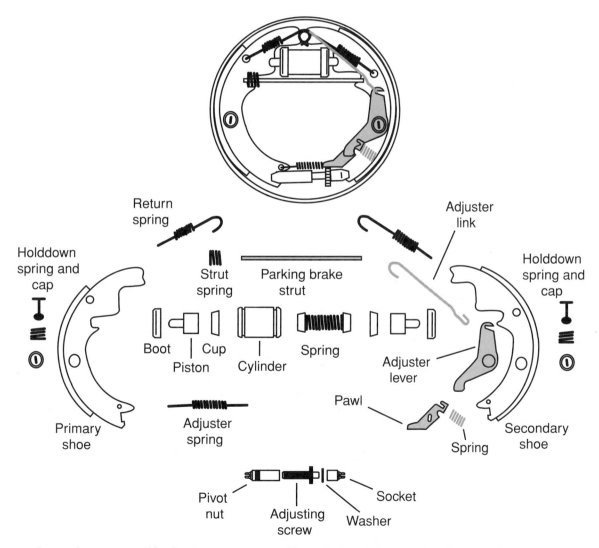

Figure 10. Some duo-servo self-adjusters are operated by a link and lever rather than a cable.

must be kept separated and not intermixed. If a starwheel adjuster is installed on the wrong side, the adjuster will not adjust at all.

Nonservo Starwheel Adjusters

Many different linkage designs are used to adjust the starwheels in nonservo brake assemblies. The starwheel self-adjusters can be operated by either the leading or the trailing shoe and can work whenever the brakes are operated in either forward or reverse.

The starwheel is usually part of the parking brake strut that is mounted between the two shoes. A pawl can be operated by either the leading or the trailing shoe to engage and rotate the starwheel as the brakes are applied or released.

Nonservo Ratchet Adjusters

A lever-type ratchet adjuster has a large lever and small latch attached to the leading shoe. Teeth on the lever and the latch form a ratchet mechanism that moves the shoes outward to take up excess clearance as the brakes are applied.

Another type of ratchet adjuster consists of a pair of large- and small-toothed ratchets and a spacer strut attached to the trailing shoe **(Figure 11)**. The strut is connected to the secondary shoe through the parking

brake lever and the inner edge of the large ratchet. As the gap between the shoes and the drums becomes greater, the strut and the primary shoe move together to close the gap. More movement will cause the large ratchet on the secondary shoe to rotate inward against the small ratchet, which is spring loaded, and reach a new adjustment position.

Other similar systems are actuated by the parking brake and adjust when the parking brake is applied and released. The adjustment lever is attached to the parking brake lever **(Figure 12)**. As the lining wears, application of the parking brake will restore the proper lining clearance.

Nonservo Cam Adjusters

Cam-type adjusters use cams with an adjuster pin that fits in a slot on the shoes **(Figure 13)**. As the brake shoes move outward, the pin in the slot moves the cam to a new position if adjustment is needed. Shoe retraction and proper lining clearance are always maintained because the pin diameter is smaller than the width of the slot on the shoe. These brakes can be adjusted even while the vehicle is at a standstill because brake pedal application is all that is needed to move the cams into proper adjustment. Because the brakes can be adjusted completely with one application of the brake pedal, this adjuster is sometimes called the one-shot adjuster.

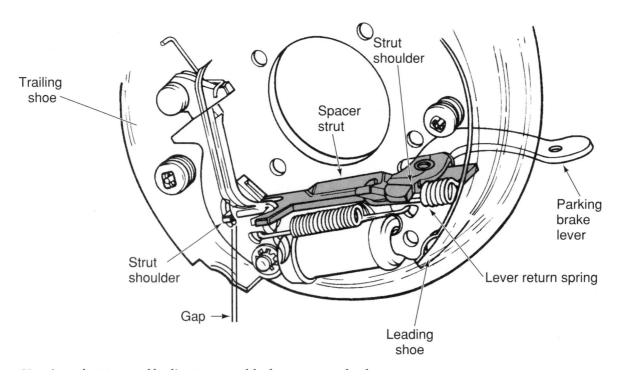

Figure 11.   A ratchet-type self-adjuster assembly for nonservo brakes.

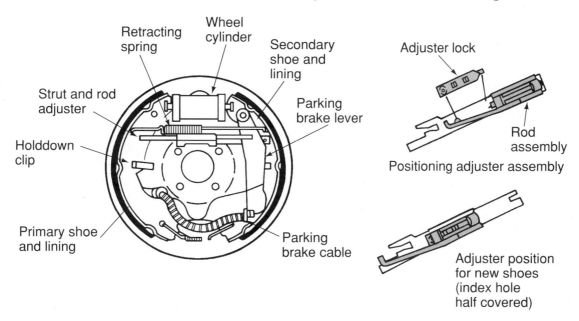

Figure 12. The self-adjuster in this setup is activated by the parking brake.

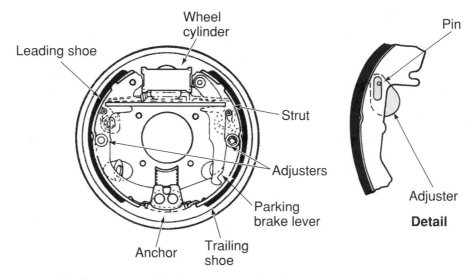

Figure 13. A cam-type self-adjuster assembly for nonservo brakes.

Summary

- Nearly all of the drum brakes installed on today's vehicles are one of two designs: nonservo brakes or duo-servo brakes.
- In nonservo brake assemblies, the brake shoes are described as leading or trailing shoes.
- In a duo-servo brake assembly, there is a primary and a secondary brake shoe.
- The wheel cylinder of a nonservo brake assembly forces the top of each shoe outward toward the drum,

and each shoe pivots on the anchor at the bottom. The leading shoe develops self-energizing action and provides most of the braking force. The force of drum rotation works against the wheel cylinder force on the trailing shoe, so it is not self-energizing.

- A duo-servo brake has a single anchor post at the top of the backing plate, and the return springs hold the shoes tightly against the anchor post. The bottoms of the shoes are not anchored to the backing plate but are linked to each other by the adjuster link.

- The lining of the primary shoe is shorter than the lining on the secondary shoe. As the wheel cylinder applies force against both brake shoes, it pushes the top of the primary shoe away from the anchor. The top edge of the primary shoe contacts the rotating drum first and develops self-energizing action that applies servo force to the bottom of the secondary shoe, which then becomes self-energized.
- Brake adjustment is made automatically and by technicians servicing the brake system.
- The bottoms of the brake shoes in a duo-servo brake assembly are connected to each other by an adjuster link. One half of the adjuster link is internally threaded, and the other is externally threaded with a starwheel near one end.
- The starwheel allows for the lengthening and shortening of the adjuster link.
- Automatic adjusters in duo-servo systems have a cable, a heavy wire link, or a lever attached to the secondary shoe and to a smaller lever, or pawl, that engages the starwheel.
- The starwheel self-adjusters in nonservo brake assemblies can be operated by either the leading or the trailing shoe and can work whenever the brakes are operated in either forward or reverse.
- Some nonservo brake assemblies have a lever-type ratchet adjuster or use cams with an adjuster pin that fits in a slot on the shoes.

Review Questions

1. Explain how a duo-servo brake assembly works to provide great braking ability.
2. When moving in reverse, the _____ shoe provides most of the stopping force in a duo-servo brake system.
3. When moving forward, the _____ shoe provides most of the stopping force in a nonservo brake system.
4. Why do drum brakes need to be adjusted on a regular basis?

5. Technician A says that most self-adjusters in duo-servo systems are activated by the secondary shoe. Technician B says that some self-adjusters actuate only when the parking brake is applied. Who is correct?
   A. Technician A only
   B. Technician B only
   C. Both Technician A and Technician B
   D. Neither Technician A nor Technician B

Chapter 26

Disc Brake Construction

Introduction

Disc brakes are used at the front wheels of nearly all cars and light trucks built since the mid-1970s and, on many vehicles, at all four wheels. Disc brakes have several important strong points and a few, less important, disadvantages. The principal advantages of disc brakes are strong fade resistance, self-adjustment, and reduced pulling and grabbing. Disadvantages are noise, poorer parking brake operation that requires some complicated linkage, and the lack of self-energizing (servo) action.

Fade Resistance

Brake operation is a process of changing kinetic energy (motion) into thermal energy (heat) through the application of friction **(Figure 1)**. When any brake installation reaches its limits of heat dissipation, brake fade sets in. Brake fade is simply the loss of braking power due to excessive heat that reduces friction between brake linings and the rotors or drums.

One major factor that makes disc brakes more fade resistant than drum brakes is that the friction surfaces are exposed to the air. Many brake rotors also have cooling passages, or fins **(Figure 2)**, to improve heat dissipation.

Disc brakes have good fade resistance because most of the swept area is cooling while only a small area is in direct contact with the frictional material at any given time **(Figure 3)**. Disc brakes do not suffer from mechanical fade because the rotor does not expand away from the pads. If anything, the rotor expands very slightly toward the pads if it becomes very hot. Disc brakes are self-cleaning, which reduces water and gas fade. Additionally, centrifugal force works to remove water and gas from the surfaces of the rotor.

Disc Brake Self-Adjustment

Brake adjustment is the process of compensating for lining wear and maintaining correct clearance between brake linings and the drum or rotor surfaces. Disc and drum brakes both require adjustment, but, unlike drum brakes, disc brakes do not need cables, levers, screws, struts, and other mechanical linkage to maintain proper pad-to-rotor clearance.

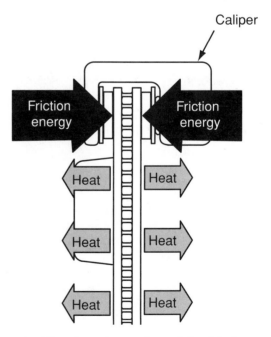

Figure 1. The clamping action of the friction material against the rotating disc or rotor generates much heat.

231

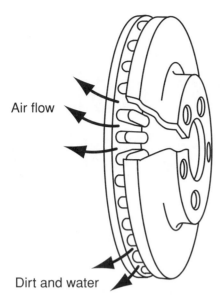

Air flow

Dirt and water

Figure 2. Cooling fins built into the rotor help to dissipate heat, and centrifugal force helps to keep the rotors clean by throwing off water and dirt.

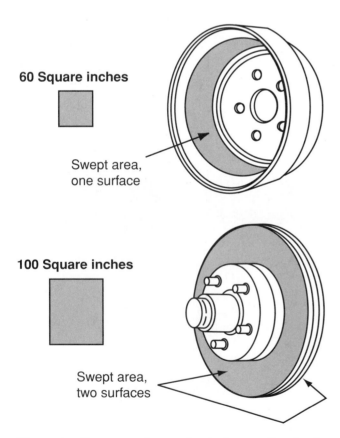

60 Square inches

Swept area, one surface

100 Square inches

Swept area, two surfaces

Figure 3. For any given wheel size, a disc brake always has 35–50 percent more swept area than a drum brake to dissipate heat.

When disc brakes are applied, the caliper pistons move out far enough to apply braking force from the pads to the rotors. When the brakes are released, the caliper pistons retract only far enough to release pressure **(Figure 4)**. The pads always have only a few thousandths of an inch clearance from the rotor, regardless of lining wear.

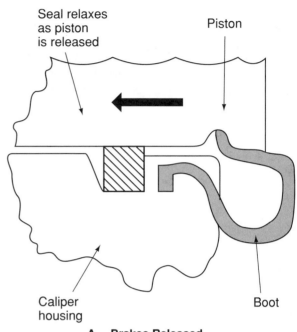

Seal relaxes as piston is released

Piston

Caliper housing

Boot

A.   Brakes Released

Piston seal extended

Piston

Caliper housing

Boot

B.   Brakes Applied

Figure 4. (A) When the brakes are released, the caliper piston seal holds the piston in a retracted position. (B) When the brakes are applied, the seal extends and then returns to the relaxed position to retract the piston and provide clearance between the pads and the rotor.

Brake Servo Action, Pulling, and Grabbing

Slight pull can occur in a front disc brake installation due to a sticking caliper piston or problems in hoses or valves. Disc brake pull, however, is usually never as severe as drum brake pull can be. Disc brakes require higher application force from the hydraulic system than drum brakes require. The greater size ratio between master cylinder pistons and caliper pistons, along with universal use of power boosters with disc brakes, provides the necessary braking force without requiring undue pedal effort by the driver.

DISC BRAKE NOISE

Disc brakes are noisier than drum brakes. The noise usually occurs when the brakes are applied and released and is commonly caused by a slight high-speed rattling of the pads. Modern disc brakes use various devices and techniques to minimize noise. Antirattle springs and clips hold pads securely in the calipers and help to dampen noise. Adhesives are also used to hold the pads to the pistons and caliper mounting points to further dampen squeaks and rattles. On some installations, metal or plastic shims between the pads, calipers, and pistons help to reduce vibration and noise.

Simple squeaks and rattles can be annoying to the driver, but they generally do not affect brake operation. When the noise becomes a continuous grinding or loud scraping sound as the brakes are applied, it often indicates that the linings have worn to the metal surfaces of the pads. These sounds mean that it is time for brake service before vehicle safety is jeopardized further.

Many disc brake pads have audible wear indicators, which are small steel pins or clips that rub on the rotor to create a constant squeal when the pads have worn to their minimum thickness. These wear indicators are designed to create a sound that will alarm the driver into having the brakes checked but will not damage the rotor surface.

DISC BRAKE CONSTRUCTION

The basic parts of a disc brake are a rotor and hub and a caliper assembly **(Figure 5)**. The rotor provides the friction surface for stopping the wheel. The rotor has a machined braking surface on each side. The wheel is mounted to the rotor hub by wheel lug nuts and studs. The hub houses wheel bearings that allow the wheel to freely rotate while keeping it centered on its axle.

The hydraulic and friction components are housed in the caliper assembly that straddles the outside diameter of

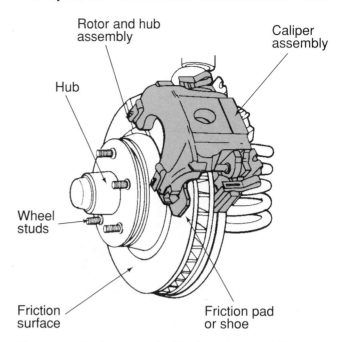

Figure 5. Basic parts of a disc brake assembly.

the rotor. When the brakes are applied, a piston inside the caliper is forced outward by hydraulic pressure. The pressure of the piston is exerted through the pads in a clamping action on the rotor. **Figure 6** illustrates basic disc brake operation.

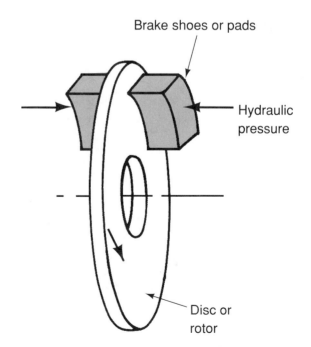

Figure 6. Disc brakes stop the wheel when the brake pads are forced against the rotating rotor.

ROTORS, HUBS, AND BEARINGS

The disc brake **rotor** has two main parts: the hub and the braking surface **(Figure 7)**. The hub is where the wheel is mounted and contains the wheel bearings. The braking surface is the machined surface on both sides of the rotor. The braking surface is carefully machined to provide a friction surface for the brake pads. The entire rotor is usually made of cast iron, which provides an excellent friction surface. The rotor side where the wheel is mounted is the outboard side. The other side, toward the center of the car, is the inboard side.

The size of the rotor braking surface is determined by the diameter of the rotor. Large cars, which require more braking energy, have rotors measuring 12 inches or more in diameter. Smaller, lighter cars can use smaller rotors. Generally, manufacturers want to keep parts as small and light as possible, while maintaining efficient braking ability.

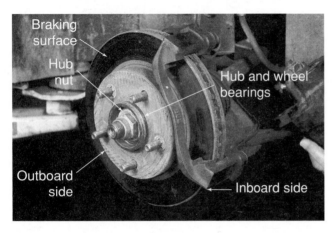

Figure 7. The major parts of a typical rotor, hub, and caliper support assembly.

> **You Should Know** *Some customers install dust shields over the brake units to reduce the buildup of dirt on the vehicle's wheels. The covers help keep the wheels clean but also help to overheat the rotors by blocking the air flow needed to cool the rotors.*

The rotor is protected from water and dirt due to road splash by a sheet metal splash shield that is bolted to the steering knuckle **(Figure 8)**. The outboard side is shielded

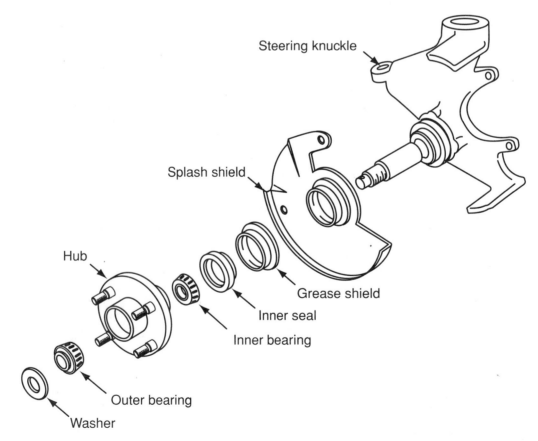

Figure 8. The splash shield is mounted on the steering knuckle inboard of the rotor.

by the vehicle's wheel. The splash shield and wheel also are important in directing air over the rotor to aid cooling.

Fixed and Floating Rotors

Rotors can be classified by the hub design as fixed (with an integral hub) or floating (with a separate hub). A **fixed rotor** has the hub and the rotor cast as a single unit **(Figure 9)**.

Floating rotors and their hubs are made as two separate parts **(Figure 10)**. The hub is a conventional casting and is mounted on wheel bearings or on the axle. The

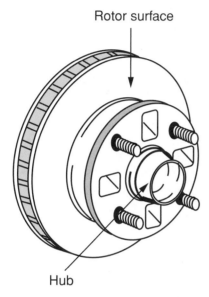

Rotor surface

Hub

Figure 9. A rotor that is cast as a single unit with its hub is called a fixed or integral rotor.

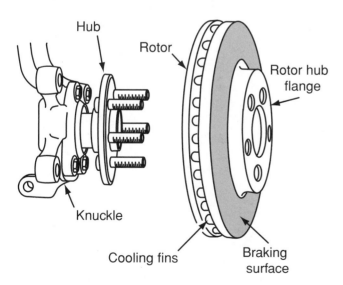

Hub

Rotor

Rotor hub flange

Knuckle

Cooling fins

Braking surface

Figure 10. A rotor that is cast as a separate part and is fastened to the hub is called a floating rotor.

wheel studs are mounted in the hub and pass through the rotor center section. This kind of rotor is called a hubless, or floating, rotor. One advantage of this design is that the rotor is less expensive and can be replaced easily and economically when the braking surface of the rotor is worn beyond machining limits.

Composite Rotors

Traditionally, brake rotors were manufactured as a single iron casting. The development of floating, two-piece rotors and the need to reduce vehicle weight led to the development of **composite rotors**. Composite rotors are made of different materials, usually cast iron and steel, to reduce weight. The friction surfaces and the hubs are cast iron, but supporting parts of the rotor are made of lighter steel stampings. The steel and iron sections are bonded to each other under heat and high pressure to form a one-piece finished assembly **(Figure 11)**. Composite rotors can be fixed components with integral hubs or they can be floating rotors, mounted on a separate hub. Because the friction surfaces of composite rotors are cast iron, the wear standards are generally the same as they are for other rotors. The procedure for refinishing composite rotors is also the same as for other rotors but requires special adapters and mounts for the lathe.

Solid and Ventilated Rotors

A rotor can be solid or it can be ventilated **(Figure 12)**. A **solid rotor** is simply a solid piece of metal with a friction surface on each side. A solid rotor is light, simple, cheap, and easy to manufacture. Because they do not have the cooling capacity of a ventilated rotor, solid rotors usually are used on small cars of moderate performance.

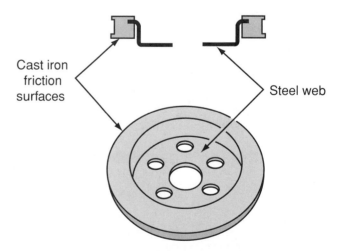

Cast iron friction surfaces

Steel web

Figure 11. The cast iron friction surface is cast onto a steel web during the manufacture of a composite rotor. The steel web of this rotor also contains the parking brake drum.

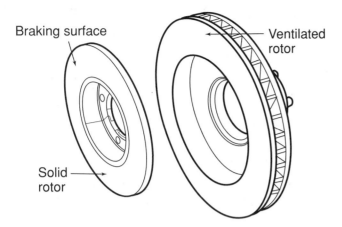

Figure 12.   At the left is a solid rotor and at the right
is a ventilated one.

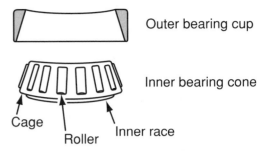

Figure 13.   In a tapered roller wheel bearing, the inner
race, the rollers, and the cage are a separate assembly from the outer race or cup.

A **ventilated rotor** has cooling fins cast between the braking surfaces to increase the cooling area of the rotor. When the wheel is in motion, the rotation of these fins in the rotor also increases air circulation and brake cooling.

Although ventilated rotors are larger and heavier than solid rotors, these disadvantages are more than offset by their better cooling ability and heat dissipation. A heavy ventilated rotor can affect wheel balance, so many rotors have balance weights installed between the fins.

Some ventilated rotors have cooling fins that are curved or formed at an angle to the hub center. These fins increase centrifugal force on rotor airflow and increase the air volume that removes heat. Such rotors are called **unidirectional rotors** because the fins work properly only when the rotor rotates in one direction. Therefore, unidirectional rotors cannot be interchanged from the right side to the left side on the car.

A few cars have solid rotors with holes drilled through the friction surfaces. These drilled rotors are not made to increase cooling as much as they are to release water and hot gases from the rotor surface that can cause water or gas fading. Drilled rotors are typically very light and can have very short service lives. Therefore, they are used mostly on racecars and dual-purpose, high-performance cars.

Rotor Hubs and Wheel Bearings

Tapered roller bearings, installed in the wheel hubs, are the most common bearings used on the front wheels of rear-wheel-drive vehicles and the rear wheels of front-wheel-drive cars. The **tapered roller bearing** has two main parts: the inner **bearing cone** and the outer **bearing cup** (Figure 13). The bearing cone is an assembly that contains steel tapered rollers. The rollers ride on an inner cone-shaped race and are held together by a **bearing cage**. The

bearing fits into the outer cup, or race, which is pressed into the hub. This provides two surfaces, an inner cone and an outer cup, for the rollers to ride on.

A large tapered roller bearing is installed between the hub and the spindle on the inboard side, and a smaller tapered roller bearing is installed on the outboard side. The bearings are held in place with a thrust washer, nut, nut lock, and cotter pin. A dust cap fits over the assembly to keep dirt out and lubricant in. A seal on the inboard side prevents lubricant from escaping at this end. **Figure 14** is a cross-sectional view of a bearing assembly.

BASIC CALIPER PARTS AND OPERATION

The disc **brake caliper** converts hydraulic pressure from the master cylinder to mechanical force that pushes the brake pads against the rotor. The caliper is mounted over the rotor   **(Figure 15)**. Although there are many design differences among calipers, all calipers have a caliper body, or housing, hydraulic lines and internal hydraulic passages, one or more pistons, piston seals, and dust boots.

Figure 16 is a cross-sectional view of a floating caliper. A brake line from the master cylinder is attached to the caliper housing. When the brakes are applied, fluid under pressure from the master cylinder enters the caliper housing. The caliper has at least one large hydraulic piston located in a piston bore. During braking, fluid pressure behind the piston increases. Pressure is exerted equally against the bottom of the piston and the bottom of the cylinder bore. The pressure applied to the piston is transmitted to the inboard brake pad to force the lining against the inboard rotor surface. Depending on caliper design, fluid pressure can be routed to matching pistons on the outboard side of the caliper, or the pressure applied to the bottom of the cylinder can force the caliper to move inboard on its mount. In either case, the caliper applies

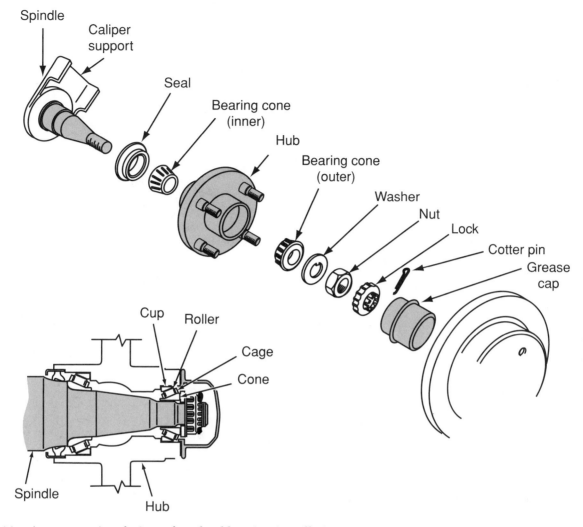

Figure 14.   A cross-sectional view of a wheel bearing installation.

Figure 15.   The typical location of a brake caliper.

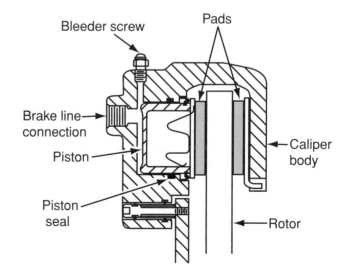

Figure 16.   A cross-sectional view of a typical caliper.

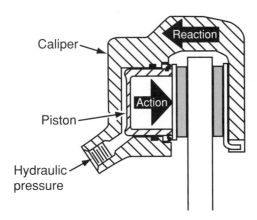

Figure 17. Hydraulic pressure in the caliper forces the piston and one pad in one direction against the rotor and the caliper body and other pad in the opposite direction.

mechanical force equally to the pads on both sides of the rotor to stop the car **(Figure 17)**.

Caliper Body

The caliper body is a U-shaped casting that wraps around both sides of the rotor. Nearly all caliper bodies are made of cast iron. Single-piston caliper bodies are usually cast in one piece, but caliper bodies with pistons on the inboard and outboard sides of the rotor are usually cast in two pieces and bolted together.

Front calipers usually are mounted on **caliper supports** that are integral parts of the **steering knuckles**. Although one-piece steering knuckle and caliper support forgings are the simplest and most economical way to manufacture these components, some front caliper supports are separate pieces that bolt to the steering knuckle. Rear calipers are mounted on caliper supports that bolt to the rear suspension or the axle housing.

Most caliper bodies have one or two large openings in the top of the casting through which the lining thickness can be inspected. Although these openings are handy inspection points, that is not their primary purpose. Caliper bodies are cast with these openings to reduce weight and, more importantly, to minimize large masses of iron that can cause uneven thermal expansion of the assembly. You will find a few caliper bodies without these openings because they were not needed structurally.

Caliper Hydraulic Passages and Lines

The fitting for the brake line is located on the inboard side of the caliper body, as is the bleeder screw. A movable caliper with one or two pistons on the inboard side has hydraulic passages cast into the inboard side of the body. A fixed caliper with pistons on both the inboard and the outboard sides requires crossover hydraulic lines to the out-

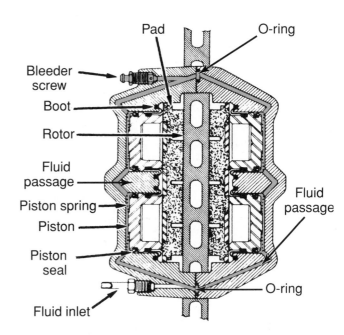

Figure 18. This cross-sectional view shows the fluid crossover and the seals in a fixed caliper.

board side. These usually are cast into the caliper halves and sealed with O-rings when the caliper body is bolted together **(Figure 18)**. Some fixed calipers, however, have external steel tubing to carry fluid to the outboard pistons.

Caliper Pistons

Depending on its design, a caliper may have one, two, three, or four pistons, with single-piston calipers being the most common by far. The piston operates in a bore that is cast and machined into the caliper body. The piston is the part that converts hydraulic pressure to mechanical force on the brake pads. To do its job, the piston must be strong enough to operate with several thousand pounds of pressure, and it must resist corrosion and high temperatures.

The inner side of the piston contacts the brake fluid in the caliper bore, and the outer side bears against the steel brake pad. The brake pads operate at temperatures above the boiling point of brake fluid, and the piston must help to insulate the fluid from this extreme heat. The piston surface that contacts the pad is hollow, or cup shaped, which reduces weight and reduces the area available to absorb heat from the pad. The ability to absorb heat and not transfer it to the fluid also is an important design consideration for the piston materials. Caliper pistons typically are made of steel, cast iron, **phenolic plastic**, or aluminum.

Chrome-plated cast iron and steel pistons are the most common. Iron pistons were used in many early disc brakes, but steel is more common in late-model designs. Steel pistons are strong and thermally stable, but they are heavier than desired in late-model brake assemblies and they can conduct excessive heat to the brake fluid.

To reduce both weight and heat transfer, manufacturers turned to phenolic plastic pistons in the mid-1970s. Phenolic pistons are strong, light, excellent insulators, and economical to make. Additionally, the plastic surface is not as slippery as chrome plating and grips the piston seals better than a steel piston does.

Phenolic pistons also have some disadvantages. Among them is a tendency to wear faster and score more easily than steel pistons. Early designs also tended to stick in their bores due to caliper bore varnish and corrosion. Improved dust boots and seals have reduced that problem, however.

Aluminum would seem to be a good material for caliper pistons because of its light weight. In fact, aluminum has been used for pistons in some high-performance brakes where weight is critical, but it has serious drawbacks that limit its use in passenger car and light truck brakes.

Aluminum expands faster than iron or steel, and aluminum pistons must be made with more clearance in their caliper bores. This, in turn, increases the possibility of leakage, as does aluminum's greater tendency toward scoring and corrosion. The major problem with aluminum caliper pistons, however, is their ability to transfer heat. Aluminum is a very poor thermal insulator and increases the danger of boiling the brake fluid.

Caliper Piston Seals

Brake calipers require seals to keep fluid from escaping between the pistons and their bores, but caliper piston seals also perform other functions. Disc brakes do not have return springs to move the pads out of contact with the rotor. This is accomplished by the caliper piston seal.

During braking, the piston seal is deflected, or bent, by the hydraulic pressure. When the pressure is released, the seals relax or retract, pulling the pistons back from the rotors. The flexing of the seal releases pressure from the rotor but maintains only very slight clearance between the pad and the rotor. The seal fits closely around the piston and holds it in position. Because the seal is installed at the outer end of the caliper bore, it keeps dirt and moisture out of most of the bore and away from the piston. This minimizes corrosion and damage to the caliper bores and pistons.

Caliper Dust Boots

A rubber boot fits around every caliper piston to keep dirt and moisture out of the caliper bore **(Figure 19)**. The opening in the center of the boot fits tightly around the outer end of the piston. The outer circumference of the boot can be attached to the caliper body by a retaining ring in the caliper or by tucking it into a groove inside the bore.

BRAKE PADS

Each brake caliper contains two **brake pads**. The brake pads are positioned in the caliper on the inboard and out-

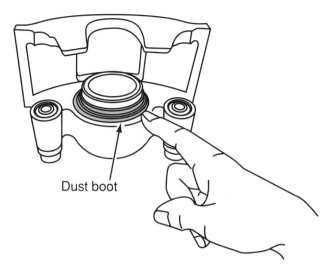

Figure 19. A dust boot keeps dirt and moisture out of the piston's bore in the caliper.

board sides of the rotor. The caliper piston, or pistons, force the brake pad linings against the rotor surfaces to stop the car.

Fundamentally, a brake pad is a steel plate with a friction material lining bonded or riveted to its surface **(Figure 20)**. The simple appearance of a brake pad, however, hides some sophisticated engineering that goes into its design and manufacturing. Friction materials are classified as organic (nonmetallic), semimetallic, fully metallic, and—in some advanced systems—synthetic.

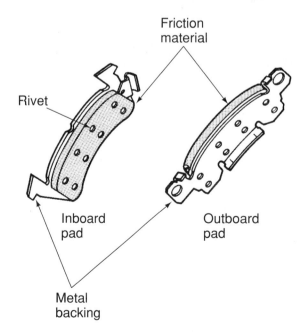

Figure 20. The parts of a disc brake pad.

Summary

- Disc brakes are more fade resistant than drum brakes are because the friction surfaces are exposed to the air and the swept area is much greater.
- Disc brakes require high application force from the hydraulic system, which means there is a greater size ratio between master cylinder pistons and caliper pistons.
- Disc brakes can be noisier than drum brakes because of a high-speed rattling of the pads. Antirattle springs and clips, adhesives, and/or plastic shims are used to minimize this rattling.
- A continuous grinding or loud scraping sound as the brakes are applied often indicates that the linings have worn to the metal surfaces of the pads.
- Many disc brake pads have audible wear indicators, which are small steel pins or clips that rub on the rotor to create a constant squeal when the pads have worn to their minimum thickness.
- The rotor of a disc brake assembly provides the friction surface for stopping the wheel.
- The hydraulic and friction components are housed in the caliper assembly that straddles the outside diameter of the rotor.
- A fixed rotor has the hub and the rotor cast as a single unit.
- Floating rotors and their hubs are made as two separate parts.
- Composite rotors are made of different materials, usually cast iron and steel, to reduce weight. The friction surfaces and the hubs are cast iron, but supporting parts of the rotor are made of lighter steel stampings. The steel and iron sections are bonded to form a one-piece finished assembly.
- A solid rotor is simply a solid piece of metal with a friction surface on each side.
- A ventilated rotor has cooling fins cast between the braking surfaces to increase the cooling area of the rotor.
- Tapered roller bearings, installed in the wheel hubs, are the most common bearings used on the front wheels of rear-wheel-drive vehicles and the rear wheels of front-wheel-drive vehicles.
- The disc brake caliper converts hydraulic pressure from the master cylinder to mechanical force that pushes the brake pads against the rotor.
- There are many design differences among calipers, but all calipers have a caliper body, or housing, hydraulic lines and internal hydraulic passages; one or more pistons; friction pads, piston seals, and dust boots.
- Front calipers usually are mounted on caliper supports that are integral parts of the steering knuckles.
- A caliper can have one, two, three, or four pistons.
- Caliper pistons typically are made of steel, cast iron, phenolic plastic, or aluminum.
- Brake calipers require seals to keep fluid from escaping between the pistons and their bores, but caliper piston seals also move the pads out of contact with the rotor when the brakes are released.
- A rubber dust boot fits around every caliper piston to keep dirt and moisture out of the caliper bore.

Review Questions

1. List the basic components of a disc brake assembly.
2. List the components of a typical caliper assembly.
3. The two different rotor assemblies used on today's vehicles are the _____ rotor and the _____ rotor.
4. Functions of a splash shield are being discussed. Technician A says that disc brakes will function normally without a splash shield in place. Technician B says that a splash shield helps to direct cooling air over the rotor. Who is correct?
   A. Technician A only
   B. Technician B only
   C. Both Technician A and Technician B
   D. Neither Technician A nor Technician B
5. Technician A says that the piston seal retracts the caliper piston when hydraulic pressure is released. Technician B says that a return spring is used to retract a caliper piston. Who is correct?
   A. Technician A only
   B. Technician B only
   C. Both Technician A and Technician B
   D. Neither Technician A nor Technician B

Chapter 27

Disc Brake Design Variations

Introduction

Although all disc brakes have the same common parts, that is, calipers, rotors, pistons, and pads, they are commonly classified into two groups in terms of caliper operation: fixed and movable. Fixed calipers are bolted rigidly to the caliper support on the steering knuckle or on the rear axle or suspension. Fixed calipers have pistons and cylinders (bores) on the inboard and outboard sides. Hydraulic pressure is applied equally to the inboard and outboard pistons to force the pads against the rotor. Movable calipers slide or float on the caliper support. Sliding or floating calipers have a piston on only the inboard side, and hydraulic pressure is applied to the piston to force the inboard pad against the rotor. At the same time, hydraulic pressure on the bottom of the caliper bore forces the caliper to move inboard and clamp the outboard pad against the rotor with equal force.

Although the caliper design dictates the classification of disc brake systems, other features such as the number of pistons in the caliper and the type of seals used in the caliper further define the brake system.

CALIPER PISTONS

Fixed calipers **(Figure 1)** have a minimum of two pistons with at least one on each side of the caliper. Probably the best-known fixed caliper brake is the four-piston Delco Moraine brake system used on Chevrolet Corvettes from 1965 through the early 1980s. Some high-performance cars are still equipped with four-piston calipers. Four-piston calipers have two pistons on each side of the caliper. In the past, a few three-piston designs were manufactured (with one large inboard and two small outboard pistons).

CALIPER PISTON SEALS

Movable calipers use **fixed seals**, also called **square-cut seals** or **lathe-cut seals**. Fixed calipers use a type of lip seal called a **stroking seal**.

Fixed Seals

A fixed seal, used with a movable caliper, is installed in a groove in the inner circumference of the caliper bore. The piston fits through the inside of the seal and is free to move

Figure 1. A movable caliper assembly with two pistons for greater braking power.

in the seal. The outer circumference of the seal remains in a fixed position in the caliper bore. Many seals are square or rectangular in cross section, but others have different cross section shapes. Because all seals are not identical, it is important to be sure that replacement seals match the shape of the originals.

During braking, the piston seal is deflected, or bent, by the hydraulic pressure. When the pressure is released, the seals relax, or retract, pulling the pistons back from the rotors **(Figure 2)**. The seal flexing releases pressure from the rotor but maintains only a very slight clearance of 0.001 inch or less between the pad and the rotor. The seal fits closely around the piston and holds it in position. Because the seal is installed at the outer end of the caliper bore, it keeps dirt and moisture out of most of the bore and away from the piston. This minimizes corrosion and damage to the caliper bores and pistons.

As the brake linings wear, the piston can move out toward the rotor to compensate for wear. The seal continues to retract the piston by the same amount, however. Thus, the piston can travel outward, but its inward movement is restricted by the flexibility of the seal. This action provides the inherent self-adjusting ability of disc brakes, and pedal height and travel remain constant throughout the life of the brake linings.

Steel pistons used with fixed seals have very close clearances in their bores. Manufacturers typically specify 0.002–0.005 inch of clearance. Phenolic pistons require slightly more clearance to allow for more expansion. They are typically installed with 0.005–0.010 inch of clearance. The close fit of the pistons in their bores and the close running clearances between pads and rotors keep pistons from cocking in the calipers and minimize piston knock back due to rotor runout or warping. Close piston-to-bore clearances also keep the seals from flexing too much and rolling in their grooves.

Low-Drag Caliper Seals

Low-drag calipers increase the clearance between the brake pads and rotors when the brakes are released. This increased clearance reduces friction, improves fuel mileage, and reduces exhaust emissions. In a low-drag caliper, the groove for the fixed seal has a tapered outer edge. This lets the seal flex farther as the brakes are applied. This increased flexing outward as the brakes are applied is matched by equal inward flexing as the brakes are released. The result is more clearance between the pads and rotors.

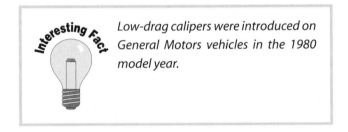

Interesting Fact *Low-drag calipers were introduced on General Motors vehicles in the 1980 model year.*

Low-drag calipers require a quick-takeup master cylinder, which provides more fluid volume with the first part of the pedal stroke to take up the greater pad-to-rotor clearance. The combination of low-drag calipers and a quick-takeup master cylinder maintains normal brake pedal height and travel.

Stroking Seals

Fixed-caliper disc brakes have piston seals that are called **stroking seals**. A stroking seal is similar to a lip seal and is installed at the rear of the piston **(Figure 3)**. Unlike a fixed seal that is installed in a groove in the caliper bore, a stroking seal is installed on the piston and moves with it. Sealing is provided by the seal lip, and the seal flexes to retract the piston similar to the way a fixed seal works. Pistons used with stroking seals are installed with slightly more clearance than pistons with fixed seals.

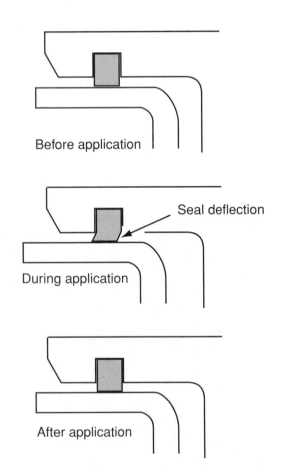

Before application

During application

Seal deflection

After application

Figure 2. These cross-sectional views show how a seal deflects during brake application and then returns to the relaxed position to retract the piston.

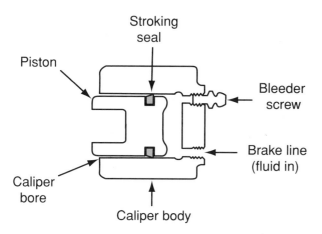

Figure 3. A stroking seal is like a lip seal and is installed toward the rear of the piston.

Because stroking seals are installed at the rear of the pistons, the caliper bore can be exposed to moisture and dirt if the dust boot fails. This can lead to scored bore walls and leakage as the piston moves out in its bore. Stroking seals also provide less support for the piston in the bore and can let the piston cock, or tilt, slightly. If a piston tilts in its bore, it is more likely to be knocked back in the bore as the rotor turns. Piston travel will then increase on the next brake application. Piston knock back can cause excessive and inconsistent pedal travel. Because of these reasons—caliper corrosion and leakage, and excessive pedal travel—stroking seals are not used in late-model brake designs.

FIXED CALIPERS

A **fixed caliper** is bolted to its support and does not move when the brakes are applied. A fixed caliper must have pistons on both the inboard and the outboard sides.

The pads for a fixed caliper brake are held in the caliper body by locating pins and have antirattle clips and springs to reduce noise and vibration. Fixed calipers must have equal piston areas on each side so that an equal amount of hydraulic pressure can apply an equal force to the pads. When the brakes are released, the pistons are retracted by their seals. Nearly all fixed caliper brakes use stroking-type piston seals.

Because the body of a fixed caliper brake is mounted rigidly to the vehicle, it must be aligned precisely over the brake rotor. Piston travel must be perpendicular to the rotor surface, and the pads must be exactly parallel to the rotor. If the caliper is misaligned, or cocked, on its support, the pads will wear unevenly and the pistons can stick in their bores. Just as importantly, the caliper must be centered over the rotor so that the inboard and outboard pistons move equal distances to apply the pads. If the caliper is offset to one side or the other, unequal piston travel can create a spongy brake pedal feel and uneven pad wear. Moreover, it can be more difficult to bleed all air from the caliper during service. Shims often are used to align fixed calipers on their supports.

Fixed calipers are large and heavy, which provides good heat dissipation and strength to resist high hydraulic pressures. Because a fixed caliper is mounted rigidly to the vehicle, it does not tend to flex under high temperature and pressure from repeated use. Fixed calipers can provide consistent braking feel under repeated, hard use.

The use of fixed calipers has declined over the past twenty years, and they are not found on many late-model, lightweight vehicles. The greater weight of a fixed caliper is its worst disadvantage, and it increases the percentage of unsprung weight on the steering knuckle. Fixed calipers also are more expensive to make than floating or sliding calipers, as well as harder to service. Because fixed calipers are built with a two-piece body, they require more time to service, and leaks have more opportunities to develop around O-rings and crossover line fittings.

MOVABLE CALIPERS

By far, most late-model cars and light trucks have disc brakes with movable calipers. Movable caliper brakes are further subdivided into floating calipers and sliding calipers, which identify the ways in which the calipers are mounted on their supports.

Movable calipers are lighter, easier to service, and cheaper to build than fixed calipers. Because of their simpler design, they also are less likely to develop leaks. The flexible mounting of a movable caliper provides some beneficial self-alignment of the caliper with the rotor. If the flexibility becomes excessive, however, the pads can wear at an angle, or become tapered, which decreases pad life.

Floating Calipers

Floating calipers began to appear on domestic and some imported vehicles in the late 1960s. A floating caliper brake has a one-piece caliper body and usually one large piston on the inboard side. Some medium-duty trucks, particularly Fords, and some late-model General Motors FWD cars have floating caliper brakes with two pistons on the inboard side **(Figure 4)**.

The caliper is mounted to its support on two locating pins, or guide pins, that are threaded into the caliper support **(Figure 5)**. The caliper slides on the pin in a sleeve or bushing. The bushing may be lined with Teflon® or have a highly polished surface for low friction. The pins let the caliper move in and out and provide some flexibility for lateral movement to help the caliper stay aligned with the rotor. The pads are attached to the piston on the inboard side and to the caliper housing on the outboard side.

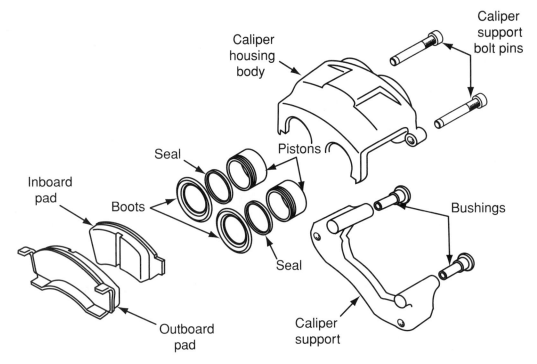

Figure 4.   Some late-model floating calipers are fitted with two pistons.

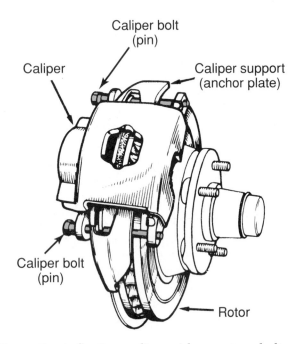

Figure 5.   A floating caliper rides on two bolts or guide pins in the caliper support.

In a floating caliper, the piston and the bottom of the caliper bore are both hydraulic pressure surfaces of equal size, and equal pressure is applied to both the piston and the caliper. Hydraulic pressure against the bottom of the caliper bore creates a reaction force that moves the caliper body inward as the piston moves outward. As a result, the rotor is clamped between the piston on one side and the caliper body on the other.

When the driver releases the brakes, the pressure behind the piston drops. The seal relaxes, or springs back, and retracts the piston. As the piston moves back, the caliper relaxes and moves in the opposite direction on the guide pins to the unapplied position. The piston seals provide the self-adjusting action and the required pad-to-rotor clearance.

Sliding Calipers

Sliding calipers operate on exactly the same principles as floating calipers, but their mounting method is different. The caliper support has two V-shaped surfaces that are called abutments, or **ways (Figure 6)**. The caliper housing has two matching machined surfaces. The caliper slides onto the caliper support, where the two parts are held together with a caliper support spring, a key, and a key retaining screw. An antirattle spring is used to prevent noise from vibration.

When a sliding caliper is replaced, the caliper ways on the caliper support should be inspected closely and caliper movement should be checked to be sure that the replacement caliper slides correctly. It may be necessary to polish the caliper ways with a fine file or emery cloth.

The pins on a floating caliper should be lubricated according to the manufacturer's instructions, but lubrication is even more important on a sliding caliper. Both the

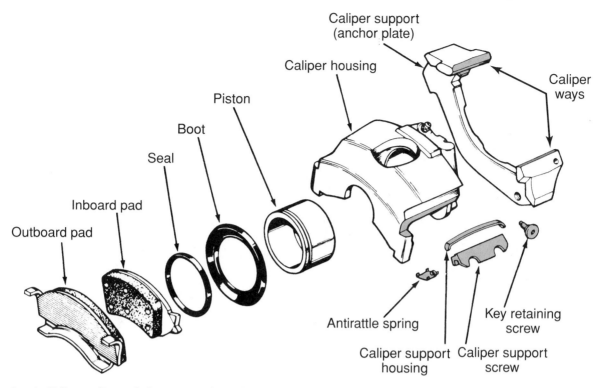

Figure 6.   A sliding caliper slides on machined ways on the caliper support.

caliper ways and the mating surfaces on the caliper should be lubricated with high-temperature brake grease. This ensures smooth operation and prevents corrosion on the sliding surfaces.

REAR WHEEL DISC BRAKES

Rear wheel disc brake calipers can be fixed, floating, or sliding, and all of these designs work in the same way as when they are used at the front wheels. The only difference between a front and a rear disc brake caliper is the need for a parking brake in the rear.

Drum brakes provide a better static coefficient of friction than disc brakes do. The brake linings grab and hold the drums more tightly than disc pads can hold a rotor. The servo action of drum brakes contributes to this feature, as does the larger area of brake shoe linings compared to the lining on disc brake pads.

Because most brake installations have discs at the front and drums at the rear, the parking brake weaknesses of disc brakes are not a problem. Four-wheel disc brake installations, however, must have some way to mechanically apply the rear brakes. With movable (floating or sliding) calipers, this is usually done with a cam-and-lever arrangement that mechanically moves the caliper's piston to develop clamping force **(Figure 7)**. Disc brakes with fixed calipers usually have small, cable-operated brake shoes that grip a small drum surface inside the center of

the rotor. Some late-model rear disc brakes with sliding calipers also have small drum-type parking brakes built into the rotors **(Figure 8)**. Late-model General Motors trucks use an expandable, single metal band covered with friction material inside the parking brake drum that is machined on the inside of the rear brake disc **(Figure 9)**.

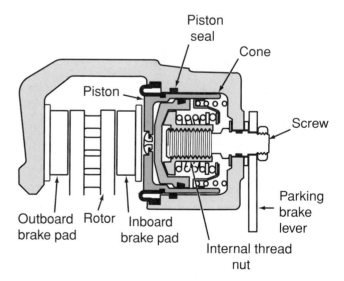

Figure 7.   Most rear disc brakes have a lever-and-screw mechanism to apply the service brakes for parking.

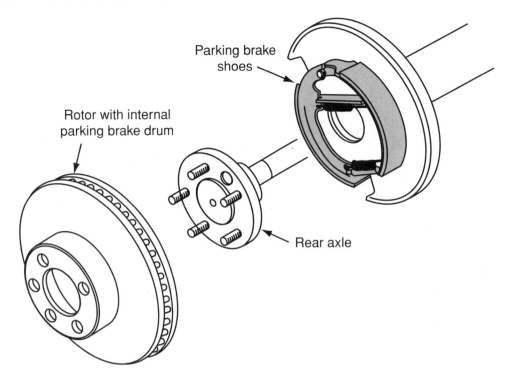

Figure 8.   Some rear disc brake assemblies have a rotor with an internal brake drum and a small set of parking brake shoes.

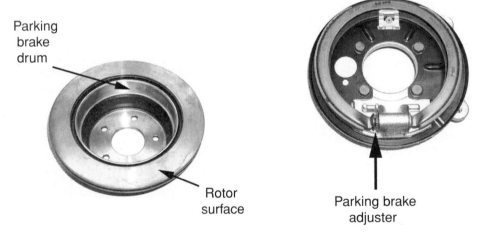

Figure 9.   Late-model General Motors trucks use a single metal band covered with friction material inside a brake drum machined inside the rotor as the parking brake.

Summary

- Disc brake systems are commonly classified into two groups in terms of caliper operation: fixed and movable.
- Fixed calipers are bolted rigidly to the caliper support and have pistons and cylinders (bores) on the inboard and outboard sides.
- Movable calipers slide or float on the caliper support and have pistons on only the inboard side.
- Fixed calipers have a minimum of two pistons with at least one on each side of the caliper.
- Movable calipers use fixed seals, also called square-cut seals or lathe-cut seals, to keep fluid from escaping

between the pistons and their bores and to pull the piston back into its bore when the brakes are released.

■ Fixed calipers use a type of lip seal called a stroking seal.

■ Low-drag calipers increase the clearance between the brake pads and rotors when the brakes are released, thus providing improved fuel mileage.

■ Low-drag calipers require a quick-takeup master cylinder.

■ Rear wheel disc brake calipers may be fixed, floating, or sliding types.

■ The only difference between a front and a rear disc brake caliper is the need for a parking brake in the rear.

■ With movable rear calipers, a cam-and-lever arrangement mechanically moves the caliper's piston to develop clamping force for the parking brake.

Review Questions

1. How does a low-drag caliper operate?
2. What is the difference between floating and sliding calipers?
3. What types of caliper seals are commonly used in movable calipers?
4. Low-drag calipers are being discussed. Technician A says that low-drag calipers require a quick-takeup master cylinder. Technician B says that the combination of low-drag calipers and a quick-takeup master cylinder maintains normal brake pedal height and travel. Who is correct?
   A. Technician A only
   B. Technician B only
   C. Both Technician A and Technician B
   D. Neither Technician A nor Technician B

5. Technician A says that fixed calipers use a piston on each side of the rotor to apply the brakes. Technician B says that sliding calipers typically use only one piston on one side of the rotor. Who is correct?
   A. Technician A only
   B. Technician B only
   C. Both Technician A and Technician B
   D. Neither Technician A nor Technician B

Chapter 28

Brake Shoe and Pad Linings

Introduction

All drum brakes have a pair of shoes at each wheel. On nonservo brakes, the lining on the leading and trailing shoes is usually the same length and positioned in the same location on each shoe. On a duo-servo brake, however, the primary shoe lining is shorter than the secondary lining and can even have a different coefficient of friction. The secondary shoe applies most of the stopping force and thus requires a longer lining. The coefficient of friction for the secondary lining is often higher than for the primary lining to provide good stopping power. On the other hand, the lower coefficient of friction for the primary lining helps to keep the brakes from applying too harshly and locking.

Lining for the secondary shoe is nearly always centered from top to bottom on the shoe table. Primary lining can be centered from top to bottom on the table, or it can be offset toward either the top or bottom depending on the desired operation of the brake assembly **(Figure 1)**.

Each brake caliper contains two brake pads **(Figure 2)**. The brake pads are positioned in the caliper on the inboard and outboard sides of the rotor. The caliper piston, or pistons, forces the brake pad linings against the rotor surfaces to stop the car.

BRAKE FRICTION MATERIALS

The friction material used for drum brake lining is generally softer than that used on disc brake pads because the friction surface is larger. Today's brake lining materials are much the same for drum brakes as for disc brakes except for differences in friction coefficients. Each of these has dif-

ferent characteristics. Some of the most important characteristics are:

1. The ability to resist fading when the brake system temperature increases.
2. The ability to resist fading when the parts get wet.
3. The ability to recover quickly from heat or water fading.
4. The ability to wear gradually without causing excessive wear to brake rotors or drum surfaces.

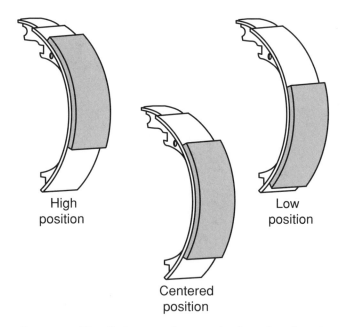

High position

Low position

Centered position

Figure 1. The lining can be attached to the shoe in different locations depending on the brake design and application.

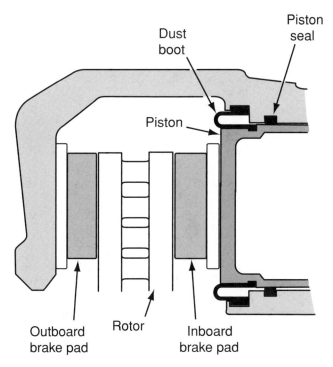

Figure 2. The placement of the brake pads in fixed and floating caliper brake designs.

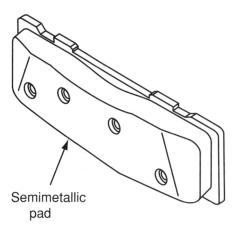

Figure 3. Asbestos-free semimetallic brake pads.

5. The ability to provide a quiet, smooth frictional contact with a rotor or drum.

Organic Linings

Organic linings are made from nonmetallic fibers bonded together to form a composite material. In the past, asbestos was the main ingredient in organic linings but is no longer used in shoes made in North America. Today's organic brake linings contain the following types of materials:

- Friction materials and **friction modifiers**, some common examples of which are graphite, powdered metals, and even nut shells.
- **Fillers**, which are secondary materials added for noise reduction, heat transfer, and other purposes.
- Binders, which are glues that hold the other materials together.
- Curing agents that accelerate the chemical reaction of the binders and other materials.

Organic linings have a high coefficient of friction. They are economical, quiet, wear slowly, and are only mildly abrasive to drums. However, organic linings fade more quickly than other materials and do not operate well at high temperatures. High-temperature organic linings are available for high-performance use but they do not work as well at low temperatures and wear faster than regular organic linings.

Semimetallic Linings

Semimetallic materials **(Figure 3)** are made from a mixture of organic or synthetic fibers and certain metals molded together; they do not contain asbestos. **Semimetallic linings** are harder and more fade resistant than organic materials are but require higher brake pedal effort.

Most semimetallic linings contain about 50 percent iron and steel fibers. Copper also has been used in some semimetallic linings and, in smaller amounts, in organic linings. Concerns about copper contamination of the water systems of the United States has led to reduced use in brake linings, however.

Semimetallic linings operate best above 200°F and must be warmed up to bring them into full efficiency. Consequently, semimetallic linings are typically less efficient than organic linings at low temperatures.

Semimetallic linings were sometimes used on older heavy or high-performance vehicles with four-wheel drum brakes. Currently, semimetallic linings are used only on front disc brakes of passenger cars and light trucks. The lighter braking loads on rear brakes, particularly on FWD vehicles, might never heat semimetallic linings to their required operating efficiency. Semimetallic linings also have a lower static coefficient of friction than do organic linings, which makes semimetallic linings less efficient with parking brakes.

Semimetallic linings often are blamed for increased rotor wear, but this is not entirely true. Early semimetallic linings were more abrasive than current materials, which may cause no more wear with the properly matched rotors than organic materials. Also, the better heat transfer characteristics of semimetallic linings can reduce rotor temperatures and help to counteract abrasiveness. Many small, FWD cars built since the early 1980s have smaller front brakes that require the better high-temperature friction characteristics and heat transfer abilities of semimetallic linings.

Metallic Linings

Fully metallic materials were used for many years in racing. **Metallic lining** is made from powdered metal that is formed into blocks by heat and pressure. These materials provide excellent resistance to brake fade but require high brake pedal pressure and create the most wear on rotors and drums. Metallic linings work very poorly until they are fully warmed. Improved high-temperature organic linings and semimetallic materials have made metallic linings almost obsolete for late-model automotive use. Metallic linings are also extremely noisy, which is something that must be considered by customers when choosing the type of brake lining to install on their vehicle.

Synthetic Linings

The goals of improved braking performance and the disadvantages of the other lining materials have led to the development of **synthetic lining** materials. They are classified as synthetic because they are made of non-organic, nonmetallic, and nonasbestos materials. Two types of synthetic materials are commonly used as brake linings for drum brakes: fiberglass and **aramid fibers**.

Fiberglass was introduced as a brake lining material to help eliminate asbestos. Like asbestos, it has good heat resistance, a good coefficient of friction, and excellent structural strength. The disadvantages of fiberglass are its higher cost and its reduced friction at very high temperatures. Overall, fiberglass linings perform similarly to organic linings and are used primarily in rear drum brakes.

Aramid fibers are a family of synthetic materials that are five times stronger than steel, pound for pound, but weigh little more than half what an equal volume of fiberglass weighs. Friction materials made with aramid fibers are made similarly to organic and fiberglass linings. Aramid fibers have a coefficient of friction similar to semimetallic linings when cold and close to that of organic linings when hot. Overall, the performance of aramid linings is somewhere between organic and semimetallic materials but with much better wear resistance and longevity than organic materials.

FRICTION MATERIAL SELECTION

Identification codes, called the **Automotive Friction Material Edge Code**, are printed on the edges of drum brake linings **(Figure 4)** and disc brake pads **(Figure 5)**. The letters and numbers identify the manufacturer of the lining material and the material used, and the last two letters identify the cold and hot coefficients of friction.

From a service standpoint, the friction codes are the most important. These codes are not the primary factor for selecting replacement linings. They do not address lining

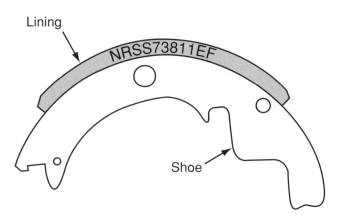

Figure 4. An example of edge coding on a brake shoe.

Figure 5. Automotive Friction Material Edge Code.

quality or its hardness, but they do indicate the coefficient of friction, as follows:

C = not over 0.15
D = over 0.15 but not over 0.25
E = over 0.25 but not over 0.35
F = over 0.35 but not over 0.45
G = over 0.45 but not over 0.55
H = over 0.55

It is important to use the recommended friction material when replacing brake shoes and pads. The incorrect type of friction material can affect the stopping characteristics of the car. These codes, however, indicate only the coefficient of friction. They do not address lining quality or its hardness.

Hard and soft are terms applied to linings within a general category of material. Thus, any particular organic lining may be considered as a hard or a soft organic material. Overall, organic linings are considered softer than semimetallic linings, and semimetallic linings are considered softer than fully metallic linings. A hard lining usually has a low coefficient of friction but resists fade better and lasts longer than a soft lining. A soft lining has a higher coefficient of friction but fades sooner and wears faster than a hard lining. A soft lining is less abrasive on drum surfaces and operates more quietly than a hard lining. It also is common to use linings with a lower coefficient of friction on the rear brakes than on the front to minimize rear brake lockup.

Although the coefficient of friction and hardness of linings can vary quite a bit for the same kind of material, approximate coefficient of friction values can be assigned to different lining materials to help you understand the performance characteristics of the different materials:

Organic—cold: 0.44; warm: 0.48
Semimetallic—cold: 0.38; warm: 0.40
Metallic—cold: 0.25; warm: 0.35
Synthetic—cold: 0.38; warm: 0.45

FRICTION MATERIAL ATTACHMENT

Brake shoes are typically made from welded steel, although some are aluminum. The outer part is called the table and is curved to match the curvature of the drum. The brake lining is riveted or bonded with an adhesive to the brake shoe table **(Figure 6)**. The web is the inner part of the shoe that is perpendicular to the table and to which all of the springs and other linkage parts attach **(Figure 7)**.

Brake pads are made in two parts **(Figure 8)**. They are constructed of a stamped metal backing with the friction lining riveted, bonded, or attached using a combination of riveting and bonding to the metal backing.

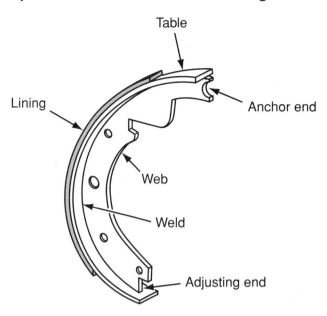

Figure 7. The parts of a typical brake shoe.

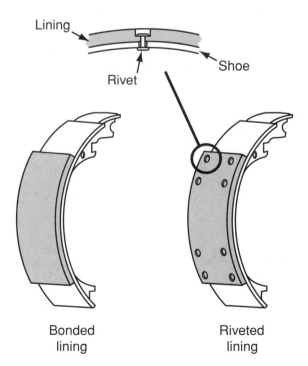

Figure 6. A brake lining can be either riveted or bonded to the shoe.

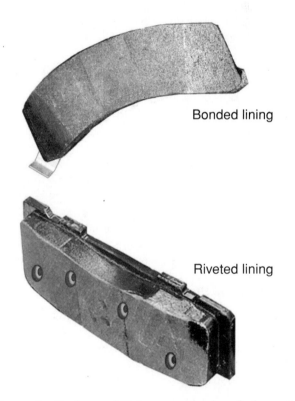

Figure 8. Brake pad linings can be bonded or riveted to the metal backing.

Riveted Linings

In a riveted brake pad assembly, the **riveted lining** is attached to the steel pad by copper or aluminum rivets. Riveting allows a small amount of flexing between the pad and lining to absorb vibration and reduce noise. Riveting also is very reliable, and rivets maintain a secure attachment at high temperature and high mileage.

Rivets require that about one-third to one-quarter of the lining thickness remain below the top of the rivet for secure attachment. This places the rivet head closer to the lining's friction surface and reduces the service life or mileage of the shoe or pad because not all of the lining is available for wear. Once the lining wears enough to allow the rivet heads to contact the drum or rotor, the drum or rotor can become damaged, and therefore the shoes or pads should be replaced. Also, the drum or rotor can get scored by abrasive particles that can collect in the holes above the rivet heads.

Historically, riveting was the most common method for lining attachment. However, since bonding has become so dependable, rivets are used more with semimetallic linings than with organic linings.

Bonded Linings

Bonding is a method of attaching the friction material to the pad with high-strength, high-temperature adhesive. **Bonded linings** have fewer cracking problems than riveted linings because they do not have the stress points created by the rivet holes. Bonded linings are immediately identified by the lack of holes in the lining material; the holes in riveted linings are for the rivets.

Bonded linings can be noisier than riveted linings because they do not have the flexibility between the pad and lining that riveted linings have. A noise complaint often can be fixed by replacing bonded linings with riveted linings that have the recommended friction characteristics.

Bonded linings can provide longer service life because more material is available for wear before any metal contacts the drum or rotor. If the steel backing of a shoe or pad contacts the drum or rotor, severe scoring will occur, which normally results in the replacement of the drum or rotor.

Mold-Bonded Linings

In a pad assembly with **mold-bonded lining**, an adhesive is applied to the pad and the uncured lining material is poured onto the pad in a mold. The assembly is then cured at high temperature to fuse the lining and adhesive to the pad. Holes drilled in the pad are countersunk from the rear so that the lining material flows through the holes and rivets itself to the pad as it cures. Many high-performance pads are made this way to avoid the stress-cracking problems associated with rivets while providing very secure pad attachment.

LINING-TO-DRUM FIT

When drum brakes are applied, a semicircular lining contacts a circular drum. The shoe does not move in a straight line toward the drum, nor does the entire lining surface contact the drum at the same time. For these reasons, the surface fit of the lining to the drum is important.

The leverage that applies the brakes and the motion of brake shoes as they are applied prevent full lining contact with the drum. Therefore, lining shape must be adjusted to overcome these natural conditions and work toward providing full lining-to-drum contact.

The hydraulic force of the wheel cylinder pistons and the rigid positions of the shoe anchors tend to force one end of each shoe into contact with the drum before the center of the shoe. Most drum brake noise complaints are caused by binding at the ends of the shoes as they contact the drum before the center does. More importantly, if the full surface of new brake linings does not contact the drum as the linings are broken in, the linings can overheat and become glazed, which reduces braking effectiveness.

To compensate for these natural characteristics of drum brakes, linings are fitted to shoes to provide more clearance at the ends than at the center. This process is known as **arcing** the shoes, and it forms either cam-ground or undersize linings. A **cam-ground lining** is thinner at the ends than at the center, and the lining surface is not a portion of a circle with a constant radius. An undersize lining has a uniform thickness and a constant radius **(Figure 9)**, but it has a smaller outside diameter than the inside diameter of the drum. A fixed-anchor arc is a variation of an undersize lining in which the overall arc is offset slightly so that one end of the lining is thicker than the other.

Arc grinding, or arcing, of new brake linings used to be a standard part of brake service. Today, however, brake shoe linings are seldom arced as part of the procedure for replacing brake shoes. In fact, most modern brake lathes do not even have the attachments necessary to arc shoes.

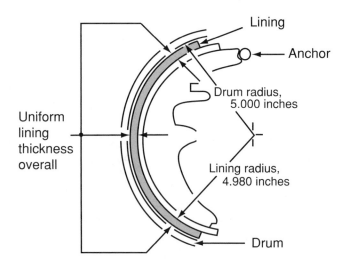

Figure 9. An undersize lining has a shorter radius than the drum radius.

Brake shoes are sold with the lining surface already contoured for proper drum fit.

Several factors led to the decline of arc grinding. Concerns about airborne asbestos and brake dust were the most dominant. Because rear drum brakes provide a smaller percentage of overall braking for late-model FWD cars with front disc brakes, brake linings and drums are less prone to overheating and lining wear is not as severe as in the past. Therefore, arcing requirements are not as great, and precontoured linings are provided by manufacturers.

PAD-TO-CALIPER ATTACHMENT

Most brake pads are held in the caliper by locating tabs formed on the end of the metal backing plate **(Figure 10)**.

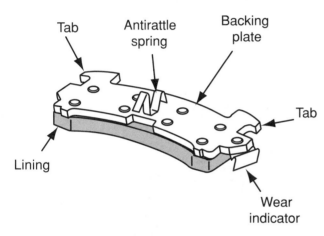

Figure 10.   Tabs on the metal backing plate locate the pad in the caliper.

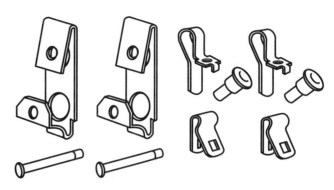

Figure 12.   Typical antirattle and retaining hardware for brake pads.

The pads also can be retained by retaining pins that go through holes in the metal pad backing.

Many pads have antirattle shims or springs or support plates or clips, which are spring steel clips that hold the pads in position to keep them from rattling when the pads are out of contact with the rotor **(Figure 11)**. These small parts are called **pad hardware** and are often replaced when the pads are replaced. **Figure 12** shows a selection of typical pad hardware.

BRAKE PAD WEAR INDICATORS

Many current brake systems have **pad wear indicators**, or sensors, to warn the driver that the lining material has worn to its minimum thickness and that the pads require replacement. The most common types are audible contact sensors and electronic sensors.

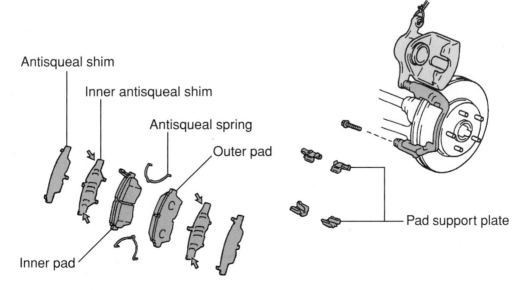

Figure 11.   A sliding caliper with support plates, antirattle shims, and springs.

NEW PAD **WORN PAD**

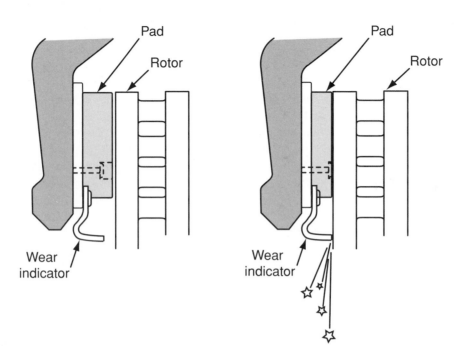

Figure 13. As the pad lining wears, the wear indicator eventually contacts the rotor and makes a noise, warning the driver that new pads are needed.

The audible sensor is the oldest type. The audible system uses small spring clips on the brake pads and lining that make a noise when the linings are worn enough to be replaced. The spring clips are attached to the edge of the brake shoe or into the shoe from the pad side. They are shaped or positioned to contact the rotor when the lining wears to where it should be replaced. When the linings wear far enough, the sensor contacts the rotor **(Figure 13)** and makes a high-pitched squeal when the brakes are not applied to warn the driver that the system needs servicing.

Electronic sensors provide a warning lamp or message on the instrument panel to inform the driver of worn brake pad linings. The electronic sensor uses pellets embedded into the friction material to complete the circuit. Early systems used a grounding logic to complete a parallel circuit and turn on the warning lamp **(Figure 14A)**. When the linings are worn enough, the contacts touch the rotor. This completes an electrical circuit to turn on a warning lamp on the instrument panel. If a wire within the circuit became grounded, the warning lamp would also light. An open circuit would not be detected and would disable the wear indicator circuit. Later systems were designed to use open logic for wear detection. Open logic allows the system to detect both opens and shorts-to-ground in the circuit, as well as indicate wear **(Figure 14B)**.

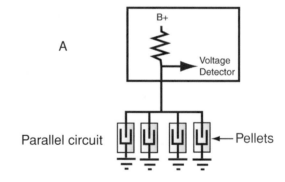

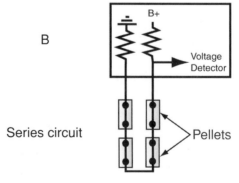

Figure 14. Electronic wear sensors can be based on parallel (A) or series (B) circuits within the pad.

Summary

- On nonservo brakes, the lining on the leading and trailing shoes is usually the same length and positioned in the same location on each shoe.
- On a duo-servo brake, the primary shoe lining is shorter than the secondary lining and can have a different coefficient of friction.
- Lining for the secondary shoe is almost always centered from top to bottom on the shoe table, and the primary lining can be centered from top to bottom on the table or offset toward either the top or the bottom.
- The friction material used for disc brake pads is generally harder than that used on drum brake linings because the friction surface is smaller and high pressures are used to push the pads into contact with the rotor.
- Organic linings contain friction materials and friction modifiers, fillers, binders, and curing agents.
- Organic linings have a high coefficient of friction for normal braking. They are economical, quiet, wear slowly, and are only mildly abrasive to drums and rotors.
- Semimetallic materials are made from a mixture of organic or synthetic fibers and certain metals molded together; they are harder and more fade resistant than organic materials but require greater brake pedal effort.
- Metallic lining is made from powdered metal and provides excellent resistance to brake fade but requires high brake pedal pressure and creates the most wear on rotors and drums.
- Synthetic linings are made of non-organic, nonmetallic, and nonasbestos materials, typically fiberglass and aramid fibers.
- Identification codes, called the Automotive Friction Material Edge Code, are printed on the edges of drum brake shoe and disc brake pad linings.
- A hard lining has a low coefficient of friction but resists fade better and lasts longer than a soft lining.
- A soft lining has a higher coefficient of friction and fades sooner and wears faster than a hard lining, but it is less abrasive on rotor surfaces and operates more quietly than a hard lining.
- The lining material for brake pads and shoes is riveted, bonded, or attached using a combination of riveting and bonding to its metal backing.
- A mold-bonded lining is manufactured with an adhesive applied to the pad, and the uncured lining material is poured onto the pad in a mold.
- Arcing brake shoes allows more clearance between the ends of the shoe and drum as compared with the clearance between the center of the shoe and the drum.
- A cam-ground lining is thinner at the ends than at the center.
- An undersize lining has a uniform thickness and a constant radius, but it has a smaller outside diameter than the inside diameter of the drum.
- A fixed-anchor arc has an overall arc that is offset slightly so that one end of the lining is thicker than the other.
- Most brake pads are held in the caliper by locating tabs formed on the end of the metal backing plate. The pads also can be retained by retaining pins that go through holes in the metal pad backing.
- Many pads have antirattle shims, springs, support plates, or spring steel clips that hold the pads in position to keep them from rattling when the pads are out of contact with the rotor. These small parts are called pad hardware.
- Pad wear indicators, or sensors, warn the driver that the lining material has worn to its minimum thickness and that the pads require replacement. The most common types are audible contact sensors and electronic sensors.

Review Questions

1. List three common materials used as friction materials and friction modifiers in organic brake linings.
2. _____ linings have been considered softer than semimetallic linings, and semimetallic linings are considered softer than _____ linings. Typically, a hard lining has a _____ coefficient of friction but resists fade better and lasts longer than a soft lining. A soft lining has a _____ coefficient of friction but fades sooner and wears faster than a hard lining. Additionally, a _____ lining is less abrasive on rotor surfaces and operates more quietly than a _____ lining.
3. What are the two most common types of brake pad wear indicators?
4. The four types of lining friction material are: _____, _____, _____, and _____.
5. Which of the following types of lining material has the lowest coefficient of friction when it is cold?
   A. Organic
   B. Semimetallic
   C. Metallic
   D. Synthetic

6. Which of the following statements about brake lining material is *not* true?
   A. All organic materials are considered soft lining materials.
   B. Semimetallic linings are considered softer than fully metallic linings.
   C. A hard lining usually has a low coefficient of friction.
   D. A soft lining fades sooner and wears faster than a hard lining.

7. While discussing duo-servo brake systems Technician A says that the lining for the secondary shoe is almost always centered from top to bottom on the shoe table. Technician B says that the primary lining can be centered from top to bottom on the table or it can be offset toward either the top or the bottom. Who is correct?
   A. Technician A only
   B. Technician B only
   C. Both Technician A and Technician B
   D. Neither Technician A nor Technician B

8. Friction material used in disc brake pads is being discussed. Technician A says that inner and outer pad frictional pads can have different material friction coefficients. Technician B says that friction pad linings with the same coefficient of friction and other characteristics as the original parts should be used when replacing brake pads. Who is correct?
   A. Technician A only
   B. Technician B only
   C. Both Technician A and Technician B
   D. Neither Technician A nor Technician B

Section 6

Wheel Brake Service

Interesting Fact

Modern disc brakes were developed from aircraft brakes used in World War II. These brake systems were originally known as "spot" brakes because the pads contacted one spot on each side of the rotor. Service to spot brakes ensures the spot and the pads are capable of generating enough friction to safely stop the vehicle.

SECTION OBJECTIVES

After you have read, studied, and practiced the contents of this section, you should be able to:

- Diagnose drum brake problems; determine and perform needed repairs.
- Remove, clean, and inspect brake shoes and lining and related hardware, including all springs, pins, clips, levers, adjusters, and backing plate; determine and perform needed repairs.
- Remove, disassemble, clean, inspect, and service a wheel cylinder.
- Properly lubricate brake shoe support pads on backing plate, adjusters, and other drum brake hardware.
- Install replacement brake shoes and related hardware.
- Adjust brake shoes and reinstall brake drums and wheel bearings.
- Preadjust brake shoes and parking brake before installing brake drums or drum/hub assemblies.
- Reinstall wheels, torque lug nuts, and make final brake system checks and adjustments.
- Diagnose disc brake problems, including problems caused by the caliper, pads, hydraulic system, or rotor.
- Remove, clean, inspect, and replace brake pads and retaining hardware.
- Remove caliper assembly from mountings; clean and inspect for leaks and damage to caliper housing; check caliper mounting and slides for wear and damage; determine necessary action.
- Reassemble, lubricate, and reinstall caliper, pads, and related hardware and seat pads and inspect for leaks.
- Adjust calipers with an integrated parking brake system.
- Remove, clean (using proper safety procedures), inspect, measure, and service brake drums and brake rotors, including machining on a brake lathe.
- Describe the components of a wheel bearing assembly.
- Service wheel bearings, including removal, cleaning, repacking, and installation.
- Troubleshoot braking problems related to wheel bearings.

Chapter 29

Drum Brake Service

Introduction

This chapter covers diagnosis, disassembly, cleaning, inspection, and reassembly of drum brake systems. When disassembling or performing any service to drum brakes, you should pay heed to the following precautions:

- When servicing drum brakes, never use an air hose or a dry brush to clean the brakes. Use OSHA-approved cleaning equipment to avoid breathing brake dust.
- Do not spill brake fluid on the vehicle; it may damage the paint. To keep fluid from spraying or running out of lines and hoses, cover the fittings with shop cloths when disconnecting them.
- Always use the type of DOT brake fluid specified by the vehicle's manufacturer.
- During servicing, keep grease, oil, brake fluid, or any other foreign material off the brake linings and drums.
- Handle brake shoes and drums carefully to avoid scratching the drums or nicking or scratching brake linings.
- Always service drum brakes and linings in axle sets.

DIAGNOSING DRUM BRAKE PROBLEMS

If you suspect problems in the drum brake system, road test the vehicle and note any abnormal operation. As you apply the brake pedal **(Figure 1)**, check for excessive travel and sponginess. Listen for noises: not just the obvious sounds of linings grinding on the drums, but mechanical clanks, clunks, and rattles. Common problems and their probable causes follow.

- *Brake drag or grab*—oil, grease, or brake fluid on linings; a damaged or improperly adjusted brake shoe; glazed or worn linings; a broken or weakened return spring; a

loose backing plate; an out-of-round drum; a scored drum; or a faulty wheel cylinder.
- *Pulling to one side*—a damaged or improperly adjusted brake shoe; a loose lining; glazed or worn linings; a broken or weakened return spring; self-adjuster malfunction; oil, grease, or brake fluid on linings; an out-of-round drum; a scored drum; a faulty wheel cylinder; a collapsed brake hose; or damaged steering or suspension parts.
- *Wheel lockup*—a damaged or improperly adjusted brake shoe; glazed or worn linings; a loose backing plate; oil, grease, or brake fluid on linings; a faulty wheel cylinder; a collapsed brake hose; or a restricted line or hose.
- *Hard pedal*—a bad booster.

Figure 1. Press down on the brake pedal and check for excessive pedal travel and sponginess.

- *Excessive pedal travel*—air in the system; a damaged or improperly adjusted brake shoe; glazed or worn linings; self-adjuster malfunction; a bad master cylinder; a cracked drum; or a bell-mouthed or distorted drum.
- *Noisy operation*—a damaged or improperly adjusted brake shoe; glazed or worn linings; worn support pads on a backing plate; self-adjuster malfunction; oil, grease, or brake fluid on linings; a scored drum; or a faulty wheel cylinder.
- *Brake chatter*—loose lining; glazed or worn linings; a loose backing plate; oil, grease, or brake fluid on linings; hot spots on a drum; or an out-of-round drum.
- *Poor braking*—glazed or worn linings; or oil, grease, or brake fluid on linings.

> **You Should Know** *Road test a vehicle under safe conditions and observe all traffic laws. Never attempt any maneuvers that would jeopardize your control of the vehicle.*

If you find any of these conditions during your test drive, remove the drums and inspect the brakes. Any wear on the shoes, shoe holddown and retracting hardware, drums, or wheel cylinder will make a complete brake overhaul necessary.

Many state vehicle inspection laws specify that drums must be pulled at regular intervals and the system inspected. Always remove the drums for inspection whenever the customer has a brake-related complaint or concern or whenever you suspect a problem.

GENERAL INSPECTION

For a complete inspection of a drum brake assembly, you must remove the wheels to inspect the brake shoes, attaching hardware, and drum. Before disassembling the brake assemblies, the wheel and brake assemblies should be checked for obvious damage that could affect brake system performance, including:
- Tires for excessive or unusual wear or improper inflation.
- Wheels for bent or warped rims.
- Wheel bearings for looseness or wear.
- The suspension system for worn or broken components.
- The brake fluid level in the master cylinder.
- Signs of leakage at the master cylinder, in brake lines or hoses, at all connections, and at each wheel.

Correct any of these problems before assuming the problem is in the drum brake assembly.

DRUM BRAKE SERVICE

To service and/or inspect a drum brake system, you must raise the vehicle on a hoist or safety stands, remove the wheels, and disassemble the brake system. To prepare for servicing the brakes, follow these general steps:
1. If the vehicle has an electronically controlled suspension, turn the suspension service switch off.
2. Raise the vehicle on a hoist or safety stands and support it safely.
3. Remove the wheels from the brakes to be serviced. Brakes are always serviced in axle sets, so you may remove both front wheels, both rear wheels, or all four.
4. Thoroughly vacuum or wet-clean the brake assembly to remove all dirt, dust, and fibers.

BRAKE DRUM REMOVAL

The brake drum must be removed for all brake service except bleeding the wheel cylinder and lines or a basic manual adjustment. Removal procedures are different for fixed and floating drums. In all cases, however, you might need to back off the parking brake adjustment (**Figure 2**) or manually retract the self-adjusters (**Figure 3**) to have

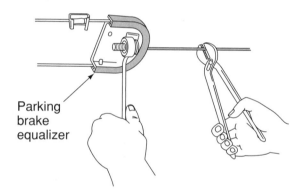

Parking brake equalizer

Figure 2. Loosen the parking cables to ensure there is enough shoe-to-drum clearance to remove the drum.

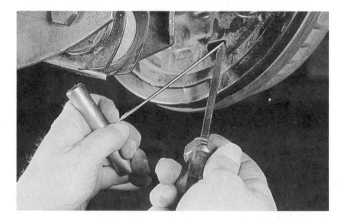

Figure 3. Use a heavy piece of wire to move the adjuster lever away from the teeth of the starwheel. Then rotate the starwheel with a brake adjustment tool to retract the shoes enough to remove the drum.

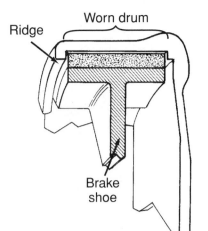

Figure 4. The ridge in this drum would prevent the drum from coming off unless the shoes were retracted.

enough shoe-to-drum clearance to remove the drum. Wear on the friction surface of the drum creates a ridge at the edge of the drum's rim. As the self-adjusters move the shoes outward to take up clearance, the shoe diameter becomes larger than the ridge diameter. If the adjuster is not retracted, the drum's ridge can jam on the shoes and prevent drum removal **(Figure 4)**. Trying to force the drum over the shoes can damage brake parts.

> ⚠ **You Should Know** *Do not step on the brake pedal while a brake drum is off. This will cause the piston in the wheel cylinder to overextend or pop apart.*

Before you remove a drum, mark it L or R for left or right, respectively, so that it gets reinstalled on the same side of the vehicle from which it was removed.

Fixed (One-Piece) Drum Removal

Brake drums that are made as a one-piece unit with the wheel hub are common as rear drums on FWD cars and on the front wheels of older vehicles with four-wheel drum brakes. The hub contains the wheel bearings and is held onto the spindle by a single large nut **(Figure 5)**. This nut also is used to adjust the wheel bearings. Remove a fixed drum mounted on tapered roller bearings as follows:

1. Release the parking brake if you are removing a rear drum.
2. Use a pair of dust cap pliers or large slip-joint pliers to remove the dust cap from the hub.
3. Remove the cotter pin from the castellated nut or nut lock on the spindle. Then remove the nut lock, if used.
4. Remove the spindle nut and washer.
5. Pull outward on the drum to slide it off the spindle. If the drum drags or catches on the brake shoes, slide it back onto the spindle and temporarily reinstall the spindle nut. Then retract the parking brake and the brake shoe adjustment.
6. When you remove the drum, be careful not to drop the outer wheel bearing on the floor and do not drag the inner bearing across the spindle, particularly the threads at the end of the spindle.

Occasionally the inner bearing race of a drum sticks on a spindle and prevents easy drum removal. In such a case, use a puller to remove the drum **(Figure 6)**. After the drum is removed, use another puller or pair of small pry bars to remove the bearing race from the spindle.

After any drum is removed, inspect the grease in the hub and on the bearings. If the grease is dirty or dried out and hard, it is a clue to possible bearing damage. Set the

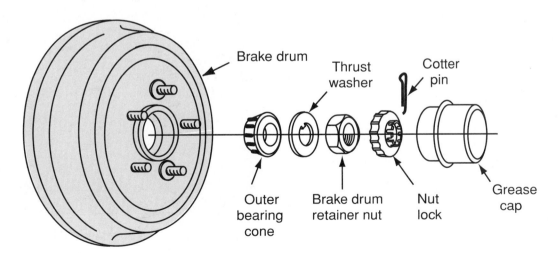

Figure 5. To remove a one-piece drum, remove the large retaining nut and slide the drum, with its bearings, off the spindle.

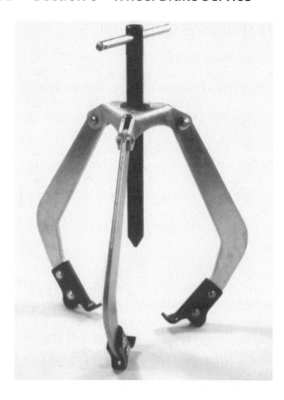

Figure 6. If a drum is stuck, it often can be removed with a puller.

drum and all bearing parts aside for cleaning and close inspection. If the grease seems to be in good condition, place the drum on a bench with the open side down. Cover the outer bearing opening with a shop cloth to keep dirt out.

Floating (Two-Piece) Drum Removal

Floating drums are separate assemblies that are not made with a wheel hub or axle. On a rear-wheel-drive (RWD) vehicle, the drums are held in place on studs in the axle flange by the wheel and wheel nuts. Some FWD cars have rear hubs that contain sealed bearings and are mounted to the spindle of the rear axle. These installations

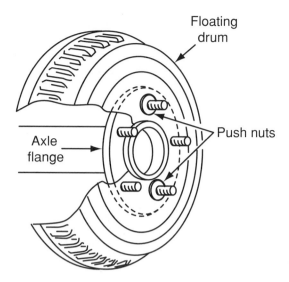

Figure 7. Push nuts are often used to retain a floating drum onto the wheel studs of an axle flange.

use a floating drum that is mounted on the hub flange by studs and held secure by the wheel's lugs.

On many floating drums, **push nuts** or speed nuts are used, during vehicle assembly, to hold the drum onto two or three studs **(Figure 7)**. Typically the push nuts do not need to be reinstalled during service. However, on some vehicles, the push nuts are used to hold the drum squarely against the axle or hub flange.

Remove a fixed drum mounted on a hub or an axle flange as follows:
1. Release the parking brake.
2. Mark the position of the drum to the axle flange so it can be reinstalled in the same location **(Figure 8)**.

Figure 8. Mark the position of the drum to the axle flange so it can be reinstalled in the same location.

3. Some drums are fastened to the axle flanges with small cap screws. Remove the cap screws with a suitable screwdriver or wrench to remove the drum. At times, an impact screwdriver might be required to loosen the screws.

4. If the drum is not rusted or stuck to its flange, lift it off and move it to a bench. If the drum drags or catches on the brake shoes, slide it back onto the flange or hub and retract the parking brake and the brake shoe adjustment.

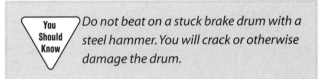
> **You Should Know** *Do not beat on a stuck brake drum with a steel hammer. You will crack or otherwise damage the drum.*

5. Penetrating oil can assist in loosening a stuck drum. If the drum is stuck to its flange **(Figure 9)**, use a large scribe or center punch to score around the joint at the drum and flange and break the surface tension. If this does not loosen the drum, place a block of hard wood against the drum and strike it with a large hammer. If the drum remains stuck, use a puller to separate it from the flange.

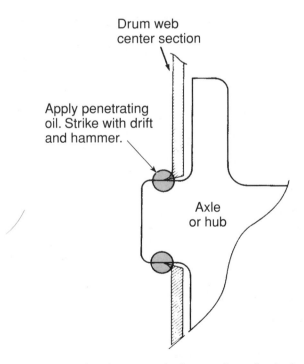

Drum web center section

Apply penetrating oil. Strike with drift and hammer.

Axle or hub

Figure 9. The drum might be stuck to the hub and make it difficult for you to remove it. Penetrating oil can help separate the drum from the hub.

DRUM BRAKE CLEANING

Dirt and dust created by drum brake lining wear stays inside the enclosure formed by the drum and backing plate. Fine metallic particles caused by drum wear combine with lining dust to create a unique brake grime, most of which accumulates on the backing plate, the shoes, the springs, and other parts of the brake assembly.

Special cleaning equipment is available for brake service and should always be used to clean brake assemblies and components. Vacuum enclosures and aqueous, or water-based, cleaning equipment are among the most popular. General-purpose parts washers also can be used to clean brake parts after they are removed from the car.

> **You Should Know** *Do not blow dust and dirt off brake assemblies with compressed air outside of a brake cleaning enclosure. Airborne dust and possibly asbestos fibers are an extreme respiratory hazard.*

Examine the rear wheels for evidence of oil or grease leakage past the wheel bearing seals. Such leakage could cause brake failure and indicates the need for additional service work. Thoroughly clean and inspect the backing plates, struts, levers, and other metal parts to be reused.

Cleaning with Parts Washers

Nearly all shops have a parts washer that is used to clean miscellaneous small and medium-sized parts that are removed from vehicles. Traditionally, parts washers have used petroleum solvent, but many current models use water-based solvents and detergents. Do not use petroleum-based products to clean brake drums. Cast iron is porous and will hold the solvent and damage newly installed shoes. A parts washer that uses water-based solvents and detergents as the only solvent is safe for brake cleaning.

DRUM BRAKE INSPECTION

Rear drum brakes wear much more slowly than front disc brakes because only 20–40 percent of the braking effort is provided by the rear brakes. On older four-wheel drum brake systems, front-to-rear wear is closer to equal, but front brakes still tend to wear more than rear brakes. Because most drum brakes today are used on the rear wheels with disc brakes on the front, it is important not to overlook rear brake inspection. After the drum is removed, set it aside for inspection and measurement. Then inspect the shoes and lining, the wheel cylinders, the springs, and other parts.

Lining and Shoe Inspection

Inspecting the lining should begin with checking its thickness. Lining thickness is the first thing—but not the only thing—that determines the need for replacement. Most vehicle manufacturers specify a minimum lining thickness of 1/32 inch (0.030 inch, or 0.75 millimeter) above the shoe table or above the closest rivet head. Use a depth gauge **(Figure 10)** or scale graduated in 1/32 or 1/64-inch increments to measure lining thickness at several points.

This thickness recommendation means that 1/32 inch is the minimum safe thickness at which the brake lining can perform to its minimum requirements. Worn linings cannot dissipate heat adequately, and the last 1/32-inch of lining will wear much faster than the first 1/32-inch of fresh linings.

Also inspect the lining for cracks, missing rivets, and looseness. Check for contamination from grease, oil, or brake fluid. A leaking wheel cylinder can deposit brake fluid on the linings. A leaking oil seal on a rear drive axle can let gear oil from the differential get onto the brakes. A less frequent, but possible, cause of lining contamination is a leaking hub grease seal for the wheel bearings. If the linings are damaged or contaminated in any way, they must be replaced. Remember also that brake linings are serviced in axle sets, so all linings for both wheels must be replaced if any are damaged.

Check the linings for unequal wear on any shoe of an axle set **(Figure 11)**. Also look for uneven lining wear on any one shoe. If one lining on a duo-servo brake is worn more than the other, be sure that the primary and second-

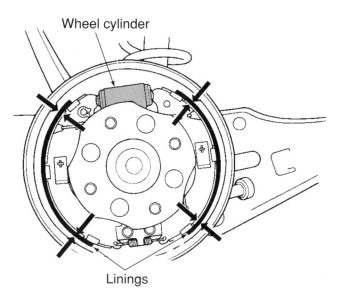

Figure 11. Inspect the linings for unequal wear.

ary shoes are installed in the right locations. If the linings on one wheel are worn more than those of the other, that drum might be scored or rough. Uneven wear from side to side on any one set of shoes can be caused by a tapered drum. Other causes for this wear are poorly adjusted shoes, a frozen wheel cylinder, damaged or distorted springs, or hydraulic problems in a diagonally split system.

If brake shoes and linings have a slight blue coloring, it indicates overheating, and the brake adjuster springs and holddown springs should be replaced. Overheated springs lose their tension and could allow a new lining to drag and wear prematurely if not replaced.

If the lining of one shoe on one wheel is worn more than the other, check the less worn shoe for binding and incomplete application. This type of problem is more likely to occur on a nonservo brake than on a duo-servo brake. If the lining is worn more in the center than at the ends or vice versa, the lining might not have been arced properly to the drum diameter or the drum is excessively worn. Check the lining-to-drum fit closely when you install new shoes with a resurfaced drum. If the linings are worn more at the end where the parking brake applies the shoes, the parking brake might be adjusted too tightly. Linings worn badly at the toe or heel also can indicate an out-of-round drum.

Wheel Cylinder and Axle Inspection

Inspect the outside of the wheel cylinder for leakage. Then pull back the dust boots and look for fluid at the ends of the cylinder **(Figure 12)**. Minor dampness or seepage and some staining are acceptable, but any noticeable fluid means that the cylinder should be overhauled or replaced. Also be sure that the pushrods engage the shoes properly.

Figure 10. Use a depth micrometer to accurately measure lining thickness.

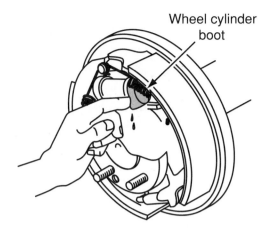

Figure 12. Inspect the wheel cylinder for leakage.

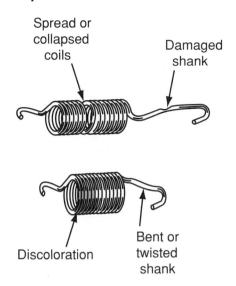

Figure 13. Carefully inspect the shoe return springs.

Check the cylinder mounting on the backing plate for looseness and missing fasteners. Most cylinders are held to the backing plate by small cap screws, but some are secured with a clip. Brake spring tension and tight clearances in the mounting hole can make a clip-mounted cylinder seem secure even when the clip is missing. Inspect this kind of cylinder mounting closely.

Inspect the brakes on a rear drive axle for contamination by gear oil leaking from the axle seal. Use a flashlight to inspect thoroughly behind the axle flange. If you catch a leaking seal in its early stages, you might be able to replace it before gear oil gets to the brake linings. Similarly, inspect the backing plate for grease leaking past the inner wheel bearing seal of a nondriving hub.

Backing Plate, Spring, and Hardware Inspection

Check for a broken or bent plate or other obvious damage. Also look for uneven wear on the shoe support pads that could indicate a bent shoe or incorrectly installed parts. Also check the support pads for etching; these imperfections can cause the brake binding. If you suspect that the backing plate is bent, place a straightedge across two of the shoe support pads as far from each other as possible. If the straightedge does not contact both pads evenly and squarely, you might need to remove the backing plate for closer inspection and measurement.

Inspect the return **(Figure 13)** and holddown springs for damage and unusual wear. Replace any damaged parts and always replace equivalent parts on both sides of the vehicle. Pry the brake shoes slightly away from the backing plate and release them. The holddowns should pull the shoes sharply back to the plate.

Inspect self-adjuster levers and pawls for wear **(Figure 14)** and replace any defective parts. Many shops make it standard practice to replace self-adjuster cables at each brake job, but you should at least inspect them for broken strands and obvious wear and stretching.

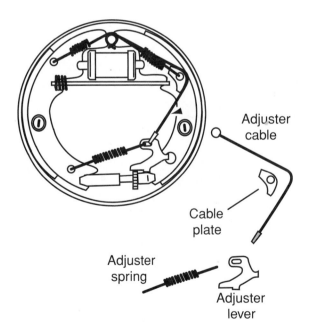

Figure 14. Inspect these self-adjuster parts for wear and damage.

Interesting Fact

Brake shoe return springs can be weakened by excessive heat and hard use but still look all right. If you have any doubt about the condition of a spring, replace it.

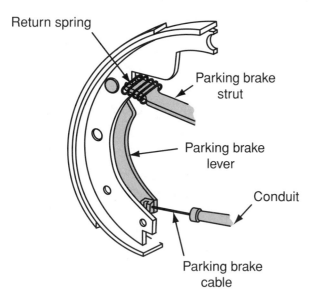

Figure 15.   Carefully inspect the parking brake hardware.

Check the parking brake linkage for damage and rust. Be sure that all parking brake levers and links are properly lubricated and free to move easily **(Figure 15)**.

Drum Inspection

Brake drums absorb heat and dissipate it to the air. As drums wear from normal use and are thinned by refinishing, the amount of metal available to absorb and release heat is reduced. As a result, the drums operate at increasingly higher temperatures. The drum structural strength also is weakened by the loss of metal. Braking forces can distort the drum's shape or lead to cracking or other problems. During your inspection of a drum, you will determine if the drum is good, needs refinishing, or should be replaced.

> **You Should Know** *Never use a drum that exceeds or will exceed its **discard limit** after machining. It is best to replace any drum that exceeds or is close to the discard limit. Also, a technician could be found liable if a known bad brake drum were installed on a vehicle that was in an accident.*

The following are conditions that you may find while inspecting a brake drum. The necessary corrective action for each of these conditions is included in the discussion.

- *Scored drum surface.* Inspect the drum braking surface for scoring by running your fingernail across the surface. Any large score marks mean that the drum must be resurfaced or replaced. The most common cause of drum scoring **(Figure 16A)** is when road grit or brake dust becomes trapped between the brake lining and drum. Glazed brake linings that have been hardened by high temperature, or inferior linings that are very hard, also can groove the drum surface. Excessive lining wear that exposes the rivet heads or shoe steel will score the drum surface. If the scoring is not too deep, the drum can be refinished.
- *Bell-mouthed drum.* Bell-mouthing is shape distortion caused by extreme heat and braking pressure **(Figure 16B)**. It is most common on wide drums that are weakly supported at the outside of the drum. Bell-mouthing makes full drum-to-lining contact impossible, so braking power is reduced. Drums must be refinished.
- *Concave drum.* A concave wear pattern **(Figure 16C)** is caused by a distorted shoe that concentrates braking pressure on the center of the drum.
- *Convex drum.* A convex wear pattern **(Figure 16D)** is caused by excessive heat or an oversized drum, which allows the open end of the drum to distort.

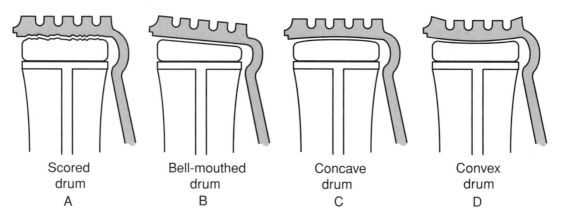

Figure 16.   Typical brake drum wear problems.

- **Hard spots**. Hard spots, or chilled spots, in the cast iron surface **(Figure 17A)** result from a change in metallurgy caused by heat. They appear as small raised areas. Brake chatter, pulling, rapid wear, hard pedal, and noise can occur. Hard spots can be ground out of the drum, but because only the raised surfaces are removed, hard spots can reappear when heat is reapplied. Drums with hard spots should be replaced.
- *Threading*. An extremely sharp or chipped tool bit or a lathe that turns too fast can literally cut a thread into the drum surface. During brake application, the shoes ride outward on the thread and then snap back with a loud crack. Threading can also cause faster lining wear and interfere with shoe alignment during braking. To avoid threading the drum surface, use a properly shaped bit and a moderate-to-slow lathe speed.
- **Heat checks**. Unlike hard spots, heat checks **(Figure 17B)** are visible on the drum surface. They are caused

by high temperatures. Heat-checked drums can have a bluish-gold tint, which is another sign of high operating temperatures. Hardened carbide lathe bits or special grinding attachments are available to service these conditions. Excessive damage by heat checks or hard spots requires drum replacement.
- *Cracked drum*. Cracks in the drum are caused by excessive stress. They can appear anywhere, but they are most common near the bolt circle or at the outside of the drum web **(Figure 17C)**. Cracks also can appear at the open edge of the braking surface. Fine cracks often are hard to see and frequently do not appear until after machining. Any crack, no matter how small, means that the drum must be replaced.
- *Out-of-round drums*. Slightly out-of-round drums usually appear good to the eye, but the problem causes pulling, grabbing, and pedal vibration or pulsation. An out-of-round or egg-shaped condition is often caused by heating and cooling during normal brake operation. To test for an out-of-round drum before the drum is removed, adjust the brake to a light drag and feel the rotation of the drum by hand. Any areas of heavy drag or no drag can indicate a problem. Remove the drum and measure it at several points to determine the amount of distortion. A brake drum that is out of round enough to cause vehicle vibration or roughness when braking should be refinished. Remove only enough metal to return the brake drum to roundness.
- *Grease or oil contamination*. If the drums have been exposed to leaking oil or grease, thoroughly clean them with a nonpetroleum solvent such as denatured alcohol or brake cleaner. Locate the source of the oil or grease leak and fix the problem before reinstalling new or refinished drums.

DRUM INSTALLATION

If the drum is a two-piece floating drum, make sure all mounting surfaces are clean. Apply a small amount of silicone dielectric compound to the pilot diameter of the drum before installing the drum on the hub or axle flange. If the drum has an alignment tang or a hole for a locating screw, make sure it is lined up with the hole in the hub or axle flange.

Install the wheel and tire on the drum and torque the wheel nuts to specifications, following the recommended tightening pattern. Failure to tighten in the correct pattern can result in increased lateral runout, brake roughness, or pulsation.

If the drum is a fixed, one-piece assembly with the hub that contains the wheel bearings, clean and repack the bearings and install the drum.

After lowering the vehicle to the ground, pump the brake pedal several times before moving the vehicle. This positions the brake linings against the drum and verifies that pedal operation is correct. If so equipped, turn the air suspension service switch back on.

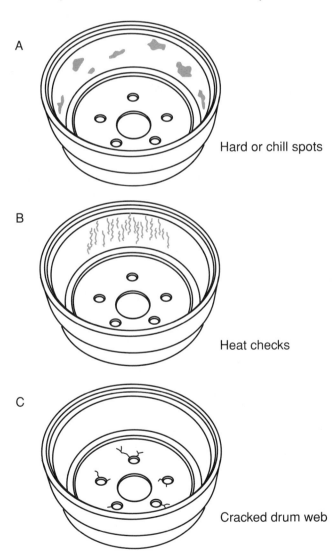

A

Hard or chill spots

B

Heat checks

C

Cracked drum web

Figure 17.   Hard spots (A), heat checks (B), and cracks (C) are common drum problems.

Summary

- During a road test, pay attention to the feel and action of the brake pedal and the behavior of the vehicle when the brakes are applied.
- Before brake work, the suspension, wheel, and brake assemblies should be checked for obvious damage.
- Vacuum enclosures and aqueous cleaning equipment are available for brake service and should always be used to clean brake assemblies and components.
- The thickness and wear pattern of the brake linings should be checked, in addition to checking the brake linings for damage.
- Inspect the outside of the wheel cylinder for leakage, then pull back the dust boots and look for fluid at the ends of the cylinder.
- Inspect the return and holddown springs for damage and unusual wear.
- Careful inspection of a brake drum will determine if the drum can be reused after refinishing or if it should be replaced. You should make sure the drum does not exceed or come close to exceeding its discard limit.

Review Questions

1. Which of the following would not cause a hard brake pedal?
   A. A leaking wheel cylinder
   B. A damaged or improperly adjusted brake shoe
   C. Glazed or worn linings
   D. A bell-mouthed or distorted drum.

2. Technician A checks the surface of a drum for scoring by running a fingernail across the surface. Technician B replaces any drum that is scored. Who is correct?
   A. Technician A only
   B. Technician B only
   C. Both Technician A and Technician B
   D. Neither Technician A nor Technician B

3. Technician A replaces all brake drums that have hard spots. Technician B replaces all drums that have heat checks. Who is correct?
   A. Technician A only
   B. Technician B only
   C. Both Technician A and Technician B
   D. Neither Technician A nor Technician B

4. Drum linings are badly worn at the ends of the linings. Technician A says that the problem is an out-of-round drum. Technician B says that the problem is a tapered drum. Who is correct?
   A. Technician A only
   B. Technician B only
   C. Both Technician A and Technician B
   D. Neither Technician A nor Technician B

5. Technician A says that weak or broken return springs can cause brake drag or pulling to one side. Technician B says that the same problems can be caused by a loose backing plate or an inoperative self-adjuster. Who is correct?
   A. Technician A only
   B. Technician B only
   C. Both Technician A and Technician B
   D. Neither Technician A nor Technician B

Chapter 30

Drum Brake Component Replacement

Introduction

The parts of a drum brake that might need to be replaced depend on the results of your diagnosis and inspection. The contents of this chapter cover the replacement of brake shoes, return springs, holddown springs, and self-adjusters. Also included are the different ways to adjust the brake shoes after the brake has been reassembled.

DRUM BRAKE DISASSEMBLY

All current designs of drum brakes have similar main components (**Figure 1**), but they can use different individual small parts and the location of the parts can be different. If you are not thoroughly familiar with the brake design you are servicing, refer to the·vehicle's service manual for guidance. It is often helpful to have an illustration of the

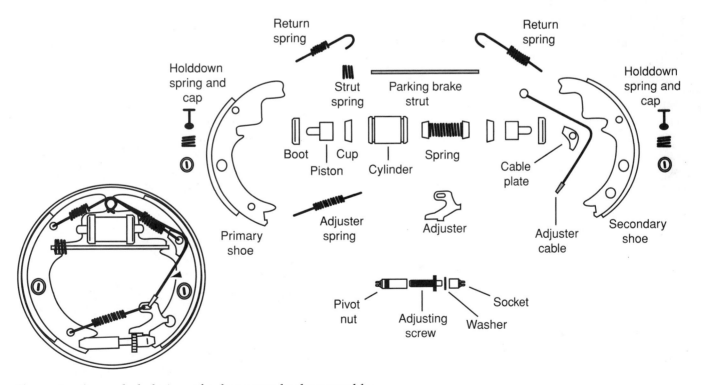

Figure 1. An exploded view of a duo-servo brake assembly.

brake assembly to refer to when you are reassembling the unit. It is also helpful to service the brake on one side of the vehicle at a time and use the brake on the opposite side as an assembly reference. Remember that left-hand and right-hand parts of many drum brakes look the same but are not. If parts, such as self-adjusters, are interchanged from one side to the other, they will not work right.

Begin your service to the drum brake by raising the vehicle on a hoist or safety stands, remove the drum, and clean the brake assembly. Then follow these general guidelines (outlined below), along with specific vehicle service procedures, to service drum brakes.

The shoe return springs are usually the first parts to be removed because their tension holds most of the other parts in place, but before you remove the springs, study their installation. Note how they are hooked over anchor posts and to which holes in the shoe webs they are attached (**Figure 2**).

Special brake spring pliers (**Figure 3**) and other tools (**Figure 4**) are available for spring removal and installation.

Figure 2. Carefully note where and how the various springs are attached.

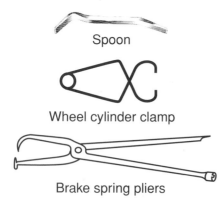

Figure 3. Three commonly used special tools for drum brakes.

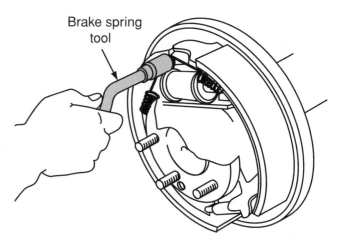

Figure 4. A brake spring tool that is used to remove and install shoe return springs.

The rear brakes on many late-model General Motors cars have a large horseshoe-shaped spring that serves as both a return spring and a holddown spring. Special lever-type pliers are used to remove and install it (**Figure 5**).

> **You Should Know** *Wear safety glasses and keep your fingers away from the return springs while you are removing and installing them. If a spring slips off a tool or mounting point its strong tension can cause injury or damage.*

After you have removed the return springs, remove the lighter holddown springs and clips. Again, special tools are available to make the job easier and to avoid damage to

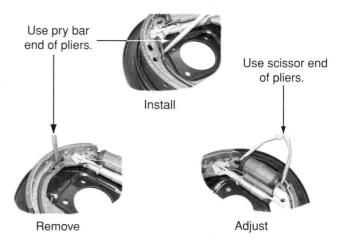

Figure 5. Use these special pliers to remove and install the unispring used in some General Motors drum brakes.

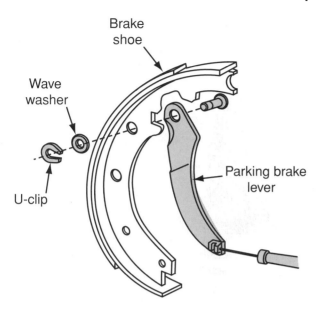

Figure 6. Remove the clip and washer at the shoe to release the parking lever from the shoe.

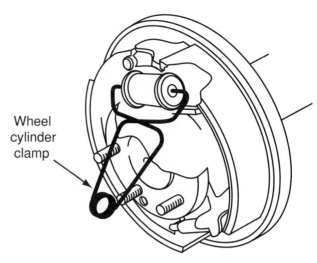

Figure 7. Wheel cylinder clamps are used to prevent the pistons from popping out of the cylinder.

parts. At this point, the shoes are loosened from the backing plate, and you can remove shoe-to-shoe springs and self-adjuster parts easily. Again, carefully note the positions in which the parts are installed so that you can reinstall them correctly.

Disconnect the parking brake linkage from rear brakes with parking brake levers **(Figure 6)** and struts. It is often easiest to disconnect the parking brake linkage after the shoes are loosened from the backing plate to release tension on the cables.

On some brakes, sections of the shoe webs fit into the ends of the wheel cylinder pistons. On other brakes, short pushrods are installed between the cylinder pistons and the shoes. If the brakes you are servicing have wheel cylinder pushrods, remove them for cleaning and reinstallation later.

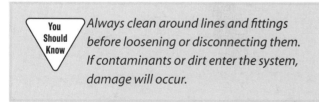

You Should Know | *Always clean around lines and fittings before loosening or disconnecting them. If contaminants or dirt enter the system, damage will occur.*

If you do not plan on removing the wheel cylinder for service or replacement, install a wheel cylinder clamp to keep the pistons from popping out of the cylinder **(Figure 7)**. On some assemblies, the clamp will make it easier to remove the brake shoes.

Remove the wheel cylinder by disconnecting the brake line and removing its mounting screws or clip

(Figure 8). Cap the brake line after the wheel cylinder is removed to keep dirt out of the hydraulic system.

After all parts are removed, inspect them *again* for damage. Clean individual parts in a parts washer and with approved cleaning equipment remove any remaining dirt from backing plates, the parking brake linkage, and other parts still attached to the vehicle.

WHEEL CYLINDER SERVICE

Typically, a bad wheel cylinder is replaced rather than rebuilt. Some wheel cylinders cannot be rebuilt and must be replaced. In this chapter, we cover the correct way to rebuild a wheel cylinder. You or the person in charge of your shop will need to decide if you will rebuild a particular wheel cylinder. Many shops make it standard practice to overhaul or replace wheel cylinders whenever new brake shoes are installed.

Wheel Cylinder Disassembly

Remove the wheel cylinder boots from the retaining grooves in the cylinder body **(Figure 9)**. Remove each boot and piston as an assembly. Next, remove the rubber cups and springs or expanders from the cylinder. Rubber parts cannot be reused; discard them to prevent accidental reuse.

Continue disassembly by removing the wheel cylinder bleeder screw from the cylinder. Wash all parts in clean, denatured alcohol and inspect the pistons for scratches, scoring, or other damage or wear. Check the cylinder bore for scoring, nicks, scratch marks, and rust. You can remove light scoring or corrosion in the bore with emery cloth lubricated with clean brake fluid or by honing in some

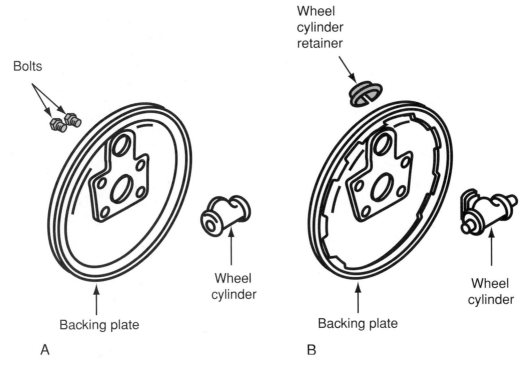

Figure 8. Two common ways that the wheel cylinder is held to the brake's backing plate: (A) by bolts; or (B) with a spring-type retainer ring or clip.

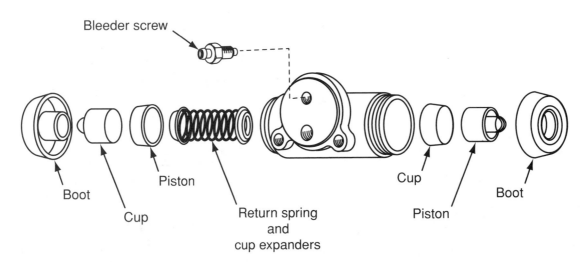

Figure 9. Parts of a typical wheel cylinder.

cylinders. Any honing performed must be light. The cylinder must not be honed more than 0.003 inch (0.08 mm) beyond its original diameter.

Wheel Cylinder Honing

An aluminum wheel cylinder should not be honed unless it has a steel insert, or liner. The bore of an unlined aluminum cylinder is rolled to close pores in the aluminum casting. This **rolled finish** is removed by honing, and the cylinder will wear prematurely. Only a cast-iron cylinder bore or steel insert can be honed to the proper finish. Keep in mind that although honing restores the finish of a cylinder, it leaves a slightly rougher finish than the original finish. The rough finish will cause the cup seals to wear faster. This is one reason why many shops do not rebuild and hone wheel cylinders. However, honing might be necessary to save a cylinder that is not easily replaced. Follow these guidelines when honing a wheel cylinder:

1. Disassemble the cylinder. If the cylinder is to be honed, clamp it in a vise by its mounting flange; do not clamp it on the bore section of the body.
2. Select a suitably sized hone and mount it in an electric drill motor or pneumatic die grinder.
3. Lubricate the cylinder bore thoroughly with fresh brake fluid and insert the hone into the bore.
4. Operate the hone at 300–500 rpm and move the hone back and forth in the bore with even strokes. While the hone is spinning, keep it moving back and forth. Holding it in one spot will result in an uneven diameter.

 Do not let the stones come completely out of the bore while the hone is rotating. The hone or the cylinder, or both, could be damaged.

5. While the hone is rotating, let it extend partly out of the open end of the cylinder. This ensures an even finish on the bore without a ridge at the open end.
6. Hone the bore for approximately 10–15 seconds while keeping the bore well lubricated with brake fluid.
7. Let the hone come to a complete stop and carefully remove it from the bore.
8. Wipe the bore with a clean rag and inspect its finish. The cylinder bore should be clean and free of rust and corrosion and have a uniform crosshatch finish **(Figure 10)**.

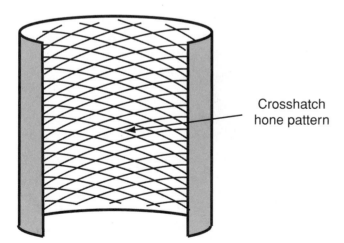

Crosshatch hone pattern

Figure 10. A crosshatch hone pattern is a uniform series of light grooves that cross each other at a 25- to 30-degree angle in the cylinder walls.

9. After inspecting the cylinder bore, clean it thoroughly with brake-cleaning solvent or alcohol. Do not use petroleum-based solvents.
10. Before assembling the cylinder, measure the piston-to-bore clearance. If honing has removed too much metal and clearance is excessive, the cylinder cannot be rebuilt.
11. Insert a narrow 0.003-inch (15-millimeter) feeler gauge into the cylinder bore and try to slide the piston into the bore, along with the gauge. If the piston fits, the bore is oversized. The cylinder must be replaced.

The 0.003-inch piston-to-bore clearance is a traditional guideline. Always check the vehicle service manual for exact specifications.

Wheel Cylinder Reassembly

Wash the body of the wheel cylinder thoroughly with denatured alcohol before reassembly. Let it air dry; do not dry it with compressed air because moisture in the air lines can allow rust to form.

During reassembly, make sure you install all of the parts in the wheel cylinder rebuild kit. Lubricate the new seals and all internal parts with clean brake fluid. Then install the spring **(Figure 11)**, the cup expander, the cup seals, and the pistons into the cylinder bore. Place a new dust boot over each end of the wheel cylinder. Thread the wheel cylinder bleeder screw into the upper threaded bore in the wheel cylinder. The lower bore is for the brake line.

DRUM BRAKE REASSEMBLY

Although drum brake reassembly is generally the opposite of disassembly, a few special steps must be followed in order to do the job right. Also, if any part of the hydraulic system was disconnected or removed, that part or all of the brake system will need to be bled.

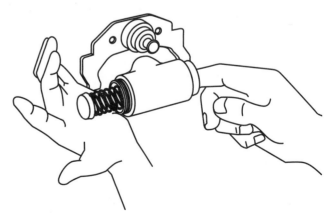

Figure 11. The spring is the first to be installed when reassembling a wheel cylinder.

Install the wheel cylinder onto the backing plate. Then remove the caps from the brake line fittings and install the line fitting into the cylinder.

Compare the new brake shoes with the old ones to be sure that the holes for springs and clips are in the same locations and that the new linings are positioned in the same locations on the shoes. This is particularly important for the primary shoes of duo-servo brakes. Also, verify that the right-hand and left-hand shoes are properly identified. Some shoes have pins for adjuster levers that will fit correctly only if installed at the correct wheel.

Remove nicks and rough spots from the raised shoe pads on the backing plate with emery cloth and then clean the area. Lightly coat the shoe pads with brake lubricant. Transfer any parking brake linkage parts from the old shoes to the new shoes. Make sure that the backing plate bolts and bolted-on anchor pins are torqued to specifications. Ensure that riveted anchor pins are secure.

Install the holddown parts using the appropriate tools to mount the shoes on the backing plate. Connect parking brake linkage. Lightly coat the surface of the parking brake pin with brake lubricant. Install the lever on the pin with a new washer and clip. Attach the parking brake cables and be sure that their movement is not restricted.

If the brake assembly has wheel cylinder pushrods, install them between the wheel cylinder pistons and the shoes. If pushrods are not used, make sure that the shoe webs fit into the ends of the wheel cylinder pistons correctly.

Once again, make sure that the left-hand and right-hand parts are on the correct side of the vehicle. This is especially important for duo-servo starwheel adjusters, which have left- and right-hand threads **(Figure 12)**. The

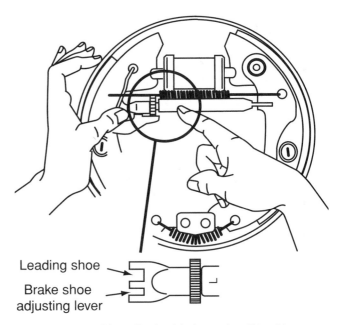

Note: Socket blade marker R and L. Install letter in upright position to ensure proper slot engagement to brake shoe adjusting lever.

Figure 13. This adjuster strut for a non-servo brake assembly has specific right-hand and left-hand positions.

adjusting mechanism in non-servo brake assemblies also has left-side and right-side only parts **(Figure 13)**.

Disassemble the adjuster and clean the parts in denatured alcohol. Clean the threads with a fine wire brush. Make sure that the adjusting screw threads into the threaded sleeve over its complete length without sticking or binding. Make sure that none of the starwheel teeth are damaged. Lubricate the screw threads with brake lubricant. Also apply brake lubricant to the inside of the socket and the socket face. Finally, apply lubricant to the open ends of the threaded sleeve and socket (end cap) when the threads are fully engaged.

Before you install the self-adjusters, turn them so that they are almost fully retracted. Setting the adjusters in a retracted position makes it easier to install the shoes and their return springs. After the shoes are installed, check the adjusters again and adjust them outward just enough to take up any slack between the adjusters and the shoes. Check the operation of the self-adjusters by prying the shoe lightly away from its anchor or by pulling the cable or link to make sure that the adjuster easily advances, one notch at a time.

Installing the return springs is usually the last step—or close to the last step—of brake reassembly. Use brake spring pliers and other special tools to install the springs

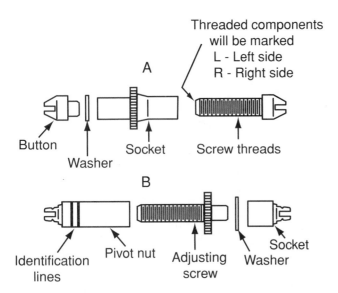

Figure 12. Locations of right and left identification on adjustment assemblies: typical Chrysler (A) and Ford (B) adjusters are shown.

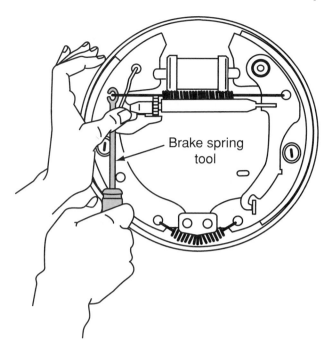

Figure 14. Use the correct special tools to install brake shoe return springs.

(Figure 14). Make sure to install each spring in the proper direction and in the proper holes in the shoe webs. Many return strings have longer straight sections at one end than at the other. If a spring is installed in the wrong hole on the shoe web, its operating tension will not be correct. If you try to stretch a spring too far to install it in a wrong hole, personal injury or damage to parts can result.

DRUM INSTALLATION

At this point, you are ready to do the initial brake adjustment and install the drum. If the brake drum is a floating type that is installed over the wheel studs on an axle flange or hub, you can install it after the initial brake adjustment.

BRAKE ADJUSTMENT

Drum brakes require manual adjustment when assembled and can require periodic manual adjustment to ensure proper lining-to-drum clearance. Correct brake adjustment ensures the proper brake pedal position and operation for safe braking.

Initial Adjustment

When you install the brake shoes and all of the brake components, the self-adjusters are retracted so that the

shoes are at the minimum diameter. Before installing the drum, you should readjust the shoes to take up most of the clearance with the drum. This will make the final adjustment, after the drum is installed, faster and more accurate.

Before making the initial adjustment of rear drum brakes, check the adjustment of the parking brake cable. New brake shoes can reduce the lining-to-drum clearance allowed by the parking brake, even when the brake is released. If necessary, *back off the parking brake adjustment* before the initial manual adjustment of the brake shoes. The parking brake must operate freely with the brake shoes and linings centered on the backing plate. Check the parking brake and readjust it if necessary as a final step after manual adjustment of the service brakes.

Initial adjustment requires a brake shoe caliper, which is a measuring tool that gauges the inside diameter of the drum and the outside diameter of the installed shoes. Place the caliper into the drum as shown in **Figure 15** and slide it back and forth to open the jaws to their widest point. Then tighten the lock screw to hold the caliper jaws in position.

Depending on the kind of caliper used, the jaw opening on its opposite side is now set to equal the drum diameter or to a specific smaller diameter. The installed diameter of the brake shoes should be adjusted to approximately 0.020–0.040 inch (0.50–1.00 millimeter) smaller than the drum diameter. Check the vehicle service manual for exact specifications.

If the caliper opening used to gauge the brake shoes equals the drum diameter, readjust it undersize by the specified amount and then place it over the widest point of

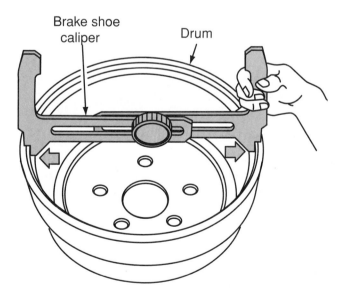

Figure 15. Set one end of the brake shoe caliper into the inside diameter of the drum.

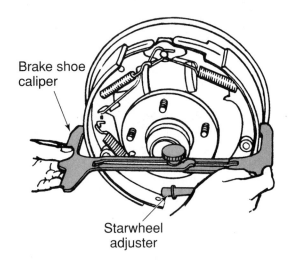

Brake shoe
caliper

Starwheel
adjuster

Figure 16. Slide the caliper, after it has been set to the drum's diameter, over the shoes and adjust the shoes until the caliper barely slides over the shoes.

the brake shoes. As an alternative, you can leave the caliper set to the drum diameter and use a feeler gauge between the caliper and one shoe to adjust the shoes. If the caliper has built-in compensation for the drum and shoe diameters, simply place it over the widest point of the shoes.

With the caliper in place over the shoes **(Figure 16)**, rotate the starwheel adjuster to expand the shoes until the caliper just slides over them without binding. Hold the self-adjuster pawl away from the starwheel with a screwdriver or a heavy wire hook while adjusting the starwheel. Some non-servo brakes have a separate starwheel or other adjuster for each shoe. For this kind of brake assembly, adjust each shoe an equal amount.

After the initial adjustment, install the brake drum and apply the brakes several times to verify that the pedal is fairly high and firm. Bleed the brakes, if necessary, to ensure that the lines are free from air. Then make a final, manual adjustment of the brakes.

Manual Shoe Adjustment

Some technicians rely on the self-adjusters to make the final adjustment automatically after new shoes are installed. This is not the best practice, however, because initial adjustment is only a rough adjustment and might not be equal on both wheels of an axle set. Subsequent self-adjustment then can remain unequal, particularly on lightly used rear drum brakes of a FWD car. A final manual adjustment lets you verify an equal and complete adjustment on both wheels.

Perform manual brake adjustments with the vehicle supported on a hoist or stands and so that the brake drum

or wheel can rotate during adjustment. Exact adjustment procedures are different for different brake designs but all are based on the principle of expanding the shoes until they contact the drum and then backing off the adjustment a specified amount. With the brakes adjusted and the drum on, pump the brake pedal once or twice. Then, recheck the adjustment. The pedal action will center the two shoes and provide for a better adjustment. Generally duo-servo brake adjustments are backed off more than non-servo brake adjustments because duo-servo brakes need more clearance for the servo action to develop proper leverage.

On some vehicles, you might feel that you can tighten the adjuster quite a bit after the linings first contact the drum. Then when you back off the adjustment, it may feel like you have to retract the shoes excessively to get free wheel rotation. Additionally, the pedal might feel springy or spongy when it is applied. All these symptoms are signs that the shoes are not properly arced to the drum. Excessive adjuster travel in either direction indicates that the shoes are bending due to incomplete lining contact with the drum. The only satisfactory long-term solution to this problem is to measure the shoe arc, remove the shoes, and either replace them with properly arced shoes or have the shoes rearced to match the drum diameter and brake installation.

Before performing any of the following brake adjustments, make sure the parking brake is fully released. If you suspect that the parking brake is holding the brake shoes off their anchors or is improperly adjusted in any way, back off the parking brake adjustment. Some noticeable slack should be present in the cables, and the linkage should not bind. Readjust the parking brake after adjusting the service brakes.

Duo-Servo Starwheel Adjustment

Duo-servo brakes use a single starwheel adjuster in the link that connects the bottoms of the brake shoes. Rotating the starwheel moves both shoes at the same time and adjusts the clearance with the drum as an assembly. Most duo-servo brakes are adjusted manually through a hole in the backing plate **(Figure 17)**, but some cars have an adjustment hole in the outboard web of the drum. This latter style requires wheel and tire removal for brake adjustment.

The adjusting hole, either in the backing plate or the drum, is usually closed with a rubber, plastic, or metal plug that you can remove easily with a screwdriver. After adjustment, the plug should be reinstalled to keep dirt and water out of the brakes. Some brakes are built without the hole in the backing plate or drum, but the location for the hole is scored or lanced. At the time of the first brake adjustment, the scored area (or knockout) must be removed with a drill

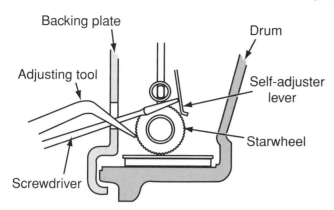

Figure 17. If the adjustment opening is in the backing plate, use a screwdriver to push the self-adjuster lever away from the starwheel during manual adjustment.

or by knocking it out with a small chisel or punch and a plug installed.

To adjust duo-servo brakes, insert a brake adjusting tool or a flat-blade screwdriver through the adjusting hole to engage the notches of the starwheel. With your other hand, use another small screwdriver or a wire hook to push or pull the self-adjuster pawl away from the starwheel. The method of disengaging the self-adjuster is different depending on brake design and on whether you reach the adjuster through the backing plate or through the drum.

It is possible to force the starwheel against the self-adjuster without disengaging it. If you do this, however, you will quickly wear down the teeth of the starwheel and the edge of the pawl. The self-adjuster will not operate properly then and might never adjust the brakes during their service life. Always disengage the self-adjuster before manually turning the starwheel.

To expand the adjuster link and decrease lining-to-drum clearance, the handle of the adjusting tool is moved upward as you turn the starwheel. To retract the adjuster link and increase lining-to-drum clearance, the tool handle is typically moved downward. This general rule can vary for different brake designs, however, and even from one side of the vehicle to the other. Check the vehicle service manual for instructions.

Rotate the vehicle's wheel by hand as you adjust the brakes. For duo-servo brakes, rotate the wheel in the direction of forward rotation as you expand the shoes. The self-energizing operation and servo action of the shoes will help to center the shoes in the drum during adjustment.

Exact adjustment instructions will vary from one manufacturer and one brake design to another. Some instructions say to adjust the brakes outward until you feel a light drag on the drum as you turn the wheel. Other instructions

call for a heavy drag or to lock the wheels. The instructions will then direct you to back off or retract the shoes a specific number of notches or clicks of the starwheel. Generally, you retract the shoes more on a duo-servo brake than on a nonservo brake.

Nonservo Starwheel Adjustment

Most nonservo brakes have a single starwheel adjuster similar to the adjuster for duo-servo brakes. The adjuster linkage is mounted higher on the shoe webs **(Figure 18)** than the adjuster link used for duo-servo brakes because leading shoes and trailing shoes are rigidly anchored to the bottom of the backing plate.

Some nonservo brakes have a separate starwheel or other adjuster for each shoe. For this type of brake assembly, adjust each shoe an equal amount. Nonservo brakes with quadrant adjusters are adjusted by manually moving the toothed quadrant. Refer to the vehicle's service manual for special procedures and tool requirements.

As with duo-servo brakes, rotate the vehicle's wheel by hand as you adjust the brakes. Rotation methods are a bit different, however. If the brake has a single starwheel, start by rotating the wheel forward until specified drum contact is made. Then rotate the wheel in reverse to verify adjustment. If the brake has separate starwheels for each shoe, rotate the wheel forward as you adjust the forward shoe, then rotate the wheel in reverse as you adjust the rear shoe.

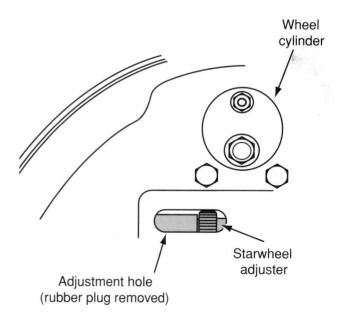

Figure 18. This starwheel for a nonservo brake is mounted higher on the brake assembly than the adjuster for a duo-servo brake assembly.

Summary

- The brake assembly should be disassembled in an orderly fashion with great attention given to the exact position of the various springs used in the assembly.
- Special tools designed for drum brake work should be used to avoid damage to the components and injury to you, the technician.
- Wheel cylinders are typically replaced when they are faulty, but most can be rebuilt.
- Light scoring or corrosion in the bore of a wheel cylinder can be removed with emery cloth lubricated with clean brake fluid or by honing.
- Before reassembly of a wheel cylinder, wash the body of the wheel cylinder thoroughly with denatured alcohol and let it air dry. Then lubricate all parts with clean brake fluid.

- Before assembly of the brake unit, compare all new parts to the old ones to ensure that they are exactly the correct size and design.
- After all brake parts have been attached to the shoe and backing plate, the unit is ready for initial brake adjustment and the installation of the drum.
- Use a brake shoe caliper to preliminarily adjust the brakes.
- After the drum is installed over the shoes, the brakes need to be adjusted again.
- Although the exact adjustment procedures are different for different brake designs, they are all based on the principle of expanding the shoes until they contact the drum and then backing off the adjustment a specified amount.

Review Questions

1. While servicing the brakes Technician A removes nicks and rough spots from the raised shoe pads on the backing plate with emery cloth. Technician B lightly coats the shoe pads with brake lubricant. Who is correct?
   A. Technician A only
   B. Technician B only
   C. Both Technician A and Technician B
   D. Neither Technician A nor Technician B

2. While discussing wheel cylinder service Technician A says that honing restores the bore of a cylinder but leaves a rougher finish on the cylinder wall. Technician B says that cost and warranty issues often make it more practical to replace wheel cylinders than to hone them. Who is correct?
   A. Technician A only
   B. Technician B only
   C. Both Technician A and Technician B
   D. Neither Technician A nor Technician B

3. While installing return springs in a drum brake assembly Technician A says that if a spring is installed in the wrong hole on the shoe web, its operating tension will not be correct. Technician B says that if you try to stretch a spring too far to install it in a wrong hole, damage to parts or personal injury can result. Who is correct?
   A. Technician A only
   B. Technician B only
   C. Both Technician A and Technician B
   D. Neither Technician A nor Technician B

4. Technician A always rotates the wheel in a forward direction when adjusting and verifying the adjustment of drum brakes. Technician B adjusts the brakes with a brake caliper during assembly and then applies the brakes many times in reverse to make the final adjustment. Who is correct?
   A. Technician A only
   B. Technician B only
   C. Both Technician A and Technician B
   D. Neither Technician A nor Technician B

5. What are the first parts to remove when you are disassembling a drum brake unit? Why?

Chapter 31

Disc Brake Diagnosis

Introduction

When diagnosing disc brake systems, it is important to remember that there are other parts of the brake system besides the disc brakes. Many problems are caused by the hydraulic system and the rear drum brakes (if the vehicle is so equipped) or the vehicle's suspension and tires. You should always check the following items and correct any problems that may exist before moving on to the brake system. You should also keep in mind how the problem(s) are affecting brake performance so you can accurately diagnose the brake system.

Check the following:

1. Tires for excessive or unusual wear or improper inflation.
2. Wheels for bent or warped rims.
3. Wheel bearings for looseness or wear.
4. The suspension system for worn or broken components.
5. Improper mounting of caliper and rotor.
6. The brake fluid level in the master cylinder.
7. Signs of leakage at the master cylinder, in brake lines or hoses, at all connections, and at each wheel.

If the disc brakes **(Figure 1)** are faulty, the repairs can be quite straightforward and rather simple. However, some repairs can be difficult. Make sure you refer to the appropriate service manual before undertaking any repair to a disc brake system.

DIAGNOSING DISC BRAKE PROBLEMS

Poorly performing disc brakes usually result from worn brake pads or other parts; poorly fitted or incorrectly

assembled parts; or rotor problems, such as grooving, distortion, or grease and dirt on the rotor surface. Worn pads increase the braking effort needed to stop the vehicle, but the same problem can be caused by a sticking or sluggish caliper piston.

The brake system should be carefully inspected with a thorough look at the calipers, hoses, lines, and master cylinder fluid level. If you suspect disc brake problems, road test the vehicle and compare your findings with those listed in **Figure 2**.

Make sure you operate the vehicle wisely and safely during the road test. Pay attention to everything. Note any unusual behavior, noise, or vibration. Also note when and

Figure 1. A disc brake assembly.

SYMPTOM   Excessive pedal effort
Possible Causes
Binding brake pedal linkage
Brake pads contaminated with grease or brake fluid
Frozen caliper piston
Glazed brake pad linings
Malfunctioning power booster
Restricted brake lines or hoses

SYMPTOM   Excessive pedal travel
Possible Causes
Air in the hydraulic system
Contaminated brake fluid
Distorted brake pad(s)
Improperly adjusted master cylinder or power booster
    pushrod
Internal master cylinder leakage
Loose or worn wheel bearings
Loose, broken, or worn caliper attachments
Pressure loss in one hydraulic system
Worn or damaged piston seals

SYMPTOM   Spongy pedal
Possible Causes
Air in the hydraulic system
Brake hose expanding under pressure
Contaminated brake fluid
Damaged power booster mounts
Distorted brake caliper
Internal master cylinder leakage
Loose master cylinder mounting bolts

SYMPTOM   Sinking pedal
Possible Causes
Hydraulic system leak
Internal master cylinder leakage

SYMPTOM   Pulsing brake pedal
Possible Causes
Excessive rotor lateral runout
Loose or worn wheel bearings
Rusted rotor
Variation of rotor thickness

SYMPTOM   One brake drags
Possible Causes
Distorted brake pad
Loose or worn wheel bearings
Restricted brake lines or hoses
Sticking caliper piston
Swollen caliper seal

SYMPTOM   Rear brakes drag
Possible Causes
Improper parking brake adjustment
Restricted brake lines or hoses
Sticking or frozen parking brake cable

SYMPTOM   All brakes drag
Possible Causes
Binding brake pedal linkage
Improperly adjusted master cylinder or power booster
    pushrod
Weak brake pedal return spring
Sticking master cylinder pistons
Swollen rubber parts due to contaminated brake fluid

SYMPTOM   One brake locks
Possible Causes
Brake pads contaminated with grease or brake fluid
Failing wheel bearing
Incorrect tire pressure
Mismatched or worn tires
Wrong brake pad for application

SYMPTOM   Front brake lockup (premature)
Possible Causes
Faulty metering valve
Faulty proportioning valve
Rear brake hydraulic circuit failing
Restricted brake lines or hoses

SYMPTOM   Rear brake lockup (premature)
Possible Causes
Faulty proportioning valve
Front brake hydraulic circuit failing
Restricted brake lines or hoses

SYMPTOM   Uneven pad wear
Possible Causes
Caliper loosely mounted
Caliper not aligned with rotor
Distorted caliper mount
Sticking caliper piston
Sticking or binding caliper mount

SYMPTOM   Rapid lining wear
Possible Causes
Master cylinder partially applied at all times
Rough rotor surface
Wrong brake pad for application

Figure 2.   A symptom-based diagnostic chart. *(continued on following page)*

SYMPTOM   Pull during braking
Possible Causes
Brake lining experiencing water fade
Brake pad contaminated with grease or brake fluid
Incorrect tire pressure
Incorrect wheel alignment
Mismatched or worn tires
Restricted brake line or hose
Unequal brake action at the sides of the vehicle
Worn suspension parts

SYMPTOM   Brake fade
Possible Causes
Contaminated brake fluid
Dragging brakes
Excessively worn rotors
Glazed brake pads
Overheated brakes
Wrong brake pad for application

SYMPTOM   Steering wheel shimmy
Possible Causes
Damaged or out-of-balance tires and wheels
Excessive rotor runout
Worn suspension parts

SYMPTOM   Brakes grab
Possible Causes
Binding brake pedal linkage
Brake pads contaminated with grease or brake fluid
Caliper loosely mounted
Failing wheel bearing

Faulty metering valve
Sticking caliper piston
Wrong brake pads for application

SYMPTOM   Clicking during brake application
Possible Causes
Caliper loosely mounted
Poor fitting brake pad(s)
Worn suspension parts

SYMPTOM   Squeal during braking
Possible Causes
Glazed brake pads
Loose or worn wheel bearings
Pad wear indicator contacting rotor
Poor fitting brake pad(s)
Weak, loose, or missing pad hardware
Worn suspension parts
Wrong brake pad for application

SYMPTOM   Grinding or scraping
Possible Causes
Excessively worn pads
Debris embedded in the lining
Rotor contacting splash shield
Rotor contacting caliper or mount
Caliper mounting bolts too long

SYMPTOM   Rattle when the brakes are not applied
Possible Causes
Excessive rotor runout
Weak, loose, or missing pad hardware

Figure 2.   A symptom-based diagnostic chart. (*continued*)

where the problems occur. These details will help you determine what the cause of the problem is.

As you apply the pedal, check for excessive travel and sponginess. Listen for noises, not just the obvious sounds of grinding pads or pad linings, but mechanical clanks, clunks, and rattles. A vehicle that pulls to one side when the brakes are applied can have a bad caliper or loose caliper at one wheel. Grease or brake fluid might have contaminated the pads and rotor, or the pad and lining might be bent or damaged. Grabbing brakes also can be caused by grease or brake fluid contamination or by a malfunctioning or loose caliper. Worn rotors or pads also can cause roughness or pedal pulsation when the brakes are applied.

INSPECTING BRAKE PADS

Disc brake pad linings should be inspected regularly at the time or mileage intervals recommended by the manu-

facturer. The time intervals vary with manufacturer and vehicle model. A convenient time to check pad wear is whenever the wheels are removed for tire rotation.

Follow these basic steps for any disc brake pad inspection:

1. Raise the vehicle on a hoist or safety stands. Be sure it is properly centered and secured on the stands or hoist.
2. Mark the relationship of the wheel to the hub to ensure proper wheel balance upon reassembly.
3. Remove the wheel and tire from the hub. Be careful not to hit the brake caliper, the rotor splash shield, and the steering knuckle or suspension parts.
4. If the rotor and hub are a two-piece assembly, reinstall two wheel nuts to hold the rotor on the hub.
5. On most disc brakes, you can inspect the pads without removing the calipers. Check both ends of the outboard pad by looking in at each end of the caliper

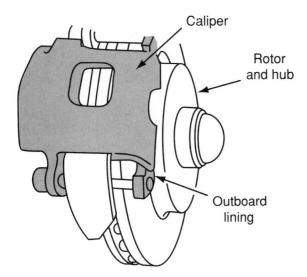

Figure 3. Inspect both ends of the outward pad by looking at the end of the caliper. These are the areas where the highest rate of pad wear occurs.

(Figure 3). These are the areas where the highest rate of pad wear occurs. Also check the lining thickness on the inboard pad to be certain it has not worn prematurely. If the caliper has an opening in the top, look through it to view the inboard pad and lining **(Figure 4)**. Some calipers do not have such openings, and you cannot inspect the lining of the pads without removing them from the caliper.

6. On vehicles with floating or sliding calipers, check for uneven wear on the inboard and outboard linings. If the inboard pad shows more wear than the outboard

Figure 4. Most calipers have an opening on the top of the casting that allows for a good view of the brake pads.

pad, the caliper should be overhauled and/or the slides replaced. If the outboard pad shows more wear, the sliding components of the assembly might be sticking, bent, or damaged. In any case, uneven brake wear is a sign that the pads and/or calipers need service.

7. If the customer's complaint is poor braking, remove the pads from the calipers and carefully check them. Replace the pads if the linings are worn, glazed (shiny and smooth), heat damaged, cracked, or contaminated with dirt or brake fluid.

8. Carefully inspect the entire brake assembly at each wheel, including the caliper, brake lines and hoses, and the rotors. Look for fluid leaks, broken or cracked brake hoses or lines, and a damaged rotor. Repair any of these problems before replacing the brake pads.

If the customer complains that the brakes are making a high-pitched squeal, immediately suspect an audible brake pad wear indicator, signaling that the system needs service. Instrument panel warning lamps also are indicators that the pads have worn beyond specifications.

Interesting Fact

BMWs and a few other vehicles have an electronic disc brake wear indicator. When the pads wear to the point where the sensor contacts the rotor, the brake warning lamp lights. After the pads are replaced, the lamp might stay lit until the system learns that new pads were installed. This normally takes a drive of about one-quarter of a mile.

Most manufacturers specify a minimum lining thickness of 1/32-inch for bonded pads. Ford, however, wants you to change brake pads when the lining is 1/8-inch thick or less. By any measure, 1/32-inch may be the minimum thickness to prevent rotor scoring, but braking efficiency is dramatically reduced with worn linings. Worn linings cannot dissipate heat adequately, and the last 1/32-inch of lining will wear and peel off much faster than the first 1/32-inch of fresh linings. When visually evaluating pad lining wear, consider that any lining worn to the thickness of the metal backing pad needs replacement **(Figure 5)**.

ROTOR INSPECTION

Inspect the brake rotors **(Figure 6)** whenever the pads or calipers are being serviced or when the tires are rotated or removed for other work. Many rotor problems might not be apparent during a visual inspection. Rotor thickness, parallelism, runout, flatness, and depth of scoring can be measured only with precision gauges and micrometers. Other

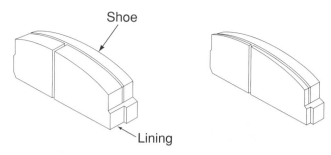

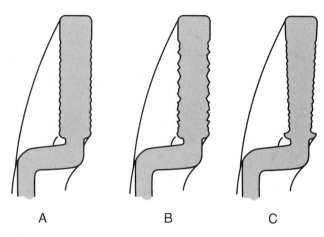

New pad and lining **Ready for replacement**

Figure 5. Any lining worn to the thickness of the metal backing should be replaced.

A   B   C

Figure 7. The small grooves on rotor A are acceptable for continued use. Rotor B has deeper scoring and requires refinishing if the thickness allows. Rotor C has extreme taper and will probably need to be replaced.

Figure 6. Carefully inspect the rotor before you remove it from the vehicle.

faced within its thickness limits, you will need a micrometer with a pointed anvil and spindle.

4. Inspect the rotor thoroughly for cracks or broken edges. Replace any rotor that is cracked or chipped, but do not mistake small surface checks in the rotor for structural cracks. Surface checks will normally disappear when a rotor is resurfaced. Structural cracks, however, will be more visible when surrounded by a freshly turned rotor surface.

5. Inspect the rotor surfaces for heat checking and hard spots **(Figure 8)**. Heat checking appears as many small interlaced cracks on the surface. Heat checking lowers the heat dissipation ability and friction coefficient of

rotor conditions should be checked with equal precision and thoroughness. Inspect a rotor as follows:

1. If the rotor is dirty enough to interfere with inspection, clean it with a shop cloth dampened in brake cleaning solvent or alcohol. If the friction surface is rusted, remove the rust with medium-grit sandpaper or emery cloth and then clean it with brake cleaner or alcohol.

2. If necessary, remove the splash shield for complete inspection of the inboard surface. Most often, turning the rotor will allow for a good view of the inboard surface.

3. Inspect both rotor surfaces for scoring and grooving. Scoring or small grooves up to 0.010 inch (0.25 millimeter) deep are usually acceptable for proper braking performance **(Figure 7)**. To determine the depth of grooves and whether or not the rotor can be resur-

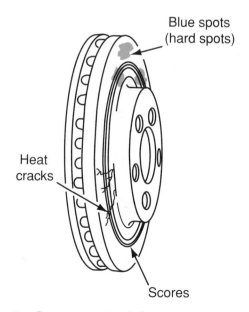

Figure 8. Common rotor defects.

Figure 9. The groove in this rotor surface aids in the cooling of the rotor and helps to reduce noise.

the rotor surface. Heat checking does not disappear with resurfacing. Hard spots appear as round, shiny, bluish areas on the friction surface. It may be possible to machine hard spots out of a rotor if the rotor has not been previously turned and has enough thickness for extended machining. Hard spots are difficult to machine, however, and might require special cutting bits for the brake lathe.

6. Inspect the fins of vented rotors for cracks and rust. Rust near the fins can cause the rotor to expand and lead to rotor **parallelism**, or thickness variations, and excessive runout. Machining the rotor might remove runout and thickness variations, but rotor expansion due to rust can cause these problems to reappear soon. Rusted rotors should be replaced.

Some rotors, particularly on General Motors vehicles, have a single deep groove manufactured into each surface **(Figure 9)**. This groove helps to keep the pads from moving outward and reduces operating noise.

ROTOR MEASUREMENT

Various micrometers and dial indicators are needed to measure a rotor. A micrometer is used to measure rotor thickness and parallelism, as well as taper. A dial indicator is used to measure rotor runout. Some rotors require an additional surface depth measurement.

Rotor Thickness and Parallelism

All brake rotors except the earliest examples from the mid-1960s should have a discard thickness dimension cast

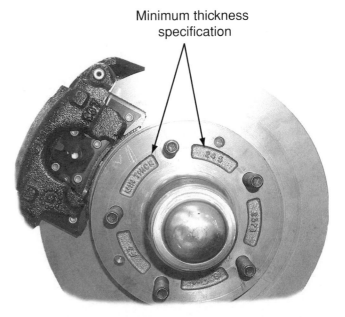

Minimum thickness specification

Figure 10. The discard dimension, or minimum thickness specification, is cast or stamped into all disc brake rotors.

into them **(Figure 10)**. If you cannot find a discard dimension on a rotor or if it is hard to read, check a service manual for thickness specifications. The **discard dimension** seems like a simple specification, but you must understand its complete meaning and how to apply it to rotor service.

Rotor discard thickness dimensions are given in two or three decimal points (hundredths or thousandths of an inch or hundredths of a millimeter), such as 1.25 inch, 1.375 inch, 0.750 inch, or 24.75 millimeters. When you measure rotor thickness, you must subtract 0.015–0.030 inch (0.40–0.75 millimeter) to allow for wear after the rotor is returned to service. If you resurface the rotor, it must similarly be 0.015–0.030 inch thicker than the discard dimension after machining to allow for wear. A rotor should not be returned to service, with or without resurfacing, if its thickness is at or near the discard dimension.

Rotor parallelism refers to thickness variations in the rotor from one measurement point to another around the rotor surface. Thickness variations and excessive rotor runout are the major causes of brake pedal pulsation. Parallelism is measured at the same time as basic rotor thickness.

Measure rotor thickness as follows. Use a micrometer graduated in ten-thousandths of an inch or hundredths of a millimeter.

1. Raise the vehicle on a hoist or safety stands and remove the wheels. If the rotor is removable from the hub (a floating, two-piece rotor), place flat washers on all wheel studs and reinstall the wheel nuts. Torque the nuts to specifications to minimize any possible runout.

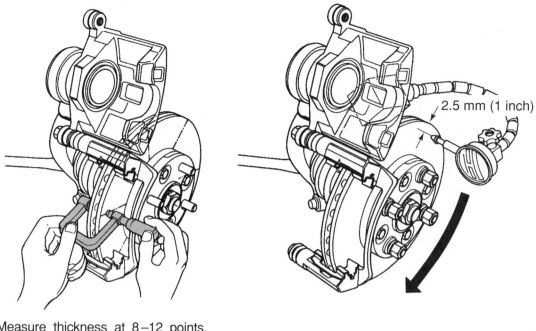

Measure thickness at 8–12 points, equally spaced around the rotor, all about 1 inch from outer edge of rotor.

Figure 11.   Use a micrometer to measure thickness and a dial indicator to measure rotor runout.

2. If the caliper must be removed to measure the rotor, hang it from the underbody or suspension with heavy wire so it will not drop.

3. Place an outside micrometer about 1 inch in from the outer edge of the rotor and measure the thickness **(Figure 11)**. Compare the measurement to specifications. Take all measurements at the same distance from the edge so that rotor taper does not affect measurement comparisons.

4. Also check the vehicle's service manual for an allowable thickness variation. Many manufacturers hold tolerances on thickness variations as close as 0.0005 inch (0.013 millimeter).

5. Repeat the measurement at about eight points equidistant (45 degrees) around the surface of the rotor **(Figure 12)**, and compare each measurement to specifications. If the rotor is thinner than the minimum thickness at any point or if thickness variations exceed limits, the rotor must be replaced.

6. If the rotor is deeply grooved, it must be thick enough to allow the grooves to be completely removed without turning the rotor to less than its minimum thickness. To measure the depth of the grooves on both sides of the rotor, use a micrometer with a pointed anvil and spindle. Measure rotor thickness to the bottom of the deepest grooves in about eight places. If rotor thickness at the bottom of the deepest grooves is at or near the discard dimension, replace the rotor.

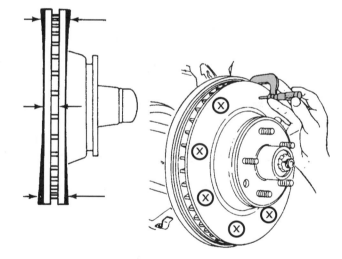

Figure 12.   Check for thickness variations by measuring the thickness around the rotor at a minimum of eight points.

Rotor Lateral Runout

Excessive **rotor lateral runout** causes the rotor to wobble from side to side as it rotates **(Figure 13)**. This wobble knocks the pads farther back than normal, which causes the pedal to pulse or vibrate as it is applied. You might also notice an increase in brake pedal travel because the pistons must move a greater distance to contact the rotor surface.

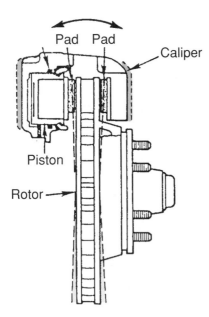

Figure 13.   Excessive rotor lateral runout knocks the piston back into its bore and causes pedal pulsation and increased pedal travel.

For best brake performance, lateral runout should be less than 0.003 inch (0.08 millimeter) for most vehicles. Some manufacturers, however, specify runout limits as small as 0.002 inch (0.05 millimeter) or as great as 0.008 inch (0.20 millimeter).

Runout measurements are taken only on the outboard surface of the rotor. Using a dial indicator and suitable mounting adapters, measure runout as follows:

1. Raise the vehicle on a hoist or safety stands and remove the wheels. If the rotor is removable from the hub (a floating, two-piece rotor), place flat washers on all wheel studs and reinstall the wheel nuts. Torque the nuts to specifications to remove looseness from runout measurements.
2. If the caliper must be removed to measure runout, hang it from the underbody or suspension with heavy wire so the weight of the caliper is not put on the brake hose.
3. If the caliper was not removed and the rotor drags heavily on the pads as you try to rotate it, open the bleeder screw and push the pistons back in the calipers **(Figure 14)**. Then bleed the system.
4. If the rotor is mounted on adjustable wheel bearings, readjust the bearings to remove bearing end play from the runout measurements. Do not overtighten the bearings. Bearing end play is the amount of preload applied to the bearings. Preload is designed to allow the bearing to spin freely while preventing wobble. On rotors bolted solidly to the axles of front-wheel-drive cars, bearing end play is not a factor in rotor runout measurement. If excessive bearing end play is noted,

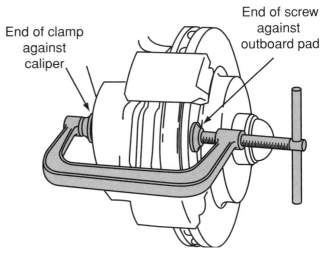

Figure 14.   Pistons can be retracted by placing a C-clamp against the back of the caliper and the outboard pad. Tighten the clamp to push the piston back into its bore.

the bearing assembly must be replaced. Bearing end play is best checked with a dial indicator.

5. Clamp the dial indicator support to the steering knuckle or other suspension part that will hold it securely as you turn the rotor **(Figure 15)**.
6. Be sure that the rotor surface, where you will place the dial indicator tip, is free of dirt and rust in the area.
7. Position the dial indicator so that its tip contacts the rotor at 90 degrees. Place the indicator tip on the fric-

Figure 15.   The dial indicator must have a stable mount in order to provide an accurate reading.

tion surface about 1 inch in from the outer edge of the rotor. Do not place the dial indicator in a grooved or scored area.

8. Turn the rotor until the lowest reading appears on the dial indicator and then set the indicator to zero.

9. Turn the rotor through one complete revolution and compare the lowest to the highest reading on the indicator. This is the maximum runout of the rotor. If the dial indicator reading exceeds specifications, resurface or replace the rotor.

If the rotor is a floating type, try repositioning it on the hub one or two bolt positions from its original location and repeat the runout measurement. If repositioning a floating rotor fixes an excessive runout condition, make index marks on the rotor and hub so that the rotor will be installed in the correct position. A few manufacturers recommend the use of a wedge-type shim placed behind the rotor to correct for runout.

If the rotor is mounted on nonadjustable wheel bearings that have any amount of end play, you must account for this end play in rotor runout measurements. To do this, press in on the rotor and turn it to the point of the lowest dial indicator reading. Set the indicator to zero and pull outward on the rotor. Read the indicator, which now shows the amount of bearing end play. Then turn the rotor to the point of the highest reading and pull outward. Subtract the end-play reading from the reading at this point to determine runout of the rotor alone.

Summary

- Begin diagnosis by checking the tires, wheels, wheel bearings, suspension system, mounting of brake parts, brake fluid level, and all brake lines or hoses for indications of needed service.
- Check both ends of the outboard pad by looking in at each end of the caliper. These are the areas where the highest rate of pad wear occurs.
- If the inboard pad shows more wear than the outboard pad, the caliper should be overhauled. If the outboard pad shows more wear, the sliding components of the assembly might be sticking, bent, or damaged.
- Most manufacturers specify a minimum lining thickness of $1/32$-inch.
- Pad replacement is the most frequent disc brake service because pads wear as they do what they are supposed to do.
- Calipers might require replacement or overhaul to replace worn parts or fix leaks and corrosion.
- Inspect disc brake rotors whenever the pads or calipers are being serviced or when the wheels are rotated or removed for other work.

- Rotors need to be resurfaced if they are grooved, scored, or otherwise badly worn or if runout and parallelism measurements are out of limits. Rotors cannot be resurfaced beyond their minimum thickness specification.
- Replace any rotor that is cracked or chipped, but do not mistake small surface checks in the rotor for structural cracks.
- Heat checking appears as many small interlaced cracks on the surface and cannot be removed by resurfacing the rotor.
- Hard spots appear as round, shiny, bluish areas on the friction surface. It might be possible to machine hard spots out of a rotor. Hard spots are difficult to machine and might require special cutting bits for the brake lathe.
- All brake rotors have a discard thickness dimension cast into them.
- Runout measurements are taken only on the outboard surface of the rotor with a dial indicator.

Review Questions

1. What is rotor parallelism and how is it checked?
2. Technician A uses a micrometer to measure rotor runout. Technician B uses a dial indicator to measure rotor runout. Who is correct?
   A. Technician A only
   B. Technician B only
   C. Both Technician A and Technician B
   D. Neither Technician A nor Technician B

3. While diagnosing a four-wheel disc brake system Technician A says that worn suspension parts can affect the operation of the brakes. Technician B says that grabbing brakes can be caused by contaminated brake linings. Who is correct?
   A. Technician A only
   B. Technician B only
   C. Both Technician A and Technician B
   D. Neither Technician A nor Technician B

4. Rotor serviceability is being discussed. Technician A says that the minimum wear thickness of a rotor is the discard thickness of the rotor. Technician B says that the thickness of a rotor after machining must be the specified minimum wear thickness or thicker. Who is correct?
   A. Technician A only
   B. Technician B only
   C. Both Technician A and Technician B
   D. Neither Technician A nor Technician B

5. Which of the following statements about inspecting disc brake pads is *not* true?
   A. If the inboard pad shows more wear than the outboard pad, the caliper should be overhauled.
   B. When the thickness of a lining is less than ½ inch, the pad should be replaced.
   C. If the outboard pad shows more wear, the sliding components of the assembly might be sticking, bent, or damaged.
   D. To inspect the lining surfaces on most disc brakes, you must remove the pads from the calipers.

Chapter 32

Brake Pad Replacement

Introduction

The most common service to disc brakes is the replacement of brake pads. Pads are not only replaced when they are worn or damaged; they should also be replaced whenever caliper and/or rotor work is completed. Pads should always be replaced in axle sets. An axle set contains four pads: the inner and outer pads for the caliper at each wheel. Even if only one pad in a set of four is badly worn, replace all four pads after fixing the problem that caused the uneven wear.

SERVICE PRECAUTIONS

The following general precautions apply to many different disc brake service operations and are presented at the beginning of this chapter to highlight their overall importance.

- When servicing disc brakes, never use an air hose or a dry brush to clean the brake assemblies. Use OSHA-approved cleaning equipment to avoid breathing brake dust.
- Do not spill brake fluid on the vehicle; it may damage the paint.
- Always use the DOT type of brake fluid specified by the vehicle manufacturer.
- During servicing, keep grease, oil, brake fluid, or any other foreign material off the brake linings, calipers, surfaces of the rotors, and external surfaces of the hubs.
- Handle brake rotors and calipers carefully to avoid damaging the rotors or nicking and scratching brake linings.

DISC BRAKE CLEANING

Disc brakes stay much cleaner than drum brakes. A disc brake is partly shrouded by the wheel, but it is largely exposed to circulating air. Dust and dirt created by pad wear and accumulated through normal driving are thrown off by the centrifugal force of the spinning rotor. Some brake dust accumulates inside the wheel and partially on the caliper. Proper and safe cleaning of brake assemblies and components is as important for disc brakes, however, as it is for drum brakes. Always use approved cleaning equipment and solutions when doing brake work.

> **You Should Know** *Do not blow dust and dirt off brake assemblies with compressed air outside of a brake cleaning enclosure. Airborne dust and asbestos fibers are an extreme respiratory hazard.*

VEHICLE PREPARATION

Follow these general preparation steps for all disc brake service:

1. Use a brake fluid siphon, or syringe, to remove approximately two-thirds of the brake fluid from the front or disc brake reservoir on a front-rear split system. On a diagonally split system, remove about half the fluid from both master cylinder reservoirs.
2. If the vehicle has electronically controlled suspension, turn the suspension service switch off.

3. Raise the vehicle on a hoist or safety stands and support it safely.
4. Remove the wheels from the brakes to be serviced. Brakes are always serviced in axle sets, so you can remove both front wheels, both rear wheels, or all four.
5. Vacuum or wet-clean the brake assembly to remove all dirt, dust, and fibers.

Removing brake fluid from the master cylinder (**Figure 1**) is an important preliminary step for all disc brake service.

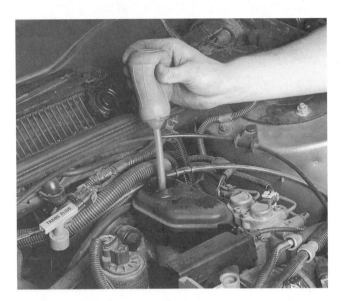

Figure 1. Remove fluid from the master cylinder before removing the calipers. This prevents an overflow of fluid as the pistons are retracted during the removal of the calipers.

If you forget to remove fluid, it may overflow the reservoir and spill when the caliper pistons are forced back into the caliper bores for caliper or pad removal. After siphoning fluid from the reservoir, replace the reservoir cover and safely discard the removed brake fluid.

Another way to prevent reservoir overflow is to open the caliper bleeder screw and run a bleeder hose into a container to catch the fluid expelled when the piston is forced back into its bore. Opening the bleeder screw also makes it easier to move the piston. This is the recommended way to retract pistons on a vehicle with ABS because it prevents damage to the hydraulic valve body. It is still good practice, however, to remove some fluid from the reservoir even if you plan to open the bleeder screw at the caliper.

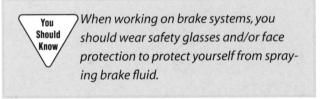

When working on brake systems, you should wear safety glasses and/or face protection to protect yourself from spraying brake fluid.

PAD REMOVAL

The details of pad replacement vary with the design of a particular caliper, but some general steps are common to all calipers. If you are unfamiliar with a particular caliper design, work on one side of the car at a time. The assembled caliper on the opposite side will be a guide for correct installation of pads and other parts. **Figure 2** is an exploded view of a typical floating caliper assembly.

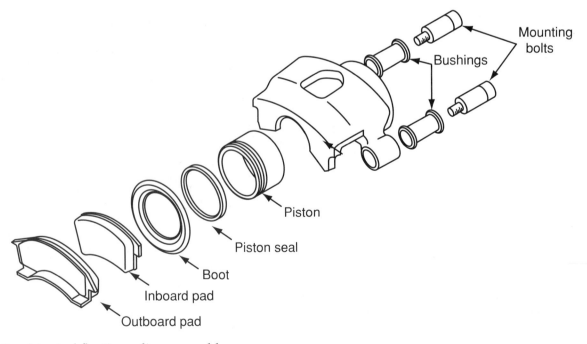

Figure 2. A typical floating caliper assembly.

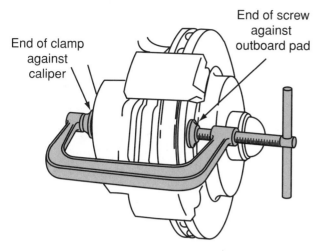

Figure 3. Place a C-clamp against the back of the outboard caliper and the pad. Tighten the clamp to retract the piston.

Several methods are used to retract caliper pistons. One method is to install a large C-clamp over the top of the caliper and against the back of the outboard pad **(Figure 3)**. Slowly tighten the clamp to push the piston into the caliper bore far enough so that the caliper can be lifted off the rotor easily.

> **You Should Know** *When the pads are excessively worn, the piston might be extended too far from its bore. Retracting the piston can damage the caliper seals. If this is the case, rebuild or replace the caliper.*

On some calipers, you can move the piston back in its bore with a large screwdriver or pry bar placed between the inboard pad lining and the rotor. On other calipers, you can use large slip-joint pliers placed over the inboard side of the caliper and a tab on the inboard pad to squeeze the piston into its bore **(Figure 4)**. On older, fixed caliper brakes, prying the pistons back into their bores is the usual way to provide clearance for pad removal.

Do not use either a C-clamp, slip-joint pliers, or a screwdriver to retract the piston of a rear caliper with a parking brake mechanism that moves the piston to apply the parking brakes. Trying to force this kind of piston back into its bore against the parking brake mechanism usually breaks the parking brake. This type of rear disc brakes requires a special tool to rotate the piston back into its bore on the automatic adjuster screw of the brake mechanism **(Figure 5)**. If you are unfamiliar with a rear caliper parking brake, refer to the vehicle service manual before trying to retract the caliper piston.

Remove the caliper bolts and sleeves from a floating caliper or the cotter pins and support keys from a sliding

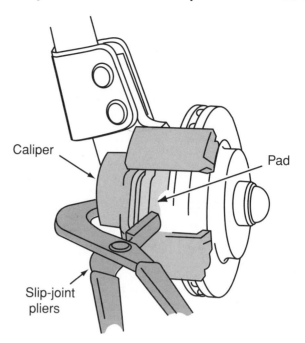

Figure 4. You can use large slip-joint pliers to retract the piston in some calipers.

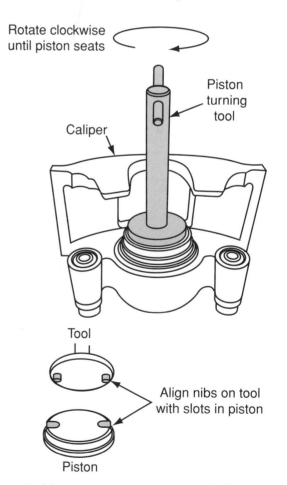

Figure 5. You must use these tools, or similar ones, to retract the piston in a rear caliper that has a parking brake mechanism attached to it.

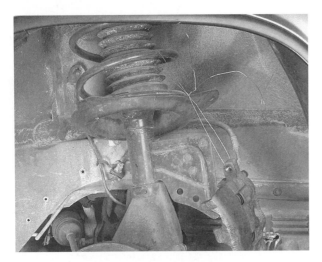

Figure 6. Suspend the caliper with wire or rope. Do not let the caliper hang from the brake hose.

caliper. Then remove the caliper from the caliper support. You do not need to disconnect the brake hose from the caliper if the caliper is not going to be removed from the vehicle. Suspend the caliper from the vehicle underbody or suspension with a heavy length of wire or rope **(Figure 6)**. Do not let the caliper hang from the brake hose.

> **You Should Know** *Never depress the brake pedal when the pads and caliper are removed. The pressure of the pedal will force the piston out of the caliper.*

Use a screwdriver to disengage the outer pad from the caliper housing. Then remove the inner pad from the piston and inner part of the caliper housing. Note the position of all springs, clips, shims, and other hardware used to attach the pads and to prevent noise. Inspect the mounting bolts and sleeves and other miscellaneous hardware for corrosion. Inspect all rubber or plastic bushings for cuts and nicks. If any part is damaged, install new parts when the caliper is reinstalled. Do not try to polish away corrosion.

PAD INSTALLATION

A complete set of springs, clips, shims and other miscellaneous pieces might be available as a pad hardware kit for popular brake assemblies. Installing such a kit is often a practical and economical choice when replacing pads or doing other caliper service.

If the pads appear serviceable, use a vernier caliper or a precision scale to measure the thickness of each brake pad lining **(Figure 7)**. Compare the pad thickness against specifications. If the lining thickness is close to or less than the service limit, replace all pads on both calipers as a set. If

Figure 7. Measure the total thickness of the pad with a vernier caliper. Then measure the thickness of just the metal backing plate. The difference between the two is the thickness of the lining.

the pads are serviceable, always reinstall the pads in their original positions. Switching pad positions can reduce braking power. The pads must also be free of grease or brake fluid if they are to be reused. Replace contaminated brake pads and wipe any excess grease off the parts.

Before installing the new pads, wipe the outside of the piston dust boot with denatured alcohol. Use a C-clamp to bottom the piston in the caliper bore, taking care not to damage the piston or the dust boot **(Figure 8)**. When the

Figure 8. Use a C-clamp to fully retract the piston. Wood between the piston and the clamp helps to protect the piston's seal and its bore.

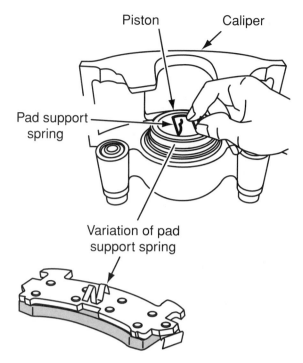

Figure 9. **Many inboard pads have retainer springs that snap into the inner opening of the piston.**

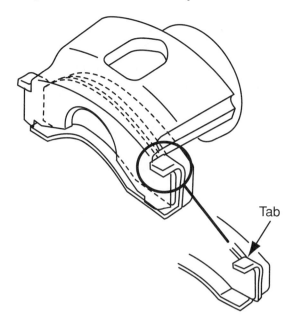

Figure 10. Many outboard pads have retaining flanges or tabs that snap around the outboard part of the caliper.

piston is bottomed in the bore, lift the inner edge of the boot next to the piston and press out any trapped air. The boot must lie flat.

Install the inboard pad and lining by snapping the pad retaining spring into the piston's inside opening **(Figure 9)**. The pad retainer spring is usually staked to the inboard pad. The pad must lie flat against the piston. After the pad is installed, verify that the boot is not touching the pad. If it is, remove the pad and reposition the boot.

Install the outboard pad in the caliper and secure any locking tabs as necessary. The exact methods and parts used for pad installation will vary with brake design **(Figure 10)**. You might need to bend the retaining flanges or tabs of some pads to ensure secure attachment to the caliper **(Figure 11)**. If you service one side of the car at a time, the opposite caliper is available as a guide for pad installation.

When specified, apply any recommended antinoise compound or shims to the backs of the pads during reassembly **(Figure 12)**. Some technicians choose to use shims or compound on all brake pads. It is never wise to use both in a single application.

If the replacement pads have audible wear sensors, the pad must be installed so that the sensor is at the leading edge of the pad in relation to wheel rotation. This does not mean at the top or front of the caliper. The rotor must contact the edge of the pad that holds the sensor first as it rotates through the caliper; this is the pad leading edge. If

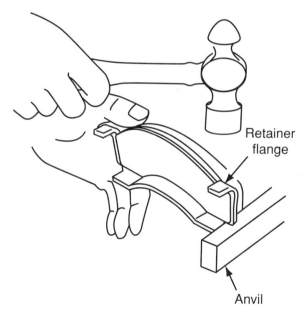

Figure 11. Bend the outboard pad retaining flange when required to do so to ensure secure attachment.

the vehicle has electronic lining wear sensors that light a warning lamp on the instrument panel, install the pads and sensors according to the procedures in the service manual.

On floating calipers, liberally coat the inside diameter of the bushings with silicone grease before installing the

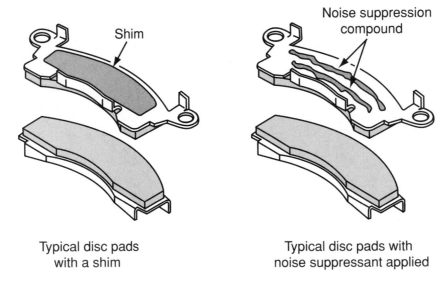

Typical disc pads
with a shim

Typical disc pads with
noise suppressant applied

Figure 12. Noise suppression compounds and shims applied or installed to the back of the pad help to reduce noise.

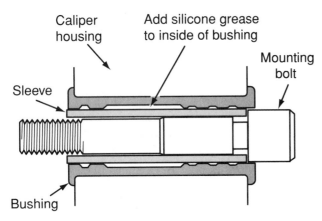

Figure 13. Lubricate the inside of the caliper mounting bushing.

mounting bolts and sleeves **(Figure 13)**. For sliding calipers, lubricate the caliper ways on the caliper support and the mating parts of the caliper housing with the recommended lubricant **(Figure 14)**.

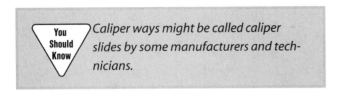

You Should Know *Caliper ways might be called caliper slides by some manufacturers and technicians.*

After the new pads are installed and the calipers remounted on the caliper supports, add fresh brake fluid to the master cylinder reservoir to bring it to the correct level. Then start the engine without moving the vehicle and apply the brake pedal. The pedal probably will go to the floor as the caliper pistons move out to take up the clearance between the new pads and the rotors. Recheck the brake fluid level and add more fluid as needed. Apply the pedal several more times until it becomes firm and verify that the fluid is at the correct level.

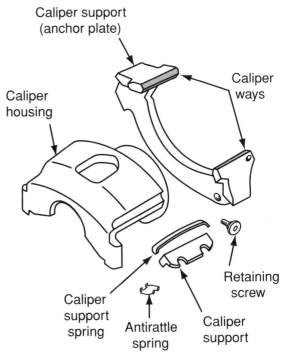

Figure 14. On sliding calipers, lubricate the caliper ways and mating areas of the caliper housing during assembly.

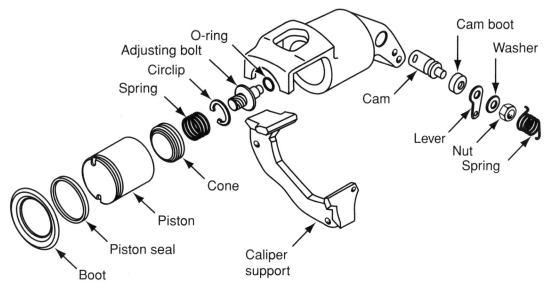

Figure 15. A typical rear disc brake caliper with a parking brake cam and lever assembly.

After proper pedal action is restored, recheck the fluid level, turn on the ignition, and release the parking brake. Verify that the brake warning lamp on the instrument panel is not lit. If it is, adjust the pressure differential valve and switch.

> **You Should Know** *Before trying to move the vehicle, press the brake pedal several times to make sure that the brakes work. Bleed the brakes as necessary and then road test the vehicle.*

Theoretically, if the brake lines were not disconnected and the pistons were not removed from the calipers, air will not have entered the system and bleeding should not be necessary. It is good practice, however, to bleed the brakes to ensure that the hydraulic system is free of air. If the pedal seems at all spongy or low after pad replacement, brake bleeding is required. If brake bleeding does not restore proper pedal action and brake performance, inspect the system thoroughly for leaks. You might need to remove and disassemble the calipers to check for corroded caliper bores and leaking seals that let air into the system.

REAR DISC BRAKE INSPECTION AND REPLACEMENT

Rear disc brakes are used on many vehicles. In most cases, the rear brakes are identical to the front disc brakes except for some type of parking brake mechanism. **Figure 15**

is an exploded view of a rear disc brake caliper that is typical of one with a parking brake mechanism.

The following is a typical procedure for inspecting and replacing rear disc brakes and pads:
1. Remove some brake fluid from the fluid reservoir at the master cylinder.
2. Raise the vehicle on the lift and remove the rear wheels. The rear caliper is often protected by a plastic shield. Remove this shield and the bolts securing the caliper.
3. Disconnect the parking brake cable from the lever on the caliper by pulling out the lockpin **(Figure 16)**.

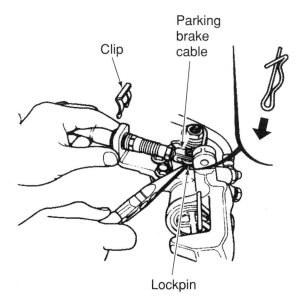

Figure 16. Disconnect the parking brake cable from the lever on the caliper by pulling out the lockpin.

4. Remove the caliper mounting bolts and lift the caliper off its support. To prevent dirt from entering the caliper body, clean the outside of the caliper before beginning disassembly.

5. To remove the brake pads, remove any pad shims and retainers and lift the pad spring out of the caliper. Then pull the pads off the caliper.

6. Check the condition of the rotor. Thoroughly clean the rotor surface and inspect it for defects and damage. Measure rotor thickness and runout.

7. To depress the caliper's piston into its bore and allow room for the new pads, use a properly sized locknut wrench and extension bar. Turn clockwise to retract the piston **(Figure 17)**. Back the piston down just enough to allow for reinstallation with the new pads. Make sure you follow the instructions given by the manufacturer. Excessive tolerances can ruin the piston's mechanisms.

8. Install the new or refinished rotor and the clean caliper bracket. Install the new brake pads, pad shims, retainers, and springs onto the caliper bracket.

9. Lubricate the dust boot with a small amount of silicone grease before installing it to reduce the chances of twisting. Rotate the piston clockwise to align the cutout in the piston with the tab on the inner brake pad by turning the piston back **(Figure 18)**. The piston boot must fit properly without twisting. If the boot becomes twisted during installation, back it out until it is seated properly.

10. Reinstall the caliper on the caliper bracket and tighten the caliper bolts to torque specifications. Then install the splash shield.

11. Reconnect the parking brake cable to the arm on the caliper. Then reinstall the caliper shield.

12. Top off the master cylinder reservoir.

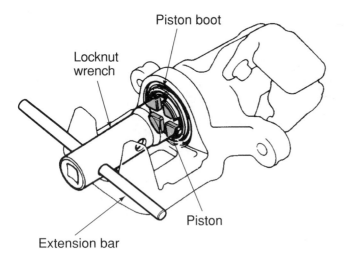

Figure 17. A locknut wrench is used to turn the piston in and out of its bore.

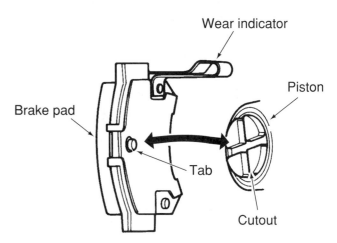

Figure 18. Rotate the piston so that the cutout in the piston is aligned with the tab on the brake pad.

13. Depress the brake pedal several times, and then adjust the brake pedal. Before making adjustments, be sure that the parking brake arm on the caliper touches the pin.

ROAD TEST AND PAD BURNISHING

Whenever new brake pads are installed, they need a short period of controlled operation that is called a **burnishing**, or **bedding-in**, period. Road testing the vehicle lets you perform this burnishing procedure and verify that the brakes work properly.

New pads require burnishing to establish full contact with the rotor and to heat and cure any resin left uncured in the friction material. Whether the rotors were refinished or not, new pads do not initially make full contact with the rotor surfaces but require a period of light wear to establish this contact. Also, when brake linings are manufactured, some of the resin materials might remain uncured until the pads are put into service. If fresh pads are subjected to hard braking, the resins can boil to the surface of the pads and cause glazing when they cool. The pads then might never operate properly.

Burnish the brake pads during the initial road test by driving at 30–35 mph (50–60 kph) and firmly but moderately applying the brakes to fully stop the car. Do this five or six times with 20–30 seconds of driving time between brake applications to let the pads cool. Then drive at highway speeds of 55–60 mph (85–90 kph) and apply the brakes another five or six times to slow the car to 20 mph (30 kph). Again allow approximately 30 seconds of driving time between brake applications to let the brake pads cool. Finally, advise the customer to avoid hard braking for the first 100 miles of city driving or the first 300 miles of highway driving.

Summary

- Brake pads should always be replaced in axle sets.
- Disc brake assemblies should be properly cleaned before disassembly.
- Removing brake fluid from the master cylinder is an important preliminary step for all disc brake service in order to prevent overflowing the reservoir when the caliper pistons are forced back into the caliper bores.
- On vehicles with ABS, the recommended way to prevent reservoir overflow is to open the caliper bleeder screw and run a bleeder hose into a container to catch the fluid expelled when the piston is forced back into its bore.
- A C-clamp, slip-joint pliers, a pry bar, or a screwdriver can be used to retract caliper pistons.
- On fixed caliper brakes, prying the pistons back into their bores is the usual way to provide clearance for pad removal.

- The piston in a rear caliper with a parking brake mechanism should be retracted only with the tools designed for that caliper and piston.
- Before installing the new pads, wipe the outside of the piston dust boot with denatured alcohol.
- If the replacement pads have audible wear sensors, the pad must be installed so that the sensor is at the leading edge of the pad in relation to wheel rotation.
- On floating calipers, liberally coat the inside diameter of the bushings with silicone grease before installing the mounting bolts and sleeves.
- On sliding calipers, lubricate the caliper ways on the caliper support and the mating parts of the caliper housing with the recommended lubricant.
- Whenever new brake pads are installed, they need a short period of controlled operation that is called a burnishing, or bedding-in, period. Road testing the vehicle lets you perform this burnishing procedure and verify that the brakes work properly.

Review Questions

1. Why is brake fluid removed from the master cylinder before working on disc brakes?
2. What is the recommended procedure for removing fluid, before servicing the brakes, from a system that has ABS?
3. While servicing disc brakes Technician A applies antinoise compound to all disc pads he installs. Technician B applies antinoise compounds and shims to all disc pads he installs. Who is correct?
   A. Technician A only
   B. Technician B only
   C. Both Technician A and Technician B
   D. Neither Technician A nor Technician B
4. While servicing the disc brakes on a vehicle Technician A says that dust and dirt created by pad wear and accumulated through normal driving are thrown off by the centrifugal force of the spinning rotor. Technician B says that proper and safe cleaning of disc brake assemblies and components is not as important for disc brakes as it is for drum brakes. Who is correct?
   A. Technician A only
   B. Technician B only
   C. Both Technician A and Technician B
   D. Neither Technician A nor Technician B
5. While installing new brake pads Technician A liberally coats the inside diameter of the bushings with silicone grease before installing the mounting bolts and sleeves of floating calipers. Technician B lubricates the caliper ways on the caliper support and the mating parts of sliding caliper housings with the recommended lubricant. Who is correct?
   A. Technician A only
   B. Technician B only
   C. Both Technician A and Technician B
   D. Neither Technician A nor Technician B

Chapter 33

Caliper Service

Introduction

Whenever brake pads are changed, the calipers **(Figure 1)** should be carefully inspected for damage, leakage, and general wear. Generally, calipers are likely to need overhaul or replacement if the vehicle is five years old or older or has more than 50,000 miles (80,000 kilometers) and the calipers have never been serviced. If the vehicle is driven hard or operated in very hot or cold temperatures, the calipers will need service sooner than calipers on a vehicle used conservatively in a moderate climate.

LOADED CALIPERS

Rebuilding a caliper and installing new pads can be a time-consuming procedure and one that allows room for error. Common mistakes include forgetting to bend brake pad locating tabs that reduce vibration and noise, leaving

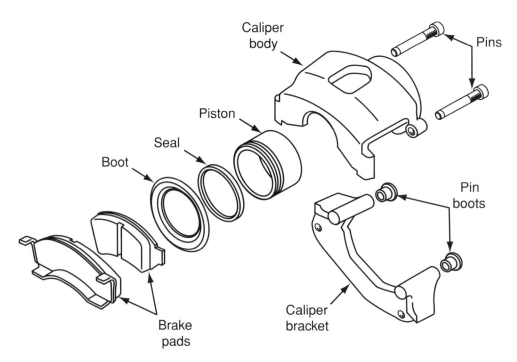

Figure 1. The parts of a caliper should be checked and/or replaced when the brake pads are replaced.

Figure 2. A loaded caliper is a rebuilt caliper with new pads, seals, and hardware.

off antirattle clips and pad insulators, or reusing worn or corroded mounting hardware that limits caliper movement and reduces pad life. To save time and frustration, many shops install **loaded calipers** when a vehicle's calipers need service. Loaded calipers are rebuilt calipers that come with brake pads and mounting hardware fully installed **(Figure 2)**.

The use of loaded calipers ensures that all components that should be replaced are replaced. When a caliper on one side of the vehicle is bad, the other side should also be replaced with the same type of loaded caliper.

Whether or not to use loaded calipers for a particular brake job is as much a business decision as a technical decision. Loaded calipers can be more economical for the customer than the labor costs of overhauling the original calipers. Additionally, loaded calipers usually have a manufacturer's warranty for all materials and labor that went into the rebuilding. On the other hand, loaded calipers for all vehicles are not available. The job of a skilled brake technician still includes the ability to properly overhaul a caliper.

If you install loaded calipers, do not overlook the necessary inspection, cleaning, and lubrication of the caliper supports on the vehicle's suspension. Also, do not overlook brake rotor inspection and service. Even the best loaded calipers cannot operate properly if installed on damaged or rusted mounting hardware or if they are forced to operate with defective rotors.

CALIPER REMOVAL

You can inspect much of the caliper while it is mounted on the vehicle, but you must remove the caliper to closely examine the dust boot, the pistons and seals, and the mounting hardware, including bushings and sleeves. Removal procedures are different for different calipers, but you should follow these general guidelines.

A brake hose can be attached to the caliper with a banjo fitting, a swivel fitting, or a rigid fitting. If the hose is attached to the caliper with a banjo or swivel fitting, disconnect it from the caliper while being careful not to twist

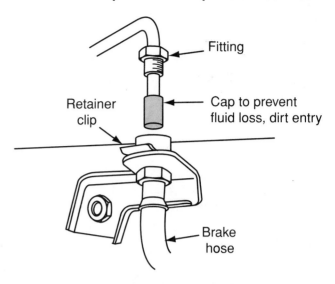

Figure 3. Cap or plug disconnected brake lines to keep dirt and moisture out.

the hose. Plug or cap the open end of the hose to keep dirt out of the brake lines. If the brake hose is attached to the caliper with a rigid fitting, disconnect the brake pipe from the hose at the hose mounting bracket and cap the end of the pipe **(Figure 3)**.

The brake pads can be removed before the caliper or along with the caliper, whichever is easier. If you remove the pads first, inspect the mounting hardware and set it aside for reinstallation or for closer inspection and replacement.

Remove the clips and keys from sliding calipers and slide each caliper off its ways on the caliper support **(Figure 4)**. Remove the mounting pins or bolts from floating calipers and similarly slide each caliper off its caliper

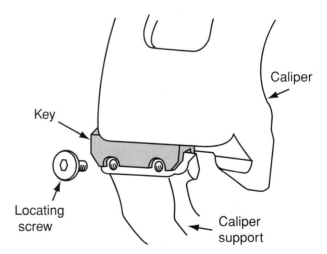

Figure 4. On this sliding caliper, the locating screw must be removed before the keys are removed.

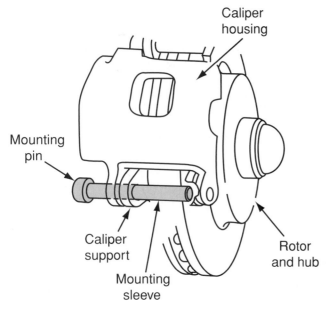

Figure 5. Remove the mounting pins or bolts to disconnect a floating caliper.

support **(Figure 5)**. Inspect the mounting hardware for excessive wear, corrosion, and other damage. Remove all bolts holding a fixed caliper to its support.

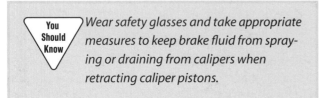

Wear safety glasses and take appropriate measures to keep brake fluid from spraying or draining from calipers when retracting caliper pistons.

Lift the caliper off the rotor and take it to a bench for further service. It is wise to work on the caliper on one side of the vehicle at a time and use the caliper on the other side for assembly and installation reference.

CALIPER INSPECTION

Look at the following areas and check for these conditions during complete caliper inspection:

1. Inspect the entire outside of the caliper body for cracks and other major damage. Replace any damaged caliper.
2. Inspect the piston dust boot closely for holes and tears. Be sure it is correctly installed in the piston and caliper and that no openings exist that could let dirt or water into the caliper's bore. If the dust boot is damaged or defective in any way, stop at this point and replace or overhaul the caliper.
3. If the dust boot is all right, inspect the caliper closely for leakage. Any sign of leakage means that a piston

seal is leaking and the caliper must be replaced or overhauled.

4. Use a C-clamp or large slip-joint pliers to force the piston to the bottom of its bore. Be careful not to damage the dust boot or seal. Note how the piston feels as it moves. If it sticks or moves unevenly, remove the piston to check for rust and scoring. Refer to specific service manual procedures to retract the pistons on rear calipers with a parking brake linkage attached to the pistons.

If a caliper passes these general checks and the vehicle is relatively new or has low mileage, it is reasonable to replace the pads without overhauling or replacing the calipers. If a caliper shows any signs of damage, excessive wear, or age, it should be replaced or overhauled. Just as pads are replaced in axle sets, calipers must always be serviced in pairs for each end of the vehicle.

BLEEDER SCREW REMOVAL

Before you begin to overhaul a caliper, try to loosen the bleeder screw **(Figure 6)**. Bleeder screws often get stuck or frozen into the caliper, particularly in aluminum caliper bodies or on vehicles driven on salted roads in the winter. A frozen bleeder screw is often a sign of extreme wear or age that immediately makes a caliper a candidate for replacement. If replacement calipers are not readily available, however, you must be able to replace the frozen screw.

If the bleeder screw does not loosen easily with a wrench, do not continue to force it until it breaks. You can loosen many frozen screws by applying penetrating oil to the outer threads and letting it work its way into the screw hole. Wait 10 minutes after applying penetrating oil and then place a deep socket over the screw so that the socket does not contact the screw but rests on the caliper body. Hit the socket several times with a hammer to break the

Bleeder cap and screw

Figure 6. Be sure you can remove the bleeder screw before beginning to service a caliper.

surface tension between the screw and the hole and to help the oil work into the threads.

Special impact sockets for loosening bleeder screws are available from tool companies. These sockets attached to an impact screwdriver are placed over the bleeder screw and struck with a hammer. This shocking effect might loosen the screw.

> *During your attempt to loosen and remove a stuck bleeder screw, never apply heat to the caliper with the piston still in the bore.*

If the bleeder screw still will not turn, heat might help to loosen it. These steps should be followed with the piston removed from the caliper and the caliper clamped in a bench vise. Heat the area of the caliper body around the screw with a torch and apply penetrating oil to the outer screw threads. Repeat this several times and try to turn the screw with a wrench. For a particularly stubborn bleeder screw, it might help to invert the caliper in the vise so that the screw is at the lowest point of the caliper. There should be some brake fluid in the caliper bore and it should cover the inner end of the bleeder. Add an ounce or two of fluid through the hose connection port if necessary. Once again, heat the caliper body. Brake fluid can work its way past the tip of the bleeder and into the inner threads to help loosen the screw.

> *Wear safety glasses whenever removing caliper pistons. Keep your hands away from the piston when using compressed air or hydraulic pressure. Cover the open hydraulic ports and the perimeter of the piston bore with clean shop cloths and take other necessary precautions to avoid spraying brake fluid. Clean up any spilled fluid immediately.*

If the bleeder screw breaks or cannot be loosened by these methods, replace the caliper. If a replacement is not available, the broken screw can be drilled out and the caliper rethreaded. Because bleeder screws are made of hard steel and because of the awkwardness of the caliper, drilling and retapping should be done by a machine shop.

CALIPER PISTON REMOVAL

Caliper pistons fit into their bores with only a few thousandths of an inch of clearance; therefore, even a small amount of rust, corrosion, or dirt can make them hard to remove. Caliper pistons are rarely loose enough in their bores to remove them by hand, so three common techniques are used for removal: compressed air, mechanical, and hydraulic.

Piston Removal with Compressed Air

Compressed air removal works best on pistons that are somewhat free or only moderately stuck in their bores.

1. Remove the caliper from the vehicle and clamp it securely in a bench vise by one of its mounting points.
2. Place a wooden block, wrapped in shop cloths, in the caliper opening. The block should be thin enough to let the piston move outward in its bore, but thick enough to keep the piston from leaving the bore completely.
3. Cover the caliper with cloths to keep brake fluid from spraying during removal.
4. Insert an OSHA-approved air nozzle with a rubber tip into the brake hose port and gradually apply air pressure **(Figure 7)**. Keep your fingers away from the piston.
5. If an OSHA-approved nozzle is not available but you can regulate the air pressure, start with about 30 psi and gradually increase line pressure. Use the lowest possible air pressure to loosen the piston. If you cannot regulate line pressure, apply air in short increments. Do not exceed normal maximum shop pressure of 90–100 psi.
6. When the piston pops free, remove the air nozzle and the wooden block and then remove the piston.

Mechanically Removing a Piston

Removing a piston mechanically involves the use of a special tool to grip the inner bore, or opening, of the piston

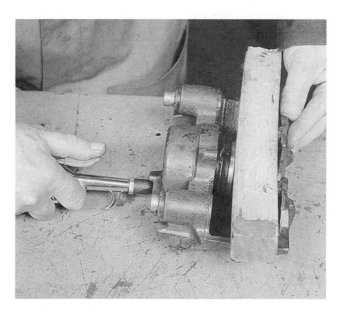

Figure 7.   Use a wooden block to limit the amount the piston can come out of its bore.

so that the piston can be twisted out of the caliper. This method works best with pistons that are only mildly stuck in their bores. It is often the quickest way to remove multiple pistons from fixed calipers.

1. Remove the caliper from the vehicle and clamp it securely in a bench vise by one of its mounting points.
2. Insert the removal tool into the piston opening and securely grip the inner surface of the piston. With a pliers-type tool, squeeze the handles tightly. With an expanding tool, turn the locking bolt until the tool grips the piston tightly.
3. Rotate the piston while working it back and forth in its bore until it loosens and slides out of the caliper.
4. If the piston cannot be removed manually, use compressed air or hydraulic pressure.

Hydraulically Removing a Piston

The vehicle's brake system can also be used to free a stuck piston. However, the vehicle's hydraulic system cannot be used for piston removal if air is present in the lines or if the master cylinder has an internal leak.

Follow these steps to use the vehicle's hydraulic system to remove a piston from a caliper:

1. Check the fluid level in the master cylinder and add fluid if necessary.
2. Raise the vehicle on a hoist or safety stands and support it securely.
3. Remove both front and/or both rear calipers from the caliper supports and hang them from the chassis or suspension with heavy wire.
4. Place a wooden block in the opening of each caliper. The block of wood should be thick enough to keep the piston from being ejected from its bore while still allowing piston movement.
5. Place large pans under the calipers to catch any fluid that may drain from the calipers.
6. Have an assistant gradually apply the brake pedal to force the pistons from their bores. If several pedal applications are needed to loosen the pistons, you might need to add fluid to the master cylinder.
7. When the pistons are loosened from their bores and pressed against the wooden blocks, disconnect and cap the brake hoses and remove the calipers from the vehicle.
8. Take the calipers to a work bench and drain brake fluid into a suitable container.
9. Remove the pistons from their bores by hand.

If hydraulic pressure cannot loosen a badly stuck caliper piston, it is probably rusted or corroded so badly that the caliper assembly should be replaced. Special hydraulic service equipment is available to remove pistons with the caliper off the vehicle. Such equipment makes piston removal a one-person operation, independent of the vehicle hydraulic system.

Potential Problems with Fixed Calipers

Mechanical removal tools can be used effectively on multiple-piston calipers, but they might not provide enough force to remove a badly stuck piston. Several methods can be used to free the piston. For example, if the caliper can be separated into two halves, the pistons can be removed more easily from each side with air or hydraulic pressure or mechanical removal tools.

At times, the loosest piston can be removed with compressed air and then reinstalled in its bore and held in place with a C-clamp. The piston seal will then retain pressure well enough to apply air pressure to loosen the other pistons.

If the pads can be removed while the calipers are installed on the vehicle, more space will exist between the pistons and the rotor. Cover the rotor with shop cloths to protect its surface and use compressed air to pop the pistons out of the caliper. They will move far enough to loosen them, but the rotor will keep them from moving out of the bores completely so that they retain enough air pressure to loosen all pistons.

DUST BOOT AND SEAL REMOVAL

After the piston is removed, remove the dust boot from the caliper. If the boot stays attached to the caliper, use a small screwdriver to pry it from its groove **(Figure 8)**. If the dust boot is attached to the piston, remove and discard it.

Use a wooden or plastic scraper or pick to remove the piston seal from the caliper bore **(Figure 9)**. To avoid scratching the caliper bore, do not use a screwdriver or metal pick. Before discarding the old piston seal, examine its shape. If it is anything but square cut, remember the direction in which it was installed so that you can install the new seal correctly.

After the caliper is disassembled, clean it with brake cleaning solvent or alcohol and a soft-bristled brush. You can use a plastic Scotchbrite® scouring pad to remove dirt

Piston dust boot

Figure 8. Use a small screwdriver to remove the dust boot from the caliper.

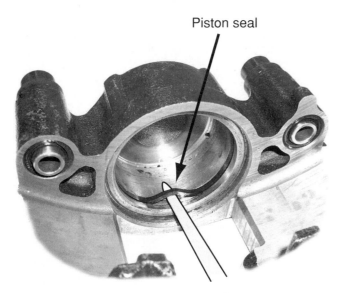

Piston seal

Figure 9. Use a wooden or plastic pick to remove the piston seal from its groove.

Figure 11. Carefully inspect the bore of the caliper.

and varnish from caliper bores. Do not use a wire brush on the caliper bore, the piston, or dust boot attaching points. A wire brush will scratch and nick caliper bore and piston surfaces, which will cause leaks after reassembly.

Use crocus cloth **(Figure 10)** lubricated with brake fluid to remove light rust, dirt, corrosion, or scratches from the caliper bore and the piston. After removing these defects, clean the piston and caliper bore with brake cleaning solvent or alcohol to remove all abrasive materials. If light corrosion cannot be removed from the caliper bore with crocus cloth, the bore can be honed. Heavy corrosion or rust or severe pitting and scratching on the piston or its bore indicate that the complete caliper should be replaced.

Figure 10. Crocus cloth can be used to polish out light corrosion.

CALIPER INTERNAL INSPECTION

If the caliper appears serviceable after it is disassembled and cleaned, closely inspect the piston and the caliper bore closely **(Figure 11)**. You can hone scratched or rusted calipers, but excessive honing can create too much clearance between the piston and the bore. In this case, the caliper must be replaced.

Inspect metal pistons for corrosion and scratching. Remember that the outside surface of the piston provides the sealing surface, so even minor scratches can create leakage. If the piston is chrome-plated, make sure that the plating is not scuffed or flaked away. Similarly, the outer anodized surface of an aluminum piston must not be worn away. Replace any piston that has a damaged surface finish.

Inspect phenolic plastic pistons for cracks and chips. Small surface chips or cracks away from sealing surfaces are acceptable, but the best choice is to replace any piston that has surface damage.

CALIPER HONING AND PISTON CLEARANCE

Honing the bore of a floating or sliding caliper is a cleaning procedure, not a machining procedure. Hone the bore only enough to remove the worst corrosion and surface wear.

1. Remove the caliper from the car and clamp it securely in a bench vise by one of its mounting points.
2. Install the proper size hone with fine-grit stones in an electric drill motor.
3. Lubricate the bore and the hone with clean brake fluid.

4. Insert the hone in the bore and operate the drill at about 500 rpm.

5. Move the hone slightly up and down in the bore, but do not let the hone come completely out of the bore while the drill motor is running.

6. After about 10 seconds, stop the motor and remove the hone.

7. Wipe the bore with a clean cloth and inspect it to see if it has been cleaned satisfactorily.

8. Repeat the honing operation until corrosion, scratches, and pits are satisfactorily removed.

9. After honing the bore, clean it thoroughly with brake cleaning solvent, alcohol, or fresh brake fluid. If you use alcohol or brake cleaning solvent, make a final cleaning pass with brake fluid. Be sure to remove all abrasive materials from the seal groove and fluid passages.

10. After cleaning the honed bore, lubricate the bore and piston with fresh brake fluid.

Piston clearance is more important than the final surface finish of the bore. Insert a feeler gauge with a thickness of the specified maximum allowable clearance into the bore. Try to insert the piston into the bore next to the feeler gauge. If the piston slides into the bore, the clearance is too great. Double-check the fit with a new piston or measure the piston and bore diameters with micrometers to be sure. If clearance exceeds specifications, replace the entire caliper assembly.

CALIPER ASSEMBLY

Assemble the caliper on a clean workbench and be sure that your hands and tools are clean and free of grease and oil. Follow these general steps to assemble most floating or sliding calipers:

1. Lubricate the piston seal and the seal groove in the caliper bore with fresh brake fluid. If the seal is any shape other than square cut, be sure that you install it in the proper direction. Square-cut seals can be installed with either side facing in either direction.

2. Insert one edge of the seal in its groove and roll it into place by hand **(Figure 12)**. Be careful not to twist, roll, or nick the seal.

3. Lubricate the piston and the caliper groove for the dust boot with fresh brake fluid **(Figure 13)**. Make sure the piston enters the caliper bore smoothly and keep it square in the bore as it is being pushed in.

4. If the dust boot attaches to the piston, install the dust boot onto the piston. Then pull it downward, turning it inside out. Now fold the boot back upward until it snaps into place.

5. Attach the dust boot to the caliper by one of the following three general methods, depending on dust boot design:

Figure 12. Carefully roll the piston's seal in place.

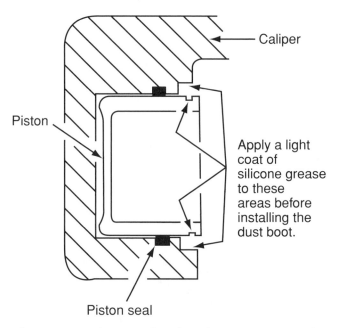

Figure 13. Lubricate the dust boot groove in the caliper.

a. If the dust boot is a press fit into the caliper, slide the piston into the dust boot and caliper bore until the boot snaps into its groove in the piston **(Figure 14)**. Then drive the boot into place with the proper driver **(Figure 15)**.

b. If the dust boot is held in the caliper by a retaining ring, slide the piston into the dust boot and caliper bore until the boot snaps into its groove in the piston. Then push the outer edge of the boot into its groove in the caliper body. Make sure that the

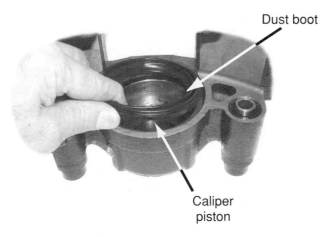

Dust boot

Caliper
piston

Figure 14.   Slide the piston into the dust boot and
caliper bore until the boot snaps into its
groove in the piston.

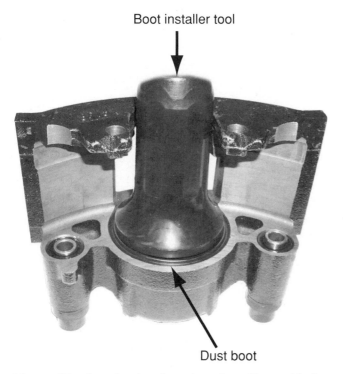

Boot installer tool

Dust boot

Figure 15.   Seat the dust boot into the caliper with the
proper driver.

boot is seated correctly and install the metal lock
ring to hold it in the groove.
c. If the dust boot is retained by a lip that fits in a
groove in the caliper bore, use the special tool
designed for this type of caliper. These tools are
installation rings of various diameters. Slide the
boot over the correctly sized ring. Install the lip of
the boot in the caliper groove, then slide the piston through the ring and boot into its bore.

Remove the ring from the piston so that the inner
lip of the dust boot snaps into the piston groove.
6. After the piston and dust boot are installed, finish
assembling the caliper by installing the brake pads.

CALIPER INSTALLATION

The procedure for installing calipers is basically the
opposite of that for removal; the exact procedures vary
with the different types of calipers. Always refer to the service manual.

Depending on the caliper design and clearance around
the caliper, it might be easier to install the pads before or
after the caliper is installed. If you install the pads first, make
sure all of the antirattle springs and clips are in place and
that any necessary antinoise compound is applied.

One of the most important points of caliper installation is lubrication of all moving parts on sliding or floating
calipers. Apply the recommended brake lubricant to contact points on caliper surfaces and on the caliper ways.
Lubricate sleeves and bushings of floating calipers according to the manufacturer's directions. Make sure you keep
lubricant off the surfaces of the rotor and pads.

Install the clips and keys on sliding calipers and slide
each caliper onto its ways on the caliper support. Install the
mounting pins or bolts for floating calipers and tighten all
bolts to the specified torque. Install all bolts holding a fixed
(stationary) caliper to its support.

Connect the brake hose to the caliper, being careful
not to twist the hose as you tighten the fitting.

REAR DISC BRAKE CALIPERS

Rear disc brake calipers with some type of parking
brake mechanism have different inspection and overhaul
procedures than front brake calipers. What follows is a typical procedure for overhauling a rear wheel caliper without
an additional drum parking brake.
1. A rear caliper is often protected by a plastic shield.
Remove this shield and the bolts securing the caliper.
2. Disconnect the parking brake cable from the lever on
the caliper by pulling out the lock pin.
3. Disconnect the brake hose from the caliper, remove
the caliper mounting bolts, and lift the caliper off its
support.
4. To remove the brake pads, remove any pad shims and
retainers and lift the pad spring out of the caliper. Then
pull the pads off the caliper.
5. To remove the caliper piston from the bore, use a properly sized locknut wrench and extension bar. Turn
them counterclockwise to back the piston out of the
bore. When the piston is free, remove the piston boot.
6. Carefully inspect the piston for wear **(Figure 16)**.
Replace the piston if it is worn or damaged in any way.

Piston

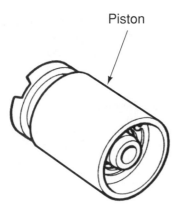

Figure 16. Carefully inspect the piston.

7. Remove the piston seal from the caliper using the tip of a screwdriver or a wooden or plastic scraper, being careful not to scratch the bore.
8. Removing the brake spring and its related parts from the caliper requires several special tools. Install a rear caliper guide in the cylinder so the cutout on the guide aligns with a tab on the brake spring cover.
9. Then install a brake spring compressor between the caliper and the rear caliper guide. Turn the shaft of the compressor tool to compress the adjusting spring and use snapring pliers to remove the circlip holding the spring in place **(Figure 17)**.
10. After removing the circlip, remove the spring compressor from the caliper. The following parts then can be

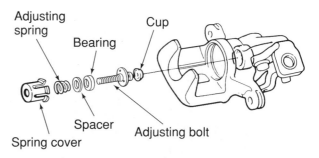

Figure 18. Once the retaining clip is removed, the spring cover, adjusting spring, spacer, bearing, adjusting bolt, and cup can also be removed.

removed: spring cover, adjusting spring, spacer, bearing, adjusting bolt, and cup **(Figure 18)**.
11. Next remove the sleeve piston and O-ring and then remove the rod from the cam.
12. Remove the return spring, the parking lever and cam assembly, and the cam boot from the caliper body **(Figure 19)**. Do not loosen the parking nut on the parking lever and cam assembly with the cam installed in the caliper. If the lever and shaft must be

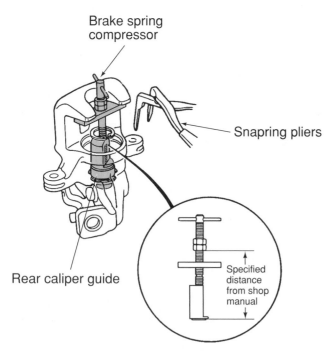

Figure 17. Spring compressor and special tool setup.

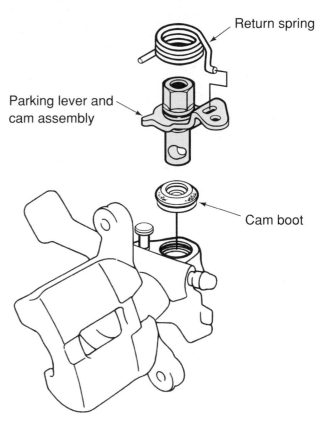

Figure 19. Remove the return spring, the parking lever and cam assembly, and the cam boot from the caliper body.

separated, secure the lever in a vise before loosening the parking nut.

13. Begin caliper reassembly by packing all cavities of the needle bearing with the specified lubricant. Coat the new cam boot with assembly lubricant and install it in the caliper. Apply assembly lubricant to the area on the pin that contacts the cam. Install the cam and lever assembly into the caliper body. Then install the return spring. If the cam and lever were separated, reassemble them before installing the cam in the caliper body.

14. Install the rod in the cam, followed by a new O-ring on the sleeve piston. Then install the sleeve piston so that the hole in the bottom of the piston is aligned with the rod in the cam, and the two pins on the piston are aligned with the holes in the caliper.

15. Install a new cup with its groove facing the bearing side of the adjusting bolt. Fit the bearing, spacer, adjusting spring, and spring cover on the adjusting bolt, then install it in the caliper bore.

16. Install the special caliper guide tool in the caliper bore, aligning the cutout on the tool with the tab on the spring cover.

17. Install the brake spring compressor and compress the spring until the tool bottoms out.

18. Make sure that the flared end of the spring cover is below the circlip groove. Then install the circlip and remove the spring compressor. Make sure that the circlip is properly seated in the groove.

19. Coat the new piston seal and piston boot with silicone grease **(Figure 20)** and install them in the caliper.

20. Coat the outside of the piston with brake fluid and install it on the adjusting bolt while rotating it clockwise with the locknut wrench.

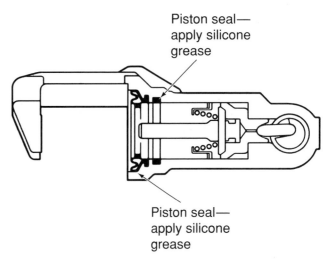

Piston seal— apply silicone grease

Piston seal— apply silicone grease

Figure 20. Coat the piston's seals with silicone grease.

21. Install the new brake pads, pad shims, retainers, and springs onto the caliper bracket.

22. Reinstall the caliper and the splash shield and tighten the caliper bolts to torque specifications.

23. Reconnect the brake hose to the caliper with new sealing washers and tighten the banjo bolt to specifications.

24. Then reconnect the parking brake cable to the arm on the caliper and reinstall the caliper shield.

25. Top off the master cylinder reservoir and bleed the brake system. Depress the brake pedal several times and then adjust the brake pedal. Before making adjustments, be sure that the parking brake arm on the caliper touches the pin.

Summary

- A loaded caliper is a rebuilt caliper with new pads, seals, and hardware.
- A brake hose can be attached to the caliper with a banjo fitting, a swivel fitting, or a rigid fitting.
- Inspect the mounting hardware for excessive wear, corrosion, and other damage.
- Inspect the entire outside of the caliper body for cracks and other major damage.
- Check the piston dust boot closely for holes, tears, and proper installation.
- Check the caliper for signs of leaks.
- Inspect all caliper mounting parts for rust and damage.
- Before you start to overhaul any caliper, try to loosen the bleeder screw.
- If the bleeder screw breaks or cannot be loosened, replace the caliper.

- Three common techniques are used to remove pistons from calipers: compressed air, mechanical, and hydraulic.
- After the piston is removed, remove the dust boot from the caliper with a small screwdriver to pry it from its groove.
- Use a wooden or plastic scraper or pick to remove the piston seal from the caliper bore.
- After the caliper is disassembled, clean it with brake cleaning solvent or alcohol and a soft-bristled brush.
- Use crocus cloth lubricated with brake fluid to remove light rust, dirt, corrosion, or scratches from the caliper bore and the piston.
- If light corrosion cannot be removed from the caliper bore with crocus cloth, the bore can be honed.
- Heavy corrosion or rust or severe pitting and scratching on the piston or its bore indicate that the complete caliper should be replaced.

- Any damaged caliper piston or one that has wear on its surface should be replaced.
- Hone the bore of a caliper only enough to remove the worst corrosion and surface wear.
- Piston clearance is more important than the final surface finish of the bore. If clearance exceeds specifications, replace the entire caliper assembly.
- During reassembly, lubricate the piston seal and the seal groove in the caliper bore with fresh brake fluid.

- Lubricate the piston and the caliper groove for the dust boot with fresh brake fluid.
- Lubricate all moving parts on sliding or floating calipers during installation of a caliper.
- Install the mounting pins or bolts for floating calipers and tighten all bolts to the specified torque.
- Install all bolts holding a fixed caliper to its support.
- Rear disc brake calipers with some type of parking brake mechanism have different inspection and overhaul procedures than front brake calipers.

Review Questions

1. When inspecting the bore of a caliper, what are the possible conditions and how are they corrected?
2. Describe the procedure for using compressed air to remove a piston from a brake caliper.
3. Loaded calipers are being discussed. Technician A says that loaded calipers are replacement calipers that come with pads and hardware already installed. Technician B says that loaded calipers should always be installed in axle sets. Who is correct?
   A. Technician A only
   B. Technician B only
   C. Both Technician A and Technician B
   D. Neither Technician A nor Technician B
4. Which of the following statements about inspecting calipers is *not* true?
   A. The entire outside of the caliper body should be inspected for cracks and other major damage.
   B. Inspect the piston dust boot closely for holes and tears.
   C. Inspect all caliper mounting parts for rust and damage.
   D. The surface of the caliper mounting plate should be inspected and measured for signs of poor parallelism.

5. When performing disc brake work Technician A works on one wheel at a time to avoid "popping" pistons out of the other caliper and to allow the other caliper to be used as a guide. Technician B never permits the caliper to hang with the weight on the brake hose. Who is correct?
   A. Technician A only
   B. Technician B only
   C. Both Technician A and Technician B
   D. Neither Technician A nor Technician B

Chapter 34

Brake Drum Service

Introduction

Many drum brake problems cause or are caused by defects on the inside surface of a brake drum. To correct these problems, brake drums are replaced or resurfaced. Most shops make drum refinishing, or turning, a standard part of complete drum brake service. However, several automobile manufacturers say that a drum should be machined only if a defect is found. Many drum problems cannot be completely identified until the drum is mounted on a lathe. Maximum braking performance cannot be obtained with a drum that is damaged or defective.

The first step in servicing a drum is to determine if it can be refinished or if it should be replaced. The following are general guidelines to help you make that decision.

- *Scored drum surface.* Any large score marks mean that the drum must be resurfaced or replaced. If the scoring is not too deep, the drum can be refinished.
- *Bell-mouthed drum.* Drums must be refinished or replaced, depending on the severity of the condition.
- *Hard spots.* Drums with hard spots should be replaced.
- *Heat checks.* Excessive damage by heat checks or hard spots requires drum replacement.
- *Cracked drum.* Any crack, no matter how small, means that the drum must be replaced.
- *Out-of-round drums.* Slightly out-of-round drums can be refinished.
- *Grease or oil contamination.* Thoroughly clean the drum with a nonpetroleum solvent, then refinish it.
- *Excessive wear.* Measure to determine if refinishing is possible; if not, replace the drum.

DRUM MEASUREMENTS

While servicing the brakes, measure each drum with a drum micrometer to make sure it is within the safe size limits. The first step of drum measurement is to check the **discard diameter**, or maximum inside diameter, that is cast or stamped on the outside of the drum **(Figure 1)**. The discard diameter is the allowable wear dimension, not the allowable machining dimension. There must be 0.030 inch (0.75 mm) left for wear *after* turning the drum. That is, the refinished drum diameter must be at least 0.030 inch less than the discard diameter. If this dimension is exceeded, the drum will wear beyond its maximum allowable diameter

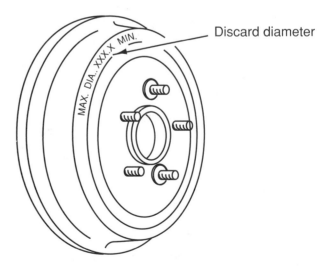

Figure 1. The discard diameter is cast or stamped on every brake drum.

during normal operation and is considered unsafe. A brake drum made before 1971 will not have a discard diameter marked on it; you will need to check a service manual for the exact specifications.

Begin measuring a drum by setting the drum micrometer to the nominal drum diameter such as 11.375 inches or 276 millimeters **(Figure 2)**. Refer to a service manual if you are unsure about the original diameter. Then insert the micrometer into the drum. Insert the end with the movable plunger first and hold it against the drum as you insert the fixed end of the micrometer.

Hold the anvil steady against the drum surface and move the dial end of the micrometer back and forth until you get the highest reading on the dial indicator **(Figure 3)**. Add the dial indicator reading to the nominal drum diameter that the micrometer was set to. For example, if you set the micrometer to 11.375 inches and the dial indicator reads 0.015 inches, the diameter of the drum is 11.390 inches **(Figure 4)**.

Compare this reading to the drum discard dimension and remember that it must be 0.030 inch (0.75 millimeter) or more *under* the discard dimension to allow for refinishing. Take at least four measurements, 90 degrees apart, around the drum opening.

Measuring the drum at four locations around its opening will also check for an out-of-round condition as well as the overall diameter. If the highest and lowest diameter measurements vary by 0.006 inch (0.15 millimeter) or more,

Dial indicator

Figure 3.   After the micrometer is in the drum, hold the fixed end firmly against the inside of the drum and carefully rock the dial indicator end until you get the highest reading.

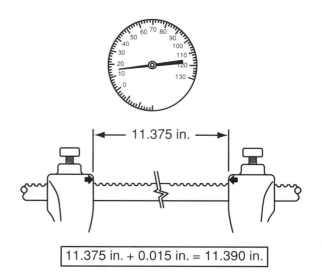

11.375 in. + 0.015 in. = 11.390 in.

Figure 4.   Add the dial indicator reading to the original micrometer setting to determine the diameter of the drum.

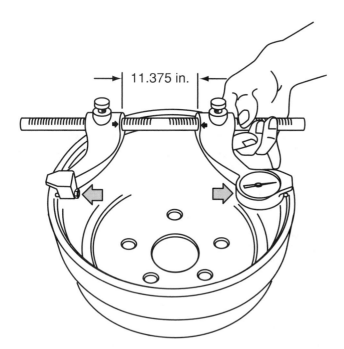

Figure 2.   Begin your measurement of the drum by setting the micrometer to the nominal diameter of the drum.

machine the drum to correct the out-of-round condition. This is true even if the drum is otherwise in good condition and within the maximum diameter limits. An out-of-round drum can cause brake chatter, grabbing, and pedal pulsation if not corrected.

The basic drum measurement will give you the drum diameter and indicate whether the drum can be refinished. If the drum is deeply scored, however, you must measure the diameter at the bottom of the deepest groove. To do this, you will need a drum micrometer adapter with pointed anvils or an inside micrometer with a long extension to span the drum diameter. Again, the diameter measured at the deepest groove must be 0.030 inch (0.75 millimeter) or more *under* the discard dimension.

Severe bell-mouth, concave, or convex wear can be visible to your eye or indicated by abnormal lining wear, but you should use an inside micrometer to measure for these conditions precisely. To check for any kind of wear or distortion across the drum surface, take several measurements in a straight line from the inside to the outside of the drum. Then repeat the measurements at about four positions. If the highest and lowest measurements taken at any straight line position across the drum vary by 0.006 inch (0.15 millimeter) or more, machine the drum to correct the problem.

If the drums are smooth and true and within safe limits, you can remove any slight scores by polishing with fine emery cloth. If scoring or light grooves cannot be removed by hand, the drum must be refinished or replaced. Even slightly rough surfaces should be turned to ensure a true drum surface and to remove any possible contamination on the surface from previous brake linings and road dust.

REFINISHING BRAKE DRUMS

Brake drums with moderate-to-severe scoring or other defects can be refinished either by turning (cutting) or grinding on a brake lathe **(Figure 5)**. Only enough metal should be removed to obtain a true, smooth friction surface. If too much metal is removed from a drum, the following unsafe conditions can result:

- Brake fade caused by the thin drum being unable to absorb heat during braking.

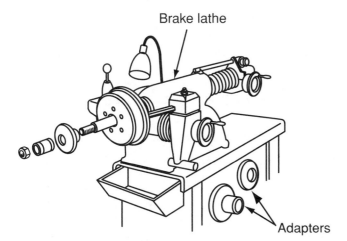

Figure 5. A drum lathe is used to refinish brake drums.

- Poor and erratic braking due to distortion of the drum.
- Noise caused by a thin drum vibrating during operation.
- Drums cracking or breaking during a very hard brake application.

If one drum must be machined, the other drum on the same axle should also be machined to the same diameter (±0.005 inch) so that braking will be equal at both wheels.

BRAKE LATHES

Brake drums are refinished on bench lathes made specifically for drum machining. Many models of drum lathes exist; therefore, always refer to the operating instructions given by the equipment's manufacturer. Nearly all brake lathes are designed to refinish both brake drums and brake rotors.

Different cutting assemblies are used for rotors and for drums. The attaching adapters, tool holders, vibration dampers, and cutting bits must be in good condition. Make sure that mounting adapters are clean and free of nicks. Always use sharp cutting tools or bits and use only replacement cutting bits recommended by the equipment manufacturer. Dull or worn bits leave a poor surface finish, which will affect braking performance.

The tip of the cutting bit should be slightly rounded, not razor sharp. A sharply pointed bit can cut a spiral groove into the drum that will cause noisy and erratic brake operation.

Mounting a Drum on a Lathe

Remove all grease and dirt from the bearing races before mounting the drum. It may be necessary to steam clean the grease out of the hub. On floating drums without bearings, clean all rust and corrosion from the hub area with emery cloth or 120-grit sandpaper.

The correct procedure for mounting a drum onto a lathe depends on whether the drum has wheel bearings mounted in its hub. When refinishing one-piece drums with bearings installed in the hub, remove the inner bearing and grease seal before mounting the drum on the lathe's arbor. Make sure that the inner bearing race is secure in the hub before you mount the drum on the lathe. If the race is not secure, the brake drum with its bearings should be replaced.

A one-piece drum with bearing races in the hub mounts to the lathe's arbor with tapered or spherical cones. A two-piece drum removed from its hub is centered on the lathe's arbor with a spring-loaded cone and clamped in place by two large cup-shaped adapters. Make sure you use the correct cones and spacers to mount the drum to the arbor shaft.

When the drum is on the lathe, install a rubber or spring-type vibration damper around the outer diameter of

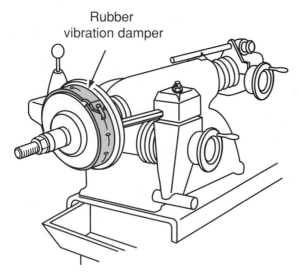

Figure 6. Install a vibration damper to keep the drum from chattering while it is being refinished.

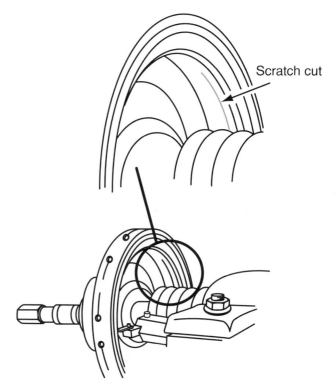

Figure 7. A pair of scratch cuts will help you check for out-of-roundness, as well as check the mounting of the drum on the lathe.

the drum **(Figure 6)** to prevent the cutting bits from chattering while they are cutting and to provide a smoother finished surface.

Machining a Drum on a Lathe

Before removing any metal from the drum, make sure that the drum is centered on the lathe's arbor and that extra runout has not been created by how the drum is mounted to the lathe. If the drum is not centered and square with the arbor, machining can add runout. To check the mounting of the drum, make a small scratch on one surface of the drum as follows:

1. Begin by backing the cutting assembly away from the drum and turning the drum through one complete revolution to be sure that there is no interference with rotation.
2. Start the lathe and advance the cutting bit until it just touches the drum surface near midpoint.
3. Let the cutting bit lightly scratch the drum, approximately 0.001 inch (0.025 millimeter) deep **(Figure 7)**.
4. Move the cutting bit away from the drum and stop the lathe. If the scratch is all the way around the drum, the drum is centered and you can proceed with resurfacing.
5. If the scratch appears intermittently, either the drum is out of round or it is not centered on the arbor. In this case, loosen the arbor nut and rotate the drum 180 degrees on the arbor; then retighten the nut.
6. Repeat steps 2 through 4 to make another scratch about ¼ inch away from the first.

7. If the second scratch matches the first scratch, the drum is significantly out of round, but it is properly centered on the lathe and you can proceed with machining.
8. If the second scratch appears opposite the first on the drum surface, remove the drum from the lathe arbor and recheck the mounting.

It is possible, although not very probable, that the lathe's arbor shaft is bent. To determine if the arbor is bent, mount a dial indicator on the lathe and disconnect lathe power. Release the pulley belt tension by moving the controlling lever; then rotate the arbor slowly by turning the drive pulleys. Observe the dial indicator needle and compare the reading to the lathe manufacturer's tolerance for arbor runout. Excessive runout indicates a bent arbor. A distorted or damaged mounting adapter is also possible.

Adjusting Lathe Settings

Before you begin to machine the drum on many lathes, you must consider and adjust three lathe settings: lathe speed (rpm), **cross-feed** (depth of cut), and spindle feed (speed of travel across the surface of the drum). On some lathes it is not possible to make all three of these adjustments.

The lathe speed usually stays constant throughout the machining operations, but most lathes have at least two or three speed settings. According to the lathe manufacturer's instructions, select the best speed for the drum you are machining. Most drums can be refinished satisfactorily at 150 rpm.

The cross-feed, or depth of cut, is the amount of metal removed by the cutting tool in each pass across the drum. You can use up to 0.015 inch (about 0.40 millimeter) for a rough cut but only about 0.005 inch (0.15 millimeter) for the finish cut.

The spindle feed is the distance the cutting bit moves across the friction surface during each lathe revolution. A spindle feed of 0.010–0.020 inch (0.25–0.50 millimeter) per revolution is good for rough cuts on most drums. Make the finish cut at a slower cross-feed of about 0.002–0.005 inch (0.05–0.15 millimeter) per revolution. **Figure 8** shows the cross-feed and spindle-feed controls for a typical lathe.

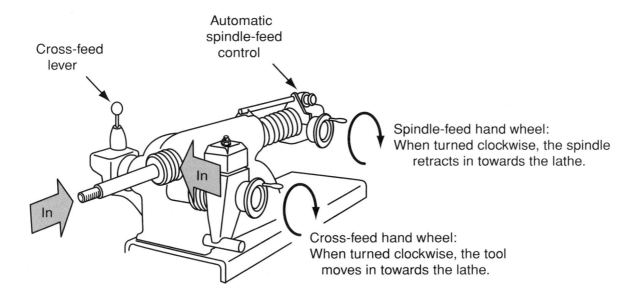

CLOCKWISE ROTATION

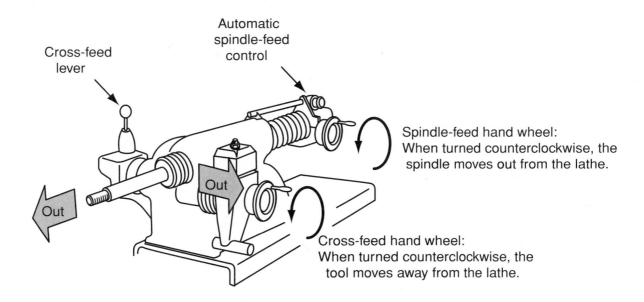

COUNTERCLOCKWISE ROTATION

Figure 8. The basic operation of the cross-feed and spindle-feed controls on a lathe.

Turning the Drum

After the drum is securely mounted and properly centered, follow these basic steps to refinish the drum:

1. Turn on the lathe and adjust it to the desired speed.
2. Advance the cutting bit to the open edge of the drum and remove the ridge of rust and metal that has formed there. Use several light cuts of 0.010–0.020 inch rather than one heavy cut.
3. Move the cutting bit to the closed edge of the drum and remove the ridge that might be present there. As you remove the ridges, note the point of the smallest drum diameter.
4. Position the cutting bit at the point of the smallest diameter and adjust the handwheel that controls the depth of cut to zero. This is the starting point for further depth-of-cut adjustments.
5. Reposition the cutting bit to the closed edge of the drum and adjust it for a rough cut as specified. The handwheel micrometer is graduated to indicate the amount of metal removed from the complete diameter. For example, if the handwheel indicates 0.010 inch, the lathe has made a cut 0.005-inch deep in the drum surface.
6. Adjust the cross-feed for a rough cut and engage the cross-feed mechanism. The lathe will automatically move the cutting bit from the inner to the outer edge of the drum. Make as many rough cuts as necessary to remove defects but stay within the dimension limits of the drum. If the cutting bit chatters as it passes over hard spots in the drum surface, grind the surface or discard the drum.
7. Complete the turning operation with a finish cut as specified.
8. After turning, keep the drum mounted on the lathe, and deburr it with 80-grit sandpaper (do not use emery cloth) to remove small imperfections in the surface.
9. If the drum does not clean up when turned to its maximum machining diameter, it must be replaced.

Some replacement brake drums are semifinished. A semifinished drum may require additional machining to obtain the proper dimensional specifications and surface finish. Fully finished drums are the most common and do not require additional machining unless it is needed to match the diameter of an old drum on the same axle set. New drums are also protected with a rust proofing coating that must be thoroughly cleaned off the friction surfaces. Use a nonpetroleum solvent, such as brake cleaner or lacquer thinner, to remove the coating.

CLEANING A REFINISHED DRUM

The surface of a freshly refinished drum contains millions of tiny metal particles. If these particles are not removed, they will become embedded in the brake lining. When the brake lining becomes contaminated with metal shavings, it becomes a fine grinding stone that will score the inside surface of the drum.

You can remove these metal particles from the drum by washing it thoroughly with hot water and wiping it down with a lint-free rag. Use compressed air to thoroughly dry the clean drum. After washing, wipe the inside of the drum (especially the newly finished surface) with a lint-free white cloth dipped in denatured alcohol or brake cleaning solvent. This operation should be repeated until dirt is no longer visible on the wiping cloth **(Figure 9)**. Allow the drum to dry before reinstalling it on the vehicle.

Figure 9.   Thoroughly clean the drum after machining.

Summary

- Many drum brake problems cause or are caused by defects on the inside surface of a brake drum.
- Most shops make drum refinishing a standard part of complete drum brake service.
- A scored drum surface requires that the drum be resurfaced or replaced.

- A bell-mouthed drum must be refinished or replaced, depending on the severity of the condition.
- Hard spots on a drum mean that the drum should be replaced.
- If the drum has excessive damage due to heat checks or hard spots, it should be replaced.

- Any crack, no matter how small, means that the drum must be replaced.
- Slightly out-of-round drums can be refinished.
- If the drum is contaminated with grease or oil, the drum should be thoroughly cleaned with a nonpetroleum solvent and then refinished.
- The first step of drum measurement is to check the drum's discard diameter.
- Use a drum micrometer to measure the diameter of the drum and to check for out-of-roundness.
- Brake drums with moderate-to-severe scoring or other defects can be refinished.
- If too much metal is removed from a drum, the following can result: brake fade, poor and erratic braking, noise, and cracking of the drum upon application of the brakes.
- Brake drums are refinished on lathes made specifically for drum machining.
- When the drum is on the lathe, install a rubber or spring-type vibration damper on the outer diameter of the drum to prevent the cutting bits from chattering while they are cutting the surface of the drum.
- Before you begin to machine the drum, you must consider and adjust three lathe settings: lathe speed, cross-feed, and spindle feed.
- The surface of a freshly refinished drum should be washed thoroughly with hot water and wiped down with a lint-free rag. Use compressed air to dry the drum.

Review Questions

1. What is the correct procedure for measuring a drum with a drum micrometer?
2. The drum discard diameter is being discussed. Technician A says that this is the maximum diameter to which the drums can be refinished. Technician B says that the drum discard diameter is the maximum allowable wear dimension. Who is correct?
   A. Technician A only
   B. Technician B only
   C. Both Technician A and Technician B
   D. Neither Technician A nor Technician B
3. When machining a drum on a brake lathe Technician A uses a spindle speed of approximately 150 rpm. Technician B sets the lathe to a fast cross-feed to obtain the finish on the surface of the drum. Who is correct?
   A. Technician A only
   B. Technician B only
   C. Both Technician A and Technician B
   D. Neither Technician A nor Technician B
4. While discussing what would happen if too much metal is removed from a drum by machining Technician A says that noise can result from the thin drum vibrating when the brakes are applied. Technician B says that the brakes could fade because the thin drum is unable to absorb heat during braking. Who is correct?
   A. Technician A only
   B. Technician B only
   C. Both Technician A and Technician B
   D. Neither Technician A nor Technician B
5. Technician A says that new drums must be cleaned to remove the rustproofing compound from the drum surface. Technician B says that refinished drums must be cleaned to remove all metal particles from the drum surface. Who is correct?
   A. Technician A only
   B. Technician B only
   C. Both Technician A and Technician B
   D. Neither Technician A nor Technician B

Chapter 35

Brake Rotor Service

Introduction

Maximum braking performance cannot be obtained with a rotor that is damaged or defective. This chapter covers replacement, resurfacing or turning, and finishing of disc brake rotors.

CLEANING ROTORS

A disc brake is partly shrouded by the wheel, but it is largely exposed to circulating air. Dust and dirt created by pad wear and accumulated through normal driving are thrown off by the centrifugal force of the spinning rotor. Some brake dust accumulates inside the wheel and partially on the caliper; therefore, proper and safe cleaning of disc brake assemblies and components before service is important.

REMOVING A ROTOR

To remove a rotor from the vehicle, raise the vehicle and remove the wheel. Then, remove the caliper from the rotor and suspend it with wire from the suspension of the vehicle. Before you remove a rotor, mark it L or R for left or right so that it gets reinstalled on the same side of the vehicle that it was removed from.

If the rotor is a two-piece floating rotor, remove it from the hub by pulling it off the hub studs **(Figure 1)**. If you cannot pull the rotor off by hand, apply penetrating oil on the front and rear rotor-to-hub mating surfaces. Strike the rotor between the studs using a rubber or plastic hammer. If this does not free the rotor, attach a three-jaw puller to the rotor and pull it off.

Whenever you separate a floating rotor from the hub flange, clean any rust or dirt from the mating surfaces of the hub and rotor. Neglecting to clean rust and dirt from the rotor and hub mounting surfaces before installing the rotor will result in increased rotor lateral runout, which will lead to premature brake pulsation and other problems.

If the rotor and hub are a one-piece assembly, remove the outer wheel bearing and lift the rotor and hub off the spindle **(Figure 2)**. Be careful not to hit the inner bearing on the spindle when removing the hub.

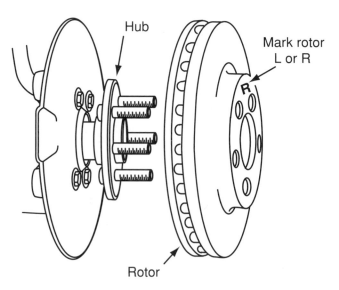

Figure 1. Remove a floating rotor by pulling it off its hub.

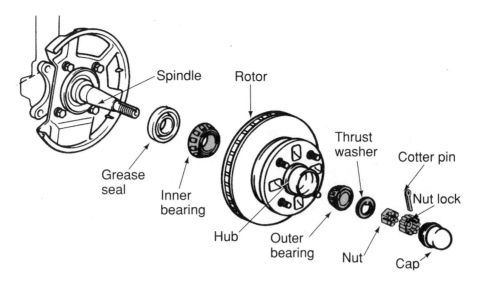

Figure 2. The bearings and mounting for a front wheel fixed rotor.

SERVICING BRAKE ROTORS

Brake rotors should be refinished only when they have minor imperfections, fail a lateral runout check, have excessive thickness variations, or have heat spots or excessive scoring that can be removed by resurfacing. However, a rotor should not be refinished if it is too thin or will be too thin after machining, regardless of the reason it needs to be turned.

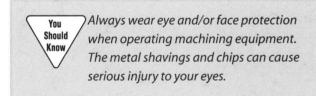

Always wear eye and/or face protection when operating machining equipment. The metal shavings and chips can cause serious injury to your eyes.

The minimum thickness specification for all brake rotors is the minimum wear dimension, not a minimum refinishing dimension. Do not use a brake rotor that does not meet these specifications. A refinished rotor must be thicker than its minimum thickness dimension.

Whenever you refinish a rotor, remove the least amount of metal possible to achieve the proper finish. This helps to ensure the longest service life from the rotor.

Never turn the rotor on one side of the vehicle without turning the rotor on the other side. Left- and right-side rotors should be the same thickness, generally within 0.002 to 0.003 inch. Similarly, equal amounts of metal should be cut off both surfaces of a rotor. Heat transfer capabilities can be reduced on a ventilated rotor if unequal amounts of metal are removed from the surfaces.

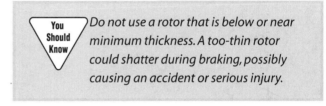

Do not use a rotor that is below or near minimum thickness. A too-thin rotor could shatter during braking, possibly causing an accident or serious injury.

Turning New Rotors

Most automobile and parts manufacturers recommend against refinishing new rotors unless runout exceeds specifications, which is not common. New rotors have the correct surface finish, which may be disturbed by their turning on a lathe. Clean any oil film off a new rotor with brake cleaning solvent or alcohol and let the rotor air dry before installing it on the vehicle.

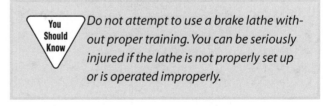

Do not attempt to use a brake lathe without proper training. You can be seriously injured if the lathe is not properly set up or is operated improperly.

BENCH BRAKE LATHES

Both off-vehicle lathes **(Figure 3)** and on-vehicle lathes are used to refinish rotors. If this is the first time you are using a particular lathe, make sure you review the operating instructions.

On a bench, off-vehicle, lathe, the rotor is mounted on the lathe's arbor and turned at a controlled speed

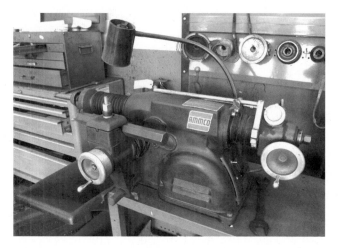

Figure 3. A typical bench brake lathe.

while a cutting bit passes across the rotor surface to remove a few thousandths of an inch of metal. The lathe turns the rotor perpendicularly to the cutting bits so that the entire rotor surface is refinished. Most lathes can operate at slow, medium, and fast speeds through a series of drive belts; however, the speed on some is not adjustable.

Most rotor cutting assemblies have two cutting bits. The rotor mounts between the bits and is pinched between them. As the cut is made, the same amount of surface material is cut from both sides of the rotor. Some lathes use only one bit for rotor refinishing, and separate cuts of equal amounts must be taken from each side.

When using a lathe with only one cutting bit, do not take the rotor off the lathe until both sides are cut. Always cut both sides of the rotor without removing it from the arbor. This ensures that both sides of the rotor will be parallel after refinishing. Check the accuracy of the cuts with a

dial indicator and an outside micrometer. Even a 0.001- to 0.002-inch variation in positioning the rotor on the lathe can add runout to a rotor.

The attaching adapters, tool holders, vibration dampers, belts, and cutting bits must be in good condition. Make sure that mounting adapters are clean and free of nicks. Always use sharp, but slightly rounded, cutting tools or bits. Dull or worn bits leave a poor surface finish, and razor-sharp bits will leave small grooves that work against the nondirectional finish needed on a rotor.

Mounting a Rotor on a Bench Lathe

A one-piece rotor with bearing races in its hub mounts to the lathe's arbor with tapered or spherical cones **(Figure 4)**. A two-piece rotor removed from its hub is centered on the lathe arbor with a spring-loaded cone and clamped in place by cup-shaped adapters **(Figure 5)**.

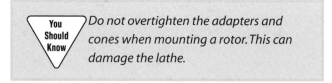

> **You Should Know** *Do not overtighten the adapters and cones when mounting a rotor. This can damage the lathe.*

Remove all grease and dirt from the bearing races before mounting the rotor. It is sometimes necessary to steam clean the grease out of the hub. When mounting rotors with their bearings installed, remove the inner bearing and grease seal before mounting the rotor on the lathe's arbor. When mounting a one-piece rotor and hub, check the inner bearing races (cones) to be sure that they are secure in the hub. If either race is loose, replace the rotor and all bearings.

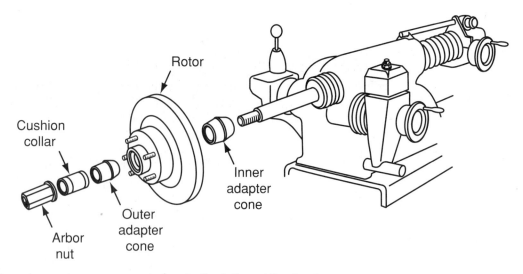

Figure 4. One-piece rotors are mounted onto the lathe with adapter cones.

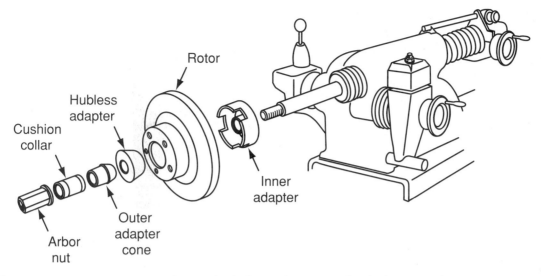

Figure 5. Two-piece rotors are mounted onto the lathe with a spring loaded cone and other adapters.

Figure 6. Use the appropriate cones and spacers to lock the rotor firmly to the arbor shaft.

Figure 7. Clean all dirt and rust from the hub-mounting area of the rotor.

Index the rotor on the wheel-bearing races to ensure that the machining is accurately indexed to the axis of the rotor. Use the appropriate cones and spacers to lock the rotor firmly to the arbor shaft **(Figure 6)**. Hubless rotors without bearings should be cleaned with emery cloth, 120-grit sandpaper, or a wire brush to remove all rust and corrosion from the hub area **(Figure 7)**.

When the rotor is on the lathe, install a rubber or spring-type vibration damper on the outer diameter of the rotor **(Figure 8)**. The vibration damper prevents the cutting bits from chattering, which results in a smoother finished surface.

Machining a Rotor on a Bench Lathe

Before removing any metal from the rotor, make sure the rotor is centered on the lathe's arbor and that extra

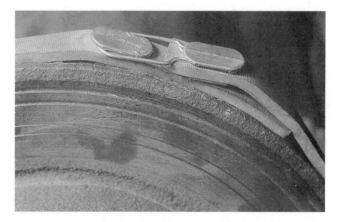

Figure 8. Install a rubber or spring-type vibration damper around the outside circumference of the rotor before machining.

runout has not been created by the mounting. If the rotor is not centered and square with the arbor, machining can actually add runout. To check rotor mounting, back the cutting assembly away from the rotor. Start the lathe and advance the cutting bit until it just touches the rotor surface near midpoint. Allow the cutting bit to lightly scratch the rotor. Move the cutting bit away from the rotor and stop the lathe. If the scratch is all the way around the rotor, the rotor is centered and you can proceed with resurfacing.

If the scratch appears as a crescent **(Figure 9)**, either the rotor has runout or it is not centered on the arbor. To determine the cause, loosen the arbor nut and rotate the rotor 180 degrees on the arbor, then retighten the nut.

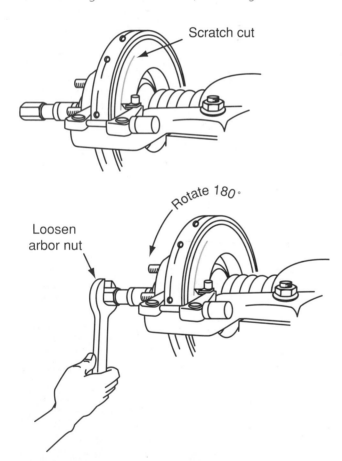

Figure 9. Check rotor runout by making a light scratch cut. Then rotate the rotor 180 degrees on the lathe's arbor and make a second cut. The position of the second cut helps to indicate the cause of any runout.

Make another scratch cut about ¼-inch away from the original scratch. If the second scratch appears at the same location as the first, the rotor has significant runout, but it is properly centered on the lathe and you can proceed with machining. If the second scratch appears opposite the first on the rotor surface, remove the rotor from the lathe arbor, recheck the mounting, and check the arbor to see if it bent.

Determining Machining Limits

To determine the approximate amount of metal to be removed, turn on the lathe and bring the cutting bit up against the rotating disc until a slight scratch is visible, as you did to verify rotor centering. Turn off the lathe and reset the depth-of-cut dial indicator to zero **(Figure 10)**. Find the deepest groove on the face of the rotor and move the cutting bit to that point without changing its depth of cut setting. Now use the depth-of-cut dial to bottom the tip of the cutter in the deepest groove. The reading on the dial now equals, or is slightly less than, the amount of metal to be removed to eliminate all grooves in the rotor. For example, if the deepest groove is 0.019 inch deep, the total amount to be removed can be 0.020 inch. For best results with cuts that have a total depth of greater than 0.015 inch, take two or more shallow cuts rather than one deep cut.

Figure 10. The depth of cut dial is graduated in thousandths of an inch to set the cutting depth.

Machining the Rotor

To make the series of refinishing cuts, proceed as follows:

1. Reset the cutting bits so that they just touch the ungrooved surface of either side of the rotor.
2. Zero the depth-of-cut indicators on the lathe.
3. Turn on the lathe and let the arbor reach full running speed.
4. Turn the depth-of-cut dials for both bits to set the first pass cut. Turning these dials moves the bits inward. The dial is calibrated in 0.001-inch increments. The first cut should be only a portion of the total anticipated depth of cut.
5. When the cutting depth is set for the first cut, activate the lathe to move the cutting bits across the surface of the rotor. After the first cut is completed, turn off the lathe and examine the rotor surface. Areas that have not yet been touched by the bits will be darker than those that have been touched **(Figure 11)**.
6. If there are large patches of unfinished surface, make another cut of the same depth. When most of the surface has been refinished, make a shallow finishing cut at a lower arbor speed. Repeat the slow finishing cut until the entire rotor surface has been refinished.
7. Do not remove any more metal than necessary for a uniform surface finish free of grooves. Make sure you do not cut the rotor thickness beyond its service limit. To ensure this, remeasure the refinished rotor with a micrometer to determine its minimum thickness and compare this measurement to the minimum refinished thickness specification.

It is very important that the rotor surface finish be a **nondirectional finish** after machining **(Figure 12)**. Dress the rotor surfaces with a sanding power tool with 120- to 150-grit aluminum oxide sandpaper. Sand each rotor surface with moderate pressure for at least 60 seconds. You

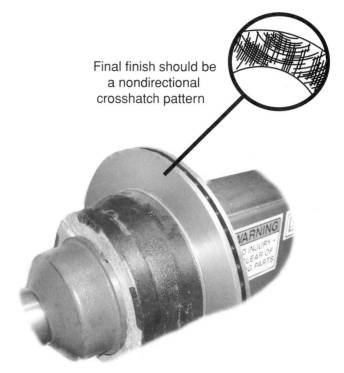

Final finish should be a nondirectional crosshatch pattern

Figure 12. A nondirectional finish on the rotor's surface is a crosshatched pattern that does not follow the arc of rotor rotation. The surface helps pad break-in and reduces brake noise.

can also do this with 150-grit aluminum oxide sandpaper. With the rotor turning at approximately 150 rpm, sand each rotor surface for at least 60 seconds with moderate pressure **(Figure 13)**. Some lathes have adapters for doing this.

Figure 11. The first cut should remove most of the worn area of the rotor.

Figure 13. Use a sander to apply a nondirectional finish.

After the rotor has been sanded, clean each surface with hot water and detergent. Then dry the rotors thoroughly with clean paper towels. If you are working in an area of high humidity or if the rotors will not be reinstalled immediately, wipe the friction surfaces with denatured alcohol and a clean cloth to be sure that all moisture is removed.

ON-VEHICLE BRAKE LATHES

One advantage of an on-vehicle lathe **(Figure 14)** is that the rotor does not need to be removed from the spindle or the hub. On-vehicle lathes also are ideal for rotors with excessive runout problems. The time and trouble needed to reproduce the exact runout condition on a bench lathe arbor is eliminated or reduced when you refinish the rotor on the vehicle.

To install the lathe, remove the wheel and then remove the caliper. To hold a two-piece floating rotor to its hub, reinstall the wheel nuts with flat washers or adapters, provided by the lathe manufacturer, against the rotor. Torque the lug nuts to prevent additional runout of the rotor. Carefully follow the manufacturer's mounting instructions and attach the lathe to the rotor. Some on-vehicle lathes mount on the caliper support; others are supported on a separate stand.

Excessive bearing end play can prevent the use of an on-vehicle lathe or require bearing replacement before an on-vehicle lathe can be used. If any end play is present in an adjustable tapered roller bearing, carefully tighten the adjusting nut by hand just enough to remove the end play before installing the lathe. After turning the rotor, readjust the bearing.

Figure 14. A typical self-powered on-vehicle brake lathe.

Vehicle-Powered Lathes

If the engine is used to turn the rotor, the lathe can be used only on drive wheels. However, a problem exists because the differential gearing in the transaxle transmits the power to the opposite wheel, not to the rotor to be resurfaced. To prevent that opposite wheel from turning, that wheel can be lowered to the floor. This may reduce floor-to-lathe clearance to the point where it is difficult to run the lathe.

> **You Should Know** *Using the engine to rotate the rotor can put a great amount of torque on the assembly, even with the engine at idle. Make sure that the caliper, hoses, and wiring harnesses are tied back away from the rotor and the lathe.*

Another way to transfer drive power to the rotor is to lock the brakes at the opposite wheel. To do this, remove and tightly plug the brake hose to the caliper on the rotor that you are refinishing. Apply the brakes so that the caliper on the opposite wheel is locked. Apply the brakes and have an assistant clamp the brake hose on the opposite side to keep that caliper applied. Release the brakes and check to see if that caliper remains applied. If it is applied, start the engine, shift the transmission into first gear or reverse (depending on the location of the lathe), and allow the engine to run at its slow idle speed. Spinning the rotor too fast will cause the tool bits to overheat and wear out faster. Excessive rotor speed during machining can damage the rotor.

During machining, the rotor must turn into the cutting edges of the bits. Depending on the design of the lathe, rotor rotation can change from one side of the vehicle to the other.

Cross-feed of an on-vehicle lathe can be automatic or manually controlled. If operated manually, advance the cutting bits slowly and steadily across the rotor. Set an automatic cross-feed to 0.003 inch (0.08 millimeter) per revolution. Depth of cut should be shallower than on a bench lathe. Make successive cuts at a depth of 0.004–0.002 inch (0.10–0.05 millimeter) until you get the desired finish.

Self-Powered Lathes

Self-powered on-vehicle lathes are more popular than lathes that use the engine to drive the rotor. Not only can a self-powered lathe be used on nondriving wheels, but rotor speed can be controlled more exactly, and you avoid dealing with engine exhaust inside the shop.

An on-car lathe can be mounted on the brake caliper support or on its own stand and indexed to the hub and the wheel studs. Each lathe has its own operating instructions, which you must follow carefully. Lathe operating procedures include most of the following steps:

1. Check for wheel bearing end play. If necessary, adjust the bearing to remove the end play. If end play exceeds specifications on a nonadjustable bearing, replace the bearing.

2. Make sure that the fluid level in the master cylinder is about half full to allow for fluid drain back when the caliper pistons are pushed back to remove the caliper.

3. Place the transmission in neutral, release the parking brake, and raise the vehicle to an appropriate working height.

4. Remove the wheel from the first rotor to be serviced. Install spacers on the studs, if necessary, for the rotor drive adapter and reinstall the lug nuts. Torque the nuts to specifications.

5. Remove the caliper from its support (push the piston back in its bore if necessary) and hang the caliper out of the way from the chassis or suspension.

6. If the lathe will be mounted on the caliper support, be sure that the area on the support around the mounting holes is free of dirt, rust, and gouges. Select the proper lathe mounting adapters and mount the lathe on the caliper support. Securely tighten all fasteners.

7. Attach the wheel drive adapters to the wheel studs and align the drive motor stand with the hub axis. Lock the stand wheels and plug in the power cord.

8. Turn on the motor before moving the cutting bits close to the rotor to be sure that the motor is turning in the right direction.

9. Then move the cutting bits until they are ½ inch in from the outer edge of the rotor. Then turn the depth-of-cut micrometer until the bits just lightly touch both rotor surfaces.

10. Move the bits outward and remove rust and dirt from the outer edge of the rotor.

11. Manually feed the bits inward until they are past the inner pad contact line. Then set the lathe stop for the inward feed, or in-feed.

12. Rotate both micrometer knobs clockwise for an initial cut of no more than 0.004 inch (0.10 millimeter).

13. Shift the lathe to automatic operation at the fast feed rate. Switch the lathe to feed outward for the first rough cut.

14. After the lathe completes the first cut, turn off the lathe and check the uniformity of the cut. Make additional cuts if necessary, with the final cut at a depth of 0.002 inch (0.05 millimeter).

PREVENTING RUNOUT AND THICKNESS PROBLEMS

Runout and thickness, or parallelism, problems can compound each other. For example, excessive runout causes the high spot of the rotor to hit the pad harder each time it rotates with the brakes off. This, in turn, causes the high spot to wear faster and leads to a parallelism problem. To avoid, or at least reduce, runout and parallelism problems, follow these guidelines during the various steps of rotor service:

- Measure rotor runout before removing the rotor from the vehicle and measure it again after mounting the rotor on a bench lathe. Runout should be the same on the vehicle and on a bench lathe. If it is not, machining can add runout to the rotor.

- Check the runout of a bench lathe arbor frequently. As little as 0.002 inch of runout in a lathe arbor can add runout to a rotor during machining.

- Do not overtighten the rotor mounting on a bench lathe arbor. Overtightening the mounting can add runout during machining; tighten by hand.

- If bearing races are to be replaced in a hub-type (one-piece) rotor, replace them before machining the rotor, not after.

- Reindexing a hubless, or floating, rotor on a wheel hub can reduce runout.

- A lightweight hub used with a hubless, or floating, rotor can warp and acquire runout. If you suspect such a problem, remove the rotor and measure the runout of the hub mounting surface while the hub is on the vehicle.

INSTALLING A ROTOR

New rotors come with a protective coating on the friction surfaces. To remove this coating, use carburetor cleaner, brake cleaner, or solvent recommended by the manufacturer.

If the rotor is a two-piece floating rotor, make sure that all mounting surfaces are clean. Apply a small amount of silicone dielectric compound to the pilot diameter of the disc brake rotor before installing the rotor on the hub. Reinstall the caliper. Install the wheel and tire on the rotor and *torque the wheel nuts to specifications,* following the recommended tightening pattern. Failure to tighten in the correct pattern can result in increased lateral runout, brake roughness, or pulsation, as well as damage to the wheels.

If the rotor is a fixed, one-piece assembly with the hub that contains the wheel bearings, clean and repack the bearings and install the rotor.

After lowering the vehicle to the ground, pump the brake pedal several times before moving the vehicle. This positions the brake linings against the rotor. If so equipped, turn the air suspension service switch back on.

Summary

- If the rotor is a two-piece floating rotor, remove it from the hub by pulling it off the hub studs.
- If the rotor and hub are a one-piece assembly, remove the outer wheel bearing and lift the rotor and hub off the spindle.
- Brake rotors should be refinished only when they have minor imperfections.
- A rotor should not be refinished if it is too thin or will be too thin after machining, regardless of the reason it needs to be turned.
- The minimum thickness specification for all brake rotors is the minimum wear dimension, not a minimum refinishing dimension.
- Never turn the rotor on one side of the vehicle without turning the rotor on the other side. Left- and right-side rotors should be the same thickness.
- Most automobile and parts manufacturers recommend against refinishing new rotors.
- Clean any oil film off the rotor with brake cleaning solvent or alcohol and let the rotor air dry before installing it on the vehicle.
- Both bench-type off-vehicle lathes and on-vehicle lathes are used to refinish rotors.
- On a bench lathe, the rotor is mounted on the lathe arbor and turned at a controlled speed while a cutting bit passes across the rotor surface.
- The tip of the cutting bit should be slightly rounded, not razor sharp.
- When the rotor is on the lathe, install a rubber or spring-type vibration damper on the outer diameter of the rotor to prevent the cutting bits from chattering during refinishing.
- It is very important that the rotor surface be made nondirectional with a sanding block or power tool after machining.
- Rotors with excessive runout can be resurfaced if the amount and position of runout is marked when the rotor is mounted on the vehicle and these same conditions can be reproduced on the bench lathe.
- One advantage of an on-vehicle lathe is that the rotor does not have to be removed from the spindle or the hub. On-vehicle lathes also are ideal for rotors with excessive runout problems.
- On-vehicle lathes can be self-powered or use the power of the vehicle to rotate the rotor. The latter of course can only be used on driven wheels.
- When mounting a two-piece floating rotor, make sure that all mounting surfaces are clean and apply a small amount of silicone dielectric compound to the pilot diameter of the disc brake rotor before installing the rotor on the hub.
- When mounting a fixed, one-piece assembly with the hub that contains the wheel bearings, clean and repack the bearings and then install the rotor.

Review Questions

1. List three conditions that dictate that a rotor should be refinished.
2. Give two main advantages of an on-vehicle lathe versus a bench lathe.
3. Rotor refinishing is being discussed. Technician A says that it is very important that the rotor surface be made nondirectional. Technician B says that rotors should be refinished as part of routine disc brake service. Who is correct?
   A. Technician A only
   B. Technician B only
   C. Both Technician A and Technician B
   D. Neither Technician A nor Technician B
4. When refinishing a rotor on a lathe, the rotor wobbles excessively. Technician A says that the lathe's arbor might be bent. Technician B says that the mounting adapters of the lathe might be distorted. Who is correct?
   A. Technician A only
   B. Technician B only
   C. Both Technician A and Technician B
   D. Neither Technician A nor Technician B

5. Which of the following is *not* a good way to reduce or avoid runout and parallelism problems?
   A. Measured rotor runout should be the same on the vehicle and on a bench lathe.
   B. Check the runout of a bench lathe arbor frequently.
   C. Do not overtighten the rotor mounting on a bench lathe arbor.
   D. Never install new bearings into a one-piece rotor before machining the rotor.

6. Which of the following statements about servicing rotors is *not* true?
   A. The minimum thickness specification for all brake rotors is the minimum wear dimension, not a minimum refinishing dimension.
   B. A refinished rotor must be thicker than its minimum thickness dimension.
   C. The overall thickness of a rotor is the important dimension, and if it is necessary, metal can be removed from only one side of a rotor to correct a problem.
   D. Left- and right-side rotors should be the same thickness, generally within 0.002 to 0.003 inch.

Wheel Bearings

Introduction

Wheel bearing problems can create braking, handling, and tire wear problems. The purpose of all bearings is to allow a shaft to rotate smoothly in a housing or to allow the housing to rotate smoothly around a shaft. Wheel and axle bearings do this for a vehicle's wheels. Typically, the bearings on an axle that drives the wheels are called axle bearings. The wheel is mounted to the hub of an axle shaft and the shaft rotates within a housing. Wheel bearings are used on nondriving axles. The wheel's hub typically rotates on a spindle. Axle bearings are typically serviced with the drive axle. Wheel bearings, however, require periodic maintenance service and are often serviced with suspension and brake work.

WHEEL BEARING MAINTENANCE

Often the front wheel hub bearing assembly for driven and nondriven wheels is two tapered bearings **(Figure 1)** facing each other. Tapered roller bearings are also used on the rear nondriven wheels of many vehicles. Tapered roller bearings must be cleaned and repacked with grease periodically and then adjusted. Each of the bearings rides in its own race. A complete tapered roller bearing assembly consists of an outer race (bearing cone), an inner race, tapered steel rollers, and a cage that holds the rollers in place. Tapered roller bearings are installed in pairs with one large bearing set on the inboard side of the hub and a smaller bearing set on the outboard side. Tapered roller bearings must be periodically cleaned and repacked with grease and then their end play must be adjusted. These bearing services are usually part of a complete brake job.

Straight roller or ball bearings are used on the drive axles of most rear-wheel-drive vehicles. Double-row ball bearings are used on the front hubs of many FWD cars. Straight roller bearings and ball bearings do not require periodic service or adjustment. Also, the front bearing arrangement often found on FWD and four-wheel-drive (4WD) vehicles is frequently nonserviceable. These types of bearings are replaced only when defective, and the procedures are outside the scope of a brake textbook.

WHEEL BEARING EFFECTS ON BRAKING

Adjustable tapered roller bearings used on nondriven wheels must be adjusted correctly. If the bearings are too

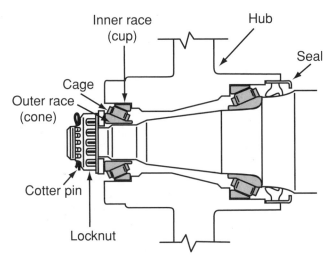

Figure 1. A typical tapered roller bearing setup on a spindle.

tight, the hub and brake assembly can overheat with accompanying problems that include brake fade. In the worst cases, the bearings can seize on the spindle due to overheating.

If wheel bearings are too loose, wheel runout can be excessive. With disc brakes, excessive runout can knock the caliper pistons farther back in their bores and increase brake pedal travel. If the bearings are loose on only one side of the car, brake application on that side will be delayed. This can cause uneven braking and a pull to the opposite side. Excessive runout also can contribute to pedal pulsation.

Whether a wheel bearing is permanently lubricated or requires periodic relubrication, the lubricant is contained in the bearing by one or more seals. If a bearing seal leaks, lubricant can contaminate brake linings and greatly reduce stopping power. Brake pull and premature lockup often are symptoms of contaminated linings.

TAPERED ROLLER BEARING TROUBLESHOOTING

Bearings rarely fail suddenly. Rather, they deteriorate slowly from dirt, lack of lubrication, and improper adjustment. Bearing wear and failure are almost always accompanied by noise and/or vibration.

Begin diagnosing possible wheel bearing problems by raising the vehicle by its frame so that the wheels can turn freely. Rotate the wheel and listen to the sounds of brake drag and bearing rotation. If the brakes are dragging slightly, they will produce a high-pitched swishing sound. Bearing noise will be a lower-pitched rumble. Low-level bearing noise and brake drag are normal. Normal bearing sounds should be uniform throughout the wheel revolutions. An uneven rumble or a grinding sound indicates possible bearing problems.

If a bearing is unusually noisy, roughness will be felt as you turn the wheel. Hold the wheel as you rotate it. Listen for a low-frequency rumbling sound and feel for rough and uneven rotation. Compare left and right wheels to determine whether a problem exists.

While rotating the wheel, try to move it in and out on the spindle and note the amount of movement. Worn or damaged bearings or bearings that need adjustment will have a noticeable amount of end play. Also, grasp the top and bottom of the tire and try to wobble the wheel and tire back and forth. You should feel little or no wobble from a properly adjusted bearing.

Bearing end play can be measured with a dial indicator placed against the wheel hub **(Figure 2)**. Set the indicator to zero, move the wheel in and out on the spindle and note the reading. Tapered roller bearings can have 0.001–0.005 inch (0.025–0.127 millimeter) of end play. Check the specifications for the exact amount of allowable end play.

Finally, inspect the wheel and the brake drum or rotor for grease that might be leaking past a bad seal. Bearing

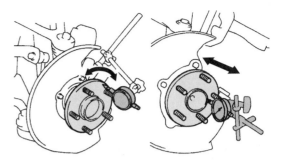

Figure 2. Wheel bearing end play can be checked with a dial indicator.

grease can contaminate brake linings, and a leaking grease seal can let dirt into the bearing. A leaking seal must be replaced, but the bearings also must be cleaned, inspected, and repacked to be sure that they have not been damaged. Whether repacking and reinstalling a used bearing or installing a new one, always install a new grease seal.

TAPERED BEARING SERVICE GUIDELINES

The procedures to remove, clean, lubricate (repack), and install tapered roller bearings are basically the same for servicing a used bearing or installing a new one. A bearing and its outer race must be replaced as a set. Installing a new bearing in a used race will cause an uneven wear pattern that can lead to premature bearing wear or failure.

When installing a new bearing, leave the bearing in its protective packaging until you are ready to lubricate it. Be sure your hands are clean before handling a bearing. Some technicians wear latex gloves or apply a light film of motor oil to their fingers before handling a bearing. Doing this prevents the acid on your fingers from damaging the bearing.

Bearing Removal

Remove the tire and wheel assembly. Then remove the inner and outer tapered roller bearings **(Figure 3)** as follows:

1. If the axle has disc brakes, remove the brake caliper and suspend it out of the way.
2. In the center of the hub is a dust (grease) cap. Using slip-joint pliers or a special dust cap removal tool **(Figure 4)**, wiggle the cap out of its recess in the hub to expose the bearing nut.
3. Remove the cotter pin **(Figure 5)** and nut lock from the bearing nut.
4. Loosen and remove the adjusting nut.
5. Place a thumb over the thrust washer of the outer bearing and carefully slide the drum or rotor off the

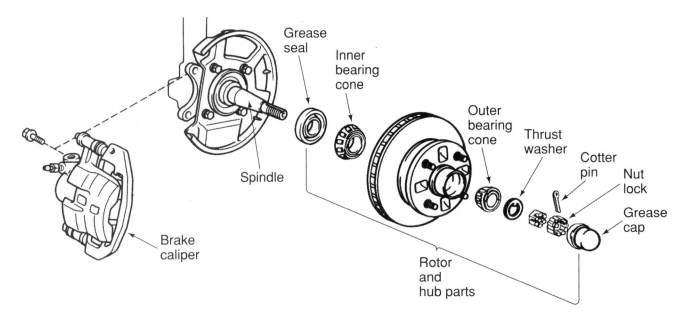

Figure 3. A typical hub and bearing assembly.

Figure 4. Remove the dust cap with dust cap pliers.

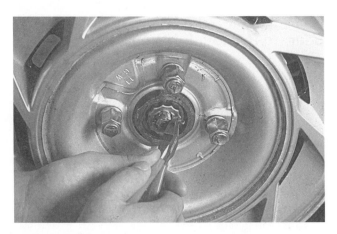

Figure 5. Use side cuts to remove the cotter pin from the nut lock and bearing nut.

spindle. Support the drum or rotor with both hands so that it does not drag on the spindle as you remove it. If a brake drum catches on the shoes, reinstall the adjusting nut and back off the brake adjustment. Then remove the drum.

6. Set the drum or rotor on a clean work surface with the inboard side down.

7. Lift the outer bearing and thrust washer from the hub and set them aside.

8. A grease seal located on the back of the hub normally keeps the inner bearing from falling out when the hub is removed. To remove the bearing assembly, the grease seal must be removed first. Remove the inner bearing and grease seal with a seal puller, large screwdriver, or a large, nonmetallic drift punch. Be careful not to damage the hub while doing this.

9. Keep the outer bearing and inner bearing separated if you plan on reusing them.

10. Wipe the bearings and races or use brake parts cleaner to clean them. While doing this, pay close attention to the condition and movement of the bearings. The bearings should rotate smoothly. Any noticeable damage means that they should be replaced.

11. Also inspect the spindle. If it is damaged or excessively worn, the steering knuckle assembly should be replaced. Slight burrs can be removed with a fine file or emery cloth.

12. Whenever a bearing is replaced, its race must be replaced with it. Remove the race by inserting a large drift punch **(Figure 6)** through the opposite side of the hub and placing it against the inner edge of the

Figure 6. Use a large drift punch and a hammer to remove the bearing race.

race. Strike the punch with a hammer at several locations around the circumference of the race until the race is driven from the hub. You also can remove a bearing race from a hub using a press and suitable adapters.

13. Once the race has been removed, wipe all grease from the inside of the hub.

Cleaning and Inspection

After the bearings are removed from the hub, wipe the old grease off the spindle with clean, lint-free shop cloths or paper towels. Examine the old grease for metal particles from the bearings or their races. Any sign of metal particles in the grease is a clue to inspect the bearings very closely for wear and damage. Also inspect the old grease for dirt, rust, and signs of moisture that could indicate a leaking grease seal.

If the outer races for the bearings are still installed in the hub, wipe old grease off them and inspect them closely for wear or damage **(Figure 7)**. Also wipe as much of the old grease as possible from the bearings and inspect them similarly for wear or damage. Inspect the grease removed from the bearings for metal particles, rust, and dirt. Turn the rollers in the cage and listen and feel for roughness.

Clean the bearings in a parts washer, using clean petroleum-based solvent. Rotate the bearings in the cage as you clean them and remove old grease with a stiff-bristled brush. Wash each bearing separately and keep the bearings sorted to be sure that they are reinstalled in the same hubs from which they were removed.

After cleaning, flush all of the cleaning solvent from the bearings with a nonpetroleum brake cleaning solvent. Let the bearings dry in the air for a few minutes and then blow

> **You Should Know** Do not spin the bearings with compressed air during cleaning. High-speed rotation of an unlubricated bearing will damage it. Compressed air also can dislodge bearings from the cage and cause additional damage or injury.

any remaining solvent from the bearings and cages with low-pressure compressed air. Direct the air through the bearing from side to side along the axis of the rollers.

Carefully inspect the bearings after cleaning. Turn each roller completely around and check the surface. Carefully inspect each of the races also. **Figure 8** shows a good and a bad bearing and race. Common roller surface damage conditions include:

- *Galling*—indicated by metal transfer or smears on the ends of the rollers. It is caused by overloading, lubricant failure, or overheating. Overheating is usually the result of adjustment that is too tight.
- *Etching*—a condition that results in a grayish-black bearing surface. It is caused by insufficient or incorrect lubricant.
- **Brinelling**—occurs when the surface is broken down or indented. It is caused by impact loading, usually because the adjustment is too loose.
- *Abrasive wear*—results in scratched rollers. It is caused by dirty or contaminated bearing lubricant.

If the roller or race shows signs of any of these conditions or if you are unsure of the bearing condition in any way, replace the bearing. A replacement bearing contains inner and outer races and the rollers in their cage. Always compare the old parts with their replacement to be sure you have the correct items.

Bearing Lubrication

Tapered roller bearings must be lubricated, or packed, with grease made especially for wheel bearing use. Disc brakes cause wheel bearings to run at higher temperatures than drum brakes do, and many wheel bearing greases are labeled for use with disc brakes. Always use the type of grease specified by the vehicle's manufacturer.

Because wheel bearing grease must operate at higher temperatures than chassis grease, it contains oils and other lubricants that have been made particularly heavy with thickening agents. Different grease manufacturers use different thickening agents and other additives, so it is best to avoid mixing greases by removing as much of the old grease as possible before repacking a bearing. Using the correct amount of lubricant is also important.

TAPERED ROLLER BEARING DIAGNOSIS

Consider the following factors when diagnosing bearing condition:
1. Consider general condition of all parts during disassembly and inspection.
2. Classify the failure with the aid of the illustrations.
3. Determine the cause.
4. Make all repairs following recommended procedures.

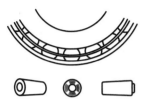

ABRASIVE STEP WEAR

Pattern on roller ends caused by fine abrasives. Clean all parts and housing. Check seals and bearings and replace if leaking, rough, or noisy.

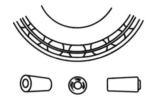

GALLING

Metal smears on roller ends due to overheating, lubricant failure, or overload. Replace bearing, check seals, and check for proper lubrication.

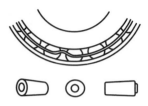

BENT CAGE

Cage damaged due to improper handling or tool usage. Replace bearing.

ABRASIVE ROLLER WEAR

Pattern on races and rollers caused by fine abrasives. Clean all parts and housings. Check seals and bearings and replace if leaking, rough, or noisy.

ETCHING

Bearing surfaces appear gray or grayish black with related etching away of material, usually at roller spacing. Replace bearings, check seals, and check for proper lubrication.

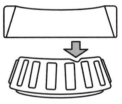

BENT CAGE

Cage damaged due to improper handling or tool usage. Replace bearing.

INDENTATIONS

Surface depressions on race and rollers caused by hard particles of foreign material. Clean all parts and housings. Check seals and replace bearings if rough or noisy.

GOOD BEARING

MISALIGNMENT

Outer race misalignment due to foreign object. Clean related parts and replace bearing. Make sure races are properly sealed.

Figure 7. Examples of good and bad bearings.

Figure 8. An excessively worn bearing and race are shown on the left and a good bearing and race are shown on the right.

Figure 9. A typical wheel bearing packer.

Tapered roller bearings can be packed with grease most efficiently and thoroughly with a bearing packer **(Figure 9)**. Place the bearing in the packer with the taper pointing down and screw the cone down on the bearing. Apply a hand-operated grease gun to the fitting on the center shaft of the packer and force grease into the bearing and out around the rollers in the cage. Some bearing packers contain a supply of grease. Pushing on the cone handle forces the bearing down into the packer and lubricant is forced into the bearing.

If a bearing packer is unavailable, force the grease with your hand **(Figure 10)** through the large end of the bearing and around the rollers. Rotate the bearing several times while packing it to ensure that grease is forced completely around each roller. Finish by spreading an even film of grease around the outside of all the rollers. After packing the bearings, set them aside on a clean sheet of lint-free paper until you are ready to install them.

Figure 10. Grease can be forced into the bearing by hand.

Bearing Installation and Adjustment

Spread a light coating of bearing grease around the entire spindle and inside the hub to a level just below the bearing races. Make sure you do not get grease on the braking surfaces of the drum or rotor.

> **You Should Know** *Always use the recommended grease. If the wrong grease is used, it might not offer the correct protection or it might liquefy from the heat and leak out of the seals.*

Install and adjust tapered roller bearings as follows:
1. If the old bearing race has been removed from the drum, install a new one as follows:
   a. Apply a thin film of grease to the back of the race and to the hub's bore.
   b. Position the race squarely at the end of the bore and use a bearing driver **(Figure 11)** or a suitably sized socket and a hammer to drive the race into the hub.
   c. Listen for a change in the sound as the race is seated in the hub.

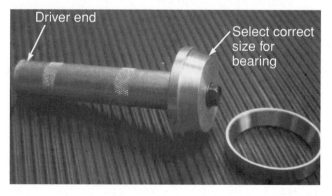

Figure 11. A bearing driver that can be used to drive in a bearing race.

Figure 12.   The lip of the grease seal must face inward when it is installed in the hub.

2. Place the drum or rotor with the outer side down on a workbench.
3. Apply a light coat of grease to the outer race for the inner bearing. Then insert the freshly packed bearing into the race.
4. Place a new grease seal into the inner bore of the hub with the lip pointing inward **(Figure 12)**.
5. Drive the seal into place with a seal driver **(Figure 13)** until the outer surface of the seal is flush with the hub.
6. Apply a thin film of bearing grease to the lip of the seal to protect it during installation.
7. Turn the hub over and apply a light coat of grease to the outer race for the outer bearing. Then insert the freshly packed bearing into the race.
8. Install the rotor or drum carefully onto the spindle in a straight line and be careful not to hit the inner bearing on the spindle threads.

Figure 13.   Drive the new seal into the hub; make sure it is fully seated.

9. Support the rotor or drum with one hand and install the outer bearing and thrust washer into the hub.
10. Install the bearing adjusting nut finger tight against the thrust washer.

> **You Should Know** *Be very careful not to touch the brake assembly with greasy hands. Clean your hands before handling the brake parts or use a clean rag to hold the brake assembly.*

11. Rotate the drum or rotor by hand and lightly tighten the adjusting nut to seat the bearings. Then adjust the bearings by one of the following methods **(Figure 14)**:
    a. Rotate the drum or rotor and lightly snug up the adjusting nut with a wrench to seat the bearings. Then back off the nut one-quarter to one-half turn or until it is just loose while continuing to rotate the drum or rotor. Tighten the nut by hand to a snug fit and lock it.
    b. Rotate the drum or rotor and tighten the adjusting nut with a torque wrench to the specified torque, which is usually 12–25 foot-pounds. Then back off the nut one-third turn and retorque it to the specified value while continuing to rotate the drum or rotor. Final torque is usually 10–15 *inch*-pounds. Lock the adjusting nut.
    c. To use a dial indicator, rotate the drum or rotor and tighten the adjusting nut with a torque wrench to 12–25 foot-pounds. Then back off the nut one-quarter to one-half turn or until it is just loose. Mount the base of the indicator as close as possible to the center of the hub. Locate the tip of the indicator's plunger on the tip of the spindle. Set the indicator to zero. Move the drum or rotor in and out and note the indicator reading. Turn the adjusting nut as necessary to obtain the specified end play, which is usually 0.001–0.005 inch (0.025–0.125 millimeter). Lock the adjusting nut.
12. Install the nut lock over the top of the bearing adjusting nut so that the slots in the nut lock align with the cotter pin hole in the spindle **(Figure 15)**.
13. Install a new cotter pin through the spindle and nut lock and bend its ends to secure it. Reinstall the dust cap in the hub.
14. If the axle has drum brakes that were backed off to allow removal of the drum, readjust the brakes.
15. If the axle has disc brakes, reinstall the caliper.
16. Reinstall the wheel and tire and lower the vehicle to the ground.

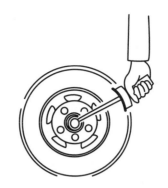

1. Hand spin the wheel.

2. Tighten the nut to 16 N•m (12 ft.-lb) to fully seat the bearings—this overcomes any burrs on threads.

3. Back off the nut until it is just loose.

4. Hand "snug up" the nut. ────────────

5. Loosen the nut until a hole in the spindle lines up with a slot in the nut. Insert the cotter pin.

6. When the bearing is properly adjusted, there will be from 0.03 to 13 mm (0.001" to 0.005") end play.

Bend end of cotter pin legs flat against nut. Cut off extra length.

Figure 14.   Typical wheel bearing adjustment procedure.

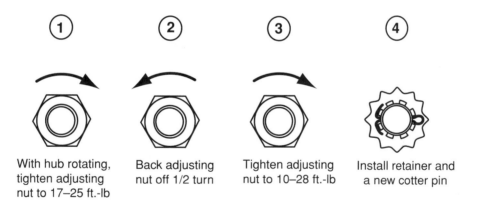

① With hub rotating, tighten adjusting nut to 17–25 ft.-lb

② Back adjusting nut off 1/2 turn

③ Tighten adjusting nut to 10–28 ft.-lb

④ Install retainer and a new cotter pin

Figure 15.   Typical spindle nut, nut lock, and cotter pin installation.

Summary

- The purpose of all bearings is to allow a shaft to rotate smoothly in a housing or to allow the housing to rotate smoothly around a shaft.
- Tapered roller bearings must be periodically cleaned and repacked with grease, and then their end play must be adjusted.

- If the bearings are too tight, the hub and brake assembly might overheat with accompanying problems that include brake fade.
- If wheel bearings are too loose, wheel runout might be excessive, and with disc brakes this can knock the caliper pistons farther back in their bores and increase

brake pedal travel. Uneven braking can result from the bearings on one side being loose.

- The causes of bearing failure can range from improper lubrication or incorrect mounting to poor condition of the shaft housing or bearing surfaces.
- The outer races, or cups, of a bearing assembly are pressed into the hub and must be driven or pressed out when installing a new bearing. A bearing and its outer race must be replaced as a set.
- Whether repacking and reinstalling a used bearing or installing a new one, always install a new grease seal.
- After the bearings are removed from the hub, wipe the old grease off the entire assembly and examine the old grease for metal particles from the bearings or their races.

- Clean the bearings in a parts washer, using clean petroleum-based solvent.
- After cleaning, flush all of the cleaning solvent from the bearings with a nonpetroleum brake-cleaning solvent.
- Common roller surface damage conditions include: galling, etching, brinelling, and abrasive wear.
- Tapered roller bearings can be packed with grease with a bearing packer or by hand.
- Before installing the bearings and the hub, inspect the spindle for wear or damage.
- Spread a light coating of bearing grease around the entire spindle and inside the hub to a level just below the bearing races.

Review Questions

1. Why is some wheel bearing grease specifically marked for disc brake use?
2. What parts make up a tapered roller bearing assembly?
3. Describe the three ways tapered bearings can be adjusted.
4. Technician A says that if the bearings are too tight, wheel runout might be excessive. Technician B says that if the bearings are loose on only one side of the car, uneven braking and a pull to the opposite side can result. Who is correct?
   A. Technician A only
   B. Technician B only
   C. Both Technician A and Technician B
   D. Neither Technician A nor Technician B

5. Wheel bearings are being discussed. Technician A says that the part that has the caged rollers is called the cone. Technician B says that the outer race part is called the bearing cone. Who is correct?
   A. Technician A only
   B. Technician B only
   C. Both Technician A and Technician B
   D. Neither Technician A nor Technician B

Section 7

Parking Brakes

SECTION OBJECTIVES

After you have read, studied, and practiced the contents of this section, you should be able to:

- Explain the function of parking brakes.
- Identify the basic types of parking brake systems.
- Identify the types of cables used to operate the parking brakes.
- Identify and explain the operation of disc brake, drum brake, and transmission drive shaft parking brakes.
- Diagnose parking brake problems; adjust as needed.
- Check parking brake cables and components for wear, rusting, binding, and corrosion; clean, lubricate, and replace as needed.
- Lubricate the parking brake system.
- Adjust calipers with integrated parking brakes.
- Adjust the parking brake and check system operation.
- Check operation of the parking brake indicator light system.

Interesting Fact

Early parking brakes were a carryover from early brakes. A lever was used to push a pad against a tire and then was locked into place. The tightness of the pad against the tire determined how well the vehicle would sit in place.

Chapter 37

Parking Brake Controls

Introduction

The purpose of a parking brake is to hold a vehicle stationary. Parking brakes are often mistakenly called emergency brakes. The parking brakes are not intended to be used as an alternative to the service brakes and to stop a vehicle. The stopping power available from parking brakes is much less than that from the service brakes. Parking brakes work only on two wheels or on the driveline; therefore, much less friction surface is available for braking energy. In the rare case of total hydraulic failure, the parking brakes can be used to stop a moving vehicle, but their application requires careful attention and skill to keep the vehicle from skidding or spinning.

The parking brake system is generally not a part of the hydraulic braking system. It is either mechanically operated by cables and levers to apply the rear brakes or it can be operated by its own hydraulic system to activate a drum brake on the transmission or drive shaft. A few vehicles use the front brakes as parking brakes.

Parking brake actuators can be operated either by hand or by foot. Many small and medium-size vehicles use a hand-operated parking brake lever mounted in the console between the front seats **(Figure 1)**. When the lever is pulled up, the parking brakes are applied. A ratchet and pawl mechanism acts to keep the brake lever applied. To release the lever and the brakes, a button on the lever is pressed and the lever is moved to unlock the ratchet.

Figure 2 shows a typical foot-operated pedal with a ratchet and pawl. Stepping on the pedal applies the brakes and engages the ratchet and pawl. A release handle and

rod or cable is attached to the ratchet release lever. When the release handle is pulled, the pawl is lifted off the ratchet to release the brakes.

The parking brake on some vehicles automatically disengages when the transmission is placed in drive or reverse; other vehicles release the brakes only when the transmission is placed in drive. The most common way to release the parking brakes automatically is with a vacuum

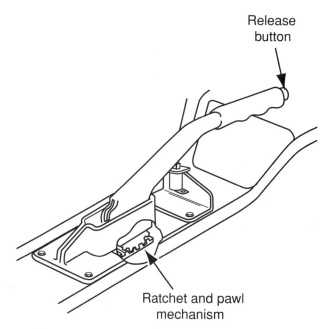

Figure 1. A typical hand-operated parking brake control.

Release button

Ratchet and pawl mechanism

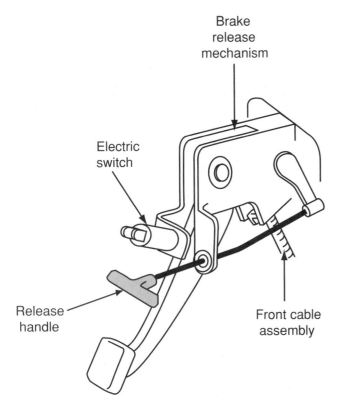

Figure 2. A typical foot-operated parking brake with a mechanical release handle.

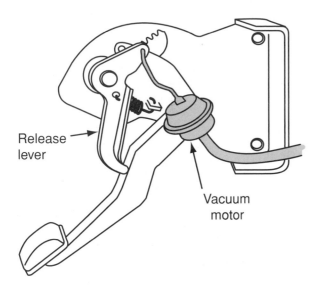

Figure 3. A typical foot-operated parking brake with vacuum release.

motor **(Figure 3)**. Vacuum is applied to the vacuum motor to move the release rod and release the brakes when the transmission is placed into gear. **Figure 4** is a simplified drawing of a typical vacuum circuit for the release of a parking brake. The parking brake release lever can be operated manually if the automatic release mechanism fails.

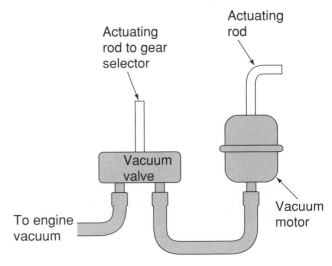

Figure 4. A simplified drawing of a typical vacuum circuit for the release of a parking brake.

Some medium-duty trucks and construction equipment use the hydraulic service brakes as the parking brakes. An electric solenoid is used to close the hydraulic lines between the wheels and master cylinder.

PARKING BRAKE CONTROLS— LEVERS AND PEDALS

The parking brakes on all late-model cars and light trucks are applied by a pedal or a lever, called the **parking brake control**. Many older vehicles and a few current medium-duty trucks have a handle under the instrument panel that is pulled to apply the parking brakes **(Figure 5)**. To

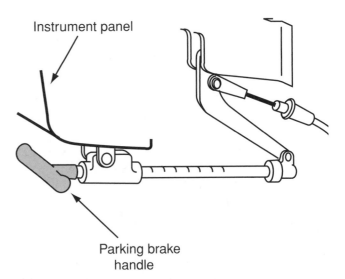

Figure 5. Older vehicles and some late-model trucks have a parking brake control handle under the instrument panel.

release the brakes, the handle must be twisted and returned to be released. Aside from the design and operation of the control handle, the linkage for this type of parking brake works the same as a lever-operated or pedal-operated brake.

Most parking brakes use the service brake shoes or pads to lock the rear wheels to hold the vehicle stationary. The parking brakes can be set most securely if the service brake pedal is pressed and held while the parking brake control lever or pedal is applied. The hydraulic system applies greater force to the shoes or pads than the parking brake mechanical linkage can apply.

Levers

The handle and lever for a lever-operated parking brake are normally installed between the two front seats. As the lever is pulled upward, the ratchet mechanism engages to keep tension on the cables and apply the brakes. The ratchet holds the lever at the height the lever was lifted to and keeps the brakes applied. To release the brakes, a spring-loaded button in the end of the lever is pressed and held while the lever is lowered.

The lever-operated parking brakes on some Chevrolet Corvettes are examples of a design in which the lever drops back to the floor after the brakes are applied. Because the parking brake control lever on these Corvettes is located between the driver's seat and the door sill, if it were to stay in the upward position with the brakes applied it would be hard to climb in and out of the car. The cables and linkage hold the brakes applied, but the lever returns to the released position. To release the parking brakes, you must pull up on the lever until you feel some resistance and then press and hold the button in the end of the lever while moving the lever back to the released position.

Pedals

In a pedal-operated parking brake system, the pedal and its release mechanism are mounted on a bracket under the left end of the instrument panel. As the pedal is pushed downward by the driver's foot, the ratchet mechanism engages to keep tension on the cables and hold the brakes applied. A spring-loaded handle or lever is pulled to release the brakes. A return spring moves the pedal to the released position.

Automatic Parking Brake Release

Automatic parking brake release systems are used only with pedal-operated parking brakes on vehicles with automatic transmissions. A vacuum motor, or servo, is attached to the release mechanism. Vacuum is applied to the servo through a solenoid-operated valve that is actuated when the engine is running and the transmission is shifted into gear from park or neutral. A rod connects the servo to the parking brake release lever. When vacuum is applied to the servo, it pulls the rod to release the brakes. Vehicles with automatic parking brake release also have a manual release handle so that the brakes can be released in case the servo or its vacuum supply fails.

WARNING LAMPS

All brake systems on vehicles built since 1967 have a warning lamp to alert the driver when one-half of the hydraulic system fails. Most vehicles also use this lamp to indicate that the parking brake is applied.

A normally open switch on the control linkage closes as the pedal is pressed or the lever is pulled. The lamp will not light, however, unless the ignition is on. Parking brake lamp switches are adjusted so that the lamp stays lit until the brake is released completely.

PARKING BRAKE LINKAGE

The parking brake linkage transmits force equally from the control pedal or lever to the shoes or pads of the parking brake.

Cables

Most parking brakes use cables to connect the control lever or pedal to the service brakes **(Figure 6)**. Parking brake cables are made of high-strength strands of steel wire that are tightly twisted together. The ends of the cables have different kinds of connectors that attach to other parts of the linkage. Some cables have threaded rods or clevises at their ends. Others have ball- or thimble-shaped connectors that fit into holes and slots on other parts of the linkage **(Figure 7)**.

The front cable connects the parking brake lever or pedal to the **equalizer**, which provides balanced braking force to each wheel. One or two rear cables run from the equalizer to the parking brake levers at each rear wheel. For example, the rear cable might pass through a guide in the equalizer and each cable end attach to each rear brake. The rear cable can slide in the guide in the equalizer if one side of the cable has greater tension. This equalizes the tension on both sides of the cable **(Figure 8)**. Some vehicles have a three-part cable installation, which includes an intermediate cable that passes through the equalizer.

Some lever-operated parking brakes have a separate cable for each rear wheel attached to the control lever. Each cable is adjusted separately, and an equalizer is not necessary.

Cable retainers and hooks maintain cable position on the rear axle, frame, and underbody of the vehicle. These

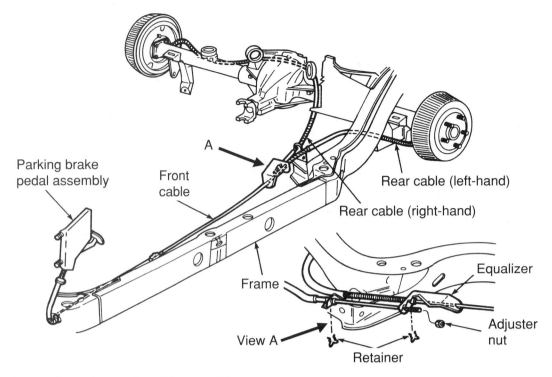

Figure 6. A typical parking brake cable assembly.

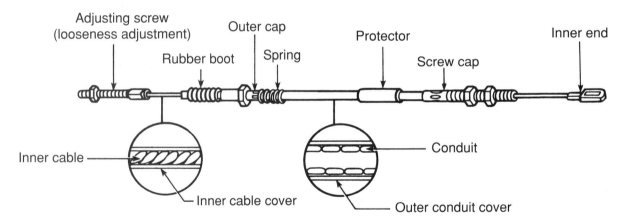

Figure 7. The parts of a typical parking brake cable assembly.

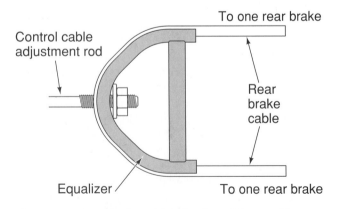

Figure 8. A typical parking brake cable equalizer.

retainers allow the cable to flex and move at their point of body attachment and help the equalizer provide its equalizing action.

Most control cables and rear brake cables are partially covered with a flexible metal **conduit**, or housing. The cable slides inside the conduit, which protects the cable from chafing or rubbing against the underside of the vehicle. Many cables are coated with nylon or plastic, which allows them to slide more easily through the conduit. One end of the conduit is fastened to a bracket on the underside of the vehicle with some type of retaining clip **(Figure 9)**, while the other end is attached to the brake.

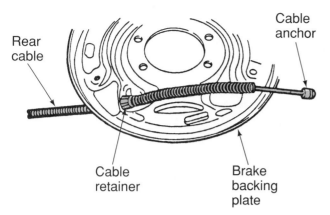

Figure 9. A cable retainer secures the conduit to the backing plate.

Rods

The most common use of solid steel rods in parking brake linkages is in lever-operated systems to span a short distance in a straight line to an equalizer or intermediate lever. The linkage rod is usually attached to the control lever by a pin. The other end of the rod is often threaded to provide linkage adjustment.

Levers

Levers are used to multiply force and are necessary in parking brake linkages to make brake application easy for the driver. Some parking brake installations have an **intermediate lever** under the vehicle body to increase the application force even more. The intermediate lever also is designed to work with the equalizer to ensure that force is applied equally to both rear wheels.

Equalizers and Adjusters

Parking brakes must apply equal force to each wheel. If the application force is unequal, the parking brakes might not safely hold the vehicle. To equalize the force at both rear service brakes, most parking brake linkage installations include an equalizer mechanism. The equalizer is also usually the adjustment point for the linkage.

The simplest example of an equalizer is a U-shaped cable guide attached to a threaded rod. The rear cable or an intermediate cable slides back and forth on the guide to balance the force applied to each wheel. In some installations, the equalizer guide is attached to a lever to increase application force.

Another kind of equalizer is installed in a long cable that runs from the driver's position to one rear wheel. A shorter cable runs from the equalizer to the other wheel. When the parking brakes are applied, the long cable applies its brake directly and then continues to move forward after the shoes or pads lock the wheel. The continued forward motion pulls the equalizer and the shorter cable to lock the brake at the other wheel.

REAR DRUM PARKING BRAKES

Parking brakes that use the rear drum brakes are the most common type of parking brake. In this type of arrangement, the parking brake cable typically runs through a conduit that goes through the backing plate. The cable end is attached to the lower end of the parking brake lever. The parking brake lever is hinged at the top of the web of the secondary, or trailing, shoe and connects to the primary, or leading, shoe through a strut. When activated, the lever and strut move the shoes away from both anchor points and into contact with the brake drum **(Figure 10)**. When tension on the cable is released, the return springs move the shoes back to their unapplied positions.

The individual parts that make up a parking brake assembly vary with the different drum brake designs, but they all work in basically the same way. Many parking brakes include various springs and clips to prevent rattles and to hold the parts in alignment.

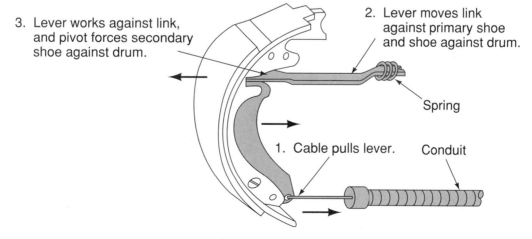

Figure 10. The operation of a parking brake lever and strut.

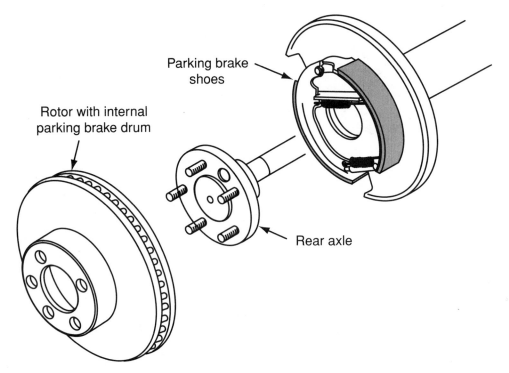

Parking brake shoes

Rotor with internal parking brake drum

Rear axle

Figure 11.   An auxiliary parking brake assembly for rear disc brakes.

REAR DISC PARKING BRAKES

Two different types of parking brakes are used with rear disc brakes: auxiliary drum parking brakes and caliper-actuated parking brakes.

Auxiliary Drum Parking Brakes

Fixed-caliper rear disc brakes and some floating or sliding caliper rear disc brakes have a small drum cast into each rotor **(Figure 11)**. A pair of small brake shoes is mounted on a backing plate that is bolted to the axle housing or the hub carrier. These parking brake shoes operate independently from the service brakes. They are applied by linkage and cables from the control pedal or lever. The cable at each wheel operates a lever and strut that apply the shoes in the same way that rear drum parking brakes work.

These auxiliary drum parking brakes must be adjusted manually with starwheels that are accessible through the backing plate or through the outboard surface of the drum. They do not have self-adjusters.

Caliper-Actuated Parking Brakes

Most floating or sliding caliper rear disc brakes have components that mechanically apply the caliper piston to lock the pads against the rotors for parking. All caliper-actuated parking brakes have a lever that protrudes from the inboard side of the caliper. These levers are operated by linkage and cables from the control pedal or lever.

The two most common types of caliper-actuated parking brakes are the screw-and-nut type and the ball-and-ramp type. A few imported cars have a third type that uses an **eccentric** shaft and a rod to apply the caliper piston. This type is not as common as the first two. An eccentric shaft acts like a cam. One portion of the shaft is oval-shaped. As the shaft rotates, the high part of the oval pushes the operating rod out to apply the brakes.

General Motors' floating caliper rear disc brakes are the most common example of the screw-and-nut parking brake mechanism **(Figure 12)**. The caliper lever is attached to an actuator screw inside the caliper that is threaded into a large nut. The nut, in turn, is splined to the inside of a large cone that fits inside the caliper piston. When the parking brake is applied, the caliper lever rotates the actuator screw. Because the nut is splined to the inside of the cone, it cannot rotate, so it forces the cone outward against the inside of the piston. Movement of the nut and cone forces the piston outward. Similarly, the piston cannot rotate because it is keyed to the brake pad, which is fixed in the caliper. The piston then applies the inboard brake pad, and the caliper slides as it does for service brake operation and forces the outboard pad against the rotor. To provide self-adjustment, an adjuster spring inside the nut and cone rotates the nut outward when the parking

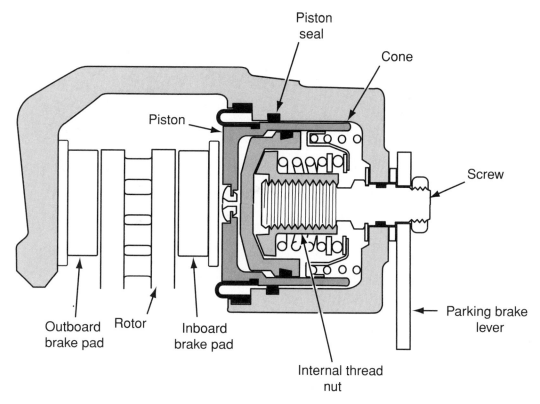

Figure 12. A General Motors screw-and-nut parking brake assembly.

brakes are released. Rotation of the nut takes up clearance as the brake pads wear.

Ford's floating caliper rear disc brakes are the most common example of the ball-and-ramp parking brake mechanism **(Figure 13)**. The caliper lever is attached to a

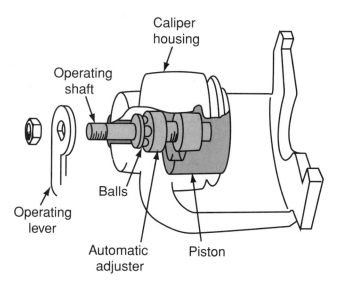

Figure 13. A Ford ball-and-ramp parking brake assembly.

shaft inside the caliper that has a small plate on the other end. Another plate is attached to a thrust screw inside the caliper piston. The two plates face each other, and three steel balls separate them. When the parking brake is applied, the caliper lever rotates the shaft and plate. Ramps in the surface of the plate force the balls outward against similar ramps in the other plate. As the plates move farther apart, the thrust screw forces the piston outward. The thrust screw cannot rotate because it is keyed to the caliper. The piston then applies the inboard brake pad, and the caliper slides as it does for service brake operation and forces the outboard pad against the rotor. When the caliper piston moves away from the thrust screw, an adjuster nut inside the piston rotates on the screw to take up clearance and provide self-adjustment. A drive ring on the nut keeps it from rotating backward.

DRIVELINE PARKING BRAKES

Driveline parking brakes can be based on an external band around a drum or expanding brake shoes inside a brake drum **(Figure 14)**. Until 1960, Chrysler used an external-band parking brake on its cars and light trucks. A brake drum was mounted on the transmission output

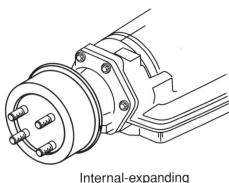

Internal-expanding
shoe type

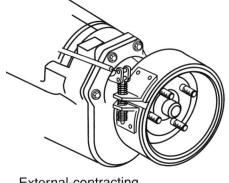

External-contracting
band type

Figure 14. Two types of driveline parking brake setups.

shaft, and linkage operated a band that contracted around the outside of the drum to lock the drive shaft.

Today, driveline parking brakes are used on some light- and medium-duty trucks, but they rely on internally expanding shoes inside a brake drum. The brake shoes and backing plate are mounted on the rear of the trans-

mission housing, and the brake drum is attached to the drive shaft. When the parking brake is applied, levers and struts attached to the brake shoes move them into contact with the brake drum. This action prevents the drum and driveshaft from turning to keep the vehicle from moving.

Summary

- Parking brakes prevent the vehicle from moving when parked.
- The control device that applies the parking brakes on today's vehicles can be either hand-lever operated or foot-pedal operated. The release mechanism can be either a manual release or an automatic release.
- Equalizers are used to balance the forces applied to the rear parking brakes during application.

- Integral disc parking brakes use the normal disc calipers as parking brakes to hold the vehicle while parked.
- Auxiliary drum parking brakes are contained inside the rotor of some rear disc brakes.
- The two most common kinds of caliper-actuated parking brakes are the screw-and-nut type and the ball-and-ramp type.

Review Questions

1. Explain the purpose of a parking brake.
2. A(n) _____ is a device that applies the same tension to each rear brake cable.
3. The parking brake cable is attached to the _____ _____ _____ that is connected to the secondary shoe web on drum brakes.
4. Technician A says that the parking brakes are mechanically operated because mechanical brakes are much more effective than hydraulic brakes. Technician B says that the parking brakes are mechanical due to the need for safety for the parking brakes to operate separately from the service brakes. Who is correct?
   A. Technician A only
   B. Technician B only
   C. Both Technician A and Technician B
   D. Neither Technician A nor Technician B

5. Technician A says that the device that releases the parking brakes whenever the transmission gear selector is in drive or reverse is the vacuum motor. Technician B says that the vacuum motor applies the parking brake when the parking brake cable is applied. Who is correct?
   A. Technician A only
   B. Technician B only
   C. Both Technician A and Technician B
   D. Neither Technician A nor Technician B

Chapter 38

Parking Brake Diagnosis and Service

Introduction

Parking brake service primarily consists of testing the system operation, adjusting the cables and linkage, and replacing components when necessary. Additionally, auxiliary drum parking brakes used with some rear disc brakes might require adjustment, and vacuum-release mechanisms and warning lamp switches might require service.

> **You Should Know** *On some ABSs, applying the parking brake while the vehicle is moving will light the ABS warning lamp and set a diagnostic trouble code in the ABS computer.*

PARKING BRAKE CHECKS

Checking brake pedal travel is the first step in diagnosing a potential parking brake problem. The next step is to inspect and test the control pedal or lever and the linkage. Finally, a performance test will verify that the parking brake can hold the vehicle stationary as required by motor vehicle safety standards.

Pedal Travel—Rear Drum Brakes

Because the brake shoes are the parking brakes for vehicles with rear drum brakes, service brake adjustment directly affects parking brake operation. If lining-to-drum clearance is excessive, the parking brake linkage might not have enough travel to apply the parking brakes completely.

If the parking brake lever or pedal must be applied to the full limit of its travel to engage the parking brakes, excessive clearance or slack exists somewhere in the system. To isolate the looseness in the system, press the service brake pedal and note its travel. If brake pedal travel seems excessive, the rear drum service brakes might need adjustment. Check and adjust the service brakes before adjusting the parking brake linkage.

> **You Should Know** *Failure to use the parking brake regularly is a leading cause of complaints about a low brake pedal. This is particularly true on some leading-trailing drum brakes that use the parking brake linkage to adjust the service brakes, and it is also a common problem with rear disc brakes. Advise your customers to use the parking brake, and do not overlook it as the cause of a low-pedal problem.*

Pedal Travel—Rear Disc Brakes

Rear disc brakes with parking brake mechanisms **(Figure 1)** include self-adjusters that keep the caliper piston from retracting too far in its bore. If the self-adjusters do not work properly, both the service brake pedal and the parking brake control might have excessive travel.

To check parking brake operation on vehicles with ball-and-ramp-type rear disc brakes **(Figure 2)**, push the parking brake lever on the caliper forward by hand. If it moves more than 20 degrees, the self-adjuster might need readjustment.

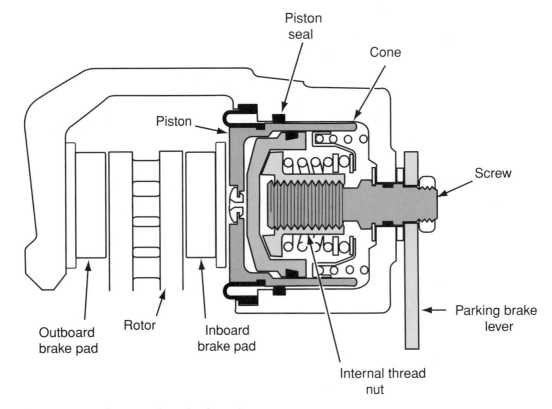

Figure 1. A GM screw-and-nut parking brake unit.

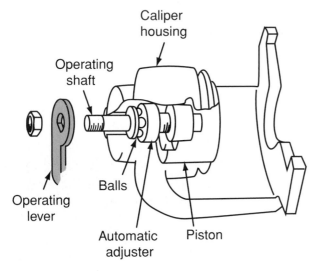

Figure 2. A typical ball-and-ramp-type parking brake assembly.

If this does not correct the problem, start the engine and apply the service brakes forty to fifty times with moderate pressure. Wait about 1 second between pedal applications. Then stop the engine and apply the pedal another thirty times with heavy foot pressure. If the travel of both the service brake pedal and the parking brake pedal or lever does not decrease, the rear calipers must be rebuilt or replaced.

Linkage Inspection and Test

The parking brakes should be fully applied when the lever or pedal is moved one-third to two-thirds of its travel **(Figure 3)**. If the parking brake lever or pedal travels more than specified, the linkage adjustment is too loose. If the parking brake lever or pedal travels less than specified, the

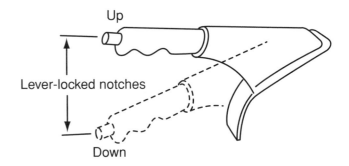

Lever-locked notches:
  Cars with rear disc brakes: 7-11
  Cars with rear drum brakes: 4-8

Figure 3. Some manufacturers specify that the parking brakes should be applied when the parking lever has traveled a certain amount.

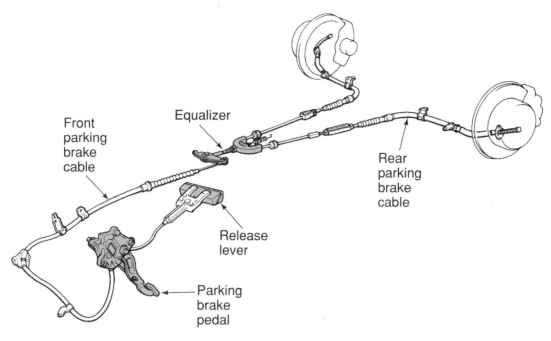

Figure 4.   A typical parking brake cable setup.

linkage adjustment is too tight. This can cause the brake linings to drag on the drum or rotor when the vehicle is moving and lead to brake fade and premature lining wear.

If the lever or pedal travel is within the specified range, raise the vehicle with the parking brakes released. Rotate the rear wheels by hand and check for brake drag. Have an assistant operate the parking brake control while you check the movement of the cables and linkage **(Figure 4)**. The pedal or lever should apply smoothly and return to its released position. The parking brake cables should move smoothly without any binding or slack.

Also inspect the cables for broken strands, corrosion, and kinks or bends **(Figure 5)**. If the brakes drag or the link-

age binds in any way, the parking brakes must be adjusted or repaired as necessary. Clean and lubricate the parking brake and noncoated metal cables with a brake lubricant or penetrating oil **(Figure 6)**. Damaged cables, conduit, and linkage must be replaced.

Parking brake cables that are coated with a plastic material do not need periodic lubrication, but you should handle these cables carefully during service. Avoid contact with sharp-edged tools or sharp surfaces on the vehicle's underbody.

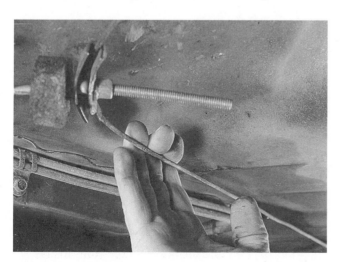

Figure 5.   Carefully inspect all cables of the parking brake cable assembly.

Figure 6.   Apply penetrating oil or brake lubricant to all exposed metal parts and cables in the parking brake assembly.

A small amount of drag is normal for rear disc brakes, but heavy drag usually indicates an overadjusted parking brake mechanism. To determine if the misadjustment is in the linkage or in the caliper self-adjusters, disconnect the cables at a convenient point so that all tension is removed from the caliper levers. If brake drag is reduced, inspect the external linkage and lubricate, adjust, or repair it as necessary. If brake drag is still excessive with the parking brake linkage disconnected, check the caliper levers to be sure that they are returning fully against their stops. If the caliper levers are operating correctly and brake drag remains excessive, the calipers must be repaired or replaced.

Performance Test

FMVSS 105 requires that the parking brake hold the vehicle stationary for five minutes on a 30 percent grade in both the forward and the reverse directions **(Figure 7)**. This standard can be the basis of a performance test for the parking brake.

Stop the vehicle on a grade of approximately 30 percent, put the transmission in neutral, apply the parking brake with

moderate force, and release the service brakes. You need not wait a full five minutes, but the vehicle should remain stationary for a reasonable amount of time. Perform this test with the vehicle facing in both directions on the grade.

CABLE AND LINKAGE ADJUSTMENT

The adjustment of the parking brake cable and linkage should be checked at particular time and mileage intervals and whenever other brake work is performed.

Most pedal-operated parking brakes have the adjustment at a point under the center of the vehicle **(Figure 8)**. Lever-operated brakes can have the adjustment under the vehicle, but it is more often located at the point where the cable, or cables, attach to the lever and is accessible from inside the car **(Figure 9)**. You might need to remove

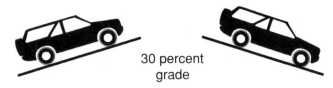

30 percent grade

Figure 7. A parking brake must be able to hold a vehicle on a 30 percent grade for five minutes in both the forward and reverse direction.

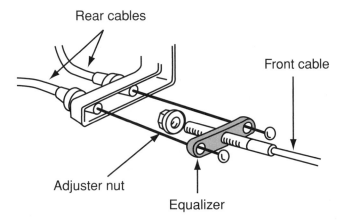

Rear cables

Front cable

Adjuster nut

Equalizer

Figure 8. Typical location for adjusting a pedal-operated parking brake linkage.

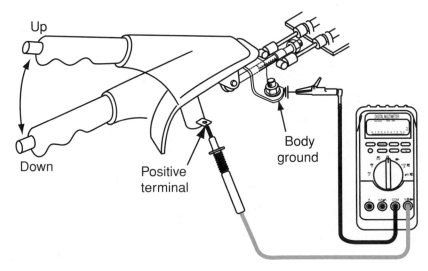

Up

Down

Positive terminal

Body ground

Figure 9. On some vehicles, the adjustment point for lever-operated parking brakes is accessible from inside the vehicle.

part of the center console or floor trim for adjustment access.

The service brakes always must be adjusted correctly before adjusting the parking brake linkage. If drum brake shoe adjustment is too loose, trying to adjust the parking brake can pull the shoes off their anchors or away from the wheel cylinder. The parking brake might appear to be adjusted correctly when it actually is not. Service brake operation and further adjustment also can be adversely affected.

If service brake adjustment is unequal from side to side and one pair of shoes is looser than the other, the parking brake linkage might engage one wheel brake more tightly than the other. The problem of unequal service brake adjustment is most critical on a lever-operated parking brake that has a separate cable for each wheel. The lever moves each cable equally; but if one wheel brake adjustment is loose, full cable travel will not fully apply that wheel brake.

A typical cable adjuster has a threaded rod at the end of a cable and two jam nuts on the rod where it passes through an equalizer or a cable anchor. If the nuts are rusted or corroded, soak them with penetrating oil before trying to adjust the linkage.

Clean the grease and dirt from the threads on either side of the adjusting nut before trying to turn the nut. Forcing the nut over dirty threads can damage the threads on the nut or the rod.

To make the adjustment, hold the adjuster nut with one wrench while loosening the locknut with another wrench. Then turn the adjuster nut to lengthen or shorten the cable as required **(Figure 10)**. If the cable tries to twist as you turn the adjuster nut, grip an unthreaded section of

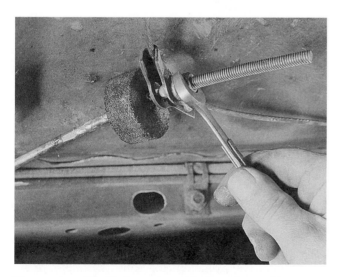

Figure 10. The linkage is typically adjusted by turning the adjustment nut at the equalizer.

the rod with locking pliers and hold the cable as you turn the adjuster nut. Finally, hold the adjuster nut in place with one wrench and retighten the locknut with another. On linkage with a separate adjustment for each cable, adjust each cable equally.

Drum Brake Cable Adjustment

To adjust a typical parking brake linkage, begin by placing the transmission in neutral and release the parking brake. Some manufacturers specify that the parking brake be partially applied during adjustment. This is intended to allow for a specific amount of slack in the linkage when the brakes are released.

1. Raise the vehicle on a hoist so that the wheels are off the ground.
2. If required, adjust the service brakes before adjusting the parking brakes.
3. Verify that the parking brake linkage is clean, properly lubricated, and operating freely.
4. Tighten the linkage until the brakes drag as you rotate the wheel by hand.
5. Then loosen the adjustment until the wheel just turns freely.
6. Apply the parking brake until the wheel locks and check the pedal or lever travel.
7. Apply and release the parking brake several times and verify that the wheels turn freely when the brake is released and are locked when the brake is applied.

Disc Brake Cable Adjustment

The parking brake cables for the rear disc brakes are adjusted similarly to the drum brake cables. The caliper levers that apply the caliper pistons are an added adjustment point. After servicing a rear brake caliper, you must operate the service brakes to position the brake pads before adjusting the parking brake. Begin by loosening the parking brake adjusting nut. Then start the engine and press the brake pedal several times to position the pads for normal operation.

Before tightening the cables, leave them loose and make sure that they exert no tension on the caliper levers. Move the caliper levers by hand to verify that they move freely and that they return fully to their unapplied positions. If the calipers have stop lugs, be sure that the levers return completely to these stops. Tighten the cable adjustment to remove all slack and to the point where the levers just start to move. Then back off the adjustment until the levers are just released but the cables have no slack. Apply and release the parking brakes several times to make sure that the levers return to the fully released positions.

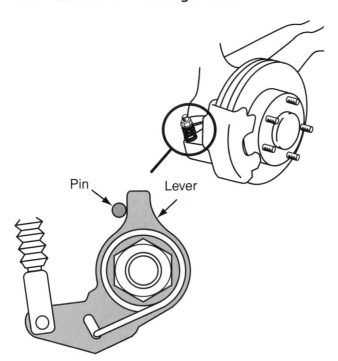

Figure 11. The adjusting point for the lever of a rear disc parking brake.

On the parking brake assembly shown in **Figure 11**, the lever of the rear brake caliper must contact the brake caliper pin. To make the adjustment, pull the parking brake hand lever up one notch and then tighten the adjusting nut until the rear wheels drag slightly when turned. Next, release the parking brake lever and verify that the rear wheels do not drag when turned. Make any readjustments necessary. When the equalizer is properly adjusted, the rear brakes should be fully applied when the parking brake lever is pulled up five to ten notches.

If the parking brake pedal or lever does not engage the brakes completely at one-third to two-thirds of its travel, the self-adjusters in the calipers are not working properly. You can try to adjust the self-adjusters by operating them, but it is more likely that the caliper will need to be repaired or replaced.

Auxiliary Drum Brake Shoe Adjustment and Replacement

Auxiliary drum parking brakes used with some rear disc brakes require two separate adjustments. Adjustment of the cables and linkage is similar to drum brake cable adjustment. The parking brake shoes must be adjusted separately, however, before any adjustments are made to the cables and linkage.

Most auxiliary drum parking brakes have manual starwheel adjusters similar to the adjusters on drum service

brakes. The adjuster is usually accessible through a hole in the outer surface in the drum portion of the rotor (**Figure 12**). On some cars, the adjuster is accessible through the hole for one of the wheel bolts. If the adjuster is accessible through a hole in the rotor, remove the wheel and then reinstall two or three wheel nuts to hold the rotor on the hub. If the adjuster is accessible through a wheel bolt hole, remove one wheel bolt. You can leave the wheel installed. Some adjusters are accessible through an opening in the inboard side of the backing plate. The starwheel adjuster on the parking brake is a simple device that can be operated with a screwdriver. Tighten the starwheel until the brakes lock and then back off the adjustment the specified number of notches or clicks.

If the shoes and linings of auxiliary drum parking brakes require replacement, the procedure is similar to the procedures for drum service brakes. The drum portion of the brake rotor cannot be resurfaced, but resurfacing should never be necessary because the parking brakes are applied and released when the vehicle is stationary.

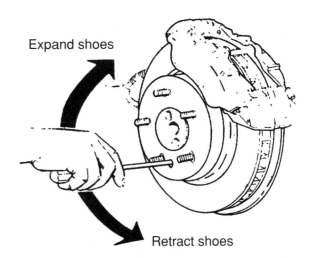

Expand shoes

Retract shoes

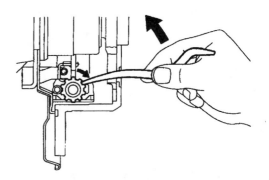

Figure 12. The adjusters for some auxiliary drum parking brakes are accessible through the outboard drum surface.

CABLE AND LINKAGE REPAIR AND REPLACEMENT

Parking brake cables and linkage can easily last the life of a vehicle if they are periodically inspected, lubricated, and adjusted. However, corrosion and rust can damage the cables and prevent proper operation. The cable can stick or seize in the conduit or individual strands can fray and rust. A sticking or seized cable can be loosened, lubricated, and adjusted. Because of the labor required, however, it is often more practical to replace the cable.

Freeing Seized Cables

All cables in a system should be checked. If a cable is only moderately corroded or sticking slightly, or if a replacement is unavailable, you can free the cable by working penetrating oil into the ends of the conduit. Let the oil soak in for a few minutes and then try to move the cable in and out of the conduit. Several applications of oil might be needed.

Most corrosion and damage are confined to the first few inches of cable inside the conduit or to the cable sections outside the conduit. You might need to disconnect one or both ends of the cable for access to corroded sections.

When the cable starts to move in the conduit, slide it in and out as far as it will go and continue to apply penetrating oil. Remove corrosion with solvent and emery cloth, steel wool, or a wire brush. You may need to apply heat to a seized cable to free it completely. In this case, disconnect and lower the cable away from the vehicle or remove it completely if possible. When the cable moves freely and dirt and corrosion have been removed, lightly lubricate the exposed sections of the cable with brake or white grease to avoid future problems.

Front (Control) Cable Replacement

If the parking brake is operated by a lever, the control cable attachment is usually accessible from inside the car. A few lever-operated cables are accessible only from under the car. If the parking brake is operated by a pedal, the control cable attachment is usually under the instrument panel. Because of the awkward access to most such installations, it is a good idea to disconnect the battery ground cable to avoid damaging any electrical system components before working under the instrument panel.

To replace the control cable, release the parking brake and raise the vehicle. Disconnect the lower end of the cable from the equalizer before trying to disconnect the upper end. The lower end is more accessible, and disconnecting it first provides slack to help disconnect the upper end and any mounting clips or grommets.

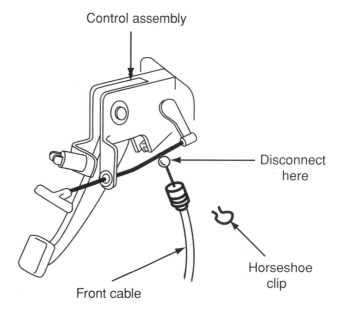

Control assembly

Disconnect here

Horseshoe clip

Front cable

Figure 13. Disconnect the upper end of the parking brake cable from the control pedal or lever.

Disconnect the upper end of the cable from the control pedal **(Figure 13)** or lever but do not pull it out of the vehicle yet. Fasten a length of cord or flexible wire to the brake cable inside the car. Then pull the cable through the floor or firewall from under the vehicle. Disconnect the cable from the cord or wire and leave the cord or wire in place through the cable opening. Fasten the cord or wire to the replacement cable and use it to help draw the cable through the opening in the floor or firewall.

> **You Should Know** *Disable the supplemental restraint system (SRS), or air bag system, before working on or near the system's electrical circuit. Accidental deployment of the air bag(s) can cause serious injuries.*

Connect the upper end of the new cable to the pedal or lever and reinstall all mounting clips, brackets, and grommets. If any attaching hardware is damaged, install new parts. Then connect the lower end of the cable to the equalizer or to an inline connector on the rear cable. Adjust the parking brakes as required.

Rear Cable Replacement

Nearly all vehicles have two rear parking brake cables connected to the equalizer or adjuster. Remove the cable adjusting nuts, then disconnect the front end of the rear cable(s) from the equalizer or from the front brake cable.

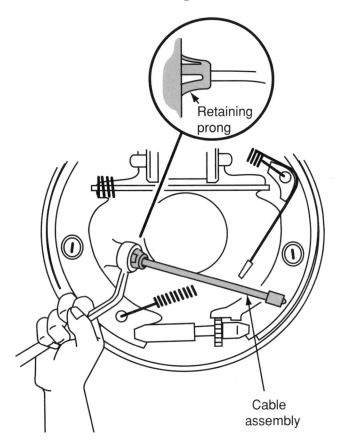

Figure 14. Use a box-end wrench or screwdriver to retract the retaining prongs of the cable's conduit.

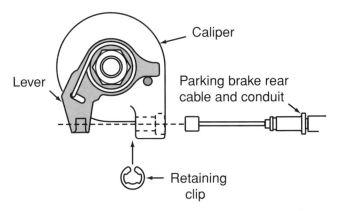

Figure 15. The end of the cable must be disconnected from the parking brake lever at the caliper.

To remove the cable from a drum brake, remove the wheel and brake drum. Disconnect the end of the rear cable from the rear parking brake lever on the rear shoe. Use the proper size offset box wrench or screwdriver to depress the conduit retaining prongs, and slide the pronged fitting out through the hole in the backing plate **(Figure 14)**.

On a typical rear disc brake, the cable conduit is secured at the rear disc brake by a clip or pin that connects the end of the cable to the parking brake lever on the caliper **(Figure 15)**. Remove the clip and pin to free the cable. It might be necessary to remove grommets and pronged retainers before the rear cable can be freed. The routing of left-hand and right-hand rear parking brake cables can be different. Refer to the service manual for specific instructions.

Attach the new rear cable to the equalizer or adjuster. Install all grommets and clips used to secure the cable. Follow the original routing pattern.

On drum brakes, insert the cable and conduit into the hole in the backing plate. Make sure that the retaining prongs lock the conduit in place where it passes through the backing plate. Hold the brake shoes in place on the

backing plate and engage the brake cable into the parking brake lever. Install the brake drum and the wheel. Install the brake cable adjusting nut. Adjust the parking brakes.

On disc brakes, insert the parking brake rear cable and conduit end into the rear disc brake caliper and install the retaining clip or pin that secures the cable to the lever. Adjust the parking brakes.

AUTOMATIC (VACUUM) RELEASE SYSTEMS

Some pedal-operated parking brakes have a vacuum release motor **(Figure 16)**. Typically, a rod connects the vacuum-actuated diaphragm inside the motor to the parking brake release handle. A hose runs from the vacuum motor to the engine intake manifold. The vacuum hose is routed through a vacuum release switch on the steering column or floor shift console. The vacuum release switch supplies engine vacuum to the parking brake release

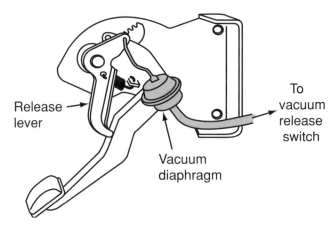

Figure 16. A typical vacuum release setup on a pedal-operated parking brake.

motor when the engine is running and the shift lever is placed in gear. This vacuum provides the power the control motor uses to release the parking brake.

The release switch vents motor vacuum when the shift lever is placed in park or neutral (and reverse on some vehicles). The vacuum release system also can be released manually at any time by pulling on the release lever, as with an ordinary parking brake.

To test the operation of the system, run the engine at idle with the transmission gear selector in neutral. Apply the parking brake and move the gear selector to drive. Watch the parking brake pedal to see if it returns to its unapplied position when the parking brake releases. If the parking brake releases, then the vacuum control is working properly.

> **You Should Know** Do not apply air pressure to the vacuum system of the parking brake release. Air pressure from the shop compressed air supply or other source will damage the actuator diaphragm in the control motor.

If the parking brake does not release, test for vacuum at the brake-release vacuum hose that connects to the release motor. Remove the vacuum hose from the motor and tee the vacuum gauge into the line. The vacuum required at the control motor is usually 10–12 inches of mercury (35–40 kilopascals). Check system vacuum with a vacuum gauge and the engine at idle. Inspect the lines between all connecting points. Low vacuum is often caused by a loose hose connection or leak in the hose. To detect a leak in the hoses, listen for a hissing sound along the hose route. Check the hoses to make sure that they are not connected to the wrong connection, kinked, or otherwise damaged.

A lack of vacuum to the motor can also be caused by a faulty release switch. Check for vacuum at the inlet of the switch and at the outlet when the transmission's gear shift is moved through its range. Vacuum should be sent to the motor when the transmission is in a forward gear range or when the vehicle is in reverse.

Parking Brake Release Switch Replacement

A faulty parking brake release switch will prevent engine vacuum from reaching the release motor. To replace a parking brake release switch mounted on the steering column (**Figure 17**), begin by disconnecting the battery ground cable and the air bag backup power supply (if so equipped). Remove the steering column cover. Then dis-

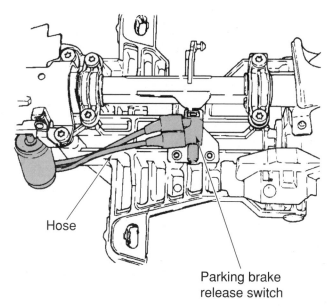

Figure 17. A typical location for a vacuum parking brake release switch.

connect the transmission's selector cable from the actuator housing and remove the retaining hardware for the steering column tube. Carefully lower the steering column tube and disconnect the vacuum hoses from the parking brake release switch. Remove the retaining screws for the switch and then remove the switch.

Begin installation of the new switch by shifting the transmission into neutral. Position the parking brake release switch over the column mounting bosses and push the switch against the turn signal canceling cam. Fasten the parking brake release switch with the retaining screws.

The new switch might have a plunger retainer that must be removed and discarded. After this retainer is removed, connect the vacuum hoses to the switch. Reassemble the steering column tube, connect the gear selector cable to the steering column, and reinstall the instrument panel trim. Complete the installation by reconnecting the air bag backup power supply and battery ground cable.

PARKING BRAKE LAMP SWITCH TEST

With the ignition on and the parking brake applied, the switch should close to light the brake warning lamp on the instrument panel. To test a typical parking brake switch, you must first gain access to the switch. For a lever-operated parking brake, you might need to remove all or part of the center console. On a pedal-operated parking brake, locate the switch on the pedal bracket under the instrument panel.

After locating the switch, disconnect the electrical harness connector and apply the brake pedal or lever. Then

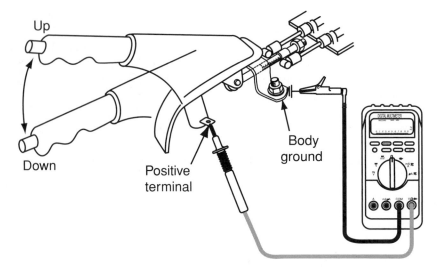

Figure 18. The action of the switch should be checked with the parking brake lever in its upward and downward positions.

use an ohmmeter or continuity tester to check the switch. The switch should be closed, and continuity should exist with the pedal or lever applied. The switch should be open, with no continuity, with the pedal or lever released.

With the brake lever pulled up, check for continuity between the positive terminal and a good ground **(Figure 18)**. Continuity should be broken when the brake lever is in the down position.

DRIVELINE PARKING BRAKE ADJUSTMENT

To adjust an internal-shoe driveline parking brake, place the transmission in neutral and release the parking brake. Raise the vehicle and rotate the drum by hand. The drum should turn freely. Remove the adjusting screw cover and loosen the clamp bolt to free the cable adjusting nut. Back off the cable adjusting nut or slide the cable sleeve in the clamp. Expand the shoes by rotating the shoe adjusting nut until the lining drags on the drum. Next, back off the nut about one notch to provide 0.010 inch of clearance between the linings and the drum. Check to make sure that the shoulders of the nut are seated in the grooves in the sleeve. As the final step, remove all slack from the parking brake cable. This can be done with the adjusting nut located at the point where the cable enters the drum or by a cable adjusting nut located under the instrument panel on the pedal control linkage.

When properly adjusted, the hand lever or foot pedal should advance four to six notches to set the parking brake firmly. The lever or pedal should not advance more than one-half of the total available stroke.

Summary

- Checking brake pedal travel is the first step in diagnosing a potential parking brake problem. The next step is to inspect and test the control pedal or lever and the linkage. Finally, a performance test will verify whether the parking brake can hold the vehicle stationary as required by motor vehicle safety standards.
- The parking brakes should be fully applied when the lever or pedal moves one-third to two-thirds of its travel.
- The parking brake cables should move smoothly without any binding or slack, and the protective conduit should be in good shape.

- Most pedal-operated parking brakes have the adjustment point at a spot under the center of the vehicle, and the adjustment point for lever-operated brakes is most often located at the spot where the cable, or cables, attach to the lever.
- A sticking or seized cable should be loosened, lubricated, adjusted, or replaced.
- Some pedal-operated parking brakes are released by a vacuum motor.
- With the ignition on and the parking brake applied, the parking lamp switch should close to light the brake warning lamp on the instrument panel.

Review Questions

1. Vacuum-operated parking brake release systems are being discussed. Technician A says that the vacuum release switch supplies engine vacuum to the parking brake release vacuum motor whenever the transaxle is placed in a forward gear. Technician B says that the release switch vents the parking brake release motor when the transaxle is placed in park, reverse, or neutral. Who is correct?
   A. Technician A only
   B. Technician B only
   C. Both Technician A and Technician B
   D. Neither Technician A nor Technician B

2. During parking brake adjustment Technician A checks and adjusts the drum-to-lining clearance before adjusting the parking brake; Technician B fully applies the parking brake lever before making the adjustment. Who is correct?
   A. Technician A only
   B. Technician B only
   C. Both Technician A and Technician B
   D. Neither Technician A nor Technician B

3. When adjusting the parking brake linkage Technician A cleans and lubricates the threads of the adjusting mechanism bolt to avoid damaging it; Technician B makes certain that the rear wheels cannot be rotated forward with the parking brake fully applied. Who is correct?
   A. Technician A only
   B. Technician B only
   C. Both Technician A and Technician B
   D. Neither Technician A nor Technician B

4. When testing a vacuum control motor Technician A checks for proper vacuum at the motor using a vacuum gauge installed with proper fittings; Technician B applies 5–10 psi of air pressure to the control motor to test the release action of the mechanical linkage. Who is correct?
   A. Technician A only
   B. Technician B only
   C. Both Technician A and Technician B
   D. Neither Technician A nor Technician B

5. Technician A says that the cable running from the parking brake control mechanism to the equalizer or adjuster is commonly referred to as the front cable. Technician B says that there might be slight routing and length differences between the left and right rear cables in a parking brake system. Who is correct?
   A. Technician A only
   B. Technician B only
   C. Both Technician A and Technician B
   D. Neither Technician A nor Technician B

Section 8

Antilock Brake Systems

SECTION OBJECTIVES

After you have read, studied, and practiced the contents of this section, you should be able to:

- Identify the components and explain the operation of a typical antilock brake system (ABS).
- Describe the differences between an integrated and a nonintegrated ABS.
- Briefly describe the major components of a four-wheel ABS.
- Identify the various types of hydraulic power units.
- Identify the major suppliers of ABSs.
- Inspect and test the brake system to determine if a complaint is related to the base brakes system or the antilock system.
- Diagnose braking problems caused by the ABS.
- Relieve high pressure from an ABS hydraulic accumulator.
- Bleed an ABS.
- Read and interpret ABS operating data displayed on a scan tool and distinguish abnormal from normal readings.
- Use manufacturers' diagnostic charts and procedures to identify the root cause of an ABS problem.
- Service, test, and adjust ABS speed sensors.
- Explain the purposes and major features of diagnostic trouble codes.
- Perform resistance and voltage waveform tests on a speed sensor and its circuit.
- Remove and install ABS electrical/electronic and hydraulic components.
- Describe the operation of traction control systems.
- Cite examples of traction control systems used by various manufacturers.
- Identify the control methods used in different traction control systems.

Interesting Fact

The United States Air Force is credited with the first practical use of antilock brakes on the B47 bomber in 1947. This brake system provided shorter and straighter stopping distances and overall safer braking on short military runways. Today, most jet aircraft have similar braking systems.

Chapter 39

Antilock Brake System Principles

Introduction

This chapter covers the basics of antilock brake systems (ABSs). Many different designs of ABSs are used on today's vehicles.

The operation of ABSs **(Figure 1)** can be thought of as electronic/hydraulic pumping of the brakes for straight-line stopping under panic conditions. Skilled drivers have always pumped the brake pedal during panic stops to avoid wheel lockup and the loss of steering control. Antilock brake systems simply get the pumping job done much faster and in a much more precise manner than the fastest human foot. ABS is typically controlled by an electronic brake control module (EBCM) that relies on inputs from several sensors. Keep in mind that a tire on the verge of slipping produces more friction with respect to the road than one that is locked and skidding. Once a tire loses its grip, friction is reduced and the vehicle takes longer to stop.

An ABS can pump the brakes up to fifteen times per second. Some systems can also control each brake separately whenever any one wheel starts to lock. An ABS can stop a car in the shortest possible distance without wheel lockup while stopping in a straight line. This condition is called **directional stability**. It allows the driver to steer the vehicle on most types of road surfaces while braking and maintain **directional control**.

An ABS is programmed to control **negative wheel slip** during braking. Negative wheel slip occurs when the speed of a wheel is less than vehicle speed. Integrated into some antilock systems is traction control, which controls **positive wheel slip** (when the wheel speed is greater than vehicle speed during acceleration).

PRESSURE MODULATION

When the driver quickly and firmly applies the brakes and holds the pedal down, the brakes of a vehicle not equipped with ABS will almost immediately lock the wheels. The vehicle slides rather than rolls to a stop. During this time, the driver also has a very difficult time keeping the vehicle straight and the vehicle will skid out of control. If the driver was able to release the brake pedal just before the wheels locked up, then reapply the brakes, the skidding could be avoided.

This release and apply of the brake pedal is exactly what an ABS does. When the brake pedal is pumped or pulsed, pressure is quickly applied and released at the wheels. This is called **pressure modulation**. Pressure modulation works to prevent wheel locking. Antilock brake systems can modulate the pressure to the brakes as often as fifteen times per second. By modulating the pressure to the brakes, friction between the tires and the road is maintained, and the vehicle is able to come to a controllable stop.

The only time reduced friction aids in braking is when a tire is on loose snow. A locked tire allows a small wedge of snow to build up ahead of it, which allows it to stop in a shorter distance than a rolling tire.

> **You Should Know** *Remind your customers that pumping the brake pedal while stopping will prevent the ABS from activating. They should always keep firm steady pressure on the brake pedal during braking.*

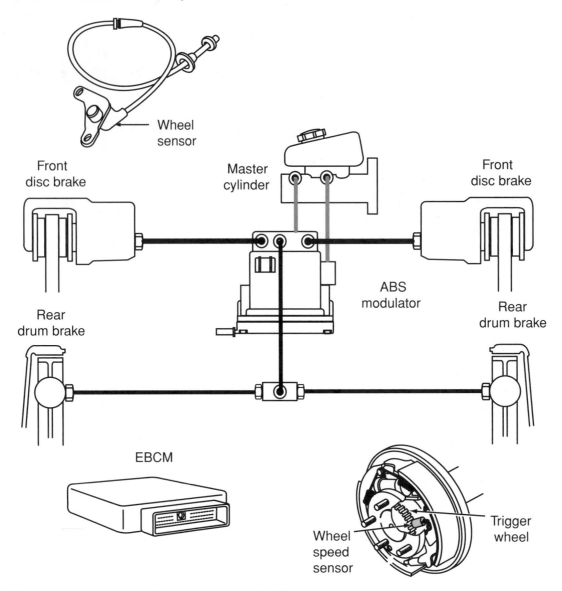

Figure 1. The main components of a commonly used ABS.

As long as a tire does not slip it goes only in the direction in which it is turned, but once it skids it has little or no directional stability. One of the big advantages of ABS is the ability to control the vehicle under all conditions.

Antilock brake systems precisely control the slip rate **(Figure 2)** of the wheels to ensure maximum grip force from the tires and thereby ensure maneuverability and stability of the vehicle. An ABS control module calculates the slip rate of the wheels based on the vehicle speed and the wheel speed, and then it controls the brake fluid pressure to attain the target slip rate.

Although an ABS prevents complete wheel lockup, an ABS allows some wheel slip in order to achieve the best braking possible. During ABS operation, the target slip rate can be from 10 to 30 percent. Zero (0) percent slip means that the wheel is rolling freely, whereas 100 percent means that the wheel is fully locked. A slip rate of 25 percent means that the velocity of a wheel is 25 percent less than that of a free-rolling wheel at the same vehicle speed. Many things are considered when determining the target slip rate for a particular vehicle. For some, the range is very low—5–10 percent; on others, it is high—20–30 percent.

PEDAL FEEL

The brake pedal on a vehicle equipped with an ABS can have a different feel than that of a conventional braking system. When the ABS is activated, a small bump followed by rapid pedal pulsations will continue until the vehicle comes to a stop or the ABS turns off. These pulsations are the result

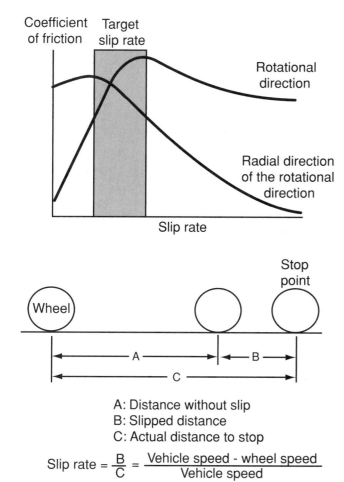

A: Distance without slip
B: Slipped distance
C: Actual distance to stop

$$\text{Slip rate} = \frac{B}{C} = \frac{\text{Vehicle speed - wheel speed}}{\text{Vehicle speed}}$$

Figure 2.   Defining slip rate.

of the modulation of pressure to the brakes and are felt more on some systems than on others. This is due to the use of dumping valves in some modulation units. With ABS, the brake pedal effort and pedal feel during normal braking are similar to that of a conventional power brake system.

COMPUTER SYSTEMS

A computer provides the precise and immediate control of an antilock brake system. Electronic control modules (ECMs) are either standalone units or are combined with a hydraulic control unit. The latter are called **electrohydraulic units**. Many newer vehicles have the antilock brake and traction control functions incorporated into the powertrain control module (PCM); therefore, a separate ABS unit is not necessary.

Computers control many systems and have been the key to providing cleaner, more fuel efficient, better performing, and safer vehicles. All computers have four stages of operation: input, processing, storage, and output.

A computer receives information in the form of voltage input signals from various sensors. It then processes, or computes, the input data to make decisions and perform output functions. Along with processing, the computer can remember, or store, an input signal for later reference or store and delay an output signal. It also can store a previous operating condition or other information. Most importantly, a computer stores its own operating instructions in the form of its program. After receiving and processing input data **(Figure 3)**, the computer sends output information or commands to display devices and actua-

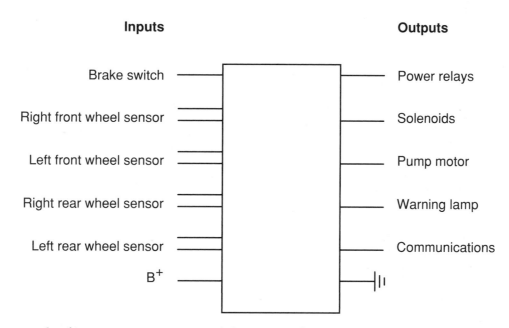

Figure 3.   An example of inputs to a computer and the outputs from it.

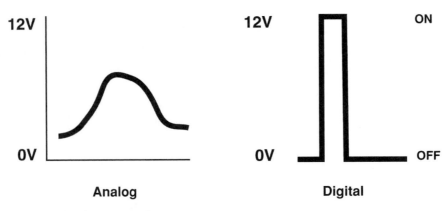

Figure 4. Analog and digital voltage signals.

tors. A display device can be an instrument panel indicator, and typical actuators are solenoids or motors that do mechanical work.

Analog and Digital Computers

Computer input signals can be either analog or digital. An **analog signal** means that a voltage signal or a processing function is infinitely variable; that is, it flows smoothly as it changes from one voltage level to another. A digital signal is one in which the voltage toggles, or steps, from one voltage to another **(Figure 4)**.

A computer needs to receive digital inputs in order to process the information. Not all sensors provide a digital signal to the computer; therefore, the computer must be able to change analog input signals to digital bits (binary digits) of information. This is accomplished with an analog-to-digital (AD) converter circuit. Based on its computations, the computer sends an output voltage to a system actuator or display device.

Computer Memory

The **microprocessor** is the **central processing unit (CPU)** of the computer. All of the mathematical operations and logical decisions occur in its integrated circuits. Other electronic integrated circuit devices provide the computer storage, or memory, function. Memory circuits can store three basic kinds of data: the computer operating system, input data from system sensors, and output data to system actuators.

Memory can be divided into two different types: memory that can be changed and memory that cannot be changed. For any computer to do its job properly, its program must be stored in memory and stay unchanged. Also, the program cannot be lost when power is turned off. This type of stored, permanent memory is **read-only memory (ROM)**. For most automotive computers, the operating program is loaded into **programmable read-only memory (PROM)**. A variation of PROM is called **erasable programmable read-only memory (EPROM)**. The EPROM differs from PROM in two ways: the computer can write information into it for permanent storage, and it can be erased and reprogrammed by the carmaker. (Other PROMs are programmed by the chipmaker and have a one-time use.)

Input and output signals that are stored temporarily are held in computer memory circuits called **random-access memory (RAM)**, where information can be written into and read. Two types of RAM are used in automobile computers: volatile and nonvolatile. Information in the volatile RAM is erased when the ignition is turned off. However, volatile RAM can be wired directly to the battery so that its information is not erased by shutting off the ignition. This type of RAM is often called **keep-alive memory (KAM)**. RAM and KAM circuits lose their memory, however, when power is disconnected from them.

Nonvolatile RAM retains its information even when power is disconnected. This type of RAM is often used for storing odometer information in an electronic speedometer. The memory chip retains the accumulated mileage of the vehicle. If the speedometer assembly is replaced, the odometer chip must be removed and installed in the new speedometer.

Computer Programs

A computer needs instructions to do its job, and these instructions are the computer's **program**, often referred to as computer **software**. The ABS computer has several programs that it follows at various times of vehicle operation. Often, the ABS ECM will communicate with another ECM or computer, sharing information. For example, the ABS ECM sends a data request to the engine's ECM to reduce torque to prevent wheel slip during acceleration. This request travels across a **network** and the engine's ECM will follow its program to reduce engine torque. Another example of this

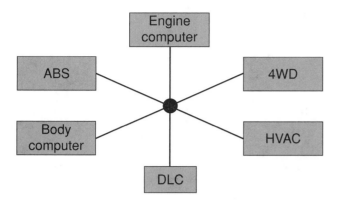

Figure 5. A computer network.

networking **(Figure 5)** is the use of data from the ABS ECM to help the engine's ECM or PCM decide if a change in operation is caused by road conditions or an engine misfire in OBD II systems.

How a computer will respond to conditions depends on its program, the input it receives, and the relative data it might have stored in its memory. All wheel sensor **frequency** inputs are compared to one another **(Figure 6)**. This allows the computer to determine if one particular wheel slows down faster than the others. This **select low** parameter causes the computer to activate the next stage of the program. During this stage, a hydraulic process called **isolate, hold, and dump** tries to equalize the speed of the wheels.

ABS COMPONENTS

The many different designs of ABSs vary by their basic layout, operation, and components. Variations also are based on the type of power-assist that is used and on whether the system is integral or not. The ABS components that can be found on a vehicle can be divided into categories: hydraulic and electrical/electronic components. Keep in mind that no one system uses all of the parts discussed here.

Hydraulic Components

Most accumulators for ABS are charged with nitrogen gas **(Figure 7)** and are an integral part of the modulator unit. This unit is typically found on vehicles with a hydraulically assisted brake system.

The antilock hydraulic **control valve assembly** controls the release and application of brake system pressure to the wheel brake assemblies. It can be an integral type, meaning that this unit is combined with the power-boost and master cylinder units into one assembly. The nonintegral type is mounted externally from the master cylinder/power booster unit and is located between the master cylinder and wheel brake assemblies. Both types generally contain solenoid valves that control the releasing, the holding, and the applying of brake system pressure.

The booster pump is an electric motor and pump that provides pressurized hydraulic fluid for the ABS system. The pump's motor is controlled by the system's ECM.

The booster/master cylinder assembly **(Figure 8)**, sometimes called the hydraulic unit, contains the valves and pistons used to modulate hydraulic pressure in the wheel circuits during ABS operation. Power brake assist is provided by pressurized brake fluid supplied by a hydraulic pump.

Different than pressure accumulators, fluid accumulators temporarily store brake fluid that is removed from the wheel brake units during an ABS cycle. This fluid is then used by the pump to build pressure for the brake hydraulic system. There are normally two fluid accumulators in a hydraulic control unit, one each for the primary and secondary hydraulic circuits.

The hydraulic control unit contains the solenoid valves, the fluid accumulators, the pump, and an electric motor. This is actually a combination unit composed of many individual components found separately in some systems. The unit can have one pump and one motor or it will have one motor and two pumps—one pump for half of the hydraulic system and the other for the other half.

The main valve is a two-position valve also controlled by the ABS control module and is open only in the ABS

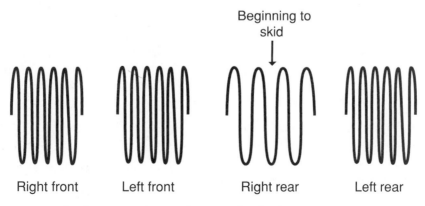

Figure 6. The ECM compares the frequency inputs from each wheel sensor.

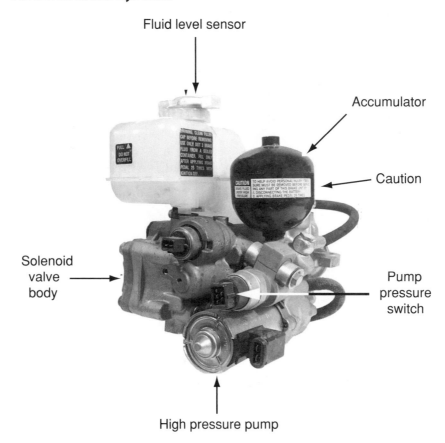

Fluid level sensor

Accumulator

Caution

Solenoid valve body

Pump pressure switch

High pressure pump

Figure 7. An integral ABS setup with a solenoid valve body, accumulator, and high-pressure pump.

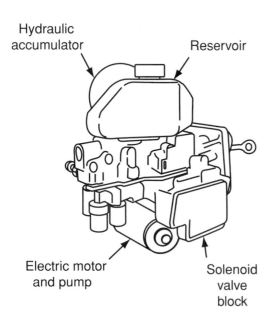

Hydraulic accumulator

Reservoir

Electric motor and pump

Solenoid valve block

Figure 8. A master cylinder assembly with an electric booster pump and solenoid valve block.

mode. When open, pressurized brake fluid from the booster circuit is directed into the master cylinder (front brake) circuits to prevent excessive pedal travel.

The modulator unit **(Figure 9)** controls the flow of pressurized brake fluid to the individual wheel circuits. Normally, the modulator is made up of solenoids that open and close valves that control the flow of fluid to the wheel brake units and electrical relays that activate or deactivate the solenoids through the commands of the ECM. The modulator unit can also be called the hydraulic actuator, hydraulic power unit, or the electrohydraulic control valve.

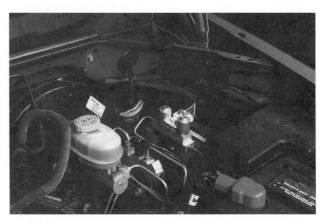

Figure 9. The modulator assembly for rear brake ABS.

The **solenoid valves** are located in the modulator unit. The ECM switches the solenoids on or off to increase, decrease, or maintain the hydraulic pressure to the individual wheel units.

The valve block assembly attaches to the side of the booster/master cylinder and contains the hydraulic wheel circuit solenoid valves. The ECM controls the position of these solenoid valves. The valve block is serviceable separately from the booster/master cylinder but should not be disassembled.

Wheel circuit valves are the solenoid valves used to control each wheel circuit or channel. When inlet and outlet valves of a circuit are used in combination, pressure can be increased, decreased, or held steady in the circuit. The position of each valve is determined by the control module. Outlet valves are normally closed, and inlet valves are normally open. Valves are activated when the ECM switches 12 volts to the circuit solenoids. During normal driving, the circuits are not activated.

Electrical/Electronic Components

The ABS control module is normally mounted inside the trunk on the wheel housing, is mounted to the master cylinder, or is part of the hydraulic control unit **(Figure 10)**. It monitors system operation and controls antilock function when needed. The module relies on inputs from the wheel speed sensors and feedback from the hydraulic unit to determine if the antilock brake system is operating correctly and to determine when the antilock mode is required. The module also has a self-diagnostic function.

The brake pedal switch **(Figure 11)** is normally closed. When the brake pedal travel exceeds the antilock brake pedal sensor switch setting during an antilock stop, the antilock brake control module senses that the antilock brake pedal sensor switch is open.

The **data link connector (DLC)** provides access and/or control of vehicle information, operating conditions, and diagnostic information through a scan tool.

A **diagnostic trouble code (DTC)** is a numeric identifier for a fault condition identified by the ABS's internal diagnostic system.

Most ABS-equipped vehicles are fitted with two different warning lights. One of the warning lights is tied directly to the ABS, whereas the other lamp is part of the base brake system. All vehicles have a red warning light. This lamp is lit when there is a problem with the brake system or when the parking brake is on. An amber warning lamp lights when there is a fault in the ABS system.

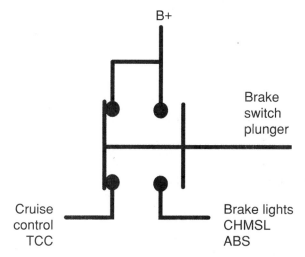

Figure 11.   A simplified schematic for a brake pedal switch.

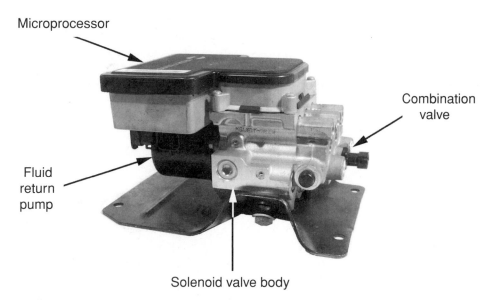

Figure 10.   A control module (microprocessor) that is integral with the solenoid valve body.

A lateral acceleration sensor (also called a lateral accelerometer) is used on a few vehicles. It monitors the sideward movement of the vehicle while it is turning a corner. This information is sent to the control module to ensure proper braking during turns.

The pressure switch controls pump motor operation and the low-pressure warning light circuit. The pressure switch grounds the pump motor relay coil circuit, activating the pump when accumulator pressure drops below 2030 psi. The switch cuts off the motor when the pressure reaches 2610 psi. The pressure switch also contains switches to activate the dash-mounted warning light if accumulator pressure drops below 1500 psi. The pressure switch unit is typically found on vehicles with a hydraulically assisted brake system.

The pressure differential switch is located in the modulator unit. This switch sends a signal to the control module whenever there is an undesirable difference in hydraulic pressures within the brake system.

Relays are electromagnetic devices used to control a high-current circuit with a low-current switching circuit. In ABS, relays are used to switch motors and solenoids. A low-current signal from the control module energizes the relays that complete the electrical circuit for the motor or solenoid.

The toothed ring can be located on an axle shaft, a constant velocity (CV) joint, a differential gear, or a wheel's hub. This ring is used to trigger the wheel speed sensor. The ring has a number of teeth around its circumference. The number of teeth varies by manufacturer and vehicle model. As the ring rotates and each tooth passes by the wheel speed sensor, an AC voltage signal is generated between the sensor and the tooth. As the tooth moves away from the sensor, the signal is broken until the next tooth comes close to the sensor. The end result is a pulsing signal that is sent to the control module. The control module translates the signal into wheel speed. The toothed ring can also be called the reluctor, tone ring, or gear pulser.

A wheel speed sensor **(Figure 12)** is comprised of a small magnet and a coil of wire inside a molded weather-sealed package that has two twisted-pair leads attached to it **(Figure 13)**.

Figure 12. A permanent magnet (PM) generator-type wheel speed sensor.

Figure 13. An example of a twisted pair.

A permanent magnet (PM) generator has an AC voltage output. Radio signals are also AC-modulated signals; therefore, PM generators can be affected by **radio frequency interference (RFI)**. The wires to and from a PM generator must be shielded from other wires in the vehicle, such as spark plug wires and those that connect to the AC generator and radiate **electromagnetic interference (EMI)**. Because the suspension must be able to move as the vehicle travels, the electrical wiring to the wheel speed sensors cannot be shielded in conduit. Instead, the wires are twisted at least one turn for every 1.75 inches. The wires are also routed to take advantage of the metal parts of the vehicle that could provide some shielding for the wires.

Summary

- An ABS can stop a car in the shortest possible distance without wheel lockup while stopping in a straight line, and it can give directional stability and control.
- An ABS is programmed to control negative wheel slip during braking. Negative wheel slip occurs when the speed of a wheel is less than vehicle speed.

- Traction control systems control positive wheel slip when the wheel speed is greater than vehicle speed during acceleration.
- An ABS allows some wheel slip in order to achieve the best braking possible.
- A driver operating the brake pedal on some vehicles equipped with ABS might experience a small bump

followed by rapid pedal pulsations that will continue until the vehicle comes to a stop or the ABS turns off. These pulsations are the result of the modulation of pressure to the brakes.

■ The ECM for an ABS can be a standalone unit or can be combined with a hydraulic control unit and called an electrohydraulic unit.

■ Many newer vehicles have the antilock brake and traction control functions incorporated into the PCM; therefore, a separate ABS unit is not necessary.

■ All computers have four stages of operation: input, processing, storage, and output.

■ Common hydraulic ABS components include: accumulator, antilock hydraulic control valve assembly, booster pump, booster/master cylinder assembly, fluid accumulators, hydraulic control unit, main valve, modulator unit, solenoid valves, valve block assembly, and wheel circuit valves.

■ Common electrical or electronic ABS components include: ABS control module, brake pedal switch, data link connector, lateral acceleration sensor, pressure switches, pressure differential switch, toothed ring, and wheel speed sensors.

Review Questions

1. What are the three stages of ABS pressure control and what does each do to the wheel brake?

2. Why are the wires leading to some wheel speed sensors twisted?

3. Technician A says that the ABS control module senses brake application from a brake switch input signal. Technician B says that the ABS control module senses wheel speed from one or more magnetic pickup sensors. Who is correct?
   A. Technician A only
   B. Technician B only
   C. Both Technician A and Technician B
   D. Neither Technician A nor Technician B

4. Technician A says that wheel speed sensors send an AC voltage signal to the control module. Technician B says that wheel speed signals create an AC signal voltage by altering a reference voltage received from the control module. Who is correct?
   A. Technician A only
   B. Technician B only
   C. Both Technician A and Technician B
   D. Neither Technician A nor Technician B

5. Technician A says that the gas-filled pressure chamber used in many antilock systems is called the accumulator. Technician B says that ABS control modules are digital computers. Who is correct?
   A. Technician A only
   B. Technician B only
   C. Both Technician A and Technician B
   D. Neither Technician A nor Technician B

Chapter 40

Common Antilock Brake System Designs

Introduction

Antilock brake systems found on today's vehicles are manufactured by one of the following companies: Bendix, Bosch, Delco Moraine, ITT Teves, Kelsey-Hayes, or Lucas Girling. Each manufacturer has a unique way to accomplish the same thing—vehicle control during braking. When working with an ABS, it is important that you identify the exact system you are working with and follow the specific service procedures for that system. Keep in mind that nearly fifty different ABSs have been used by the industry in recent years.

BASIC TYPES OF ABSs

The exact manner in which hydraulic pressure is controlled depends on the exact ABS design. The great majority of the earlier ABSs were **integrated** or **integral ABSs**. They combine the master cylinder, hydraulic booster, and ABSs hydraulic circuitry into a single hydraulic assembly.

Other ABSs are **nonintegral**. They use a conventional vacuum-assist booster and master cylinder. The ABS hydraulic control unit is a separate mechanism **(Figure 1)**. In some nonintegrated systems, the master cylinder supplies brake fluid to the hydraulic unit. Although the hydraulic unit is a separate assembly, it still uses a high-pressure pump/motor, accumulator, and fast-acting solenoid valves to control hydraulic pressure to the wheels. Both integral and nonintegral systems operate in much the same way; therefore, an understanding of one system will lend itself to the understanding of the other systems.

General Motors' electromagnetic ABS is a different type of nonintegral system that uses a conventional vacuum power booster and master brake cylinder. It does not use a

high-pressure pump/motor, accumulator, and fast-acting solenoid valves to control hydraulic pressure. Instead, it uses a hydraulic modulator.

In addition to being classified as integral and nonintegral ABSs, systems can be broken down into the level of control they provide. Antilock brake systems can be one-, two-, three-, or four-**channel** two- or four-wheel systems. A channel is merely a hydraulic circuit to the brakes.

One-Channel Systems

These systems offer antilock brake performance to the rear wheels only. They do not provide antilock performance to the steering wheels. Two-wheel systems are most often found on light trucks and some sport utility vehicles.

In one-channel systems, the rear brakes on both sides of the vehicle are modulated at the same time to control skidding. These systems rely on the input from a centrally located speed sensor. The speed sensor is normally positioned on the ring gear in the differential unit **(Figure 2)**, transmission, or transfer case.

Two-Channel Systems

Two-channel systems can be found on some diagonally split brake systems. These systems use two speed sensors to provide wheel speed data for the regulation of all four wheels. One sensor has input that controls the right front wheel; the other sensor performs likewise for the left front wheel.

Brake hydraulic pressure to the opposite rear wheel is controlled simultaneously with its diagonally located front wheel. For example, the right rear wheel receives the same pumping instructions as the left front wheel. This system is an upgrade from the two-wheel system because it does provide steering control. However, it can have shortcomings under certain operating conditions.

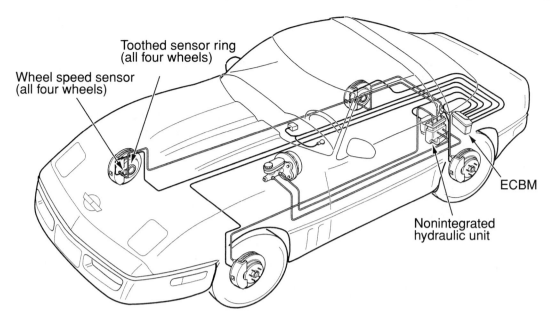

Figure 1.   An example of a nonintegrated (nonintegral) ABS with a separate hydraulic unit.

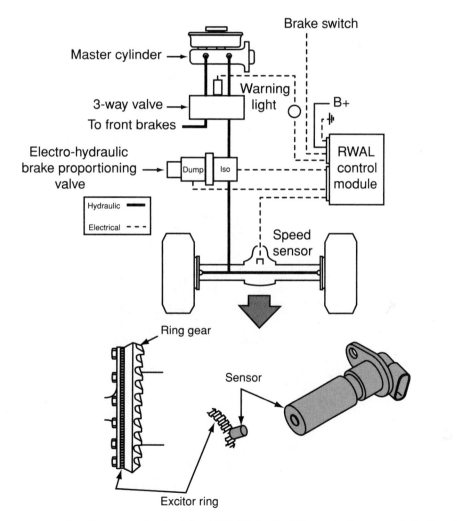

Figure 2.   The main components of a rear wheel–only ABS. Note that the speed sensor is in the differential at the ring gear.

Three-Channel Systems

Some hydraulic systems that are split from front to rear use a three-channel circuit and are called four-wheel ABSs. These systems have individual hydraulic circuits to each of the two front wheels and a single circuit to the two rear wheels.

Four-Channel Systems

The system that is the most effective and most common ABS available is a four-channel system, in which sensors monitor each of the four wheels. With this continuous information, the ABS control module ensures that each wheel receives the exact braking force it needs to maintain both antilock and steering control.

BASIC OPERATION

The exact operation of an ABS depends on its design and manufacturer. It would take many pages to explain the operation of each, and as soon as you read the explanations, there would be two or more new systems that would not have been explained. The exact operation of any system can be easily understood if you comprehend the basic operation of a few. The primary difference in operation among them all is based on the components used by the system. Therefore, the following systems were chosen as examples of how certain systems operate with the components they have.

One-Channel Systems

These systems are primarily used to prevent rear wheel lockup on pickup trucks and sport-utility vehicles (SUVs),

especially under light payload conditions. These systems consist of a standard power brake system, an ECM, and an isolation/dump valve assembly. The valve assembly is attached to the master cylinder at the rear brake line. Both rear wheel brake assemblies are controlled by the valve assembly under ABS conditions.

Under normal braking, pressure will pass through the valve assembly **(Figure 3)**. The control module receives a signal from the brake switch when brakes are applied and begins to monitor the vehicle speed sensor (VSS) signal at speeds over 5 mph. If the control module detects a deceleration rate from the VSS that would indicate probable rear wheel lock-up, it will activate the isolation valve **(Figure 4)**, which will stop the buildup of pressure to the rear wheels. If further deceleration occurs that would indicate lockup, the ECM will rapidly pulse the dump valve **(Figure 5)** to release brake pressure into the accumulator. The ECM will continue to pulse the dump valve until rear wheel deceleration matches the vehicle's deceleration rate or the desired slip rate. When wheel speed picks up, the ECM will turn off the isolation valve, allowing the fluid in the accumulator to return to the master cylinder and normal braking control to resume.

The ECM has three distinct functions: it performs self-test diagnostics, it monitors the ABS action and system, and it controls the ABS solenoid valves. When the ignition switch is turned on, the ECM checks its ROM and RAM. If an error is detected, a DTC is set in memory. A DTC will also be set if the control module senses a problem during ABS operation. The ECM continuously monitors the speed of the differential ring gear through signals from the rear wheel speed sensor. The ECM also receives signals from the brake light switch, brake warning lamp switch, reset

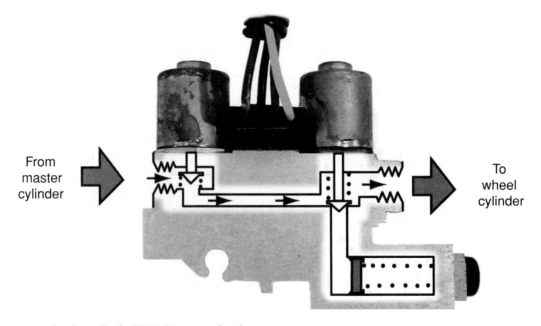

From master cylinder →   ← To wheel cylinder

Figure 3.   A rear wheel antilock (RWAL) control valve at rest.

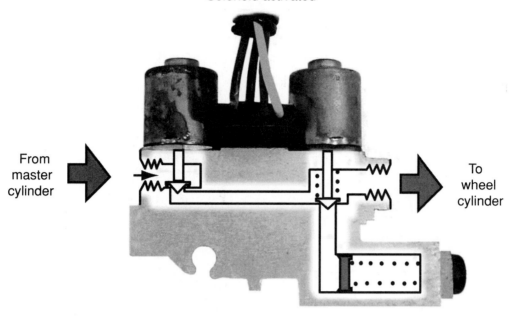

Figure 4.   The isolate/hold position of an RWAL control valve.

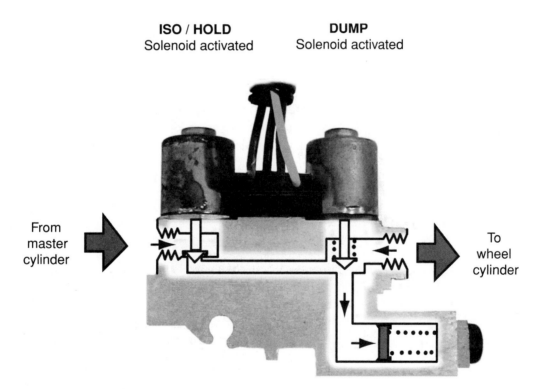

Figure 5.   The dump position of an RWAL control valve.

switch, and the 4WD switch. Preventing wheel lockup is the primary responsibility of the ECM. It does this by controlling the operation of the isolation and dump solenoid valves.

The one-channel is disabled on four-wheel drive vehicles when in the four-wheel drive mode due to transfer case operation. Switching the transfer case into two-wheel drive mode will reenable the ABS system.

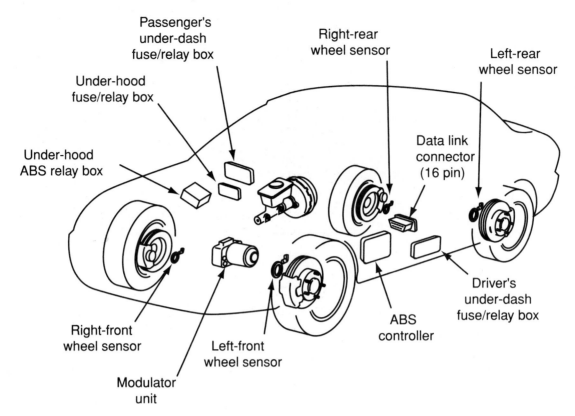

Figure 6. The basic electrical and hydraulic components for a four-wheel ABS.

Four-Wheel Systems

The hydraulic circuit for this type of system is an independent four-channel type, one channel for each wheel **(Figure 6)**. The hydraulic control unit is a separate unit. In the hydraulic control unit, there are two valves per wheel; therefore, a total of eight valves is used. Some systems have three channels, one for each of the front wheels and one for the rear axle. Obviously, these systems have only three pairs of solenoids **(Figure 7)**. Normal braking is accomplished by a conventional vacuum power-assist brake system.

By modulating brake pressure, the system prevents wheel lockup during an emergency stop. It allows the driver to maintain steering control and stop the vehicle in the shortest possible distance under most conditions. During ABS operation, the driver will sense a pulsation in the brake pedal and a clicking sound.

ABS OPERATION

The **ABS control module** calculates the slip rate of the wheels and controls the brake fluid pressure to attain the target slip rate. If the ECM senses that a wheel is about to lock, based on input sensor data, it pulses the normally open inlet solenoid valve closed for that circuit. This prevents any more fluid from entering that circuit. The ABS

control module then looks at the sensor signal from the affected wheel again. If that wheel is still decelerating faster than the other three wheels, it opens the normally closed outlet solenoid valve for that circuit. This dumps any pressure that is trapped between the closed inlet valve and the brake back to the master cylinder reservoir. Once the affected wheel returns to the same speed as the other wheels, the control module returns the valves to their normal condition, allowing fluid flow to the affected brake.

Based on the inputs from the vehicle speed and wheel speed sensors, the control module calculates the slip rate of each wheel, and it transmits a control signal to the modulator unit solenoid valve when the slip rate is high.

Wheel speed at each wheel is measured by variable-reluctance sensors and sensor indicators (toothed ring or wheel). The sensors operate on magnetic induction principles. As the teeth on the brake sensor indicators rotate past the sensors, AC current is generated. The AC frequency changes in accordance with the wheel speed. The ABS control unit detects the wheel sensor signal frequency and thereby detects the wheel speed.

A few systems are equipped with a lateral acceleration switch to change the system's programming when hard braking while cornering. The switch might be no more than a left-hand and right-hand mercury switch located in a common housing that responds to forces generated by

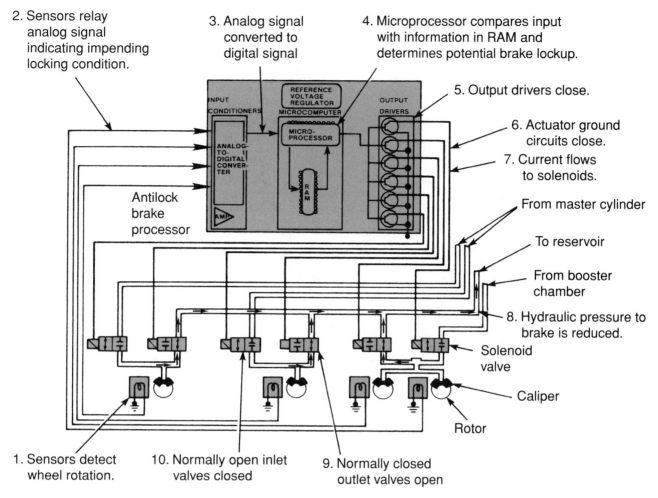

2. Sensors relay analog signal indicating impending locking condition.

3. Analog signal converted to digital signal

4. Microprocessor compares input with information in RAM and determines potential brake lockup.

5. Output drivers close.

6. Actuator ground circuits close.

7. Current flows to solenoids.

From master cylinder

To reservoir

From booster chamber

8. Hydraulic pressure to brake is reduced.

Solenoid valve

Caliper

Rotor

Antilock brake processor

1. Sensors detect wheel rotation.

10. Normally open inlet valves closed

9. Normally closed outlet valves open

Figure 7. ABS operation when there is the potential for wheel lockup.

left-hand/right-hand turns and cornering movements. Late-model lateral acceleration sensors use a Hall effect switch.

Modulator Assembly

The ABS modulator assembly consists of the inlet solenoid valve, outlet solenoid valve, reservoir, pump, pump motor, and the damping chamber. The modulator reduces fluid pressure to the calipers. The hydraulic control has three modes: pressure reduction (decrease), pressure retaining (hold), and pressure intensifying (increase).

While in the pressure reduction decrease mode, the inlet valve is closed and the outlet valve is open. During this mode, fluid pressure to the caliper is blocked and the existing fluid in the caliper flows through the outlet valve back to the master cylinder reservoir. During the pressure-intensifying mode, the inlet valve is open and the outlet valve is closed. Pressurized fluid is pumped to the caliper. To keep the pressure at the caliper, during the pressure-retaining mode, the inlet and outlet valves are closed.

The pump/motor assembly provides the extra fluid required during an ABS stop. The pump is supplied fluid that is released to the accumulators when the outlet valve is open during an ABS stop. The accumulators provide temporary storage for the fluid during an ABS stop. The pump is also used to drain the accumulator circuits after the ABS stop is complete. The pump is run by an electric motor that is controlled by a relay controlled by the ABS control module. The pump is continuously on during an ABS stop and remains on for about five seconds after the stop is complete.

Keep in mind that the activity of the solenoid valves changes rapidly, several times each second. This means that the fluid under pressure must be redirected quickly; this is the primary job of the pump.

Self-Diagnosis

The control module monitors the electromechanical components of the system. A malfunction of the system will cause the control module to shut off or inhibit the sys-

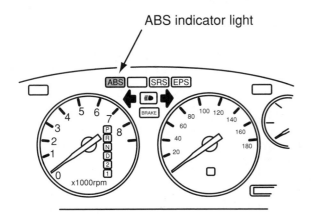

Figure 8. A typical ABS warning lamp.

tem. However, normal power-assisted braking remains. Malfunctions are indicated by a warning indicator in the instrument cluster **(Figure 8)**. The system is self-monitoring. When the ignition switch is placed in the run position, the ABS control module will perform a preliminary self-check on its electrical system indicated by a second illumination of the amber ABS indicator in the instrument cluster. During vehicle operation, during normal and antilock braking, the control module monitors all electrical ABS functions and some hydraulic functions. With most malfunctions of

the ABS, the amber ABS indicator will be illuminated and a DTC recorded.

General Motors' Electromagnetic Systems

Beginning in 1991, General Motors began equipping certain small and mid-size vehicles with an ABS called the Delco-Moraine ABS-VI. This system uses a conventional vacuum power booster and master brake cylinder. It does not use a high-pressure pump/motor and accumulator and fast-acting solenoid valves to control hydraulic pressure. Instead, it uses a hydraulic modulator **(Figure 9)** that operates using a principle called electromagnetic braking.

As in integrated systems, wheel speed is monitored using individual speed sensors. When one wheel begins to decelerate faster than the others while braking, the control module signals the hydraulic modulator assembly to reduce pressure to the affected brake.

The ABS-VI modulator contains three small screw plungers—one for each front brake circuit and one for the rear brake circuit. These plungers are driven by electric motors. At the top of each plunger cavity is a check ball that controls hydraulic pressure within the brake circuit.

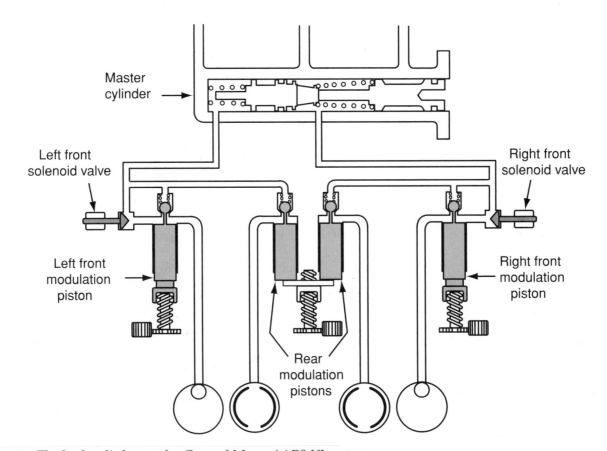

Figure 9. The hydraulic layout for General Motors' ABS-VI system.

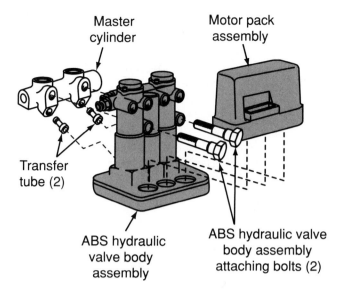

Master cylinder

Motor pack assembly

Transfer tube (2)

ABS hydraulic valve body assembly

ABS hydraulic valve body assembly attaching bolts (2)

Figure 10. The mounting of the hydraulic modulator assembly to the master cylinder in an ABS-VI system.

The hydraulic valve body (modulator) assembly and motor pack is mounted to the master cylinder **(Figure 10)**. The hydraulic brake circuit for each front wheel is controlled by a motor, a gear-driven ball screw, a solenoid, a piston, and a check valve. The rear wheel circuit is controlled by check balls and a single motor; therefore, both rear brakes are modulated together.

The motors are high-speed, bidirectional motors that quickly and precisely position the ball screws. Each motor has a brake that allows its ball screw to maintain its position against hydraulic pressures. The front motors have an electromagnetic brake (EMB), and the rear motor uses an expansion spring brake (ESB).

Other Brake System Controls

A few automobiles are now equipped with an electronic brake-assist system that can recognize emergency braking and automatically apply full-power brake force for shorter stopping distances. This system is activated only in emergency braking situations and does not affect normal brake operation.

The system recognizes emergency braking by the speed at which the brake is depressed, and the brakes are automatically applied under full power. The system is driver-adaptive as it learns the driver's braking habits by using sensors to monitor every movement of the brake pedal. When the sensors detect an emergency stop, an electronic valve at the power brake booster is turned on. This supplies full braking power to the wheels. The ABS system prevents the wheels from locking in spite of the full-power braking force.

Summary

- A great number of early antilock brake designs were integrated or integral systems, which means that they combined the master cylinder, hydraulic booster, and ABS hydraulic circuitry into a single hydraulic assembly.
- General Motors' electromagnetic ABS is a different type of nonintegral system that uses a conventional vacuum power booster and master brake cylinder but does not use a high-pressure pump/motor and accumulator and fast-acting solenoid valves to control hydraulic pressure.
- Antilock brake systems can be one-, two-, three-, or four-channel, two- or four-wheel systems. A channel is merely a hydraulic circuit to the brakes.
- In one-channel systems, the rear brakes on both sides of the vehicle are modulated at the same time to control skidding.
- A three-channel system has individual hydraulic circuits to each of the two front wheels and a single circuit to the two rear wheels.
- The most effective and common ABS is a four-channel system, in which each wheel receives the exact braking force it needs to maintain both antilock and steering control.
- One-channel systems are used to prevent rear wheel lockup on pickup trucks and SUVs and consist of a standard power brake system, an ECM, and an isolation/dump valve assembly attached to the master cylinder at the rear brake line.
- The hydraulic circuit for a four-channel system includes one channel for each wheel. In the hydraulic control unit, there are two valves per wheel; therefore, a total of eight valves is used. The ABS control module calculates the slip rate of the wheels and controls the brake fluid pressure to attain the target slip rate.
- A typical ABS modulator assembly operates in three different modes to control the fluid pressure to the individual wheel brakes: pressure reduction (decrease), pressure retention (hold), and pressure intensification (increase).
- General Motors' ABS-VI uses a hydraulic modulator that operates using a principle called electromagnetic braking.
- The ABS-VI modulator contains three small screw plungers—one for each front brake circuit and one for the rear brake circuit. These plungers are driven by electric motors. At the top of each plunger cavity is a check ball that controls hydraulic pressure within the brake circuit.

Review Questions

1. Explain the principal differences between integral and nonintegral ABS.
2. What are the functions of an isolation and a dump-release solenoid valve?
3. In a three-channel antilock system, the front brakes are controlled _____ and the rear brakes are controlled _____.
4. In a nonintegral ABS installation, the antilock components are _____ to the base brake system.

5. Technician A says that four-wheel ABS can operate with three-channel output control. Technician B says that four-wheel ABS can operate with four-channel output control. Who is correct?
   A. Technician A only
   B. Technician B only
   C. Both Technician A and Technician B
   D. Neither Technician A nor Technician B

Chapter 41

Manufacturers' Antilock Brake Systems

Introduction

Most antilock systems currently in use are built by the following manufacturers:

- Bendix
- Bosch (and Delco-Bosch)
- Delphi Chassis (formerly the Delco Moraine Division of General Motors)
- Kelsey-Hayes
- Nippondenso
- Nissan
- Sumitomo
- Teves

Each design has a designation to distinguish its features and applications. This chapter looks at some of the different designs built by the brake manufacturers.

Keep in mind that ABSs are primarily defined by the number of channels they use and whether or not the system is an integral part of the brake system **(Figure 1)**. For the most part, these variations determine the components of the system and their location.

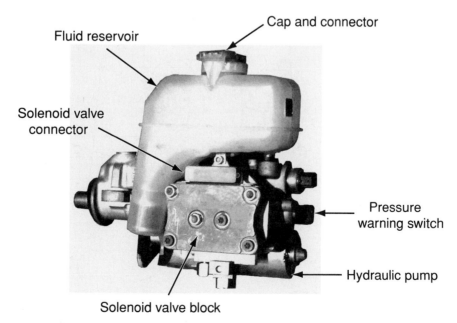

Figure 1. An integrated ABS combines the master cylinder, brake booster, and ABS components into a single unit.

Ford Motor Company was the first United States carmaker to experiment with ABS on a production car when it offered an antiskid option on the 1954 Lincoln Continental Mark II. The system added too much weight to the car and cost too much, so it was put aside. Then a decade later, Ford and Kelsey-Hayes jointly developed a rear-wheel antilock system that appeared on 1969 Thunderbirds and Lincolns. Called "Sure-Track," this system used an analog control module and a vacuum-controlled actuator to control rear brake pressure. Each rear wheel had a magnetic pickup speed sensor that was remarkably similar to those with modern antilock systems. The analog control module and vacuum actuator limited the pulse rate for the rear hydraulic circuits to only four cycles per second. Sure-Track was a good idea before its time. It went away after a year or two until digital electronics made ABS practical in the mid-1980s.

BOSCH SYSTEMS

Several different Bosch brake systems are used on domestic and import cars. The following designs are presented in the order of their introduction:

- Bosch 3 is an integrated four-channel system that was used by Cadillac from 1987 through 1992 and by Chrysler from 1988 and later **(Figure 2)**.
- Bosch 2, 2E, 2U, and 2S Micro ABS are nonintegrated, four-wheel systems. Most are three channel; some are four channel. The Bosch 2 was first used on United States–specification BMW and Mercedes models. This system and its related versions (2S, 2E, 2U, and 2S Micro) have been used on a wide variety of domestic and import vehicles through the 2000 model year.
- Bosch ABS/ASR adds traction control, which Bosch calls **acceleration slip regulation (ASR)**, to the ABS. The ABS/ASR system is a nonintegrated, three-channel, four-wheel system; its ASR component can operate the rear brakes independently of the ABS. The ABS/ASR system is used on some 1992 and later General Motors models.
- Bosch 5 is a nonintegrated, four-channel, four-wheel ABS with traction control. It is used on some 1995 and later Corvettes.

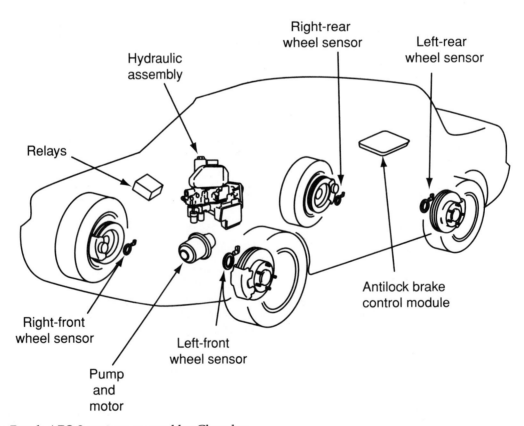

Figure 2. The Bosch ABS-3 system as used by Chrysler.

Bosch 2 Systems

Variations of the Bosch 2 system are the 2E, 2S, 2U, and 2S Micro. These have been used by all three domestic manufacturers on various models, as well as by Toyota and Nissan, from 1986 through the late 1990s.

Most Bosch 2 systems use three channels, two at the front and one at the rear, to provide four-wheel antilock capability **(Figure 3)**. An exception is the 2U system used on the 1993 and later Mercury Villager, which has a separate hydraulic channel for each wheel. All are nonintegrated; the master cylinder and hydraulic modulator assembly are separate units.

The modulator valve assembly contains three solenoid-operated valves, two accumulators, a pump, and a pump motor. Wheel sensor input is received by the control module at separate terminals **(Figure 4)**. If a sensor malfunctions, the control module disables the antilock function when the vehicle reaches a road speed between 3 and 11 mph. If all four sensors (three sensors on 2E and rear-wheel-drive 2U systems) should malfunction or their signals not reach the control module for some other reason, the module has no way of sensing that the vehicle is moving. Because the module thinks that the car is at rest, the antilock function will be disabled without lighting the indicator lamp.

When the brakes are applied and the wheel sensor data is normal, no current is sent to the modulator valves, and the control module continues to monitor the system. If the wheel sensor signal indicates a possible wheel lockup condition, the module uses the brakes-applied signal to

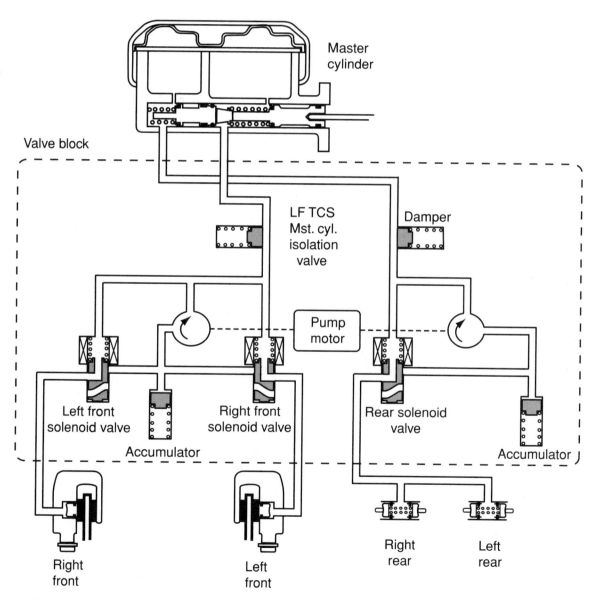

Figure 3. Most versions of the Bosch 2U/2S ABS have three channels.

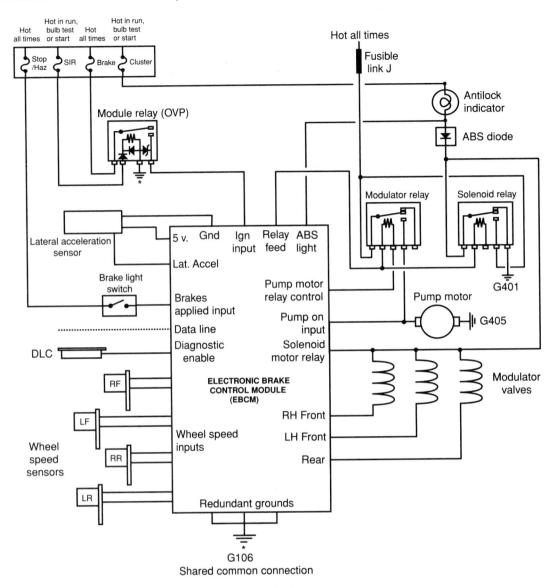

Figure 4. An electrical schematic for a Bosch 2S-micro ABS.

start the control phase. The control phase consists of 4 to 10 cycles per second. Each cycle consists of three modes:

1. The module signals the proper modulator valve to isolate the caliper. This maintains fluid pressure in the caliper.
2. If excessive wheel deceleration continues, the module signals the valve to reduce fluid pressure at the caliper.
3. The reduction in pressure allows the wheel to accelerate. When wheel acceleration reaches a specified point, the module signals the valve to increase fluid pressure.

Bosch ABS with ASR

This system is a version of the Bosch 2U/2S design that includes traction control. The ASR system controls engine output and vehicle braking to reduce excessive wheel spin during acceleration or cornering. The electronic brake control module (EBCM) uses a combination of inputs—wheel slip from the wheel sensors and cornering force from the lateral acceleration sensor—to determine when wheel spin becomes excessive **(Figure 5)**.

ASR operation is a good example of how traction control includes several aspects of power train management. The ASR system's first step in reducing wheel spin is to reduce engine power. This is done when the EBCM signals the powertrain control module (PCM) to retard ignition timing. If more power reduction is needed, the throttle valve opening is reduced. If power reduction alone is not enough, the EBCM will apply one or both rear brakes, using the hydraulic pump to supply fluid pressure and the hydraulic valves to control it.

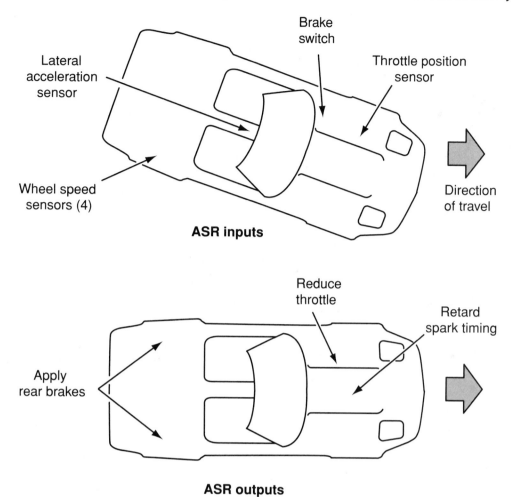

Lateral acceleration sensor

Brake switch

Throttle position sensor

Wheel speed sensors (4)

Direction of travel

ASR inputs

Reduce throttle

Retard spark timing

Apply rear brakes

ASR outputs

Figure 5. Inputs and outputs of a Bosch ABS/ASR system.

TEVES SYSTEMS

Three Teves systems in recent use are the Mark II, Mark IV and IVg (some with traction control), and the Mark 20. Versions of these three Teves systems are used by Ford, General Motors, and Chrysler.

Teves Mark II

The Teves Mark II system was used on many domestic vehicles from the mid-1980s through the early 1990s. The Mark II is an integrated system, incorporating the master cylinder into a single hydraulic control unit (HCU) with the hydraulic pump, motor, modulator, and accumulator **(Figure 6)**. The pump in the hydraulic control unit (HCU) provides power brake operation as well as ABS. The brake pedal pushrod acts on the power booster, which in turn acts on the master cylinder. The ABS portion of the system controls all four wheels with four sensors and three hydraulic channels.

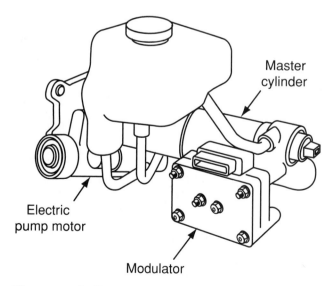

Master cylinder

Electric pump motor

Modulator

Figure 6. A Teves master cylinder and booster assembly.

Teves Mark IV

The Mark IV system is similar in operation to the Mark II system and replaces it in many applications. However, this system uses a conventional master cylinder and power booster and a separate ABS solenoid and pump assembly to control the ABS hydraulic action **(Figure 7)**. Like the Mark II, the Mark IV ABS uses four input channels and three output channels.

The low-speed traction control (LSTC) version of Teves IV includes two isolation valves or traction control valves. These are normally open, allowing flow and return of brake fluid from the master cylinder to the front calipers. When traction control is activated, the HCU delivers brake pressure to the caliper at whichever wheel is spinning. The isolation valves close to block the return of brake fluid to the master cylinder and allow brake pressure to build up at the spinning wheel.

The most important difference between the Mark IV and Mark IVg designs is that fluid for the HCU in the Mark IVg is drawn from a pair of accumulators **(Figure 8)**, rather than from the master cylinder reservoir. The accumulators are part of the HCU. One supplies the primary hydraulic circuit, and one supplies the secondary circuit during brake operation. The pump operates continuously during an ABS stop; extra fluid is stored in the accumulators. Two additional, smaller accumulators store fluid used to protect lip seals and to reduce pump noise. Because the master cylinder reservoir is not used to store fluid for ABS operation, the master cylinder is of the conventional design.

Teves Mark 20

This system is used on 1997 and later Chrysler vehicles. It is similar to previous Teves systems (Mark IV and IVg). The minivan version of the system can include LSTC.

The version used on Jeeps includes an **acceleration sensor (Figure 9)**. Also known as a G-switch, this sensor provides information about the rate of forward or reverse deceleration to the electronic control module (CAB).

The sensor includes three mercury switches. The switches work as follows:

- In their normal positions, all three mercury switches are closed.
- Under hard forward braking on a high-traction surface, such as dry pavement, both the G1 and G2 switches open.
- On a medium-traction surface, such as gravel, the G2 switch opens.

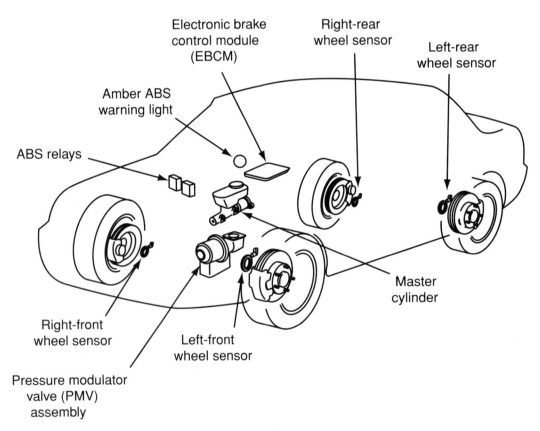

Figure 7. A Teves four-wheel integrated ABS.

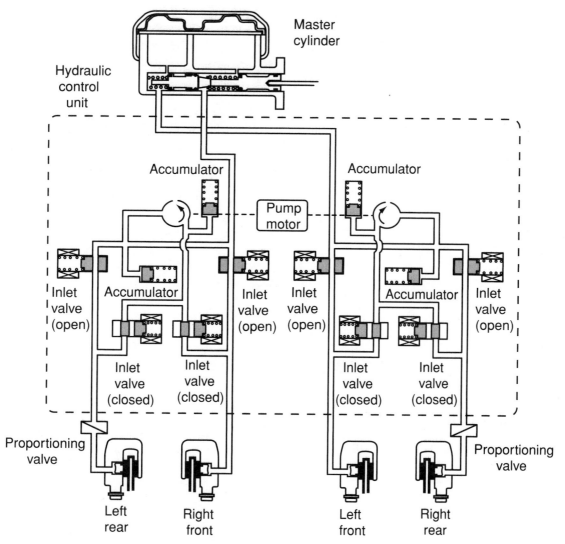

Figure 8. A Teves Mark IVg ABS with two fluid accumulators.

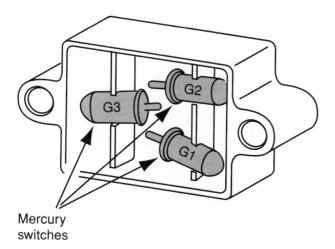

Figure 9. An acceleration sensor used in Jeeps with the Teves Mark 20 system.

- On a low-traction surface, such as ice, neither switch opens.
- Under hard braking in reverse, the G3 opens.

The angles at which the mercury switches are mounted in the vehicle determine the amount of deceleration required to open them. For this reason, the acceleration sensor must be installed correctly and is clearly marked for proper mounting.

DELPHI CHASSIS (DELCO MORAINE) SYSTEMS

The Delphi Chassis division of General Motors (formerly Delco Moraine) has manufactured two ABSs: the Delco Moraine III (PowerMaster) and the Delphi Chassis VI (Delco ABS VI).

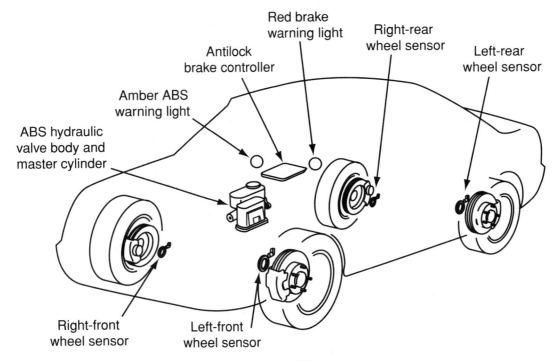

Figure 10. The major components of a Delphi Chassis ABS III.

Delco Moraine ABS III System

The Delco Moraine ABS III system **(Figure 10)** is a four-wheel integrated system that consists of a base brake system, an antilock brake electronic control module (ECM), wheel speed sensors at each wheel, two indicator lamps, and front and rear enable relays.

For the Delco ABS III, a PowerMaster-III booster and master cylinder assembly is used. The PowerMaster-III unit is a single assembly containing a dual-piston master cylinder, power brake booster, electric pump, and accumulator. It also contains a bleeder valve, a pressure relief valve, a pressure switch, and three solenoids. The unit applies hydraulic pressure through the accumulator when the brake pedal is depressed. Power braking is provided by the power brake booster. However, rear power assist and hydraulic pressure for rear braking are produced by the electric pump. Hydraulic pressure for the front brakes comes from the master cylinder. The primary piston operates the right front caliper, and the secondary piston operates the left front caliper.

The pump supplies the accumulator with pressurized fluid for rear braking. The accumulator stores the pressurized fluid so that the pump does not need to run all the time. A rubber diaphragm in the accumulator isolates brake fluid from the nitrogen gas that pressurizes the fluid.

Delphi Chassis (Delco Moraine) ABS VI

Beginning in 1991, General Motors began equipping some small and midsize vehicles with an ABS called the Delphi Chassis VI system. This system uses a vacuum power booster and master cylinder. To control hydraulic pressure, it has a hydraulic modulator **(Figure 11)** that is driven by three electric motors.

The hydraulic modulator and motor pack assembly are mounted to the master cylinder. The hydraulic brake circuit for each front wheel is controlled by a motor **(Figure 12)**, a

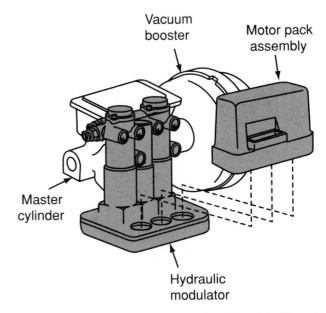

Figure 11. The components of a Delphi Chassis ABS VI.

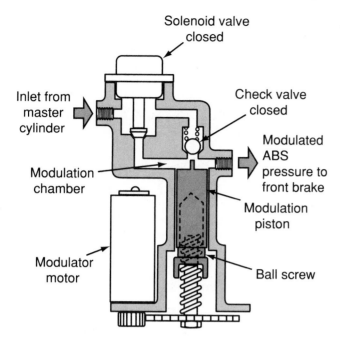

Figure 12. An ABS VI front brake modulator.

gear-driven ball screw, a solenoid, a piston, and a check valve. The rear wheel circuit is controlled by check balls and a single motor **(Figure 13)**; therefore, both rear brakes are modulated together.

KELSEY-HAYES SYSTEMS

Kelsey-Hayes two-wheel systems are nonintegrated, or add-on, systems that use a traditional vacuum power brake booster.

Ford Rear Antilock Brake System

The Ford **rear antilock brake system (RABS)** was used on light-duty pickup trucks and SUVs from 1987 through 1997. During normal braking, the hold valve remains open while the dump valve blocks off the accumulator. Brake fluid travels unrestricted to the rear wheel cylinders to apply the brakes.

If the control module detects a lockup condition, it energizes the isolation solenoid valve to close off brake fluid flow to the wheel cylinders. The dump solenoid valve is pulsed open, and fluid in the line between the wheel cylinders and isolation valve enters the accumulator. This reduces pressure at the rear wheels. A pressure-maintain signal from the control module closes the isolation valve, which allows fluid in the accumulator to return to the line between the valve and wheel cylinders. This sequence is pulsed several times a second until the danger of wheel lockup has passed. The control module then de-energizes the hold valve, allowing the brake fluid to pass to the wheel cylinders without restriction.

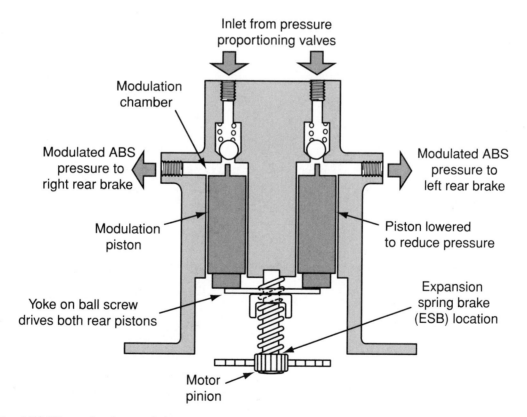

Figure 13. An ABS VI rear brake modulator.

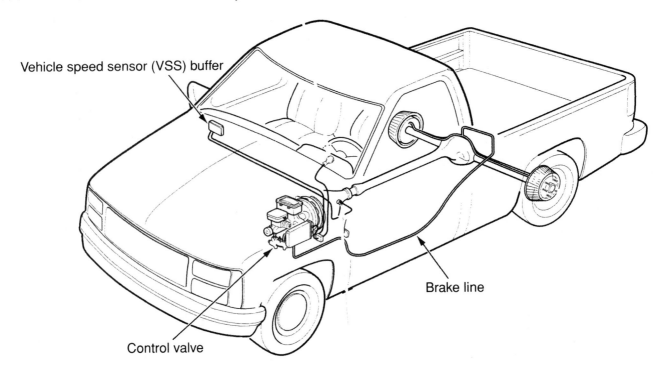

Figure 14. A Kelsey-Hayes RWAL system.

General Motors Rear Wheel Antilock Systems

The **rear wheel antilock (RWAL)** system used on Chevrolet and GMC light-duty pickup trucks and SUVs **(Figure 14)** is a bit more complex than the RABS. The basic components are the same, except for sensor type and location. The wheel speed sensor signal is passed through a stand-alone **digital ratio adapter circuit (DRAC)** or one integrated into the instrument cluster where the signal is processed before being sent to the control module. This is necessary because the same speed sensor signal also is used by the cruise control module, the engine control module, and the speedometer and odometer.

The speed sensor for RWAL systems is in the transmission on two-wheel-drive (2WD) models and in the transfer case of 4WD models. Because the sensor takes its signal from the transmission or transfer case output shaft, changing the axle ratio or rear tire size will affect the signal. If such modifications are made, the stand-alone DRAC must be changed or the ratio adapter in the instrument cluster must be recalibrated in order for the systems that rely on the speed sensor signal to function properly.

Dodge Truck Rear Wheel Antilock Systems

The Chrysler version of the Kelsey-Hayes RWAL system was used on Dodge trucks from 1988 through 1997. Dodge's RWAL system is nearly identical to RABS with the principal differences being:

- The Dodge electronic control module contains a poly-fuse, which is a thermal fuse that melts to prevent system operation if the module cannot shut down the system when a malfunction occurs.
- The Dodge diagnostic connector is inside the passenger compartment, under the glove box or behind the right-hand kick panel.
- Like the RABS control module, the RWAL module can store only one trouble code at a time. Unlike RABS, the Dodge module retains the code in memory after the ignition is turned off.

Four-Wheel Antilock Systems

The Kelsey-Hayes four-wheel antilock (4WAL) brake system was introduced as an option on some General Motors vans, light-duty trucks, and SUVs. Similar versions also have been used on Ford and Dodge trucks. The 4WAL system is also known as the Kelsey-Hayes EBC4 system. (EBC stands for electronic brake control.)

The 4WAL is a nonintegral system. Most installations have a vacuum power booster and separate master cylinder. EBC4 systems on Dodge diesel trucks are combined with Hydro-Boost power brake systems.

The system's EBCM is combined in one housing with the brake pressure modulator valve to form the electrohydraulic control unit, or ETC **(Figure 15)**.

Wheel speed sensor input signals can be separate signals from all four wheels or a three-channel input with separate front wheel signals and a combined signal for both

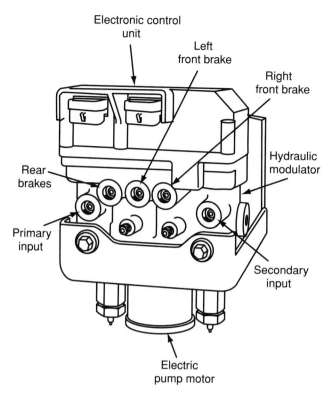

Figure 15.   The major components of a 4WAL system.

rear wheels. Earlier systems use four-wheel speed sensors and are called select-low systems, because the control module selects the rear wheel with the slowest speed during ABS braking to control the brakes of both rear wheels. Most later versions of this system use the vehicle speed sensor signal instead of two rear wheel sensors to control rear wheel braking. The system uses three isolation valves, three pulse-width-modulated (PWM) dump valves, a high-pressure accumulator for the front brakes and one for the rear, and a low-pressure accumulator for the front and one for the rear.

Later Kelsey-Hayes and Lucas-Varity Systems

In 1995, General Motors began using the EBC310 system. It operates in the same way as earlier three-channel 4WAL systems but has a more compact control unit. The same system was also used by Chrysler.

In 1998, Chrysler began using the Lucas-Varity EBC2 system to replace earlier RWAL systems as standard equipment on all Dodge trucks. (The Kelsey-Hayes Company was sold to a British firm, Lucas-Varity, and antilock systems built by the company since then are known as Lucas-Varity systems.) The EBC2 system is a one-channel, one-sensor rear-wheel system with a separate rear hydraulic control unit and a control module. The EBC2 system uses one isolation valve, one dump valve, and a single accumulator to modulate the rear brakes similarly to the operation of earlier RWAL systems.

The Lucas-Varity EBC325 four-wheel system is available as an option on all 1998 and later Dodge trucks. This system is an evolutionary development of earlier three-channel Kelsey-Hayes 4WAL systems.

BENDIX SYSTEMS

Through the years, many automobile manufacturers have used Bendix systems. What follows is a discussion of the most common designs.

Interesting Fact

Chrysler introduced a four-wheel ABS called "Sure-Brake" on some 1966 vehicles; but like other early attempts at ABS, it did not perform up to expectations and was not well received by the public. Chrysler then entered in a joint venture with Bendix and produced the first production four-wheel ABS as an option on the 1971 Chrysler Imperial.

Bendix Antilock-6

The Antilock-6 system is a diagonally split brake system with ABS components added to it. The system consists of a modulator, four-wheel speed sensors, and an antilock controller. The modulator **(Figure 16)** consists of an electric pump, two accumulators, a pressure differential valve, and six solenoid control valves from which the system takes its name—Antilock-6.

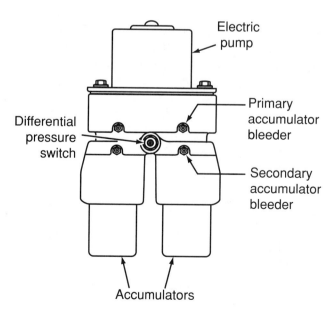

Figure 16.   The hydraulic modulator of a Bendix Antilock-6 system.

Bendix Antilock-9 and Antilock-10

These Bendix systems are integral, four-wheel, three- or four-channel ABS installations. The Antilock-10 system was used on various Chrysler passenger cars and mini vans. The Antilock-9 system was used on Jeeps **(Figure 17)**. Both consist of a brake system that uses the ABS pump and pressure controls for power assist, an electronic control unit (ECU), an integral master cylinder and booster assembly, a pressure modulator, pressure switches, an electric boost pump, two accumulators, a speed sensor at each wheel, and two warning lamps.

The systems derive their names from the number of solenoid valves in the hydraulic unit: nine in the case of Antilock-9 and ten in the case of Antilock-10. The hydraulic unit is attached to the master cylinder and power booster. In the Antilock-9 system, it controls the pressure to three channels: one for both rear brakes, one for the right front, and another for the left front. In the Antilock-10 system, it controls the pressure to four channels, one for each wheel.

The master cylinder is a diagonally split type. It provides fluid for braking as well as for the modulator during ABS operation. The power booster is a hydraulic type and is fed high-pressure fluid by an electric boost pump.

These systems use two accumulators, one inside the booster pump and the other next to the master cylinder. Both accumulators are charged with high-pressure nitrogen and store pressurized fluid at 1700–2000 psi for power brake operation. The booster accumulator is charged to 450 psi and the other is charged to 1000 psi.

Bendix LC4

The Bendix LC4 is Chrysler's designation for the Bendix III system, which replaced the Bendix Antilock-6 system in some Chryslers. The system is nonintegrated and uses a conventional vacuum booster and master cylinder that operate two diagonally split hydraulic circuits. The system uses four channels to control each wheel independently of the others.

Isolation valves are not used. During normal braking, the normally open build-and-decay solenoids let the hydraulic fluid pass from the master cylinder to each brake. During ABS braking, fluid flows to sump in the hydraulic modulator, where it is stored, then pumped into the system's two accu-

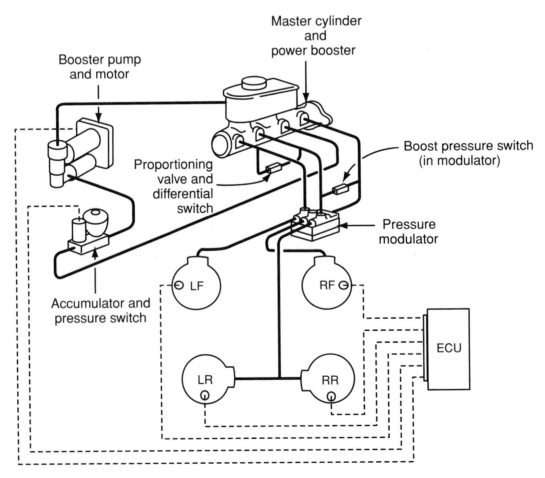

Figure 17. The major components of a Bendix Antilock-9 system.

mulators. The accumulators, which are spring loaded but not gas charged, provide pressurized hydraulic fluid as needed. Four shuttle assemblies regulate the build rate by restricting the flow of fluid during ABS stops.

Bendix ABX4

The Bendix ABX4 system differs from other ABS systems in several significant ways. There are four hydraulic channels, with the two rear channels controlled simultaneously to prevent yawing that might be caused by uneven deceleration of the rear wheels.

The hydraulic control unit is mounted beneath the master cylinder **(Figure 18)**. The ABS braking is accomplished through four decay solenoids and four shuttle valves, one of each for each brake. The decay solenoids reduce hydraulic pressure at the brake units.

Bendix Mecatronic

The Bendix Mecatronic is a nonintegrated, four-channel system used on some 1995 and later Ford vehicles. A version of the system includes traction control. The control module contains four solenoids for ABS, plus two more on systems with traction control. The control unit contains two separate microprocessors, one to monitor the wheel sensors and the other to control the solenoids and pump.

IMPORTED VEHICLE ABSs

Antilock brake system manufacturers for Asian and European vehicles include Bosch, Teves, Nissan, Nippondenso,

Sumitomo, Lucas-Varity, Bendix, and others. Most imported antilock systems are the nonintegral designs. Some imported vehicles have integral ABS, and the systems are typically variations of Bosch and Teves systems.

Toyota Rear-Wheel ABS

Unlike the Kelsey-Hayes RWAL systems, the Toyota system uses hydraulic pressure from the power steering system to apply additional brake pressure to the rear wheels during ABS braking. The Toyota system uses a single solenoid to isolate the two rear brakes in case of wheel lockup and then bleeds down the pressure to those brakes. When higher pressure is required for the rear brakes during an ABS stop, the power steering system acts as a hydraulic booster to increase pressure faster than the driver's foot can.

Sumitomo ABS

The Sumitomo ABS is a nonintegral, four-wheel, three-channel system based on Bosch designs and is similar in operation to the Bosch 2 systems. The three- and four-channel versions of the Sumitomo system have three or four solenoids, respectively, inside the hydraulic control assembly, or modulator **(Figure 19)**. The electric pump and the nitrogen-charged accumulator generate ABS pressure. On Honda and Acura models, the pump and accumulator also supply hydraulic pressure for power-assisted normal braking. The pump and accumulator are part of the hydraulic unit, except on Honda and Acura versions, where they are mounted separately. **Figure 20** is a simplified hydraulic diagram showing the basic operation of a Sumitomo system.

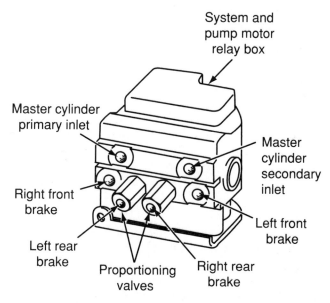

Figure 18. The hydraulic control unit for a Bendix ABX4 system.

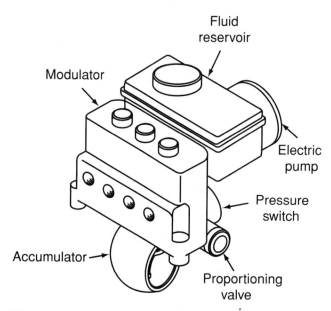

Figure 19. A typical hydraulic control unit or modulator used in Sumitomo systems.

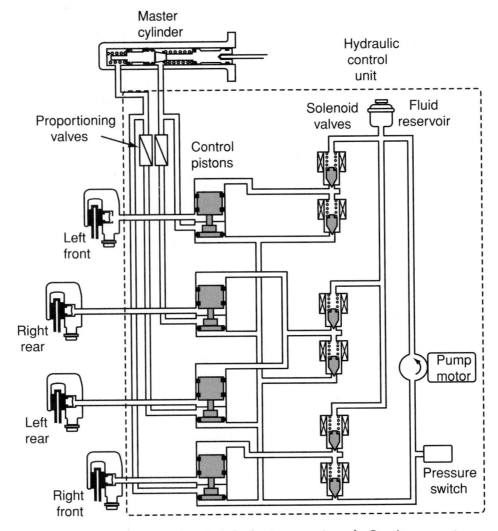

Figure 20.   A simplified hydraulic diagram showing the basic operation of a Sumitomo system.

Summary

- Most antilock systems currently in use are built by Bosch, Teves, Kelsey-Hayes, Delphi Chassis, Bendix, Nippondenso, Nissan, and Sumitomo.
- The Bosch 3 is an integrated four-channel system.
- Bosch 2, 2E, 2U, and 2S Micro ABS are nonintegrated, four-wheel systems. Most are three channel; some are four channel.
- Bosch ABS/ASR adds traction control, which Bosch calls acceleration slip regulation (ASR), to the ABS. ABS/ASR is a nonintegrated, three-channel, four-wheel system.
- Bosch 5 is a nonintegrated, four-channel, four-wheel ABS with traction control.
- The Teves Mark II is an integrated system that controls all four wheels with four sensors and three hydraulic channels, using the select-low principle at the rear.

- The Mark IV system is nonintegral and has four input channels and three output channels.
- The Delco Moraine ABS III system is a four-wheel integrated system with a PowerMaster-III booster and master cylinder assembly.
- The Delphi Chassis VI is a nonintegral system that uses a hydraulic modulator driven by three electric motors.
- All Kelsey-Hayes two-wheel systems (RABS and RAWL) are nonintegrated systems.
- The Kelsey-Hayes 4WAL brake system is a three-channel, nonintegral system.
- The Bendix Antilock-6, Antilock-9, and Antilock-10 systems are integral, four-wheel, three- or four-channel ABS installations.

- The Bendix ABX4 system has four hydraulic channels, with the two rear channels controlled simultaneously to prevent yawing that might be caused by uneven deceleration of the rear wheels.
- The Bendix Mecatronic system is a nonintegrated, four-channel system, and some versions include traction control.

- Most imported antilock systems are the nonintegral designs. Some imported vehicles do have integral ABS and most of these are variations of Bosch and Teves systems.
- The Sumitomo ABS is a nonintegral, four-wheel, three-channel system based on Bosch designs and is similar in operation to the Bosch 2 systems.

Review Questions

1. List at least four major ABS manufacturers.
2. RABS and RWAL systems control the _____-brakes in _____-_____.
3. An integral ABS typically incorporates the master cylinder into a single hydraulic control unit with the _____ _____, _____, _____, and _____.
4. Four-wheel antilock systems can provide either _____-channel control or _____-channel control.
5. Technician A says that the Bendix 9 ABS on Jeep vehicles has a speed sensor on each wheel. Technician B says that the Bendix 9 modulator contains nine solenoid valves. Who is correct?
   A. Technician A only
   B. Technician B only
   C. Both Technician A and Technician B
   D. Neither Technician A nor Technician B
6. Technician A says that the Kelsey-Hayes RABS and RWAL systems on Ford and Dodge trucks have the speed sensor in the differential and the sensor ring on the ring gear. Technician B says that on General Motors trucks with RWAL ABS, the speed sensor is in the trans-mission tail shaft housing and the sensor ring is on the transmission output shaft. Who is correct?
   A. Technician A only
   B. Technician B only
   C. Both Technician A and Technician B
   D. Neither Technician A nor Technician B
7. Technician A says that the Kelsey-Hayes RABS and RWAL antilock systems are all basically the same. Technician B says that the Kelsey-Hayes RABS and RWAL systems cannot provide acceleration traction control because they have no way to develop pressure except by the driver's foot on the brake pedal. Who is correct?
   A. Technician A only
   B. Technician B only
   C. Both Technician A and Technician B
   D. Neither Technician A nor Technician B
8. Technician A says that during ABS operation, solenoid valves release hydraulic pressure to the brakes. Technician B says that solenoid valves hold hydraulic pressure applied to the brakes. Who is correct?
   A. Technician A only
   B. Technician B only
   C. Both Technician A and Technician B
   D. Neither Technician A nor Technician B

Chapter 42

General Antilock Brake System Service and Diagnosis

Introduction

An ABS is more likely to experience basic brake problems than problems within its own control circuitry or components. The ABS will not work properly, if at all, if the brakes do not work properly in the nonantilock mode. Some ABS problems will be electrical, but many will result from a hydraulic system malfunction. When such problems are detected by the ABS control module, it will disable the antilock function and light the antilock indicator lamp on the instrument panel **(Figure 1)**. The cause can be as simple as a low hydraulic fluid level or a leaking hose, line, or connection.

BASIC ABS TESTING

The best way to diagnose an ABS is to approach solving the problem in a logical way. To do this, you should follow these seven steps, in order:
1. Verify the customer's complaint.
2. Conduct preliminary inspections and checks.
3. Check all service information for information about the complaint and the system, including service bulletins and recall notices.
4. Follow the diagnostic procedures outlined in the service manual for the specific complaint.
5. Interpret and respond to all diagnostic codes.

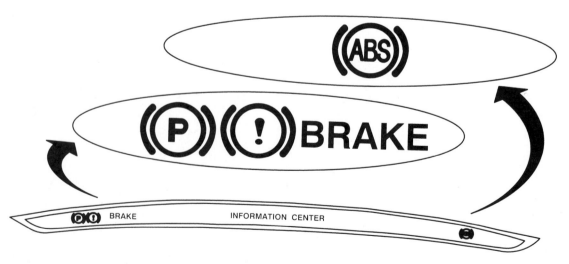

Figure 1. When the ignition switch is turned to the on position, the ABS and brake lamps should light for a short time.

6. Define and isolate the cause of the complaint or problem.
7. Fix the problem and verify the repair.

VERIFY THE CUSTOMER'S COMPLAINT

It is important that you totally understand what the complaint or problem is before you venture in and try to find the cause. This is the purpose of the first four steps. This may include an interview or road test with the customer to thoroughly define the complaint and to identify when and where the problem occurs.

Listen to the customer's complaint carefully and get as much information as possible. Avoid asking closed-end questions, such as: "Does it happen when the engine is hot?" Ask open-ended questions, such as: "What temperature is the car when this happens?"

Finally, check the vehicle yourself to verify that the problem exists as described. Try to re-create the conditions that the customer describes. You might not always be able to duplicate the conditions exactly, but try to come as close as possible. You might have to road test the vehicle or you might have to leave it standing overnight to re-create a cold-operating problem. During the test drive, you might also note another symptom that the customer missed. Any information you can gather will help.

BASIC INSPECTION AND VEHICLE CHECKS

Because many ABS problems are caused by the basics, it is wise to conduct all of the preliminary checks required for normal, hydraulic brake systems. Plus, you should conduct a thorough inspection of the electronic system. Whenever diagnosing a brake system, remember that suspension, steering, wheel, and tire problems will cause the brakes to operate, or appear to operate, abnormally.

Look for obvious faults and try to eliminate simple problems first. Look for loose or broken wires, connectors, and hydraulic lines or hoses. Check for leaks. Check for mechanical and electrical tampering and collision damage.

Troubleshooting with the Brake Pedal

You can determine the probable cause of a wide variety of brake system problems by identifying apparent symptoms as you apply the pedal. **Figure 2** lists symptoms and probable causes that can be determined in this way.

Check the System's Warning Lamp

Vehicles with ABS have an amber, or yellow, instrument panel lamp that lights to indicate major system problems. The ABS lamp should light when the ignition key is turned on without starting the engine. This is a basic bulb check similar to the bulb check for alternator or other brake sys-

tem warning lamps. The ABS lamp might go out within one or two seconds of turning on the ignition or it might stay lit for a longer time. If the lamp does not light when the ignition is turned on, the computer probably will not go into its self-test mode. The problem might be as simple as a burned out bulb or it can be a problem with the computer itself.

To identify the cause for the bulb not illuminating, begin by checking the bulb. Then test the lamp circuit for correct power and ground. Follow the manufacturer's instructions for checking the lamp to verify that the system can perform its self-test functions.

If the warning lamp lights steadily with the engine running, it indicates that a system problem exists. The light might indicate that the computer has set a trouble code. On many vehicles, some system problems will set a trouble code in the computer memory but will not light the warning lamp.

If the warning lamp does not light with the key on or if the computer will not go into a diagnostic mode, you might need to conduct some voltage tests at a diagnostic connector. Nearly all diagnostic connectors have a ground terminal that is used for one or more test modes. Use a voltmeter or ohmmeter to check the continuity between the diagnostic ground terminal and the battery negative terminal. High ground resistance or an open circuit can keep the computer out of the self-test mode and can be a clue to other system problems.

Various other terminals on the diagnostic connector might have other levels of voltage applied to them at different times. Some might have battery (system) voltage under certain conditions, whereas others can have 5 volts, 7 volts, or a variable voltage applied to them.

> **You Should Know** *Dirty or damaged wheel speed sensors and damaged sensor wiring harnesses are leading triggers to turn on ABS warning lamps. Do not rush to condemn the ABS computer or hydraulic module before checking the speed sensors and tone wheels.*

SERVICE INFORMATION

Often accurately defining the problem and locating related information in TSBs and other materials can reveal the cause of the problem. No manufacturer makes a perfect vehicle, and as the manufacturer recognizes common occurrences of a problem, it will issue a statement regarding the fix of the problem. Also, for many diagnostic trouble codes (DTCs) and symptoms, service manuals will give a simple diagnostic chart or path for identifying the cause of

Troubleshooting at the Brake Pedal

SYMPTOM	PROBABLE CAUSE
Pedal surging, brake chatter, vehicle surge during braking	Front discs out of round; excessive disc thickness variation; bearings out of adjustment
	Rear drums out of round; hard spots caused by overheating
Brakes grab	Hard spots on front discs or rear drums; cracked pads or shoe linings
Car pulls to one side	Misaligned front end; drum brake components malfunctioning; frozen caliper pistons or contaminated front brake pads; pinched lines or leaking seals
Excessive pedal effort	Insufficient engine vacuum; defective booster; vacuum leak; frozen piston; contaminated or glazed linings
Rear brakes drag	Misadjusted parking brake; rear brakes out of adjustment; weak shoe return springs; frozen wheel cylinder pistons
All brakes drag	Frozen brake pistons; misadjusted stoplamp switch; restricted pedal return; defective master cylinder; contaminated brake fluid
Low-speed disc brake squeak	Worn pad linings
Scraping noise when brakes are applied	Brake linings completely worn out
Intermittent chirp when drum brakes are applied	Insufficient backing plate pad lubricant
Intermittent clunk when drum brakes are applied	Threaded drums
Rear wheel lockup	Contaminated linings; front calipers frozen; defective combination valve
Pedal low and spongy with excessive pedal travel	Insufficient fluid in system; air in hydraulic system
Pedal low and firm with excessive pedal travel	Brakes out of adjustment
Brakes release slowly and pedal does not fully return	Frozen caliper or wheel cylinder pistons; defective drum brake return springs; binding pedal linkage
Brakes drag and pedal does not fully return	Contaminated brake fluid; defective master cylinder; defective vacuum booster or vacuum check valve; binding pedal linkage or lack of lubrication

Figure 2. Use the feel of the brake pedal to help you diagnose brake problems.

the problem. These are designed to be followed step by step and will lead to a conclusion if you follow the path matched exactly to the symptom. Check all available information before moving on in your diagnostics.

Sometimes the symptom will not match any of those described in the service manual. By eliminating those circuits and components that are working correctly from the list of possible causes of the problem, you can identify what might be causing the problem and what should be tested further.

TROUBLE CODES

Self-test programs and diagnostic modes operate differently on different vehicles, but all provide the same basic type of information. Most ABS computers allow a scan tool to read system trouble codes.

It is important to remember that diagnostic trouble codes (DTCs) can be set by out-of-range signals. This does not mean that the sensor or sensor circuit is bad. It could mean that the sensor is working properly but there is a mechanical or hydraulic problem causing the abnormal signals. Problems such as loose connections, broken wires, corrosion, and poor grounds will affect the signals in that circuit.

If the computer recognizes a condition that is not right, it records a DTC. A DTC is a two-, three-, four-, or five-digit numeric or alphanumeric code that represents a particular problem. Most systems illuminate the ABS warning lamp for many, but not all, trouble codes. Additionally, the electronic control module (ECM) for most ABSs will store the code in long-term memory.

A DTC can indicate a problem in a particular circuit or subsystem, but it does not always pinpoint the exact cause of the problem. Checking DTCs is an overall, or area, test of the system. Manufacturers have different names for codes and categorize them in different ways.

With the introduction of the second-generation on-board diagnostic systems (OBD II) for engine control systems, the term "diagnostic trouble codes" has been widely applied to codes for all automotive control systems. Two general terms we can use for all codes are hard and soft codes.

A **hard code** indicates a failure that is present at the time of testing and will be permanent until you fix it. If you turn off the ignition and clear the codes, a hard code will reappear immediately or within a few minutes because the problem still exists until you correct it. A **soft code**, or memory code, indicates an intermittent problem: one that comes and goes. Soft codes are the computer's way of remembering a problem that occurred some time in the past, before testing, but that is not present now. The problem might not reappear if you clear the codes and retest the system.

Because soft codes indicate intermittent problems, diagnostic charts and pinpoint tests usually do not immediately isolate the problem. To identify the cause of problems that cause soft codes, you should not open or disconnect electrical connectors until you have checked them in normal operation or by doing a wiggle test. Disconnecting and reconnecting a connector might temporarily solve a problem without revealing the basic cause.

After retrieving trouble codes, write down any codes that are present. Remember that if you clear the codes, soft codes might not reappear right away. Some antilock systems will display only one code at a time, even when several faults exist. On these systems, you must repair and clear each code in sequence and then retest the system until no more codes appear.

If you determine that a code is hard, check the service manual for pinpoint tests to identify the cause of that code. If you determine that a code is soft, use the intermittent diagnosis procedures to help pinpoint the problem. All manufacturers recommend that trouble codes be diagnosed and serviced in a basic order: hard codes first, followed by soft codes.

OPERATING RANGE TESTS

The signal from an analog sensor can drift out of range as the sensor ages or wears. Some sensors can develop an erratic signal, or dropout, at one point in the signal range. A loose or corroded ground connection for a sensor also can force the signal out of limits.

These and similar problems can cause definite problems without setting a code. You can check the operation of many sensors, however, by using the operating range charts provided by the manufacturers. These charts list signal range specifications for voltage, resistance, frequency, or temperature that the sensor provides under varying conditions.

Use a digital volt-ohmmeter (DVOM), a frequency counter, or other appropriate instrument to test the sensor signal at the sensor connector and, if necessary, at the main connector to the computer. You can back probe many sensor connectors or install jumper wires to provide connection points for your meter. You might need a breakout box or harness to check sensor signals at the main computer connector. If possible, operate the sensor through its full range and check the signal at several points.

Use Computer Pin Voltage Charts

A computer pin voltage, or pinout, chart identifies all the connector terminals at the main computer connector by number, circuit name, and function. The voltage or resistance levels that should be present under various conditions also are often listed. Some circuits have different voltage specifications with the key on and the engine off,

Antilock brake control module connector (end view)

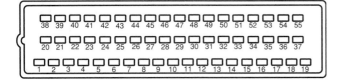

EXAMPLES

Pin number	Circuit	Circuit function
1	530 (LG/Y)	Ground
2	498 (PK)	ABS valve assembly
3	532 (O/Y)	ABS power relay
4	—	Not used
5	549 (BR/W)	ABS pedal sensor switch
6	—	Not used
26	535 (LB/R)	ABS switch No. 2
27	524 (PK/BK)	RR brake sensor-LO
28	519 (LG/BK)	LR brake sensor-LO
29	516 (Y/BK)	RF brake sensor-LO
30	522 (T/BK)	LF brake sensor-LO
31	462 (P)	Pump motor speed

Figure 3.   A segment of a pin voltage chart from a service manual. These are used to test circuit functions in an ABS circuit.

during cranking, and when the engine is running. Use the pin voltage charts **(Figure 3)** to check input and output signals at the computer. Checking signals at the computer is closely related to sensor operating range tests.

Check Ground Continuity

With the ignition switch on, the circuit energized, and current flowing, use a digital voltmeter to check the voltage drop across the main computer ground connection and across the ground connection of any sensor that you think might be causing a problem **(Figure 4)**. Low-resistance ground connections are critical for electronic control circuits.

Voltage drop across the ground connection for any electronic circuit should be 0.1 volt or less. The voltage drop across a high-resistance ground connection in series with a sensor circuit reduces the signal voltage of the sensor. This ground resistance can offset the signal voltage enough to cause serious problems. For example, a 0.5-volt drop across the ground connection on a sensor that operates on a 5-volt reference equals a 10 percent measurement error.

TROUBLESHOOTING INTERMITTENT PROBLEMS

The cause for soft codes or intermittent problems can be the hardest to diagnose. If you are lucky, the intermittent

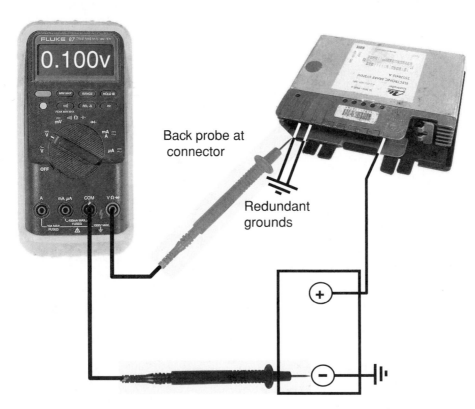

Figure 4.   With current flowing through the circuit, check the voltage drops at the various grounds of the circuit. There should never be more than 0.100 volt dropped across any ground circuit.

problem will set a soft code. This gives you a clue, at least, about the general area in which to start testing. You might need to simulate the conditions that caused the problem or road test the vehicle to catch the intermittent fault.

Wiggle Tests and Special Actuator or Sensor Tests

Most control systems have long-term memory that will record soft codes. A wiggle test puts the computer and a scan tool in communication so that the scan tool will indicate when a soft code is being set. Tap or wiggle the system's wiring and connectors in an attempt to cause the problem. If the problem occurs, remember what you did when the code was set. Use the scan tool to read the codes from the car again to verify the fault.

Check Connectors for Damage

Many intermittent problems are caused by damaged connectors and terminals. Unplug the connectors in the problem circuit and inspect them carefully for bent or broken terminals; corrosion; terminals that have been forced back in the connector shell, causing an intermittent connection; and loose, frayed, or broken wires in the connector shell. Most connectors on vehicle control systems are repairable, and, often, spraying a small amount of electronic contact cleaner on the terminals will help to remove corrosion and moisture.

> **You Should Know** *Never use spray electrical contact cleaner on a harness connector unless the wiring harness has been disconnected from the ECM. Electrostatic discharge (EDS) can damage the ECM.*

Road Test and Record Data

If the vehicle's ECM is capable of transmitting data during a road test (and most ABS computers can), drive the vehicle and try to duplicate the problem. Use the snapshot or data-recording function of the scan tool to record and store all computer activity when the problem occurs. Then analyze the data in the shop to try to locate the cause of the problem.

ABS HYDRAULIC BRAKE SYSTEM SERVICE

An integrated ABS with a pressurized accumulator is similar to a Hydro-Boost power brake system in that it uses

power steering fluid to operate both the steering gear and the brake booster. In both designs, the hydraulic fluid is under very high pressure. Opening a hydraulic line in this system without first discharging the pressure can result in a messy and potentially dangerous situation.

Depressurizing the System (Relieving Accumulator Pressure)

To depressurize a typical system, turn the ignition switch off and then disconnect the negative cable at the battery. Pump the brake pedal at least twenty-five times with about 50 pounds of pedal force. As the accumulator pressure discharges, you will notice a change in pedal feel. When you feel an increase in pedal effort, pump the pedal a few more times. This will remove all hydraulic pressure from the system. Some systems may require up to fifty pedal applications to relieve accumulator pressure completely.

Fluid Level Check and Refill

Some systems require pressurizing or depressurizing the system before checking and refilling the fluid reservoir. An example of needing to pressurize the system is the early Teves Mark II system. Turn the ignition on and pump the brake pedal until the hydraulic pump motor starts. When the pump shuts off, you can visually check the level through the translucent reservoir. Under certain conditions, the fluid level might be above the max fill line on the reservoir. In this case, shut the ignition off and then turn it back on. Pump the brake pedal again to start the hydraulic pump. When the pump stops, the fluid level should be accurate. If the fluid is below the max fill line **(Figure 5)**, remove the cap and add enough fluid to bring it to the correct level.

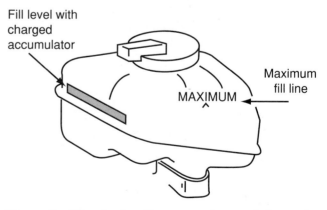

Fill level with charged accumulator

Maximum fill line

MAXIMUM

Figure 5. The hydraulic reservoir used in early Teves Mark II systems has two fill marks. One indicates the normal level and the other the maximum fill amount.

To check the fluid level on later Teves systems, depressurize the system and look at the fluid through the translucent reservoir. The level should be at the full mark. With the system pressurized, the level will be somewhere below the full mark.

Never overfill the reservoir in any system. This will cause the fluid to overflow when the accumulator discharges during its normal operation.

BLEEDING THE SYSTEM

Bleeding an ABS is fundamentally the same as bleeding a non-ABS hydraulic system. However, some systems require additional steps. Always check the manufacturer's instructions before bleeding a system, even if you have worked on a similar system before. Some manufacturers suggest either pressure bleeding or manual bleeding for their various ABS designs. Others specify either one method or the other.

> **You Should Know** *If a system requires that a high-pressure accumulator be charged to bleed the brakes, follow the manufacturer's instructions and precautions exactly when working with hydraulic components.*

When bleeding an ABS hydraulic system, it is good practice to flush the system completely to ensure that all old fluid and possible contamination are removed. The following paragraphs summarize the special instructions for bleeding some of the most common systems but should not be substituted for the manufacturers' specific procedures. Before bleeding any ABS, correct any conditions that would set diagnostic trouble codes and clear all codes from the computer's memory.

> **You Should Know** *If a brake-pull problem develops after bleeding an ABS, completely flush and bleed the system again. If sludge or dirt of any kind gets trapped in ABS solenoids or valves, it can unbalance hydraulic pressure at the wheels—even for non-ABS braking. Flushing the dirt out of the system often solves an ABS pulling problem.*

Bendix 9 (Jeep)

The Bendix 9 system can be bled manually or with pressure equipment. The entire system, including the accumulator, pump, and master cylinder, must be bled if any hydraulic connection is opened. To bleed these, loosen the fittings, one at a time, on the side of the hydraulic modulator while an assistant holds steady pressure on the brake pedal. Bleed the accumulator, pump, and master cylinder before bleeding the wheel brakes.

To bleed the wheel brakes, turn the ignition off and pump the brake pedal thirty to forty times to exhaust accumulator pressure. Leave the ignition off and the accumulator depressurized while bleeding the brakes.

Bendix 10

The Bendix 10 system can be bled with pressure equipment or manually. Before bleeding the wheel brakes, turn the ignition off and pump the brake pedal thirty to forty times to exhaust accumulator pressure. Leave the ignition off and the accumulator depressurized while bleeding the brakes.

Bendix 6

If the hydraulic modulator is not removed or otherwise exposed to air, the bleeding procedure for the Bendix 6 system is the same as for the Bendix 10 system. If air has entered the modulator, special procedures beyond the scope of this text are needed to purge the air from the system.

Bendix LC4

The Bendix LC4 system can only be manually bled. Pressure-bleeding equipment does not produce pressure high enough to remove all air from the system. Bleeding the LC4 system is a two-stage process that requires the use of a scan tool. Follow the instructions given with the scan tool.

Bendix ABX-4

Unless air has entered the modulator, you can manually or pressure bleed the Bendix ABX-4 system as you would a nonantilock system. However, Bendix 4 and ABX-4 systems are best bled manually. If air has entered the hydraulic modulator, bleed the wheel brakes first and then bleed the modulator, using a scan tool to control the operation of the system.

After bleeding the modulator, repeat the bleeding operations at all four wheels. To thoroughly purge these systems, however, you also must cycle certain solenoid valves while holding down the brake pedal.

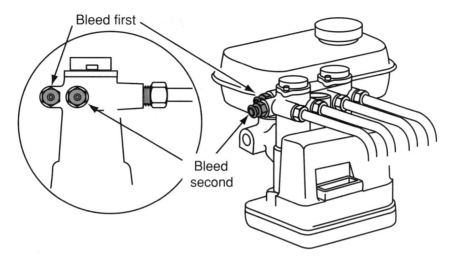

Figure 6.   Bleed these two points on the ABS VI modulator before moving to the rest of the system.

Bosch 2, 2E, 2S, 2U, Micro, and ABS/ASR

The Bosch 2 systems can be bled either manually or with pressure bleeding methods, just as you would a nonantilock system. Because this design has been used by many manufacturers for many years it is best to check the service manual for the exact bleeding procedures, as they vary with year and model.

Delco Moraine III

Before bleeding the wheel brakes, turn the ignition off and depressurize the accumulator. Manually bleed the front brakes. To bleed the rear brakes, turn the ignition on to let the pump recharge the accumulator. Have an assistant press the brake pedal slowly while you open the right rear bleeder screw. Hold the bleeder screw open for approximately fifteen seconds while your assistant maintains pedal pressure. Then close the bleeder screw, release pedal pressure, refill the fluid reservoir, and repeat the procedure at the left rear wheel.

Delphi Chassis (Delco Moraine) ABS-VI

To bleed an ABS-VI system, use a scan tool to move the pistons in the front and rear modulators to their upper positions to unseat the check balls in the hydraulic circuits. If the appropriate scan tool is not available, the pistons can be manually moved. Start the engine and run it for ten seconds. Verify that the ABS lamp is off and listen for several sharp clicks that indicate the pistons are being driven to their upper positions. If you do not hear these distinct clicks, drive the car at 5 mph and again listen for the clicking of the ABS control motors. When you hear the clicks, care-

fully drive the car to the service bay without activating ABS operation.

Before bleeding the wheel brakes, bleed the modulator at its rear bleeder screw **(Figure 6)**. Then repeat this procedure at the forward bleeder screw. After bleeding the modulator, loosen all of the brake line fittings on the outboard side of the modulator and manually bleed air from the upper ends of the brake lines **(Figure 7)**.

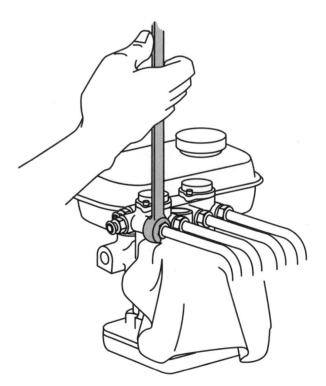

Figure 7.   Bleed air from the brake line fittings on the side of the ABS VI modulator.

The last stage of bleeding an ABS-VI system is to bleed the wheel brakes, either manually or with pressure equipment. Always refer to the appropriate service manual for the specific wheel bleeding sequence.

Kelsey-Hayes EBC2 (RABS and RWAL)

These rear-wheel ABSs can be bled manually or with pressure equipment in much the same way as a nonantilock system. On most Kelsey-Hayes systems, the combination valve must be held open for pressure bleeding. The modulator must be bled separately, if air has entered it. Some modulators have bleeder screws; others require that a brake line fitting be loosened for bleeding.

Kelsey-Hayes EBC410

If no air has entered the hydraulic lines of this ABS design, non-ABS manual and pressure bleeding methods can be used on this system. If air has entered the ABS valve assembly, special bleeding procedures that require a scan tool must be used.

Teves Mark II

You can bleed a Teves Mark II system manually or with pressure equipment. The front brakes can be bled with or without system accumulator pressure, but the accumulator must be fully charged to manually bleed the rear brakes. Because of the high pressure at the bleed screws, you must be very careful when opening a screw to bleed the line. Within ten seconds after opening the screw, the fluid flow should be free of air bubbles and the screw can be closed. Refill the reservoir after bleeding each wheel.

Teves Mark IV

Standard non-ABS manual or pressure bleeding methods can be used for this system. Unlike the Teves Mark II system, the accumulator of the Mark IV system must be depressurized before bleeding the rear brakes. Proper bleeding for these systems consists of first manually bleeding the base brakes, then bleeding the hydraulic control unit, and finally repeating a manual base brake bleeding sequence.

Summary

- An ABS will not work properly if the brakes do not work properly in the nonantilock mode.
- When system problems are detected by the ABS control module, it will disable the antilock function and light the antilock indicator lamp on the instrument panel.
- Before suspecting and testing the ABS, thoroughly check the base brake system.
- When the ignition switch is turned on, both lamps should light. After a short period of time, the ABS lamp should turn off, as should the brake lamp if the parking brake is not engaged.
- The feel and behavior of the brake pedal can help to determine the probable cause of a wide variety of brake system problems.
- Begin troubleshooting the ABS by getting as much information as possible about the problem from the customer.
- Inspect all mechanical components, vacuum and hydraulic lines, and the wiring of the system.
- Check the vehicle to verify that the problem exists as described by the customer.
- An ABS ECM will record a DTC if it recognizes a condition that is not right.

- A hard code indicates a failure that is present at the time of testing and will be permanent until it is corrected.
- A soft code, or memory code, indicates an intermittent problem, one that comes and goes.
- The operation of many sensors can be checked by using the operating voltage, resistance, frequency, and temperature range charts provided by the manufacturers.
- All ground connections and circuits should be checked because a bad ground will cause incorrect signals from the sensors.
- Perform a wiggle test and a thorough inspection of the wires and connectors to help identify the cause of an intermittent electrical problem.
- Some systems must be depressurized before checking brake fluid levels, bleeding the brakes, and other procedures.
- Bleeding an ABS is fundamentally the same as bleeding a non-ABS hydraulic system. However, some systems require additional steps, and those steps should be identified in the service manual before proceeding.

Review Questions

1. Besides indicating ABS faults for some systems, the red brake warning lamp on the instrument panel can indicate failure of the _____ brake system, application of the _____-brake, or low fluid level in the _____.

2. When road testing a car with ABS, several _____ should be felt through the brake pedal during heavy braking.

3. Technician A says that an ABS problem can cause the electronic control module to shut off, or inhibit, the system. Technician B says that a loss of hydraulic fluid or power booster pressure will disable the ABS. Who is correct?
   A. Technician A only
   B. Technician B only
   C. Both Technician A and Technician B
   D. Neither Technician A nor Technician B

4. Before working on the brakes of a vehicle with antilock brakes, Technician A relieves system pressure by slightly opening a bleeder screw and allowing fluid to spray into a wide-mouth container; Technician B turns off the ignition and pumps the brake pedal thirty to forty times until she feels an increase in pedal pressure. Who is correct?
   A. Technician A only
   B. Technician B only
   C. Both Technician A and Technician B
   D. Neither Technician A nor Technician B

5. Bleeding an ABS is being discussed. Technician A says that cracking open the brake line connections at the hydraulic module and bleeding the air into a shop rag is recommended procedure for some systems. Technician B says that pressure bleeding is required for bleeding ABSs. Who is correct?
   A. Technician A only
   B. Technician B only
   C. Both Technician A and Technician B
   D. Neither Technician A nor Technician B

6. Technician A says that it is normal for the amber antilock lamp to light when the parking brake is engaged and the engine is running. Technician B says that it is normal for the red brake lamp to light when the ignition is first turned on and the engine is off. Who is correct?
   A. Technician A only
   B. Technician B only
   C. Both Technician A and Technician B
   D. Neither Technician A nor Technician B

Chapter 43

Testing Common Antilock Brake Systems

Introduction

Each ABS has its own diagnostic procedure involving use of the brake and antilock warning lamps, special testers, breakout boxes, scan tools, troubleshooting charts, and wiring diagrams. You must have the manufacturer's procedures for the year, make, and model of the vehicle you are testing. Systems vary in the number of codes they can store and whether they will retain the code in memory when the ignition is turned off.

> **You Should Know** *Before driving a vehicle with an ABS complaint, especially if the red brake warning lamp is lit, test the brakes at low speed to make sure that the car will stop normally. An illuminated red brake warning lamp can indicate reduced braking ability.*

GENERAL MOTORS ABS TESTING

On most 1988 and later General Motors systems, the ABS control module will illuminate an ABS indicator lamp on the instrument panel if a problem occurs in the system. Most systems will store trouble codes and flash them on the warning lamp or transmit them to a scan tool. Many systems also can transmit system operating data to a scan tool. Late-model General Motors vehicles might not have a second brake warning lamp. The red lamp is used for the basic brake system and the ABS. The only indication of an ABS fault is shown by the intensity of the warning lamp.

Because of the great variety of systems used by General Motors, ABS testing usually requires that you enter the vehicle identification number (VIN) into the scan tool **(Figure 1)**. The scan tool might ask additional questions before you can begin testing.

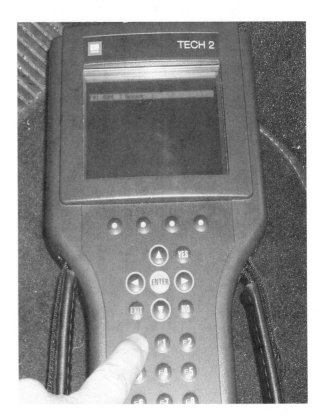

Figure 1. Normally, the vehicle's VIN must be entered into the scan tool before diagnostics can begin.

Follow these steps to connect a scan tool to the vehicle for ABS testing:

1. Turn the ignition off.
2. Select the correct scan tool adapter for the vehicle being tested and attach it to the data communication cable using the two captive screws.

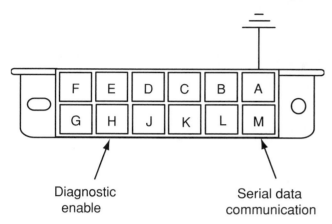

Figure 2.   A typical ALDL connector.

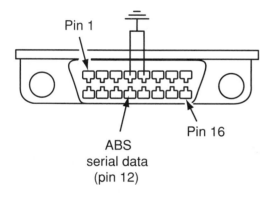

Figure 3.   A typical DLC.

3. Attach the power cable to the test adapter and connect the power cable to the vehicle. (Skip this step for OBD II adapter connections.)
4. Connect the communication cable with the adapter to the vehicle's test connector. The connector on early-model General Motors vehicles is called the assembly line data link (ALDL). Typically, the ALDL connector is under the instrument panel, to the right or left of the steering column **(Figure 2)**. On late-model vehicles, the 16-pin test connector is called the data link connector (DLC) and is the same connector used for OBD II engine control system testing **(Figure 3)**. For either connector, the tester's cable adapter is keyed so that it fits into the connector in only one way. Be sure it is installed securely.

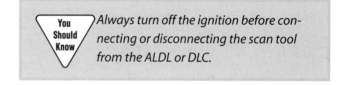 *Always turn off the ignition before connecting or disconnecting the scan tool from the ALDL or DLC.*

5. Turn on the scan tool and then enter all information requested by the tool.
6. Turn the ignition on before selecting the desired test from a scan tool's test menu.
7. You can now choose data list mode to monitor system conditions such as wheel speed sensors and brake switch input. Trouble codes may be viewed **(Figure 4)** or, if required, the special functions mode will operate ABS components such as pumps and relays. The snapshot mode can be used to capture a fault as it occurs during a test drive.
8. After the problem has been isolated and repaired, clear all stored trouble codes and take the vehicle on a road test.

ABS code	System	ABS code	System	ABS code	System
11	EBCM	36	RF WSS	54	Rear Outlet Valve
12	EBCM	37	RR WSS	55	LF WSS
21	Main Valve	38	LR WSS	56	RF WSS
22	LF Inlet Valve	41	LF WSS	57	RR WSS
23	LF Outlet Valve	42	RF WSS	58	LR WSS
24	RF Inlet Valve	43	RR WSS	61	EBCM 'Loop' Circuit
25	RF Outlet Valve	44	LR WSS	71	LF Outlet Valve
26	Rear Inlet Valve	45	2 Sensors (LF)	72	RF Outlet Valve
27	Rear Outlet Valve	46	2 Sensors (RF)	73	Rear Outlet Valve
31	LF WSS	47	2 Sensors (Rear)	74	Rear Outlet Valve
32	RF WSS	48	3 Sensors	75	LF WSS
33	RR WSS	51	LF Outlet Valve	76	RF WSS
34	LR WSS	52	RF Outlet Valve	77	RR WSS
35	LF WSS	53	Rear Outlet Valve	78	LR WSS

Figure 4.   Examples of some of the ABS-related DTCs that can be displayed on non–OBD II General Motors vehicles.

Some General Motors vehicles have ABS systems that cannot transmit codes or other test data to a scan tool. Most of these vehicles, however, are able to indicate system

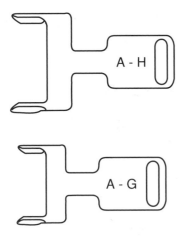

Figure 5. An ALDL A-to-H shorting key for General Motors AWAL systems (*top*) and an A-to-G shorting key for Teves Mk II systems (*bottom*).

problems by flashing the ABS warning lamp to display trouble codes. Some vehicles require a dedicated ABS tester for diagnostics, whereas a few can neither transmit codes nor flash them on an instrument panel lamp.

Flashing ABS Trouble Codes on a General Motors ABS Lamp

The Teves Mk II and 4WAL systems are examples of General Motors antilock systems that can be made to flash DTCs on the antilock warning lamp. Connect a jumper wire or insert a special shorting key between terminals H and A of the ALDL connector (**Figure 5**), and then turn on the ignition and count the flashes on the warning lamp.

General Motors ABS Trouble Codes and Data Guidelines

ABS trouble codes and operating data are available for some Bosch, Delco Moraine (Delphi Chassis), Kelsey-Hayes 4WAL and RWAL, and Teves IV systems on General Motors vehicles. Most 1994 and later General Motors antilock systems use a four-digit DTC format similar to OBD II engine and emission DTCs (**Figure 6**). When troubleshooting an

DIAGNOSTIC TROUBLE CODE AND SYMPTOM TABLE	
CHART	**SYMPTOM**
A	ABS (amber) indicator light "ON" constantly; no DTCs stored
B	ABS (amber) indicator light "ON" intermittently; no DTCs stored
C	ABS (amber) indicator light "OFF" constantly; no DTCs stored
D	Scan tool displays undefined DTCs
DIAGNOSTIC TROUBLE CODE	**DESCRIPTION**
A011	ABS indicator lamp circuit open or shorted to ground
A013	ABS indicator lamp circuit shorted to battery
A014	Enable relay contacts or fuse open
A015	Enable relay contacts shorted to battery
A016	Enable relay coil circuit open
A017	Enable relay coil circuit shorted to ground
A018	Enable relay coil circuit shorted to battery or coil shorted
A021	Left front wheel speed = 0
A022	Right front wheel speed = 0
A023	Left rear wheel speed = 0
A024	Right rear wheel speed = 0
A025	Left front excessive wheel speed variation
A026	Right front excessive wheel speed variation
A027	Left rear excessive wheel speed variation
A028	Right rear excessive wheel speed variation
A036	Low system voltage

Figure 6. An example of General Motors ABS DTCs in a format similar to OBD II engine and emission DTCs.

ABS problem, you must refer to the manufacturer's test procedures for the specific vehicle.

FORD ABS SCAN TOOL TESTING

When an ABS malfunction occurs, the ECM of Ford Motor Company products will light the ABS warning lamp on the instrument panel. In some systems, the module will store service codes in memory for most system malfunctions. The scan tool can read these codes through the vehicle's ABS test connector or OBD II data link connector.

A 1994 or later vehicle with OBD II diagnostic capabilities has a 16-pin DLC inside the passenger compartment (usually under the instrument panel, near the steering column). On pre-1994 Ford vehicles with ABS, the diagnostic connector is the unique Ford self-test connector, identified for ABS use. This test connector can be located in the trunk but is most often in the engine compartment **(Figure 7)**. **Figure 8** shows the test terminals of the test connector.

Some late-model ABS installations provide data stream information. A scan tool can display this information. The types of scan tool tests vary, depending on the model and year. In addition, there may be systems that a scan tool cannot test.

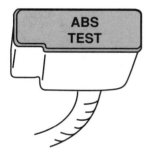

Figure 7. A typical Ford ABS test connector is identified by the labeling of "ABS TEST" on the cover of the connector.

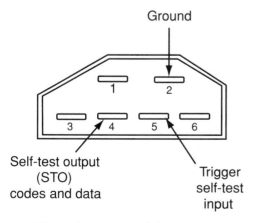

Figure 8. These three pins of the test connector are used for scan tool communication.

Ford ABS Data

In the ABS data diagnostic mode, the scan tool displays system operating data available on the ABS data stream. The ABS data selection operates similarly to data test selections for engine control testing. The data selection from the menu of a scan tool requires that the tool be able to communicate with the ABS control module. To ensure efficient communication with the ABS module, follow these steps when entering and exiting the ABS trouble codes and operating data test selections:

1. Be sure that the ignition is off when entering vehicle identification.
2. Turn the ignition on before selecting ABS codes and data from the scan tool menu.
3. Turn the ignition off after completing the ABS testing.

Once the scan tool establishes communication, the data display will appear. If it does not, disconnect the scan tool and check the vehicle's test connector for damaged terminals and open wiring. The cause of the problem can be as simple as a blown fuse or wiring fault on the vehicle. In some cases, lack of communication can indicate an ABS problem. Other causes are more common, however.

Ford ABS Service Codes

Ford refers to ABS service codes as on-demand codes and continuous codes. On-demand codes are hard codes that occur during a key on–engine off (KOEO) self-test. Continuous codes are memory codes from the ABS control module. These indicate intermittent problems that have occurred in the past, during normal vehicle operation.

On most late-model systems, continuous memory codes are read by a separate test selection on the scan tool menu. Some earlier systems transmit memory codes automatically at the end of the KOEO self-test.

Key On–Engine Off Tests

The KOEO test displays the hard codes present in the computer's memory. These are usually electrical problems, such as open and short circuits, and they must be corrected before any memory codes. On some early-model ABSs, the KOEO test also displays continuous memory codes of intermittent faults. For these vehicles, the memory codes mode selection does not appear on the scan tool's ABS test menu. Memory codes should be serviced after any other hard codes generated during the KOEO self-test.

Ford Truck RABS Service Codes

The rear ABSs used on many older Ford light trucks did not provide scan tool access but can flash service codes on the instrument panel's rear ABS warning lamp. To display codes on the warning lamp, locate the RABS diagnostic connector under the instrument panel **(Figure 9)**. Turn the ignition on, disconnect the connector, and ground the RABS module side of the harness for only one to two sec-

REAR OF INSTRUMENT PANEL

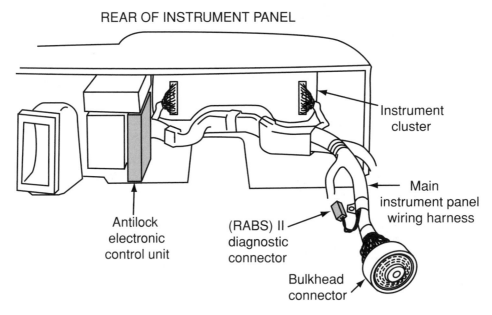

Instrument cluster

Main instrument panel wiring harness

Antilock electronic control unit

(RABS) II diagnostic connector

Bulkhead connector

Figure 9. The typical location for a RABS diagnostic connector under the vehicle's instrument panel.

onds. The rear ABS lamp on the instrument panel will flash a code for any faults that are present. The system will flash only one code at a time, so you must fix the problem, clear the module memory, and drive the truck to see if other faults occur.

> **You Should Know** *The speed sensor for a Ford RABS is in the rear axle differential housing, with the tone ring on the back of the ring gear. A leading cause of trouble codes on these installations is an accumulation of iron filings on the pickup coil's magnet. Metal particles accumulate in final drive assemblies during normal operation and present no problems, but when they get onto a speed sensor magnet, they can cause problems.*

Figure 10 is a typical troubleshooting chart for a RABS-equipped vehicle. Charts such as these can be very helpful if they are followed from top to bottom and followed precisely.

> **You Should Know** *Ground only the RABS module side of the diagnostic connector. Do not ground the B+ power side of the diagnostic connector harness. Doing so will blow the RABS memory fuse.*

DAIMLERCHRYSLER ABS SCAN TOOL TESTING

If the ABS control module detects a fault, the amber lamp will light until the fault is corrected or the trouble code cleared or, for some faults, when the ignition is turned off. Serious system problems can cause the amber and red lamps to be lit at the same time. When this occurs, ABS is disabled by the control module.

ABS Tests Available

Scan tool menus for Chrysler ABS testing are similar to the menus for Chrysler engine control testing. Most scan tools provide the following general functions, but the actual titles of the selections listed in the menus will vary.

- Trouble codes and operating data selections let you read codes and view information from the ABS control module on some vehicles.
- Actuator test mode (ATM) selections allow you to perform specific actuator operating tests on some vehicles.
- Functional test selections allow you to perform specific operating tests on some systems of some vehicles with ABS.

The test selection modes for ABS trouble codes and operating data are available on most scan tools for many Chrysler and Jeep vehicles with ABS. In these modes, a scan tool reads trouble codes and all data available on the ABS data stream. To ensure efficient communication with the ABS module, make sure that the ignition is off when entering vehicle identification. Then turn it on before selecting ABS codes and data from the scan tool menu. When finished with testing, turn the ignition off.

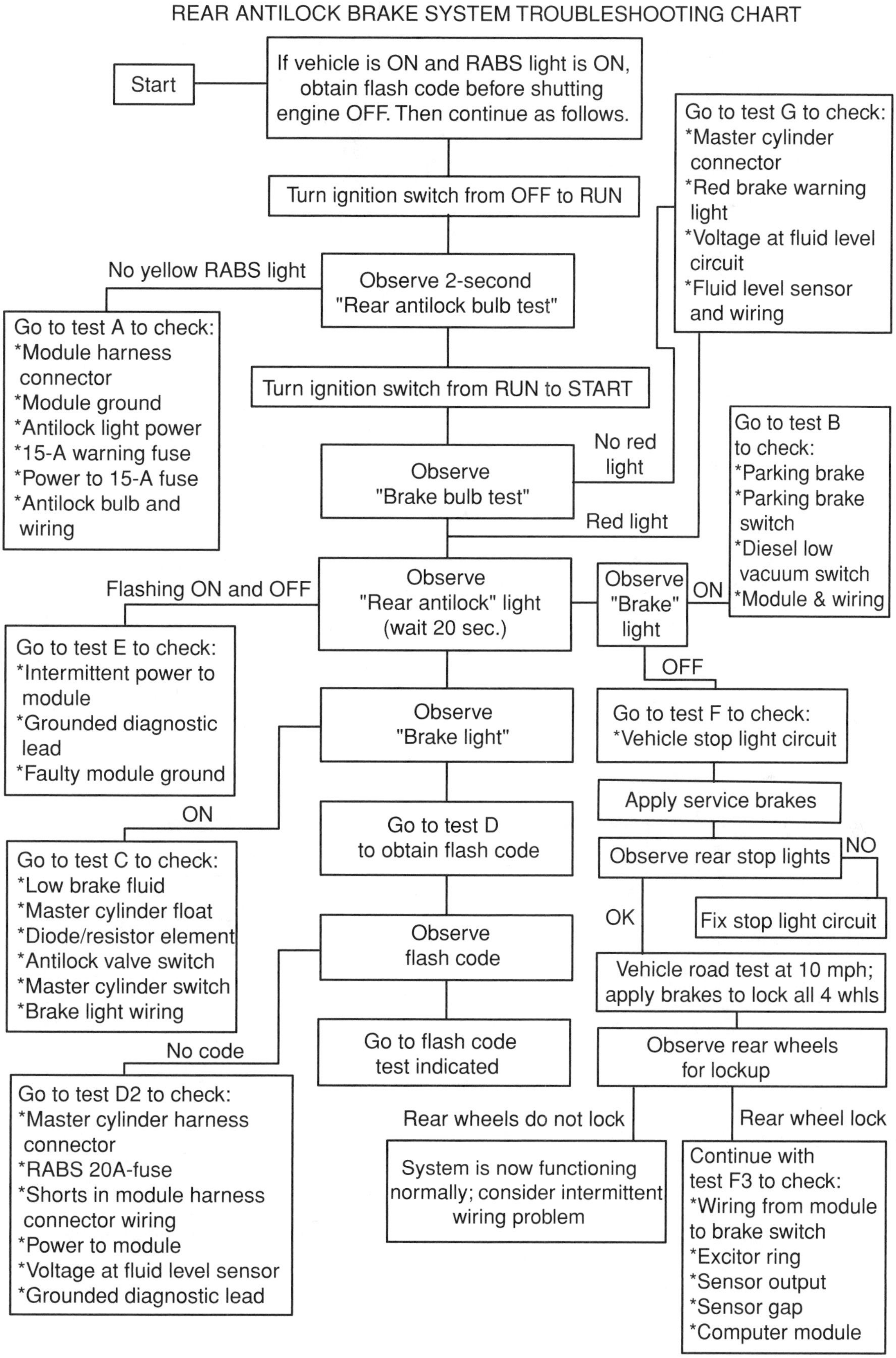

REAR ANTILOCK BRAKE SYSTEM TROUBLESHOOTING CHART

Start

If vehicle is ON and RABS light is ON, obtain flash code before shutting engine OFF. Then continue as follows.

Go to test G to check:
*Master cylinder connector
*Red brake warning light
*Voltage at fluid level circuit
*Fluid level sensor and wiring

Turn ignition switch from OFF to RUN

No yellow RABS light

Observe 2-second "Rear antilock bulb test"

Go to test A to check:
*Module harness connector
*Module ground
*Antilock light power
*15-A warning fuse
*Power to 15-A fuse
*Antilock bulb and wiring

Turn ignition switch from RUN to START

No red light

Observe "Brake bulb test"

Red light

Go to test B to check:
*Parking brake
*Parking brake switch
*Diesel low vacuum switch
*Module & wiring

Flashing ON and OFF

Observe "Rear antilock" light (wait 20 sec.)

Observe "Brake" light

ON

OFF

Go to test E to check:
*Intermittent power to module
*Grounded diagnostic lead
*Faulty module ground

Observe "Brake light"

Go to test F to check:
*Vehicle stop light circuit

Apply service brakes

ON

Go to test C to check:
*Low brake fluid
*Master cylinder float
*Diode/resistor element
*Antilock valve switch
*Master cylinder switch
*Brake light wiring

Go to test D to obtain flash code

Observe rear stop lights

NO

OK

Fix stop light circuit

Observe flash code

Vehicle road test at 10 mph; apply brakes to lock all 4 whls

No code

Go to flash code test indicated

Observe rear wheels for lockup

Go to test D2 to check:
*Master cylinder harness connector
*RABS 20A-fuse
*Shorts in module harness connector wiring
*Power to module
*Voltage at fluid level sensor
*Grounded diagnostic lead

Rear wheels do not lock

Rear wheel lock

System is now functioning normally; consider intermittent wiring problem

Continue with test F3 to check:
*Wiring from module to brake switch
*Excitor ring
*Sensor output
*Sensor gap
*Computer module

Figure 10.   A RABS troubleshooting chart.

The types of diagnostic connectors and their locations vary with the different DaimlerChrysler and Jeep vehicles. Older Chrysler vehicles have the ABS communication circuits in the body computer test connector under the instrument panel **(Figure 11)**. Most 1995 and later systems have the ABS test points in the 16-pin OBD II DLC, also under the instrument panel. **Figure 12** shows the designated pins for ABS testing in both an older and a newer diagnostic connector.

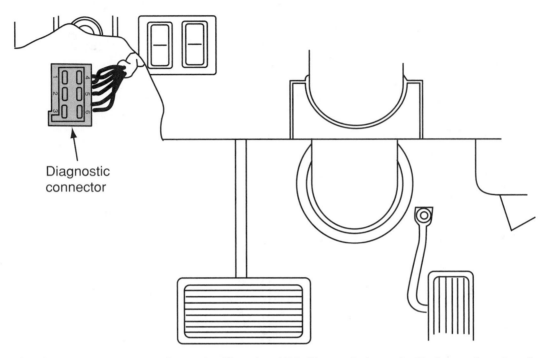

Figure 11. The diagnostic connector for early Chrysler ABS diagnosis is typically located under the instrument panel.

Cavity	Function
1	SCI receive
2	Fused B+
3	CCD bus (-)
4	CCD bus (+)
5	SCI transmit
6	Ground

Cavity	Function
3	CCD bus (+)
4	Ground
5	Ground
6	SCI receive
7	SCI transmit/ISO 9141K
11	CCD bus (-)
14	SCI receive (TCM)
16	Fused B+

Figure 12. The designated pins for ABS testing in both a pre–OBD II and an OBD II diagnostic connector.

HONDA AND ACURA ABS TESTING

Honda does not provide ABS access for aftermarket scan tools, but its antilock brake (ALB) checker can stimulate all system functions and operating conditions. It cannot be used to check system operation during test drives. The ALB checker does not display numerical trouble codes. Instead, Honda uses the amber ABS lamp and red brake lamp to flash trouble codes. On some Honda and Acura models, the codes are flashed by two light-emitting diodes (LEDs) on the electronic control module.

Reading Trouble Codes on the ABS Warning Lamp

A typical procedure for reading trouble codes in systems that use the ABS warning lamp to flash trouble codes

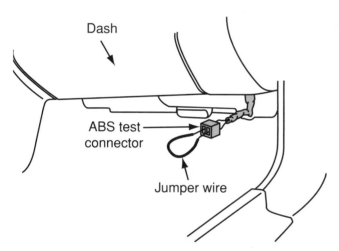

Figure 13. Jump across the terminals of the ABS test connector to begin the system's self-test.

begins with removing the connector cover from the test connector and jumping across the two terminals of the connector with a jumper wire **(Figure 13)**.

> ⚠️ **You Should Know** *Disconnect the jumper wire at the connector before starting the engine. If you start the engine with the jumper wire connected to the service connector, the malfunction indicator (MIL) will light with the engine running.*

The ignition is turned on and the ABS lamp observed. Count and record its blinking frequency. To interpret the flashing code, refer to the appropriate service manual for instructions. **Figure 14** is an example of main code and subcode numbers separated by a one-second pause. This system can display up to three separate trouble codes at one time. Each code number is separated by a five-second pause. After retrieving the codes, refer to the applicable troubleshooting chart in the service manual **(Figure 15)**. To repeat the trouble code display, turn the ignition off and repeat the procedure. Erase the codes by disconnecting the ABS fuse for at least three seconds.

ALB Checker

The Honda ALB checker can simulate each ABS function and operating mode to prove whether system operation is normal or defective. Make sure that the ABS warning lamp is not indicating an ABS system problem before using the ALB checker. When the ignition is turned on, the vehicle

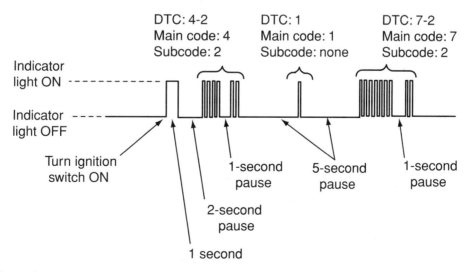

Figure 14. Honda and Acura ABS trouble codes flash according to this pattern.

| DIAGNOSTIC TROUBLE CODE (DTC) | | Problem component or system | Affected | | | | Other component |
Main code	Sub-code		Right front	Left front	Right rear	Left rear	
1		Pump motor overrun	—	—	—	—	Solenoid Pump motor Pressure switch
	2	Pump motor circuit problem	—	—	—	—	ABS motor relay ABS unit fuse ABS motor fuse
	3	High-pressure leakage	—	—	—	—	Solenoid
	4	Pressure switch	—	—	—	—	
	8	Accumulator gas leakage	—	—	—	—	Pump motor
2	1	Parking brake switch-related problem	—	—	—	—	Brake fluid level switch Brake system light
3	1	Pulser(s)	O				Wheel sensor installation
	2			O			
	3				O	O	
4	1	Wheel sensor	O				
	2			O			
	4				O		
	8					O	
5		Wheel sensor(s)			O	O	Modulator Rear brake drag
	4				O		
	8					O	
6		Fail-safe relay(s)	O		O		
	1			O			
	4					O	
7	1	Solenoid-related problem	O				ABS B3 fuse
	2			O			ABS B1 fuse Front fail-safe relay
	4				O	O	Rear fail-safe relay

Figure 15.   A typical Honda chart for interpreting trouble codes.

ABS warning lamp should light and then go off after one second. Modes 1 through 5 on the ALB checker can be used to verify proper ABS operation after the repair of suspension or body components that might have affected the wheel speed sensors or related wiring, replacement of ABS components, the replacement of the brake fluid, or after brake system bleeding.

When the ABS warning lamp on the instrument panel lights during any of the test modes, the ABS has an electrical defect. When this happens, proceed with the diagnosis of trouble codes. If there is no brake pedal kickback in modes 2 through 5 and the ABS warning lamp does not light, check for air in the ABS high-pressure lines, a restricted high-pressure line, or a faulty modulator unit.

Summary

- Each ABS has its own diagnostic procedure involving use of the brake and antilock warning lamps, special testers, breakout boxes, scan tools, troubleshooting charts, and wiring diagrams.

- Most General Motors systems will store trouble codes and flash them on the warning lamp or transmit them to a scan tool.

- When an ABS malfunction occurs, the control module of Ford Motor Company products will light the ABS warning lamp on the instrument panel.

- In some Ford systems, the module will store service codes in memory for most system malfunctions, and a scan tool can read these codes through the vehicle's ABS test connector or OBD II data link connector.

- Scan tool menus for Chrysler ABS testing are similar to the menus for Chrysler engine control testing and include trouble codes and operating data, an actuator test mode, and functional tests.

- Honda/Acura does not provide ABS access for aftermarket scan tools, but its ALB checker can stimulate all system functions and operating conditions.

- On some Honda and Acura models, the codes are flashed by two LEDs on the electronic control module.

Review Questions

1. Which of the following statements about retrieving ABS trouble codes on General Motors vehicles is *not* true?
   A. All General Motors vehicles have ABS systems that cannot transmit codes or other test data to a scan tool.
   B. Most General Motors vehicles are able to indicate system problems by flashing the ABS warning lamp to display trouble codes.
   C. The Teves Mk II system is one example of a General Motors antilock system that can be made to flash DTCs on the antilock warning lamp.
   D. Most 1994 and later General Motors antilock systems use a four-digit diagnostic trouble code format similar to OBD II engine and emission DTCs.

2. Which of the following statements about Chrysler ABS warning lights is true?
   A. DaimlerChrysler uses two brake indicator lamps: an amber brake warning lamp and a red antilock or ABS warning lamp.
   B. Both lamps light when the engine is first started and go out when the ABS self-test programs determined that the system is all right.
   C. If the ABS control module detects a fault, the red lamp will light.
   D. Serious system problems can cause the red lamp to be lit and ABS to be disabled by the control module.

3. Technician A says that each ABS has its own diagnostic procedure involving use of the brake and antilock warning lamps, special testers, breakout boxes, scan tools, troubleshooting charts, and wiring diagrams. Technician B says that ABSs vary in the number of codes they can store and whether they will retain the code in memory when the ignition is turned off. Who is correct?
   A. Technician A only
   B. Technician B only
   C. Both Technician A and Technician B
   D. Neither Technician A nor Technician B

4. While diagnosing an ABS problem with a scan tool on a Chrysler vehicle Technician A says that trouble codes and operating data selections let you read codes and view information from the ABS control module on some vehicles. Technician B says that functional test selections allow you to perform specific operating tests on some systems of some vehicles with ABS. Who is correct?
   A. Technician A only
   B. Technician B only
   C. Both Technician A and Technician B
   D. Neither Technician A nor Technician B

5. Technician A says that many late-model Ford antilock systems are tested through the OBD II test connector. Technician B says that many late-model General Motors antilock systems are tested through the OBD II test connector. Who is correct?
   A. Technician A only
   B. Technician B only
   C. Both Technician A and Technician B
   D. Neither Technician A nor Technician B

6. Shorting pin H to pin A at the ALDL connector of some General Motors trucks will flash trouble codes on the ABS instrument panel lamp for what systems?
   A. Delco Moraine III
   B. Delphi Chassis VI
   C. RABS
   D. 4WAL

Chapter 44

Testing and Replacing Antilock Brake System Components

Introduction

ABS control systems receive relatively simple input signals compared with powertrain control systems. Input signals for an ABS computer, or control module, come principally from switches and speed sensors. This chapter covers troubleshooting guidelines for these common devices and the replacement of ABS components. Vehicle manufacturers provide specific test procedures for specific components in their service manuals.

ELECTROSTATIC DISCHARGE

Some manufacturers mark certain components and circuits with a code or symbol to warn technicians that they are sensitive to electrostatic discharge. Static electricity can destroy or render a component useless.

When handling any electronic part, especially those that are static sensitive, follow these guidelines to reduce the possibility of electrostatic buildup on your body and the inadvertent discharge to the electronic part:

1. Always touch a known good ground before handling the part. This should be repeated while handling the part and more frequently after sliding across a seat, sitting down from a standing position, or walking a distance.
2. Avoid touching the electrical terminals of the part, unless you are instructed to do so in the service procedures. Keep your fingers off all electrical terminals because the oil from your skin can cause corrosion. Use an antistatic strap, if one is available.
3. When you are using a voltmeter, always connect the negative meter lead first.

4. Do not remove a part from its protective package until it is time to install the part.
5. Before removing the part from its package, ground yourself and the package to a known good ground on the vehicle.
6. When replacing a PROM, ground your body by putting a metal wire around your wrist and connect the wire to a known good ground.

SWITCH TESTING

The brake switch, cruise control switch, and brake fluid warning switch are examples of simple ABS input switches that send a signal to the system's ECM. When used as a control system sensor, a switch provides a digital, on/off, high-low voltage signal. Such a signal indicates that the brakes are released or applied.

To provide this type of input signal, the switch is installed between the computer and ground. A **reference voltage** is applied to the switch circuit inside the computer, across a fixed resistor **(Figure 1)**. The fixed resistor is often called a **pull up resistor** because it pulls the reference voltage up to the open-circuit level when the switch is open and drops the voltage when the switch is closed. The input signal for the computer is taken from a point between the resistor and the switch.

The reference voltage for a switch is most often the 5-volt reference used for other computer circuits. In some circuits, the reference voltage may be full system voltage (approximately 12 volts) or some other voltage level.

Testing a switch circuit is done by placing the switch in a known operating position (open or closed) and using

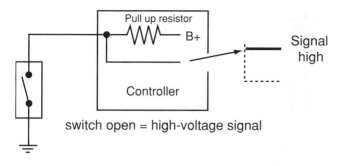

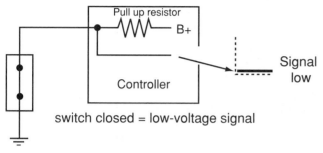

switch open = high-voltage signal

switch closed = low-voltage signal

Figure 1. A basic switch circuit used to provide an on/off input signal to a computer.

a voltmeter to check the voltage signal received by the computer. One common way to do this is to back probe the switch's wire terminal at the computer harness connector with the positive (+) lead of your voltmeter. **Figure 2** shows the basic hookup for conducting this test.

SPEED SENSOR TESTING

Speed sensors for ABSs are voltage-generating sensors called pickup coil sensors, reluctance sensors, or **permanent magnet (PM) generators**. All voltage-generating sensors operate in the same way. As a toothed sensor ring rotates past the stationary sensor, it produces a voltage frequency and intensity directly proportional to wheel speed. As the teeth of a rotating trigger wheel pass by the magnet, the magnetic field expands and collapses to generate an alternating current (AC) voltage in the coil **(Figure 3)**. This signal is sent to the ECM, which processes it to determine rotational speed. Whether the speed sensor is mounted on a wheel **(Figure 4)** or inside the trans-

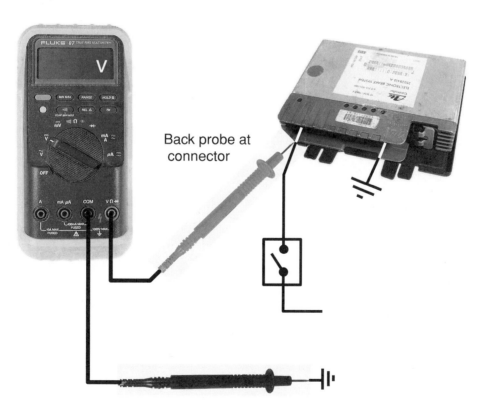

Back probe at connector

Figure 2. Back probe a switch circuit at the computer's harness connector and operate the switch to test the input voltage signal.

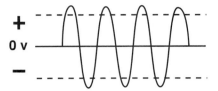

Figure 3.   A typical AC voltage sine wave signal produced by a wheel speed sensor.

Figure 4.   A typical front wheel speed sensor.

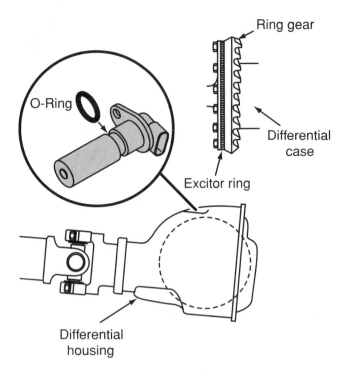

Figure 5.   A speed sensor mounted at the differential unit.

mission or differential **(Figure 5)**, it works on these same principles.

The tip of the sensor is located near a toothed ring or rotor. The toothed ring typically is part of the outer CV-joint or axle assembly or is mounted next to the differential ring gear.

Typically, the trigger wheels are replaced, as part of the assembly, with the outer CV-joint or the wheel hub and bearing assembly. Others are individually replaceable. These have a "slip-fit" and must be installed by hand or by even pressure on a hydraulic press. Each trigger wheel should be inspected for broken or damaged teeth. If any defects are found, the trigger wheel should be replaced.

Most wheel sensors are mounted on an adjustable mounting or have slotted bolt holes for adjustment. Others are mounted solidly and have no means for adjustment. Adjusting the clearance between the sensor and the trigger wheel is critical for the proper operation of ABS. If the sensor is adjustable, always follow the procedure recommended by the manufacturer. Some nonadjustable sensors have a polyethylene or paper spacer that provides for the correct gap. When reusing this type of sensor, make sure this spacer is in good condition.

ABS speed sensors produce an AC sine wave voltage signal, which is typical of any pickup coil sensor. The computer converts the AC sine wave voltage to a digital square wave signal so that it can read the sensor frequency, or speed **(Figure 6)**. The computer simply counts the trailing, or falling, edge of the square wave signal to determine wheel speed.

The AC sine wave signal from the speed sensor must have a certain **amplitude** and a smooth frequency within its specified operating range. The amplitude is the strength of the signal, or the positive and negative height of the voltage waveform. Although the signal frequency, or speed, varies with the wheel speed, it must be smooth and operate at a uniform

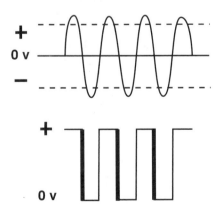

AC voltage waveform converted
to a square wave signal

Figure 6.   The computer converts the analog signal from a speed sensor to a digital signal.

amplitude. **Figure 7** compares a normal waveform with one that has low amplitude and another that has a low frequency.

A variation in speed signal frequency or a speed difference between the wheels is the normal trigger for antilock operation. An uneven, or erratic, frequency signal also can trigger a system trouble code. A normal speed sensor signal should be a repeatable AC voltage, with the frequency varying smoothly with wheel speed. A missing signal, a low-amplitude signal, or an erratic frequency signal can cause a system trouble code.

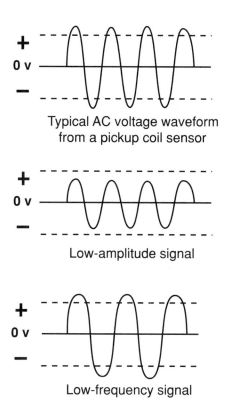

Figure 7. A typical waveform from a speed sensor (*top*). A speed sensor waveform with a low-amplitude signal (*center*). A speed sensor waveform with a low-frequency signal (*bottom*).

Sensor Resistance Testing

Most manufacturers list resistance and voltage output specifications. The resistance of the sensor is measured across the sensor's terminals. The typical range for a good sensor is 800–1400 ohms of resistance. To check the resistance of the sensor, disconnect it from its circuit and measure its resistance by connecting an ohmmeter across the two terminals of the sensor wiring harness. If resistance is out of limits, either high or low, replace the sensor. Damage to the sensor trigger wheel (tone ring), for example, can produce an uneven signal even when the pickup coil is electrically in good condition.

Sensor Voltage Waveforms

The best way to test and evaluate the signal from a speed sensor is to view its output signal on a digital storage scope or a graphing multimeter. Disconnect the sensor harness from the vehicle's wiring harness and connect the oscilloscope or graphing meter leads across the two terminals of the sensor harness. Then rotate the wheel by hand or by using the vehicle's powertrain, as appropriate.

Observing the signal while the sensor is operating lets you look for intermittent problems, such as erratic frequency changes or unstable waveforms. Damage to the trigger wheel can produce an uneven signal that you could not see without operating the sensor. A low-amplitude signal also will be very clear when viewed on an oscilloscope or a graphing multimeter.

Checking Speed Sensor Bias Voltage

Most ABS speed sensors receive a bias voltage from the system's computer. The bias voltage allows the computer to detect an open or a short in the sensor circuit before the wheel turns. The bias voltage also elevates the sensor signal off the common ground plane of the vehicle electrical system to reduce signal interference.

Figure 8 shows a simple speed sensor bias voltage circuit that uses a pull up resistor inside the computer. The bias voltage varies from manufacturer to manufacturer. It

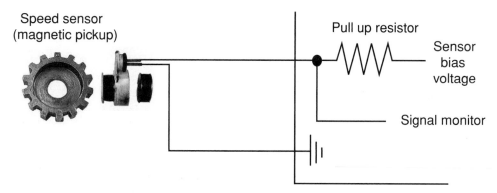

Figure 8. A simplified speed sensor circuit showing bias voltage and the signal monitor point.

can be the 5 volts used for other computer circuits or it can be a different value such as 1.5 or 1.8 volts. Check the manufacturer's specifications to determine the required bias voltage when troubleshooting a speed sensor circuit.

The ECM monitors the sensor signal at a point between the fixed pull up resistor and the pickup coil. When power is applied to the circuit, current flows through the pull up resistor and through the pickup coil to ground. The voltage drop at the signal monitor point is a predetermined portion of the reference voltage and a known value that is part of the computer program.

If an open circuit exists, no current flows through the circuit and no voltage is dropped across the pull up resistor. The signal monitor voltage will be high **(Figure 9)**. In this case, the computer will immediately set a trouble code for an open circuit fault.

If a shorted circuit exists, all or nearly all of the bias voltage is dropped across the pull up resistor. The signal monitor voltage will be lower than the programmed signal monitor voltage **(Figure 10)**.

The simple voltage divider circuit shown in these illustrations allows the computer to detect an electrical fault as soon as the ignition is turned on. The wheel does not need to turn even one revolution. You can verify an open- or short-circuit fault by connecting a voltmeter between the high-voltage side of the pickup coil circuit and ground. Depending on the circuit fault, the meter should read close to full bias voltage or close to zero volts with the ignition on. Some speed sensors receive a bias voltage to raise the signal above the common ground plane of the vehicle electrical system, as well as to detect open- and short-circuit problems.

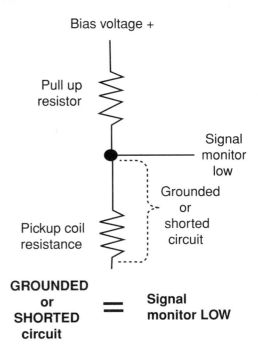

GROUNDED or SHORTED circuit = **Signal monitor LOW**

Figure 10.   If the sensor circuit is shorted, signal monitor voltage will be low and the appropriate DTC will be set.

ABS COMPONENT REPLACEMENT

When a component requires replacement, the methods are generally straightforward mechanical procedures. Some special tools might be needed to service some ABS components.

WHEEL SPEED SENSOR REPLACEMENT

A problem in a wheel speed sensor can require replacement of the sensor pickup coil and its harness or the tone ring, or both. Some wheel sensors have an adjustable air gap between the sensor head and the teeth on the tone ring. An equal number of sensors are nonadjustable. If a sensor is adjustable, follow the manufacturer's adjustment procedure. Some nonadjustable sensors use a lightweight plastic or paper spacer on the mounting surface to set the gap **(Figure 11)**. A new spacer of the correct thickness should be used whenever a sensor is reused.

Manufacturers differ in their requirement for servicing the wiring harnesses on speed sensors. The short, two-wire harness that is part of the sensor assembly on General Motors vehicles is made with very fine wire strands to provide maximum flexibility with minimum circuit resistance. General Motors specifies that the sensor harness not be repaired by any method. If it is damaged, the complete sensor assembly must be replaced. Chrysler, on the other hand, says that the harnesses on most of its wheel speed sensors

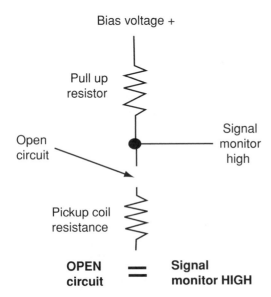

OPEN circuit = **Signal monitor HIGH**

Figure 9.   If the sensor circuit is open, signal monitor voltage will be high and the appropriate DTC will be set.

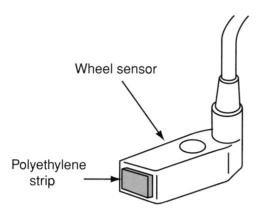

Figure 11. Some wheel speed sensors use a plastic spacer to set the air gap correctly; others use a paper spacer. Most speed sensors are not adjustable.

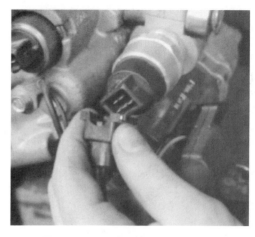

Figure 12. Disconnect the electrical connectors from the pressure switch and the motor.

can be repaired by soldering and reinsulating with heat-shrink tubing.

Because of the importance of signal accuracy from a wheel speed sensor, it is preferable to replace a sensor assembly, including the harness, instead of trying to repair the harness. Consult the manufacturer's specifications before deciding how to service these components.

Sensor tone rings that are pressed on the inside of the rotor or on the axle shaft often can be replaced. If the sensor ring is an integral part of the wheel bearing assembly, hub assembly, or the outer constant velocity joint on the axle, the entire component must be replaced if the sensor ring is damaged.

Observe these additional guidelines when servicing ABS wheel speed sensors:

- Unplug sensor electrical leads when replacing suspension components.
- If a sensor tone ring (trigger wheel) is replaceable, fit the new one in place by hand. Do not hammer or tap the sensor in place.
- If a wheel sensor ring or tone wheel is replaceable and is pressed into place, use a hydraulic press to remove the old one and to install the new one.
- Some wheel sensors require an anticorrosion coating before installation to prevent galvanic corrosion. Never substitute grease unless the manufacturer specifies its use.

Sensor assemblies that are a permanent part of the wheel bearing and hub assembly are used on many late-model General Motors cars. They require no adjustment and plug directly into the vehicle wiring harness.

PUMP AND MOTOR REPLACEMENT

Many ABSs have a pump and motor assembly, which should be replaced if they are faulty. The following is a typical procedure for removing the pump and motor from a General Motors Teves system. Reinstallation of the assembly is the reverse procedure of removal. Again, always refer to the appropriate service manual before removing and reinstalling the pump and motor assembly of an ABS.

Begin by disconnecting the battery ground cable. Then depressurize the accumulator. Now disconnect the electrical connectors from the pressure switch and the motor **(Figure 12)**. Use a clean syringe to remove about half of the brake fluid from the reservoir. Then unscrew the accumulator from the hydraulic module **(Figure 13)**. Then remove the O-ring from the accumulator. Disconnect the high-pressure hose from the pump. Now, disconnect the wire-retaining clip and pull the return hose out of the pump body. Then remove the bolt that attaches the pump and motor to the hydraulic module and remove the pump

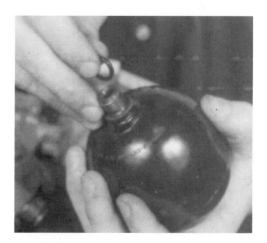

Figure 13. Unscrew the accumulator from the hydraulic module.

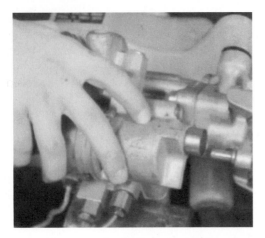

Figure 14. Remove the bolt that attaches the pump and motor to the hydraulic module and remove the pump and motor assembly by sliding it off the locating pin.

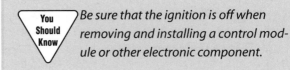

Be sure that the ignition is off when removing and installing a control module or other electronic component.

and motor assembly by sliding it off the locating pin **(Figure 14)**.

COMPUTER (CONTROL MODULE) REPLACEMENT

Before replacing a computer, check the TSBs for specific information on revised computer part numbers and the problems that they were designed to correct. Before replacing an ECM, check these items:

- *Battery voltage supply to the computer and the main system ground.* Make sure that the battery is fully charged and provides at least 9.6 volts during cranking. Make sure that the charging system is maintaining correct battery charge. Most computers receive battery volt-

age though a fuse or fusible link. Make sure that battery voltage is available at the specified terminals of the computer's main connector. Most computers are grounded remotely through several wires in the harness. Trace and check the ground connections to ensure good continuity.

- *Operation of system power relay.* Some computers receive power through a system power relay. If the vehicle is so equipped, check the relay operation.
- *Sensor reference voltage and ground circuits.* Many sensors share a common reference voltage supply from the computer and a common ground. Incorrect or erratic reference voltage or a bad common ground can affect operation of several sensors simultaneously. The symptoms can appear as if the computer has a major system problem. Repairing a wiring connection might correct the problem.
- *Resistance and current flow through all computer-controlled solenoids and relays.* Every output device controlled by a computer has a minimum resistance specification. The actuator resistance limits the current through the computer output control circuit. If the actuator is shorted, current can exceed the safe maximum and damage the computer. In most cases, current through a computer-controlled output device should not exceed 0.75 ampere (750 milliamperes). Before replacing a computer, check all output circuits for shorts or low resistance that could damage the computer.

Summary

- Electrostatic discharge is static electricity that can destroy the components of electronic circuits.
- When used as a sensor, a switch provides a digital, on/off, high-low voltage signal.
- The switch is installed between the computer and ground, and reference voltage is applied to the switch circuit inside the computer, across a fixed resistor. The input signal for the computer is taken from a point between the resistor and the switch.
- The ABS speed sensors produce an AC sine wave voltage signal, and the computer converts the AC sine wave voltage to a digital square wave signal so that it can read the sensor frequency, or speed.

- All pickup coil sensors have resistance specifications.
- The best way to test and evaluate the signal from a speed sensor is to view its output signal on a digital storage scope or a graphing multimeter.
- Most ABS speed sensors receive a bias voltage from the system computer, which allows the computer to detect an open or a short in the sensor circuit before the wheel turns.
- Some wheel sensors have an adjustable air gap between the sensor head and the teeth on the tone ring.
- Sensor tone rings that are pressed on the inside of the rotor or on the axle shaft often can be replaced. If the

sensor ring is an integral part of the wheel bearing assembly, hub assembly, or the outer constant velocity joint on the axle, the entire component must be replaced if the sensor ring is damaged.
■ As a general rule, the system's computer should be at the bottom of the list of things to replace.

■ Before replacing a computer, check battery voltage supply to the computer and the main system ground, the operation of system power relay, sensor reference voltage and ground circuits, and the resistance and current flow through all computer-controlled solenoids and relays.

Review Questions

1. When checking the waveforms of a speed sensor, what should specifically be looked at?
2. What four things should be checked before you replace the computer in an ABS?
3. When inspecting wheel speed sensors, check for all of the following *except:*
   A. Correct bias voltage
   B. Proper contact between the pole piece and tone ring
   C. Secure sensor mounting
   D. Condition of the tone ring teeth
4. When removing a vehicle's ABS electronic control module Technician A relieves hydraulic pressure in the system; Technician B disconnects the battery ground cable. Who is correct?
   A. Technician A only
   B. Technician B only
   C. Both Technician A and Technician B
   D. Neither Technician A nor Technician B
5. Technician A says that wheel speed sensors generate a square-wave frequency signal. Technician B says that a wheel speed sensor has a permanent magnet. Who is correct?
   A. Technician A only
   B. Technician B only
   C. Both Technician A and Technician B
   D. Neither Technician A nor Technician B

Chapter 45

Traction Control

Introduction

Automakers use the technology and hardware of ABSs to control tire traction and vehicle stability. An ABS pumps the brakes when a braking wheel attempts to go into a locked condition. Automatic traction control **(ATC)** systems apply the brakes when a drive wheel attempts to spin and lose traction **(Figure 1)**.

The system works best when one drive wheel is working on a good traction surface and the other is not. The system also works well when the vehicle is accelerating on slippery road surfaces, especially when climbing hills.

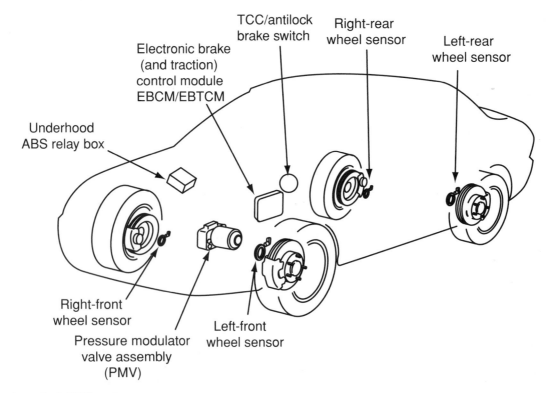

Figure 1. A typical ATC system.

An ATC system is most helpful on four-wheel or all-wheel drive vehicles where loss of traction at one wheel could hamper driver control. It is also desirable on high-powered front-wheel-drive vehicles for the same reason. Often, if **traction control** is fitted to a FWD vehicle, the ABS modified system is a three-channel system because ATC is not needed at the rear wheels. On RWD and 4WD vehicles, the system is based on a four-channel ABS system.

During road operation, the ATC system uses an electronic control module to monitor the wheel speed sensors. If a wheel enters a loss-of-traction situation, the module applies braking force to the wheel in trouble. Loss of traction is identified by comparing the vehicle's speed with the speed of the wheel. If there is a loss of traction, the speed of the wheel will be greater than expected for the particular vehicle speed. Wheel spin is normally limited to a 10 percent slippage. Some traction control systems use separated hydraulic valve units and control modules for the ABS and ATC, whereas others integrate both systems into one hydraulic control unit and a single control module. The pulse rings and wheel speed sensors remain unchanged from the ABS to the ATC.

Some ATC systems function only at low road speeds of 5–25 miles per hour. These systems are designed to reduce wheel slip and maintain traction at the drive wheels when the road is wet or snow covered. The control module monitors wheel speed. If during acceleration the module detects drive wheel slip and the brakes are not applied, the control module enters into the traction control mode. The inlet and outlet solenoid valves are pulsed and allow the brake to be quickly applied and released. The pump/motor assembly is turned on and supplies pressurized fluid to the slipping wheel's brake.

More advanced systems work at higher speeds and usually integrate some engine control functions into the control loop. For example, if the ATC system senses a loss of traction, it not only cycles the brakes but also signals the engine control module to retard ignition timing and partially close the throttle as well. Timing and throttle reduce engine output. On these vehicles, there might be no mechanical connection between the gas pedal and the throttle plates. The pedal is nothing more than an electrical switch. This is sometimes referred to as "drive by wire." If wheel slippage continues, the ATC control module might also cut fuel delivery to one or more engine cylinders. This action lowers engine output and prevents the driver from overspeeding the engine in extremely slippery conditions.

WHEEL SLIP

Controlling wheel slip is the goal of both ABS and the traction control system (TCS). ABS controls negative wheel slip by modulating (decreasing and increasing) the hydraulic pressure to the wheel, or wheels, that is skidding. The ABS control unit electronically "pumps the brakes."

One way to counteract positive wheel spin is to modulate hydraulic brake pressure to the wheel that is spinning and use the brake to slow the wheel. Most TCSs or acceleration slip regulation (ASR) systems use the method of partially applying one or both brakes on the drive wheels. Many systems use other methods before applying brake pressure.

WHEEL SPIN CONTROL STRATEGIES

Different TCS installations use an electronic brake and traction control module (EBTCM), the PCM, or an integrated vehicle control module (VCM). Besides partially applying the brakes on a wheel that is spinning, a TCS can use the following slip control methods:

- Retard ignition timing.
- Reduce or cut off fuel injection pulses to one or more cylinders.
- Increase exhaust gas recirculation (EGR) flow.
- Momentarily upshift the transmission to a higher gear.

If all of those methods do not eliminate wheel spin, the system might gradually apply the brakes at the driving wheels.

In order to use the brakes to control wheel spin, the ABS must have a pump to develop hydraulic pressure and an accumulator to store reserve pressure. During braking, the driver's foot applies pressure to the brake pedal. During acceleration, the driver's foot is nowhere near the brake pedal; so if brake pressure is to be used to stop wheel spin, the brake system must have an independent source of hydraulic pressure. The Kelsey-Hayes RABS and RWAL systems used on many pickup trucks are examples of minimal ABS designs that could not be used for traction control because they have no source of hydraulic pressure independent of the driver's foot.

As traction control has grown in popularity, manufacturers have used various combinations of these control methods. The systems have been called automatic traction control (ATC) systems, traction control systems (TCS), and acceleration slip regulation (ASR) systems. All of these names mean exactly the same thing.

The methods used to control wheel spin are a matter of engineer's choice. Most General Motors systems are examples of traction control that use several methods. General Motors systems rely primarily on engine torque management by retarding timing, cutting off fuel injection, and other tactics before using the brakes to control wheel spin. General Motors engineers believe that engine torque management is the fastest and most efficient way to control wheel spin. At the other extreme, Chrysler traction control systems rely almost exclusively on brake application to control wheel spin.

Instrument
cluster

Indicator locations vary with the vehicle options

Message
center

Figure 2. The style and location of indicator lamps vary with the different systems, but the TCS always adds one or two more lamps to the array.

DRIVER CONTROLS AND INDICATORS

Regardless of the system manufacturer or the vehicle's manufacturer, TCS adds one or two more indicator lamps to the instrument panel **(Figure 2)**. TCS also customarily has a manual override switch by which the driver can turn off the system. An amber TCS lamp will light on the instrument panel if the driver disengages the system. The same, or another, amber lamp will light as a fault occurs in the system. In this case, the TCS is automatically disabled by the system control unit. Some vehicles also have a green TCS lamp to indicate when the TCS is actively controlling wheel spin while driving.

SPECIFIC TRACTION CONTROL SYSTEMS

Traction control systems are simply add-ons to ABSs. The following paragraphs summarize the features of some other popular TCSs.

Bosch 3 with TCS

The Bosch 3 ABS with traction control claimed the distinction of being the first use of a TCS on an FWD car. Because the Bosch 3 system is an integral ABS, it has a high-pressure hydraulic pump and an accumulator. A TCS was added by including what Bosch calls a TCS plunger **(Figure 3)**. The plunger assembly is mounted below the ABS hydraulic unit and is a pair of electronically controlled mas-

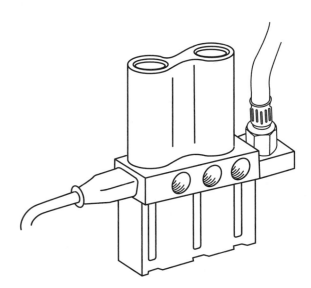

Figure 3. The TCS plunger for the Bosch 3 with traction control is an electronically controlled master cylinder.

ter cylinders that can apply brake pressure to the drive wheels when wheel spin is detected. **Figure 4** is a block diagram of the traction control hydraulic system that shows high-pressure fluid supply from the pump and accumulator to the plunger for one wheel.

Before the TCS applies brake pressure to control a spinning drive wheel, the ABS-TCS control unit signals the PCM

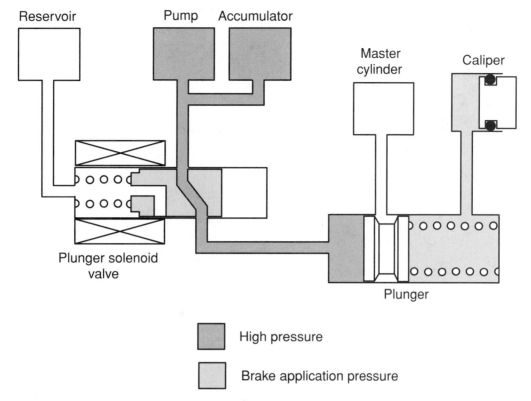

Figure 4. Traction control hydraulic system for the Bosch 3 with TCS.

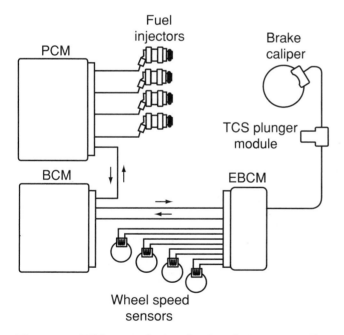

Figure 5. TCS controls for the Bosch 3 system. The PCM can cut off one to four fuel injectors on a V-8 engine to manage engine torque output.

through the BCM to shut down one to four fuel injectors to limit engine torque (**Figure 5**).

Bosch 2U and 2S with ASR

The ASR feature was added to Bosch ABS on 1992 and later Corvettes. A lateral acceleration sensor is a unique feature of this system. It measures cornering force and sends a signal to the ABS-TCS control module. When wheel spin is detected during acceleration or cornering, the brake control module signals the PCM to retard ignition timing. If the wheels continue to slip, the PCM signals a separate throttle actuator module to mechanically close the throttle. As a last resort, the brake control module can apply brake pressure to the slipping wheels.

Bosch 5 and 5.3 with TCS

TCS was offered with the nonintegral Bosch 5 and 5.3 ABS versions, starting in 1995. These systems use a lateral accelerometer and a **yaw** rate sensor to measure vehicle-cornering forces. Yaw is defined as swinging motion to the left or right of the vertical centerline or rotation around the vertical centerline. The yaw rate sensor operates like a tun-

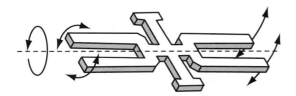

Figure 6. A yaw rate sensor as used on Bosch 5 and 5.3 systems.

ing fork with a fixed input frequency **(Figure 6)**. Lateral yaw forces on the fork change the frequency, which is translated to an input voltage signal for the TCS control module.

As with the Bosch 2U and 2S systems, Bosch 5 with TCS uses a throttle actuator module to mechanically close the throttle for engine torque management. The system also uses other engine control tactics, such as retarded timing and fuel cutoff, before it applies the drive-wheel brakes.

Teves Mk IV and Mk 20 with TCS

A TCS was added to the Teves Mk IV ABS on some General Motors cars in 1992. In 1993, Chrysler also began using the low-speed traction control system with Teves Mk IV on some models. Unlike the Bosch systems, the Teves Mk IV with TCS controls the brakes only. Engine torque management is not used.

The late-model Teves Mk 20 ABS is very similar to the earlier Mk IV system, and a TCS was introduced as an option on 1997 Chrysler mini vans.

Delphi Chassis VI with TCS

General Motors introduced the Delco Moraine ABS VI as an economical ABS. In mid-1993, the TCS was added to ABS VI, and in 1995 the system was renamed Delphi Chassis ABS/TCS VI. Since its introduction, this system has been built in several different configurations with both front-to-rear and diagonally split hydraulic systems and three- and four-channel hydraulic output control.

The traction control features of ABS/TCS VI are based primarily on engine torque management. The Delphi Chassis ABS VI system uses electric motors to control ball-screw-operated valves to modulate brake pressure. For traction control, another pair of motor-driven valves is installed in a TCS modulator, downstream from the ABS hydraulic modulator **(Figure 7)**. **Figure 8** is a hydraulic diagram of the ABS/TCS VI installation on a RWD vehicle. Note that the TCS modulator is in series between the ABS modulator and the rear driving wheels. **Figure 9** is an exploded view of the TCS modulator.

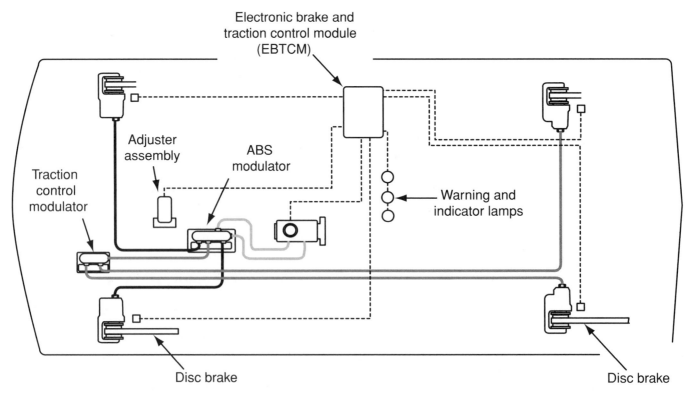

Figure 7. A typical Delphi Chassis VI ABS with TCS.

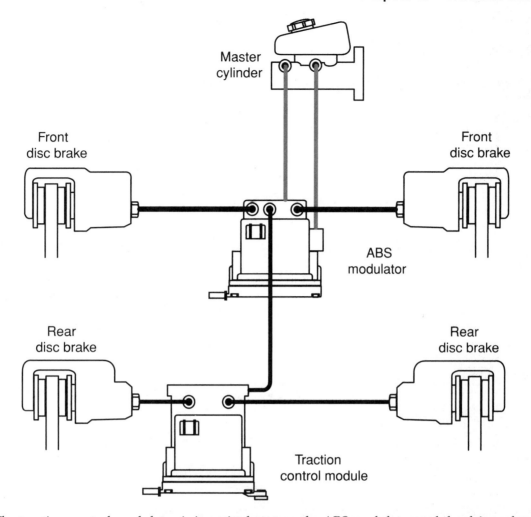

Figure 8. The traction control modulator is in series between the ABS modulator and the drive wheels on a rear-wheel-drive vehicle.

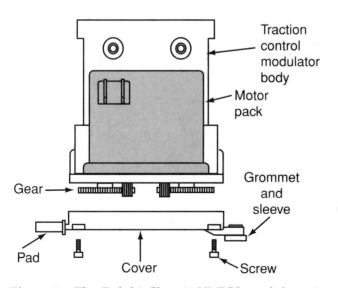

Figure 9. The Delphi Chassis VI TCS modulator is similar to the ABS modulator.

RWD Camaros and Firebirds with V-8 engines also have a throttle adjuster or relaxer assembly that uses an electric motor to close the throttle for engine torque management **(Figure 10)**.

Toyota Traction Control System

The Toyota traction control system (TRAC) is typical of the systems offered by many imported vehicle manufacturers and is quite similar to domestic TCSs. Since the early 1990s, Toyota TRAC systems have used the ABS wheel speed sensors to provide wheel speed signals to the traction control module. The TRAC system is part of an integrated ABS, but it uses engine torque management as the primary method of traction control. In this way, the Toyota system is similar to most General Motors traction control systems.

The TRAC system is basically a subsystem of the ABS. Both systems send input information to the ABS ECM and to the engine PCM. If engine torque management fails to

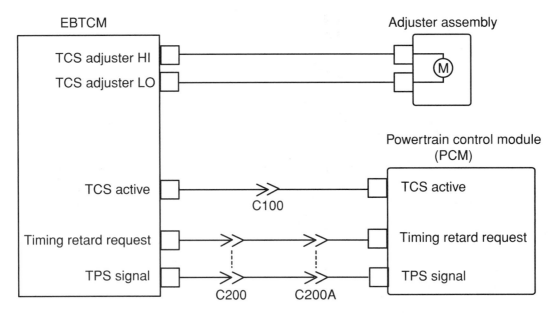

Figure 10. A block diagram of the PCM, electronic brake and traction control module, and the throttle adjuster of some Delphi Chassis VI systems.

control wheel spin, the ABS module tells the TRAC module to apply brake pressure to the slipping drive wheel or wheels. The TRAC system **(Figure 11)** has a separate pump and actuator to develop hydraulic pressure and apply it to the drive wheels.

AUTOMATIC STABILITY CONTROL

Various stability control systems are found on today's vehicles. Like TCSs, stability controls are based on and linked to the ABS **(Figure 12)**. On some vehicles, the stability con-

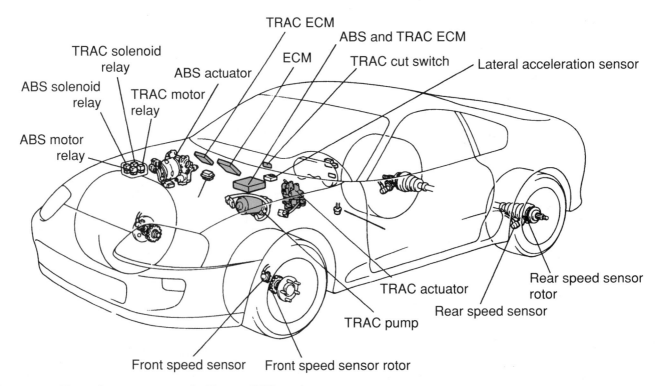

Figure 11. Typical components of a Toyota TCS.

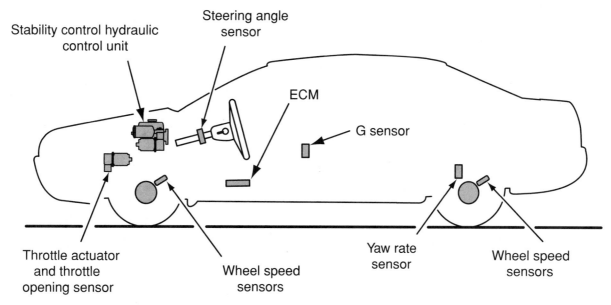

Figure 12.   The components of a typical vehicle stability control (VSC) system.

trol system is also linked to the electronic suspension system. Also on a few cars, a switch allows the driver to disable the traction control but maintain the stability controls.

Stability control systems momentarily apply the brakes at any one wheel to correct oversteer or understeer **(Figure 13)**.

The control unit receives signals from the typical sensors plus yaw, lateral acceleration (G-force), and steering angle sensors.

Understeer is a condition in which the vehicle is slow to respond to steering changes. Oversteer occurs when the

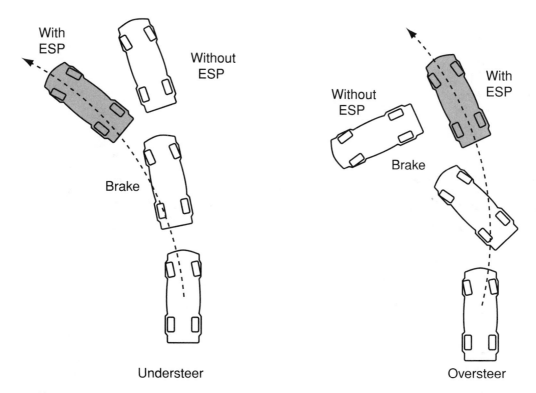

Figure 13.   The effects of oversteer and understeer and how they are affected by stability control (electronic stability program [ESP]) systems.

rear wheels try to swing around or fishtail. When the system senses understeer in a turn, the brake at the inside rear wheel is applied. During oversteer, the outside front brake is applied.

Relying on the input from the sensors and computer programming, the system calculates whether the vehicle is going exactly in the direction it is being steered. If there is the slightest difference between what the driver is asking and what the vehicle is doing, the system corrects the situation by applying one of the right or left side brakes.

The system uses the angle of the steering wheel and the speed of the four wheels to calculate the path chosen by the driver. The system then looks at lateral G-forces and vehicle yaw to measure where the vehicle is actually going.

Stability control systems can control the vehicle during acceleration, braking, and coasting. If the brakes are already applied but oversteer or understeer is occurring, the fluid pressure to the appropriate brake is increased **(Figure 14)**.

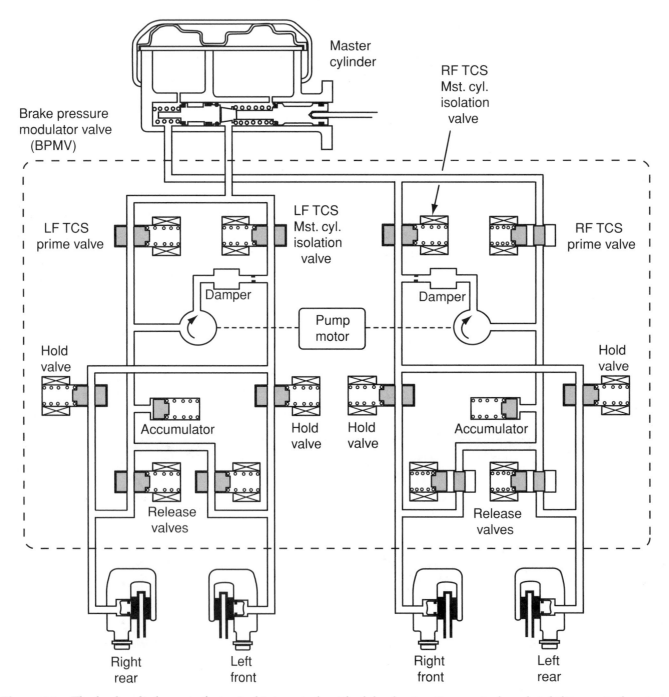

Figure 14.   The hydraulic layout of a typical integrated antilock brake, traction control, and stability control system.

TESTING TRACTION CONTROL SYSTEMS

When troubleshooting a traction control system, it is important to remember that the system will be automatically disabled by the control unit if a fault occurs. Because the TCS is combined with ABS and shares many input signals and output functions, TCS also will be disabled if an ABS fault occurs.

Traction control adds very few, if any, extra components to a vehicle. Traction control is accomplished principally by altering the control programs of the brake and engine control systems. It follows, then, that TCS testing is principally a process of using the self-diagnostic capabilities that the manufacturers include in the control program's software. Aftermarket scan tools and manufacturers' special test equipment can read TCS trouble codes and operating system data. As for other electronic tests, you must have the manufacturer's troubleshooting procedures for the vehicle you are servicing.

Summary

- Automatic traction control (ATC) systems apply the brakes when a drive wheel attempts to spin and lose traction.
- The ATC system uses an electronic control module to monitor the wheel speed sensors. If a wheel enters a loss-of-traction situation, the module applies braking force to the wheel in trouble.
- To control wheel slip, the TCS can partially apply the brakes on a wheel that is spinning, retard ignition timing, reduce or cut off fuel injection pulses to one or more cylinders, increase EGR flow, and/or momentarily upshift the transmission to a higher gear.
- Most General Motors systems rely primarily on engine torque management by retarding timing, cutting off

fuel injection, and employing other tactics before using the brakes to control wheel spin.
- Chrysler traction control systems rely almost exclusively on brake application to control wheel spin.
- A TCS adds one or two more indicator lamps to the instrument panel and typically has a manual override switch by which the driver can turn off the system.
- Stability controls are based on and linked to the ABS and some are also linked to the electronic suspension system.
- Stability control systems momentarily apply the brakes at any one wheel to correct oversteer or understeer.
- TCS testing is principally a process of using the self-diagnostic capabilities that the manufacturers include in the control program's software.

Review Questions

1. Explain the differences between positive and negative wheel slip. *negative is slippin on deceleration Positive is the opposite*
2. List the various methods used by TCSs to eliminate wheel spin. *abs electronicly Pumps Brakes or apply Brake to wheel with pos. slip.*
3. Technician A says that some TCSs eliminate wheel spin by applying the brakes on the driving wheels. Technician B says that some TCSs eliminate wheel spin by retarding ignition timing and cutting off fuel injection. Who is correct?
   A. Technician A only
   B. Technician B only
   C. Both Technician A and Technician B
   D. Neither Technician A nor Technician B
4. Technician A says that TCSs add several new components to a vehicle and require testing through their own diagnostic connectors. Technician B says

that TCSs are examples of adding new features to a vehicle primarily through computer software. Who is correct?
   A. Technician A only
   B. Technician B only
   C. Both Technician A and Technician B
   D. Neither Technician A nor Technician B
5. Technician A says that some TCSs have separate hydraulic valve units and control modules for ABS and ATC. Technician B says that some integrate both systems into one hydraulic control unit and a single control module. Who is correct?
   A. Technician A only
   B. Technician B only
   C. Both Technician A and Technician B
   D. Neither Technician A nor Technician B

Appendix A

ASE PRACTICE EXAM FOR BRAKE SYSTEMS

1. The brake system specifications for a car call for DOT 4 brake fluid. Technician A says that DOT 5 fluid can be used. Technician B says that DOT 3 fluid can be used. Who is correct?
   - A. Technician A only
   - B. Technician B only
   - C. Both Technician A and Technician B
   - D. Neither Technician A nor Technician B

2. While servicing a dual reservoir master cylinder Technician A removes the secondary piston stop bolt in order to remove the primary piston assembly; Technician B cleans the master cylinder with a degreasing solvent. Who is correct?
   - A. Technician A only
   - B. Technician B only
   - C. Both Technician A and Technician B
   - D. Neither Technician A nor Technician B

3. When multiple trouble codes are present in an antilock brake system, look for:
   - A. a weak connection at a common ground
   - B. an open circuit
   - C. low-voltage signals
   - D. high-voltage signals

4. Which of the following is a true statement?
   - A. Excessive amounts of air in the system will cause the brakes to drag.
   - B. Dragging brakes are typically caused by a stuck wheel cylinder or caliper piston.
   - C. A brake pedal return spring with too much tension can cause the brakes to drag.
   - D. All of the statements are true.

5. Technician A says that the master cylinder should be bench bled before being installed on the vehicle. Technician B says that after installing the master cylinder on the vehicle, the entire system should be bled at each wheel. Who is correct?
   - A. Technician A only
   - B. Technician B only
   - C. Both Technician A and Technician B
   - D. Neither Technician A nor Technician B

6. Two technicians were discussing brake lines. Technician A says that damaged sections of brake line can be replaced with a piece of tubing and compression fittings. Technician B says that the entire line that is damaged should be replaced and the new one bent to the right shape with a tube bender. Who is correct?
   - A. Technician A only
   - B. Technician B only
   - C. Both Technician A and Technician B
   - D. Neither Technician A nor Technician B

7. Which of the following is not likely to cause a pulsating brake pedal?
   - A. Loose wheel bearings
   - B. Worn brake pad linings
   - C. Excessive lateral runout
   - D. Nonparallel rotors

8. While servicing an ABS system Technician A uses a pressure tester to check the operation of the accumulator, hydraulic pump, and controls; Technician B tests the system by recalling the codes that are held in the control computer's memory. Who is correct?
   - A. Technician A only
   - B. Technician B only
   - C. Both Technician A and Technician B
   - D. Neither Technician A nor Technician B

9. Which term refers to variations in the thickness of the rotor?
   - A. torque
   - B. lateral runout
   - C. parallelism
   - D. pedal pulsation

10. Two technicians were discussing resurfacing brake drums. Technician A says that drums should never be cut beyond the maximum diameter designated on the drum. Technician B says that a brake drum with more than 0.005 inch out-of-roundness should be turned true on a lathe. Who is correct?
    - A. Technician A only
    - B. Technician B only
    - C. Both Technician A and Technician B
    - D. Neither Technician A nor Technician B

11. Air in the hydraulic system is being discussed. Technician A says that air can enter the system whenever a line or component is disconnected. Technician B says that air can enter the system when the fluid level in the master cylinder reservoir is low. Who is correct?
    - A. Technician A only
    - B. Technician B only
    - C. Both Technician A and Technician B
    - D. Neither Technician A nor Technician B

12. Which of the following is a true statement?
    - A. Excessive brake pedal travel is typically caused by a leaking wheel cylinder piston.
    - B. A brake pedal that moves too far before engaging the brakes is typically the result of excessive clearance between the shoes and the drum.
    - C. Excessive brake pedal travel can be caused by loose wheel bearings.
    - D. All of the statements are true.

13. Two technicians were discussing the procedure for bleeding the individual brakes on a car after the master cylinder has been bled. Technician A says that bleeding should begin at the wheel with the longest brake line. Technician B says that the proper bleeding sequence for a car with front-to-rear split is different than that for a car with a diagonal split. Who is correct?
    - A. Technician A only
    - B. Technician B only
    - C. Both Technician A and Technician B
    - D. Neither Technician A nor Technician B

14. While adjusting the front wheel bearings of a rear-wheel-drive car Technician A tightens the adjusting nut until there is no tire wobble; Technician B tightens the adjusting nut to the appropriate torque, then tightens it slightly more to align the castellations of the locking nut with the cotter pin bore. Who is correct?
    - A. Technician A only
    - B. Technician B only
    - C. Both Technician A and Technician B
    - D. Neither Technician A nor Technician B

15. Two technicians were discussing brake shoe wear. Technician A says that it is normal for the shoes to have excessive wear in the center of the shoe. Technician B says that if the lining is less than 1/32 of an inch above the rivets, it should be replaced. Who is correct?
    - A. Technician A only
    - B. Technician B only
    - C. Both Technician A and Technician B
    - D. Neither Technician A nor Technician B

16. The basic frictional parts of a brake system are being discussed. Technician A says that the harder the frictional parts are pushed together, the higher the friction. Technician B says that the harder the frictional parts are pushed together, the more heat is developed. Who is correct?
    - A. Technician A only
    - B. Technician B only
    - C. Both Technician A and Technician B
    - D. Neither Technician A nor Technician B

17. Two technicians were discussing the cause of a spongy brake pedal. Technician A says that there must be air in the system. Technician B says that this can be caused by an internal master cylinder leak. Who is correct?
    - A. Technician A only
    - B. Technician B only
    - C. Both Technician A and Technician B
    - D. Neither Technician A nor Technician B

18. While discussing how to remove the piston from a brake caliper, Technician A says that the dust boot should be removed, and then a large dull screwdriver should be inserted into the piston groove to pry the piston out. Technician B says that air pressure should be injected into the bleeder screw's bore to force the piston out of the caliper. Who is correct?
    - A. Technician A only
    - B. Technician B only
    - C. Both Technician A and Technician B
    - D. Neither Technician A nor Technician B

19. Two technicians were discussing the possible causes for a car pulling to one side during braking. Technician A says that a stuck caliper piston can cause this. Technician B says that a broken return spring in a rear drum brake can cause this. Who is correct?
    A. Technician A only
    B. Technician B only
    C. Both Technician A and Technician B
    D. Neither Technician A nor Technician B

20. While servicing a front brake rotor, Technician A uses a micrometer to check for thickness variations before turning the rotor; Technician B uses a micrometer to check the thickness of the rotor before turning the rotor. Who is correct?
    A. Technician A only
    B. Technician B only
    C. Both Technician A and Technician B
    D. Neither Technician A nor Technician B

21. Technician A says that when the driver hears the audible brake pad wear indicator, the driver should immediately park the car and not use it until the brakes are replaced. Technician B says that the tactile feedback alerts the driver to worn brake pads through pulsations felt in the brake pedal while braking. Who is correct?
    A. Technician A only
    B. Technician B only
    C. Both Technician A and Technician B
    D. Neither Technician A nor Technician B

22. Two technicians were discussing vacuum power brake systems. Technician A says that braking effort will be affected by engine or electrical problems. Technician B says that excessive pedal effort can be caused by a plugged booster air breather. Who is correct?
    A. Technician A only
    B. Technician B only
    C. Both Technician A and Technician B
    D. Neither Technician A nor Technician B

23. As part of the brake system inspection and diagnosis, Technician A checks for correct wheel alignment, inspects the tires, and notes any unbalanced loading of the vehicle. Technician B performs a test drive on a smooth, level road, testing the brakes at various speeds. Who is correct?
    A. Technician A only
    B. Technician B only
    C. Both Technician A and Technician B
    D. Neither Technician A nor Technician B

24. Which of the following is a true statement?
    A. Brake chatter can be caused by a loose metering valve.
    B. Insufficient clearance between the rotor and brake pads will cause brake chatter.
    C. A brake fluid–soaked brake pad will cause the brakes to chatter.
    D. All of the statements are true.

25. Before attempting to remove a brake drum for inspection and servicing, Technician A backs off the brake shoe adjuster; Technician B takes up all slack in the parking brake cable. Who is correct?
    A. Technician A only
    B. Technician B only
    C. Both Technician A and Technician B
    D. Neither Technician A nor Technician B

26. Two technicians were discussing parking brake systems. Technician A says that if a car has self-adjusting brakes, the parking brake should never need adjustment. Technician B says that the parking brakes are adjusted by turning the starwheel adjuster. Who is correct?
    A. Technician A only
    B. Technician B only
    C. Both Technician A and Technician B
    D. Neither Technician A nor Technician B

27. While road testing a car with an ABS system, Technician A says that during heavy braking, one or several chirps should be heard from the rear wheels as they try to lockup. Technician B says that a spongy pedal during braking is normal. Who is correct?
    A. Technician A only
    B. Technician B only
    C. Both Technician A and Technician B
    D. Neither Technician A nor Technician B

28. While servicing a parking brake, Technician A lubricates the pivot points and cables of the entire system; Technician B pulls the parking lever to its stop, then adjusts the cable so that the rear wheels are locked. Who is correct?
    A. Technician A only
    B. Technician B only
    C. Both Technician A and Technician B
    D. Neither Technician A nor Technician B

29. While servicing the front-axle disc brakes on an FWD vehicle, Technician A determines that the right wheel pad is worn and replaces both the right and left wheel pads; Technician B determines that the pads are not worn but rotates their position to ensure even pad wear. Who is correct?
    A. Technician A only
    B. Technician B only
    C. Both Technician A and Technician B
    D. Neither Technician A nor Technician B

30. While diagnosing a brake warning light that stays on, Technician A says that this can indicate a problem in the hydraulic circuit; Technician B says that the problem might be an electrical short in the circuit. Who is correct?
    A. Technician A only
    B. Technician B only
    C. Both Technician A and Technician B
    D. Neither Technician A nor Technician B

31. Which of the following is a true statement?
    A. A spongy pedal can be caused by a warped brake drum.
    B. Overheated or boiling brake fluid can cause the brake pedal to feel spongy.
    C. A soft pedal can be caused by too much vacuum in the power brake booster.
    D. None of these statements is true.

32. Two technicians were discussing installing brake shoes. Technician A says that a small amount of brake grease should be put on the heel and toe of the shoes. Technician B says that grease should be put on all areas of the backing plate where there is metal-to-metal contact. Who is correct?
    A. Technician A only
    B. Technician B only
    C. Both Technician A and Technician B
    D. Neither Technician A nor Technician B

33. Technician A says that a malfunction of the ABS causes the control module to shut off or inhibit the system. Technician B says that a loss of hydraulic fluid or power booster pressure disables the antilock brake system. Who is correct?
    A. Technician A only
    B. Technician B only
    C. Both Technician A and Technician B
    D. Neither Technician A nor Technician B

34. While replacing brake pads, Technician A removes brake fluid from the master cylinder before loosening the caliper to remove the pads; Technician B uses a C-clamp to push the piston back into its bore before removing the brake pads. Who is correct?
    A. Technician A only
    B. Technician B only
    C. Both Technician A and Technician B
    D. Neither Technician A nor Technician B

35. While road testing a car equipped with an ABS, Technician A says that during heavy braking, several pulses might be felt through the brake pedal. Technician B says that a spongy pedal during normal braking is normal. Who is correct?
    A. Technician A only
    B. Technician B only
    C. Both Technician A and Technician B
    D. Neither Technician A nor Technician B

36. Two technicians were discussing self-adjusting drum brakes. Technician A says that the brake shoe holddown springs stop the adjusters from setting the shoes too tightly against the drums. Technician B says that the operation of a cable-type self-adjuster can be checked by prying the secondary shoe away from the anchor plate and watching the action of the cable. Who is correct?
    A. Technician A only
    B. Technician B only
    C. Both Technician A and Technician B
    D. Neither Technician A nor Technician B

37. When refinishing a rotor on a lathe, the rotor wobbles excessively. Technician A says that the lathe arbor might be bent. Technician B says that the mounting adapters of the lathe might be distorted. Who is correct?
    A. Technician A only
    B. Technician B only
    C. Both Technician A and Technician B
    D. Neither Technician A nor Technician B

38. While diagnosing the cause of a vibration during braking, Technician A checks the tires and suspension before checking the brakes; Technician B checks the adjustment of the wheel bearings. Who is correct?
    A. Technician A only
    B. Technician B only
    C. Both Technician A and Technician B
    D. Neither Technician A nor Technician B

39. Two technicians were discussing wheel cylinders. Technician A says that if the cylinder bore is greater than specifications, a larger set of seals should be used. Technician B says that flat spots or grooves in the cylinder bore are normal. Who is correct?
    A. Technician A only
    B. Technician B only
    C. Both Technician A and Technician B
    D. Neither Technician A nor Technician B

40. When removing an ABS actuator, Technician A relieves hydraulic pressure in the system; Technician B disconnects the battery ground cable. Who is correct?
    A. Technician A only
    B. Technician B only
    C. Both Technician A and Technician B
    D. Neither Technician A nor Technician B

41. Two technicians were discussing proportioning valves. Technician A says that all proportioning valves are adjustable. Technician B says that only proportioning valves that are height-sensing can be adjusted. Who is correct?
    A. Technician A only
    B. Technician B only
    C. Both Technician A and Technician B
    D. Neither Technician A nor Technician B

42. The basic frictional parts of a brake system are being discussed. Technician A says that the harder the frictional parts are pushed together, the higher the friction. Technician B says that the harder the frictional parts are pushed together, the more heat is developed. Who is correct?
    A. Technician A only
    B. Technician B only
    C. Both Technician A and Technician B
    D. Neither Technician A nor Technician B

43. While diagnosing a squealing brake problem, Technician A checks to see if the antirattle spring is properly placed on the pads; Technician B checks the wear indicator on the front pads. Who is correct?
    A. Technician A only
    B. Technician B only
    C. Both Technician A and Technician B
    D. Neither Technician A nor Technician B

44. Technician A says that checking brake pedal travel is best learned through experience and that good technicians develop a "feel" over time for a good travel range. Technician B says that brake pedal travel is a set specification that is found in the service manual and measured using a special gauge and tape measure. Who is correct?
    A. Technician A only
    B. Technician B only
    C. Both Technician A and Technician B
    D. Neither Technician A nor Technician B

45. Two technicians were discussing the procedures for measuring a brake drum. Technician A says that the diameter is measured with a brake drum micrometer and that the measurement should be taken in several locations. Technician B says that the wall thickness of a drum should be measured with a micrometer. Who is correct?
    A. Technician A only
    B. Technician B only
    C. Both Technician A and Technician B
    D. Neither Technician A nor Technician B

46. While servicing the front wheel bearings on a front-wheel-drive car, Technician A removes the hub and presses a new bearing into the hub; Technician B removes the bearing and repacks the bearing with a bearing packer. Who is correct?
    A. Technician A only
    B. Technician B only
    C. Both Technician A and Technician B
    D. Neither Technician A nor Technician B

47. Which of the following is a true statement?
    A. A restricted brake line can cause a car with disc brakes to pull to one side during braking.
    B. Low inflation in one front tire can cause the car to pull during braking.
    C. A seized caliper piston can cause the car to pull to one side during braking.
    D. All of the statements are true.

48. Two technicians were discussing the causes of a rattle that occurs only when the brakes are not applied. Technician A says that the disc brake antirattle springs are incorrectly positioned or missing. Technician B says that the caliper mounting bolts are probably loose. Who is correct?
    A. Technician A only
    B. Technician B only
    C. Both Technician A and Technician B
    D. Neither Technician A nor Technician B

49. Technician A says that the piston seal retracts the caliper piston when hydraulic pressure is released. Technician B says that a return spring is used to retract a caliper piston. Who is correct?
    A. Technician A only
    B. Technician B only
    C. Both Technician A and Technician B
    D. Neither Technician A nor Technician B

50. While road testing a car equipped with an antilock brake system, Technician A says that during heavy braking several pulses might be felt through the brake pedal. Technician B says that a spongy pedal during normal braking is normal. Who is correct?
    A. Technician A only
    B. Technician B only
    C. Both Technician A and Technician B
    D. Neither Technician A nor Technician B

Appendix B

USCS AND METRIC CONVERSIONS

Linear Measurements
    1 meter (m) = 39.37 inches (in.)
    1 centimeter (cm) = 0.3937 inch
    1 millimeter (mm) = 0.03937 inch
    1 inch = 2.54 centimeters
    1 inch = 25.4 millimeters
    1 mile = 1.6093 kilometers (km)

Area (Square) Measurements
    1 square inch = 6.452 square centimeters
    1 square centimeter = 0.155 square inches

Volume Measurements
    1 cubic inch = 16.387 cubic centimeters
    1000 cubic centimeters = 1 liter (l)
    1 liter = 61.02 cubic inches
    1 gallon (gal.) = 3.7854 liters

Weight Measurements
    1 ounce (oz.) = 28.3495 grams (g)
    1 pound (lb) = 453.59 grams
    1000 grams = 1 kilogram
    1 kilogram = 2.2046 pounds

Temperature Measurements
    1 degree Fahrenheit (°F) = 9/5(°C + 32 degrees)
    1 degree Celsius (°C) = 5/9(°F − 32 degrees)

Pressure Measurements
    1 pound per square inch (psi) = 0.07031 kilograms (kg) per square centimeter
    1 kilogram per square centimeter = 14.22334 pounds per square inch
    1 bar = 14.504 pounds per square inch
    1 pound per square inch = 0.06895 bars

Torque Measurements
    10 pounds per foot = 13.558 Newton (N) meters
    1 N•m = 0.7375 ft.-lb
    1 ft.-lb = 0.138 kg m
    1 kg cm = 7.233 ft.-lb
    10 kg cm = 0.98 N•m

Appendix C

FRACTIONS OF INCHES TO DECIMAL AND METRIC EQUIVALENTS

Fraction	Decimal	Metric	Fraction	Decimal	Metric
1/64	0.015625	00.39688	33/64	0.515625	13.09687
1/32	0.03125	00.79375	17/32	0.53125	13.49375
3/64	0.046875	01.19062	35/64	0.546875	13.89062
1/16	0.0625	01.58750	17/16	0.5625	14.28750
5/64	0.078125	01.98437	37/64	0.578125	14.68437
3/32	0.09375	02.38125	19/32	0.59375	15.08125
7/64	0.109375	02.77812	39/64	0.609375	15.47812
1/8	**0.125**	**03.1750**	**5/8**	**0.625**	**15.8750**
9/64	0.140625	03.57187	41/64	0.640625	16.27187
5/32	0.15625	03.96875	21/32	0.65625	16.66875
11/64	0.171875	04.36562	43/64	0.671875	17.06562
3/16	0.1875	04.76250	11/16	0.6875	17.46250
13/64	0.203125	05.15937	45/64	0.703125	17.85937
7/32	0.21875	05.55625	23/32	0.71875	18.25625
15/64	0.234375	05.95312	47/64	0.734375	18.65312
1/4	**0.250**	**06.35000**	**3/4**	**0.750**	**19.05000**
17/64	0.265625	06.74687	49/64	0.765625	19.44687
9/32	0.28125	07.14375	25/32	0.78125	19.84375
19/64	0.296875	07.54062	51/64	0.796875	20.24062
5/16	0.3125	07.93750	13/16	0.8125	20.63750
21/64	0.328125	08.33437	53/64	0.828125	21.03437
11/32	0.34375	08.73125	27/32	0.84375	21.43125
23/64	0.359375	09.12812	55/64	0.859375	21.82812
3/8	**0.375**	**09.52500**	**7/8**	**0.875**	**22.22500**
25/64	0.390625	09.92187	57/64	0.890625	22.62187
13/32	0.40625	10.31875	29/32	0.90625	23.01875
27/64	0.421875	10.71562	59/64	0.921875	23.41562
7/16	0.4375	11.11250	15/16	0.9375	23.81250
29/64	0.453125	11.50937	61/64	0.953125	24.20937
15/32	0.46875	11.90625	31/32	0.96875	24.60625
31/64	0.484375	12.30312	63/64	0.984375	25.00312
1/2	**0.500**	**12.7000**	**64/64=1**	**1.000**	**25.40000**

Fraction	Decimal	Metric

Appendix D

GENERAL TORQUE SPECIFICATIONS

Note: The values in this chart should be used only when manufacturer's specifications are *not* available. Also, the values are valid only when SAE 10 oil is used to lubricate the threads of the bolt.

Bolt Diameter in Inches	Torque: ft.-lb		
	SAE 2	SAE 5	SAE 8
1/4	7	10	14
5/16	14	21	30
3/8	24	37	52
7/16	39	60	84
1/2	59	90	128
9/16	85	130	184
5/8	117	180	255
3/4	205	320	450
7/8	200	515	730
1	300	775	1090

Bolt Diameter in Millimeters	Torque: kg cm* kg m Property Class:									
	4.6	4.8	5.6	5.8	6.6	6.8	6.9	8.8	10.9	12.9
6	49*	63*	61*	79*	74*	95*	103*	126*	172*	206*
8	119*	153*	148*	178*	178*	230*	250*	306*	417*	500*
10	235*	303*	294*	379*	353*	455*	495*	606*	8.2	10
12	411*	529*	427*	662*	616*	7.9	8.6	10.5	14	17
14	654*	8.4	8.2	10.5	10	12	13	17	23	27
16	10	13	12	16	15	20	21	26	36	43
18	14	18	17	23	21	27	30	36	49	59
22	27	35	34	44	41	52	57	70	95	114

Bilingual Glossary

ABS event A rapid reduction in speed with the brakes applied and when one or more wheels begin to lock up.
Evento ABS *Reducción rápida de velocidad con los frenos aplicados durante la cual una o más ruedas se empiezan a bloquear.*

Acceleration sensor This sensor provides information about the rate of forward or reverse acceleration or deceleration. Also known as a G-switch.
Sensor de aceleración *Este sensor proporciona información acerca de la velocidad de aceleración o desaceleración hacia adelante o en reversa. También se conoce como interruptor G.*

Acceleration slip regulation (ASR) ASR is the name of the traction control system manufactured by Bosch.
Regulación de deslizamiento de aceleración (ASR) *Nombre del sistema de control de tracción fabricado por Bosch.*

Accumulator A container that stores hydraulic fluid under pressure. It can be used as a fluid shock absorber or as an alternate pressure source. A spring or compressed gas behind a sealed diaphragm provides the accumulator pressure.
Acumulador *Recipiente que almacena fluido hidráulico bajo presión. Se puede usar como amortiguador o como una fuente de presión alterna. Un resorte o gas comprimido detrás de un diafragma sellado proporciona la presión del acumulador.*

Actuator Any device that receives an output signal, or command, from a computer and does something in response to the signal.
Actuador *Cualquier dispositivo que recibe un comando o señal de salida de una computadora y que hace algo en respuesta a la señal.*

Air bag system A system that uses impact sensors, a vehicle's on-board computer, an inflation module, and a nylon bag in the steering column and dash to protect the driver and passenger during a head-on collision.
Sistema de bolsa de aire *Sistema que usa sensores de impacto, la computadora a bordo del vehículo, el módulo de inflado y una bolsa de fibra sintética en la columna de dirección y el tablero para proteger al conductor y al pasajero durante una colisión de frente. Sistema que usa sensores de impacto, la computadora a bordo, el módulo de inflado y una bolsa de fibra sintética en la columna de dirección y el tablero para proteger al conductor y al pasajero durante una colisión de frente. Sistema que usa sensores de impacto, la computadora a bordo del vehículo, el módulo de inflado y una bolsa de fibra sintética en la columna de dirección y el tablero para proteger al conductor y al pasajero durante una colisión de frente.*

Allen wrenches Allen wrenches, or hex-head wrenches, are used to tighten and loosen setscrews and fit into a machined hex-shaped recess in the bolt or screw.
Llaves Allen *Las llaves Allen, o de cabeza hexagonal, se usan para apretar y aflojar tornillos de presión y ajustarlos en un hueco maquinado con forma hexagonal en el perno o tornillo.*

Alternating current (AC) Electrical current that changes direction between positive and negative.
Corriente alterna *Corriente eléctrica que cambia de dirección entre positivo y negativo.*

Ammeter The instrument used to measure electrical current flow in a circuit.
Amperímetro *Instrumento que se usa para medir el flujo de corriente eléctrica en un circuito.*

Ampere The unit for measuring electrical current; usually called an amp.
Amperio *Unidad para medir la corriente eléctrica. Generalmente se conoce como amp.*

Amplitude Signal strength, or the maximum measured value of a signal.
Amplitud *Fuerza de señal o el valor medido máximo de una señal.*

Analog signal A voltage signal that varies within a given range (from high to low, including all points in between).
Señal analógica *Señal de voltaje que varía dentro de un rango (de alto a bajo, incluidos todos los puntos intermedios).*

Annular   Shaped like a circle or ring.
Anular   Con forma circular o anular.

Antilock brake control module   The computer that controls the ABS operation; can be used on some systems for traction control.
Módulo de control de frenos antibloqueo   Computadora que controla la operación del ABS. Se puede usar en algunos sistemas para control de tracción.

Antilock brake system (ABS)   A brake system that modulates hydraulic pressure to one or more wheels to prevent those wheels from locking during braking.
Sistema de frenos antibloqueo (ABS)   Sistema de frenos que modula la presión hidráulica hacia una o más ruedas y evita que se bloqueen durante el frenado.

Aqueous   Water based.
Acuoso   Con base de agua.

Aramid fibers   Refers to a family of synthetic materials that are stronger than steel but weigh little more than half of what an equal volume of fiberglass would weigh.
Fibras de aramida   Se refiere a una familia de materiales sintéticos más fuertes que el acero pero que pesan apenas un poco más de la mitad de lo que pesaría un volumen igual de fibra de vidrio.

Arcing   The process of grinding or forming brake shoe linings to conform to the diameter of the drum and to provide clearance between the shoe and the drum where it is needed.
Formación de arcos eléctricos   Proceso de esmerilar o formar el revestimiento de las zapatas de freno al diámetro del tambor y para proporcionar espacio libre entre la zapata y el tambor donde sea necesario.

Asbestos   A silicate compound with excellent heat dissipation abilities and high coefficient of friction containing millions of small, linked fibers that give it both strength and flexibility but are a health hazard.
Asbesto   Compuesto de silicato con excelente capacidad de disipación de calor y alto coeficiente de fricción que contiene millones de pequeñas fibras enlazadas que le proporcionan fuerza y flexibilidad pero que son un peligro para la salud.

Asbestosis   A progressive lung disease caused by asbestos fibers continually lodging in the lungs and inflaming the lungs' air sacs.
Asbestosis   Enfermedad pulmonar progresiva ocasionada por fibras de asbesto que se alojan en los pulmones y que inflaman los alvéolos pulmonares.

ASE   National Institute for Automotive Service Excellence.
ASE   Instituto nacional de excelencia en el servicio automotor

Aspect ratio   Expressed as a percentage, the ratio of the cross-section height to the cross-section width of a tire.
Relación de aspecto   Relación entre la altura de la sección transversal y el ancho de la sección transversal de un neumático, expresada como un porcentaje.

ATC   The acronym for automatic traction control.
ATC   Acrónimo para control de tracción automática.

Atmospheric pressure   The weight of the air that makes up the Earth's atmosphere.
Presión atmosférica   Peso del aire que constituye la atmósfera terrestre.

Atmospheric suspended   A term that describes a type of power brake vacuum booster in which atmospheric pressure is present on both sides of the diaphragm when the brakes are released; an obsolete type of vacuum booster.
Suspendido atmosférico   Término que describe un tipo de reforzador de vacío de frenos de potencia en el cual la presión atmosférica está presente en ambos lados del diafragma cuando los frenos se liberan; tipo obsoleto de reforzador de vacío.

Automotive Friction Material Edge Code   A series of codes on the side of a brake lining (disc or drum) that identifies the manufacturer, the lining material, and the coefficient of friction. These codes are for lining identification and comparison; they do not indicate quality.
Código automotor del borde de material de fricción   Serie de códigos en el lado del revestimiento del freno (disco o tambor) que identifica al fabricante, el material de revestimiento y el coeficiente de fricción. Estos códigos son para identificación y comparación del revestimiento; no indican calidad.

AWG (American Wire Gauge System)   The system used to specify wire size (conductor cross-sectional area) by a series of gauge numbers. The lower the number, the larger the cross section.
AWG (sistema estadounidense de calibre de alambre)   Sistema para especificar el tamaño del alambre (área de sección transversal del conductor) mediante una serie de números de calibre. Mientras menor sea el número, mayor será la sección transversal.

Backing plate   The mounting surface for all other parts of a drum brake assembly except the drum.
Placa de apoyo   La superficie de montaje de todas las otras piezas de un ensamble de freno de tambor, excepto el tambor.

Banjo fitting   A round, banjo-shaped tubing connector with a hollow bolt through its center that enables a brake line to be connected to a hydraulic component at a right angle.
Ajuste de guitarra   Conector redondo entubado con forma de guitarra, con un perno hueco que atraviesa el centro y que permite que una tubería de freno se conecte a un componente hidráulico en ángulo recto.

Bearing cage   The steel component that holds the rollers together in a tapered roller bearing.
Jaula de cojinete.   El componente de acero que sostiene los rodillos juntos en un cojinete de rodillos cónicos.

Bearing cone   The inner race of a tapered roller bearing; usually an integral assembly with the rollers and the cage.
Rodillo cónico de cojinete.   Anillo guía interno de un cojinete de rodillos cónicos. Habitualmente un montaje integral con los rodillos y la jaula.

Bearing cup   The outer race of a tapered roller bearing; usually pressed into the wheel hub.
Cubeta de cojinete   Anillo guía de un cojinete de rodillos cónicos; generalmente se presiona sobre el cubo de la rueda.

Bearing end play The designed looseness in a bearing assembly.
Juego longitudinal de balero *La holgura planeada en un ensamble de baleros.*

Bedding-in See burnishing.
Rasqueteado *Véase bruñido.*

Belted bias ply tire Tire construction that incorporates the belts used in radial ply tires with the older bias ply construction.
Neumático de capa diagonal estabilizadora *Construcción de neumático que incorpora los cinturones que se usan en neumáticos de capa radial con la construcción anterior de capa diagonal.*

Bias ply tire Tire construction in which the cords in the body plies of the carcass run from bead to bead at an angle from 26 to 38 degrees instead of 90 degrees, as in a radial ply tire.
Neumático de capa diagonal *Construcción de neumático en el cual las cuerdas en las capas del cuerpo de la carcasa corren de ceja a ceja en un ángulo de 26 a 38 grados en lugar de 90 grados, como en un neumático de capa radial.*

Bimetallic drum A composite brake drum made of cast iron and aluminum.
Tambor bimetálico *Tambor de freno fabricado con aleación de acero fundido y aluminio.*

Binary system The mathematical system that uses only the digits 0 and 1 to present information.
Sistema binario *Sistema matemático que usa solo los dígitos 0 y 1 para presentar información.*

Binder Adhesive, or glue, used in brake linings to bond all the other materials together.
Aglomerante *Adhesivo o pegamento que se usa en revestimientos de frenos para adherir todos los demás materiales.*

Bleeder screw A screw that opens and closes a bleeding port in a caliper or a wheel cylinder.
Tornillo purgador *Tornillo que abre y cierra un puerto de purgado en una mordaza o cilindro de rueda.*

Bolt head The part of a bolt that the socket or wrench fits over in order to torque or tighten the bolt.
Cabeza de perno *Parte del perno sobre la cual se ajusta el casquillo o la llave de tuercas para aplicar torsión o apretar el perno.*

Bonded lining Brake lining attached to the pad or shoe by high-strength, high-temperature adhesive.
Revestimiento adherido *Revestimiento de freno que se une a la pastilla o zapata mediante un adhesivo muy fuerte de alta temperatura.*

Brake bleeding A procedure that pumps fresh brake fluid into the brake hydraulic system and forces out air bubbles and the old aerated fluid through bleeding ports in calipers and wheel cylinders.
Purgado de frenos *Procedimiento que bombea líquido de frenos nuevo en el sistema hidráulico de los frenos y obliga a que salgan burbujas de aire y el líquido ventilado anterior a través de los puertos de sangrado en las mordazas y cilindros de las ruedas.*

Brake caliper The part of a disc brake system that converts hydraulic pressure back to mechanical force that applies the pads to the rotor. The caliper is mounted on the suspension or axle housing and contains a hydraulic piston and the brake pads.
Mordaza de frenos *Parte del sistema de frenos de disco que convierte de nuevo la presión hidráulica a fuerza mecánica que aplica las pastillas al rotor. La mordaza se monta en la suspensión o caja del eje y contiene un pistón hidráulico y las pastillas de frenos.*

Brake fade The partial or total loss of braking power caused by excessive heat, which reduces friction between the brake linings and the rotors or drums.
Pérdida de efectividad de freno *Pérdida parcial o total de la potencia del freno causada por calor excesivo, lo cual reduce la fricción entre los revestimientos del freno y los rotores o tambores.*

Brake pad The part of a disc brake assembly that holds the lining friction material that is forced against the rotor to create friction to stop the vehicle.
Pastilla de freno *Parte del ensamble de freno de disco que sostiene el material de fricción del revestimiento que se aplica por la fuerza contra el rotor para crear la fricción que detiene al vehículo.*

Brake shoe A metal assembly onto which the frictional lining is attached for drum brake systems.
Zapatas de Freno *Ensamble de metal en el cual el revestimiento friccionante está unido al sistema de freno de disco.*

Brinelling Denting of a bearing race following a shock load.
Brinelado *Indentación permanente en el anillo guía de cojinete debido a cargas excesivas.*

Buffer An isolating circuit used to avoid interference between a driven circuit and its driver circuit; also a storage device, or circuit, that compensates for a difference in the rate of data transmission.
Búfer *Circuito aislado que se usa para evitar la interferencia entre un circuito controlado y su circuito controlador. También un dispositivo de almacenamiento o circuito que compensa la diferencia en la velocidad de transmisión de datos.*

Burnishing To smoothen or polish a material with a frictional material under pressure.
Bruñido *Suavizar o pulir un material con un material friccionante bajo presión.*

Bushing A cylindrical lining used as a bearing assembly made of steel, brass, bronze, nylon, or plastic.
Buje *Revestimiento cilíndrico que se usa como montaje de cojinete hecho de acero, latón, bronce, fibra sintética o plástico.*

Caliper support The bracket or anchor that holds the brake caliper.
Apoyo de mordaza *Soporte o tirante de amarre que sostiene la mordaza de frenos.*

Cam-ground lining A brake shoe lining that has been arced or formed so that it is thinner at the ends than at the center and the lining surface is not a portion of a circle with a constant radius.
Revestimiento rectificado de leva. *Revestimiento de zapata de freno que se ha arqueado o formado de manera que*

sea más delgada en los extremos que en el centro y la superficie de revestimiento no es una porción de un círculo con radio constante.

Carcass The steel beads around the rim and layers of cords or plies that are bonded together to give a tire its shape and strength.
Carcasa Cejas de acero alrededor del rin y capas de cuerda que se adhieren entre sí para darle al neumático su forma y fuerza.

Casing Layers of sidewall and undertread rubber added to a tire carcass.
Casco Capas de pared y caucho cojín que se agregan a la carcasa del neumático.

Center high-mounted stop lamp (CHMSL) A third stop lamp on vehicles built since 1986, located on the vehicle centerline no lower than 3 inches below the rear window (6 inches on convertibles).
Luz de freno montada arriba y al centro (CHMSL) Tercera luz de freno en vehículos fabricados desde 1986, ubicada en la línea central del vehículo no inferior a 3 pulgadas debajo de la ventana trasera (6 pulgadas en automóviles convertibles).

Central processing unit (CPU) The calculating part of any computer that makes logical decisions by comparing conditioned input with data in memory.
Unidad central de procesamiento (CPU) Parte calculadora de cualquier computadora que toma decisiones lógicas al comparar los datos que se ingresan con datos en la memoria.

Channel Individual legs of the hydraulic system that relay pressure from the master cylinder to the wheel cylinder.
Canal Tramos individuales del sistema hidráulico que retransmiten la presión del cilindro maestro al cilindro de la rueda.

Check valve A valve that allows fluid or air flow in one direction but not in the opposite direction.
Válvula unidireccional Válvula que permite el flujo de aire en un sentido pero no en el sentido opuesto.

Chlorinated hydrocarbon solvents Colorless solvents with a strong odor of ether or chloroform and whose vapors can cause drowsiness or loss of consciousness.
Solventes clorinados de hidrocarburo Solventes incoloros con un fuerte olor a éter o cloroformo y cuyos vapores pueden causar adormecimiento o pérdida de la conciencia.

Circuit breaker Resettable circuit protection device that automatically opens in response to high current.
Disyuntores Dispositivo reajustable de protección de circuito que se abre automáticamente en respuesta a una corriente alta.

Coefficient of friction A numerical value that expresses the amount of friction between two objects, obtained by dividing tensile force (motion) by weight force. A coefficient of friction can be either static or kinetic.
Coeficiente de fricción Valor numérico que expresa la cantidad de fricción entre dos objetos, que se obtiene al dividir la fuerza tensora (movimiento) entre la fuerza de peso. El coeficiente de fricción puede ser estático o cinético.

Cold inflation pressure The tire inflation pressure after a tire has been standing for three hours or driven less than one mile after standing for three hours.
Presión de inflado en frío Presión de inflado del neumático después de que el neumático ha estado inmóvil durante tres horas o se le ha conducido por menos de una milla después de estar inmóvil durante tres horas.

Collet A collar or band that something is pushed or forced into.
Boquilla Collar o banda en la cual se empuja o se fuerza la entrada de algo.

Combination valve A hydraulic control valve with two or three valve functions in one valve body.
Válvula de combinación Válvula de control hidráulico con dos o tres funciones de válvula en un cuerpo de válvula.

Complete circuit An electrical circuit that includes a path that connects the positive and negative terminals of the electrical power source.
Circuito completo Circuito eléctrico que incluye una ruta que conecta las terminales positivas y negativas de la fuente de alimentación eléctrica.

Composite drums A drum made of different materials, usually cast iron and steel or aluminum, to reduce weight.
Tambores de aleación Tambor hecho de materiales diferentes, habitualmente hierro y acero fundido o aluminio, para reducir el peso.

Composite rotor A rotor made of different materials, usually cast iron and steel, to reduce weight.
Rotor de aleación Rotor hecho de materiales diferentes, habitualmente hierro y acero fundido, para reducir el peso.

Conductors Materials with a low resistance to the flow of current.
Conductores Materiales con una baja resistencia al flujo de corriente.

Conduit A flexible metal housing or jacket that houses the parking brake cables to protect them from dirt, rust, abrasion, and other damage.
Conducto Alojamiento o forro de metal flexible que alberga los cables del freno de estacionamiento para protegerlos de la suciedad, la herrumbre, la abrasión y otro tipo de daño.

Control valve assembly The block of metal that contains the hydraulic passages and electric solenoids used to direct brake fluid during an ABS event.
Montaje de válvula de control Bloque de metal que contiene los pasajes hidráulicos y solenoides eléctricos que se usan para dirigir el líquido de frenos durante un evento ABS.

Controller antilock brake (CAB) The computer that controls the ABS operation.
Controlador de frenos antibloqueo (CAB) Computadora que controla la operación del ABS.

Controllers Devices, such as switches or relays, that direct the flow of electrons.
Controladores Dispositivos, como los interruptores o relés, que dirigen el flujo de electrones.

Crocus cloth Fine polishing cloth, or emery cloth.
Cañamazo Trapo fino para pulir o trapo de esmeril.

Cross-feed The distance the cutting bit of a brake lathe moves across the friction surface during each lathe revolution.
Avance transversal *Distancia a la que se mueve la broca de corte de un torno de freno sobre la superficie de fricción durante cada revolución del torno.*

Crosshatch A crisscross pattern formed by a cylinder hone.
Cuadrícula *Patrón cruzado que forma un rectificador de cilindro.*

Cup expander A metal disc that bears against the inner sides of wheel cylinder seals to hold the seal lips against the cylinder bore when the brakes are released.
Extensor de copa *Disco de metal que se inclina contra los lados interiores de los sellos del cilindro de la rueda para retener los bordes del sello contra el barreno del cilindro cuando se liberan los frenos.*

Cup seal A circular rubber seal with a depressed center section surrounded by a raised sealing lip to form a cup. Cup seals often are used on the front ends of hydraulic cylinder pistons because they seal high pressure in the forward direction of travel but not in the reverse.
Sello de copa *Sello circular de caucho con una sección central hundida rodeada de bordes selladores que se elevan para formar una copa. Los sellos de copa con frecuencia se usan en los extremos delanteros de los pistones de cilindros hidráulicos porque sellan presión alta en la dirección de desplazamiento hacia adelante pero no a la inversa.*

Curing agent A class of materials used in brake linings to accelerate the chemical reaction of the binders and other materials.
Agente endurecedor *Clase de material que se usa en revestimientos de frenos para acelerar la reacción química de los aglomerantes y otros materiales.*

Current Relates to the number of electrons flowing past a given point in a given amount of time.
Corriente *Se relaciona con el número de electrones que fluyen más allá de un punto determinado en una cantidad de tiempo determinada.*

Cycle The microprocessor action of turning solenoids on or off.
Ciclo *La acción del microprocesador de encender o apagar solenoides.*

Cylinder hone An abrasive tool used to refinish the bore of a wheel cylinder or caliper.
Rectificador del cilindro *Herramienta abrasiva que se usa para reacabar el diámetro interior del calibrador del cilindro de la rueda.*

Data link connector (DLC) The attachment point for connecting a scan tool to the system.
Conector de enlace de datos (DLC) *El punto de unión para conectar al sistema una herramienta examinadora de diagnostico.*

Deployed The condition of an air bag when it has been released and expanded.
Desplegada *Condición de una bolsa de aire cuando se haya liberado y expandido.*

Diagnostic trouble code (DTC) A numerical code generated by an electronic control system to indicate a problem in a circuit or subsystem or to indicate a general condition that is out of limits.
Código de problema de diagnóstico (DTC) *Código numérico generado por un sistema de control electrónico para indicar un problema en un circuito o subsistema o para indicar una condición general que está fuera del límite.*

Dial caliper A versatile measuring instrument, accurate to 0.001 inch, capable of taking inside, outside, depth, and step measurements.
Calibrador de esfera *Versátil instrumento de medición, con una precisión de hasta 0.001 pulgadas, capaz de tomar medidas interiores, exteriores, de profundidad y escalonadas.*

Dial indicator A measuring tool used to measure component movement down to 0.001 inch. The clearance is read on a dial.
Indicador de esfera *Herramienta de medición que se usa para medir el movimiento de un componente hasta 0.001 pulgadas. La tolerancia se lee en la esfera.*

Diaphragm A flexible membrane, usually made of rubber, that isolates two substances or areas from each other.
Diafragma *Membrana flexible fabricada generalmente con caucho que aísla dos sustancias o áreas una de otra.*

Die A tool used to restore external threads on a fastener.
Dado *Herramienta que se usa para restaurar cuerdas externas en un dispositivo de fijación.*

Digital signal A voltage signal that has only two values—on or off.
Señal digital *Señal de voltaje que solo tiene dos valores: encendido o apagado.*

Digitized The process of converting an analog voltage signal to a digital equivalent that the computer can understand.
Digitalizar *Proceso de convertir una señal de voltaje análoga a un equivalente digital que la computadora pueda entender.*

Diode A semiconductor that allows current to flow through it in one direction only.
Diodo *Semiconductor que permite que la corriente fluya directamente solo en una dirección.*

Direct current (DC) A type of electrical power used in mobile applications; a unidirectional current of substantially constant value.
Corriente directa *Tipo de energía eléctrica que se usa en aplicaciones móviles; corriente unidireccional de valor sustancialmente constante.*

Directional control The ability to steer the automobile while stopping.
Control direccional *Habilidad de dirigir el automóvil mientras se hace una parada.*

Directional stability The ability to maintain a straight-line stopping action.
Estabilidad direccional *Habilidad de mantener una acción de parada en línea recta.*

Disc brake A brake in which friction is generated by brake pads rubbing against the friction surfaces on both sides of a brake disc, or rotor, attached to the wheel.
Freno de disco *Freno en el cual la fricción se genera por el frotamiento de las pastillas de freno contra las superficies de fricción en ambos lados de un freno de disco o rotor unido a la rueda.*

Discard diameter The allowable wear dimension, not the allowable machining dimension.
Diámetro de descarte *La dimensión de desgaste permitido, no la dimensión de maquinado permitido.*

Discard dimension The minimum thickness of a brake rotor or the maximum diameter of a drum.
Dimensión de descarte *Grosor mínimo de un rotor de freno o el diámetro máximo de un tambor.*

Discard limit The maximum allowable inside diameter for a brake drum.
Límite de descarte *El diámetro interior máximo permisible para un tambor de freno.*

DMM A digital multimeter.
DMM *Multímetro digital*

DOT 3, 4, and 5 United States Department of Transportation specification numbers for hydraulic brake fluids.
DOT 3, 4, y 5 *Números de especificación del Departmento de Transporte de los Estados Unidos para líquido de frenos hidráulicos.*

Double flare A type of tubing flare connection in which the end of the tubing is flared out and then is formed back on to itself.
Boquilla doble *Tipo de conexión de boquilla entubada en la cual el extremo de la tubería se acampana hacia afuera y luego se curva sobre sí mismo.*

DRAC An acronym for digital ratio adapter controller. It is used by the microprocessor to change analog signals to digital signals.
DRAC *Acrónimo de controlador adaptador de relación digital. Se usa por el microprocesador para cambiar señales análogas a señales digitales.*

Drum brake A brake in which friction is generated by brake shoes rubbing against the inside surface of a brake drum attached to the wheel.
Freno de tambor *Freno en el cual la fricción se genera por el frotamiento de las zapatas de freno contra la superficie interior de un tambor de freno unido a la rueda.*

Drum web The closed side of a brake drum.
Entramado de tambor *El lado cerrado de un tambor de freno.*

Duo-servo brake A drum brake that develops self-energizing action on the primary shoe, which in turn applies servo action to the secondary shoe to increase its application force; also called a dual-servo or a full-servo brake.
Freno duoservo *Freno de tambor que desarrolla acción autoenergizante en la zapata principal, que a su vez aplica servoacción a la zapata secundaria para aumentar su fuerza de aplicación; también se conoce como freno servodual o servocompleto.*

Duty cycle The percentage of time that a solenoid is energized during one complete on/off cycle during pulse-width modulation.
Ciclo de trabajo *Porcentaje de tiempo en el cual se energiza un solenoide durante un ciclo completo de encendido y apagado durante la modulación de ancho de pulso.*

Dynamic pressure The pressure of the fluid while it is in motion.
Presión dinámica *Presión del líquido mientras está en movimiento.*

Dynamic range The operating range of a sensor.
Rango aerodinámico *Rango operativo de un sensor.*

Eccentric Not round or concentric.
Excéntrico *Que no es ni redondo ni concéntrico.*

Electricity The type of energy caused by the flow of electrons from one atom to another. It is the release of energy as one electron leaves the orbit of one atom and jumps into the orbit of another.
Electricidad *Tipo de energía que se debe al flujo de electrones de un átomo a otro. Es la liberación de energía que ocurre cuando un electrón deja la órbita de un átomo y salta a la órbita de otro.*

Electrohydraulic unit The microprocessor and hydraulic unit combined in one unit.
Unidad electrohidráulica *El microprocesador y la unidad hidráulica se combinan en una unidad.*

Electrolyte A material whose atoms become ionized, or electrically charged, in solution. Automobile battery electrolyte is a mixture of sulfuric acid and water.
Electrolito *Material cuyos átomos se ionizan o se cargan eléctricamente al estar en solución. El electrolito de la batería de un automóvil es una mezcla de ácido sulfúrico y agua.*

Electromagnetic induction The generation of voltage in a conductor by relative motion between the conductor and a magnetic field.
Inducción electromagnética *Generación de voltaje en un conductor por movimiento relativo entre el conductor y un campo magnético.*

Electromagnetic interference A magnetic force field that influences a signal being sent to the microprocessor.
Interferencia electromagnética *Campo de fuerza magnética que influencia la señal que se envía al microprocesador.*

Electromotive force (EMF) Another name for voltage; the force or pressure that exists between a positive and negative point within an electrical circuit. This force is measured in units called volts.
EMF (Fuerza electromotriz) *Otro nombre para el voltaje; fuerza o presión que existe entre un punto positivo y un punto negativo dentro de un circuito eléctrico. Esta fuerza se mide en unidades conocida como voltios.*

EPA Environmental Protection Agency.
EPA *Agencia de protección del medio ambiente.*

Equalizer   Part of the parking brake linkage that balances application force and applies it equally to each wheel. The equalizer often contains the linkage adjustment point.
Ecualizador   *Pieza del mecanismo articulado del freno de estacionamiento que balancea la aplicación de fuerza y la aplica de manera equitativa a cada rueda. El ecualizador con frecuencia contiene el punto de ajuste del sistema articulado.*

Erasable programmable read-only memory (EPROM)   Computer memory program circuits that can be erased and reprogrammed. Erasure is done by exposing the integrated circuit's chip to ultraviolet light.
Memoria programable de solo lectura que se puede borrar (EPROM)   *Circuitos de programa de memoria de la computadora que se pueden borrar y volver a programar. El borrado se hace al exponer el chip del circuito integrado a luz ultravioleta.*

Eyeball   The eyeball is a voltage monitoring circuit.
Dispositivo vigilador   *Circuito de monitoreo de voltaje.*

Feeler gauge   A feeler gauge is a thin strip of metal or plastic of known and closely controlled thickness used to measure clearances and gaps.
Calibrador de espesor   *Tira delgada de metal o plástico de grosor conocido y estrechamente controlado que se usa para medir los espacios y huecos.*

Field of flux   A magnetic field that exists around every magnet and consists of imaginary lines along which a magnetic force acts.
Campo de flujo   *Campo magnético que existe alrededor de todos los imanes y que consiste en líneas imaginarias sobre las cuales actúa la fuerza magnética.*

Filler   A class of materials used in brake linings to reduce noise and improve heat transfer.
Rellenador   *Clase de material que se usa en revestimientos de frenos para reducir el ruido y mejorar la transferencia de calor.*

Fillet   The smooth curve where the shank flows into the bolt head.
Filete   *Curva uniforme donde el vástago se une a la cabeza del perno.*

Fixed caliper   A brake caliper that is bolted to its support and does not move when the brakes are applied. A fixed caliper must have pistons on both the inboard and the outboard sides.
Mordaza fija   *Mordaza de freno que se sujeta con pernos y que no se mueve cuando se aplican los frenos. Una mordaza fija debe tener pistones en los lados hacia el interior y hacia el exterior.*

Fixed rotor   A rotor that has the hub and the rotor cast as a single part.
Rotor fijo   *Rotor que tiene el cubo y el rotor coladas como una sola pieza.*

Fixed seal   A seal for a caliper piston that is installed in a groove in the caliper bore and that does not move with the piston.
Sello fijo   *Sello fijo para un pistón de mordaza que se instala en una ranura en el diámetro interior de la mordaza y que no se mueve con el pistón.*

Flat rate   A pay system in which technicians are paid for the amount of work they do. Each job has a flat-rate time.
Tarifa fija   *Sistema de pago en el cual se le paga a los técnicos por la cantidad de trabajo que realizan. Cada trabajo tiene un tiempo de tarifa fija.*

Flat-rate manuals   Literature containing figures dealing with the length of time specific repairs are supposed to require. Flat-rate manuals often contain a parts list with prices as well.
Manuales de tarifas fijas   *Literatura que contiene cifras que tienen que ver con la cantidad de tiempo que se supone toma hacer reparaciones específicas. Los manuales de tarifas fijas con frecuencia contienen también listas de piezas con precios.*

Floating caliper   A caliper that is mounted to its support on two locating pins, or guide pins. The caliper slides on the pin in a sleeve or bushing.
Mordaza flotante   *Mordaza que se monta en el soporte sobre dos pasadores de ubicación o pasadores guía. La mordaza se desliza sobre el pasador en una manga o buje.*

Floating drum   A brake drum that is separate from the wheel hub or axle. A floating drum usually is held in place on studs in the axle flange or hub by the wheel and wheel nuts.
Tambor flotante   *Freno de tambor que está separado del eje o cubo de la rueda. El tambor flotante con frecuencia se sujeta sobre remaches en el reborde del eje o cubo mediante la rueda y las tuercas de la rueda.*

Floating rotor   A rotor and hub assembly made of two separate parts.
Rotor flotante   *Ensamble de rotor y cubo fabricado con dos partes separadas.*

Flux lines   Flux lines are lines of magnetism.
Líneas de flujo   *Líneas de magnetismo.*

FMVSS   Federal Motor Vehicle Safety Standards.
FMVSS   *Normas federales de seguridad de vehículos automotores*

Force   Power working against resistance to cause motion.
Fuerza   *Potencia que trabaja contra la resistencia para causar movimiento.*

Frequency   The number of times, or speed, at which an action occurs within a specified time interval. In electronics, frequency indicates the number of times that a signal occurs, or repeats, in cycles per second. Cycles per second are indicated by the hertz (Hz) unit of measure.
Frecuencia   *El número de veces o velocidad a la cual ocurre una acción dentro de un intervalo específico de tiempo. En electrónica, indica el número de veces que ocurre o se repite una señal, en ciclos por segundo. Los ciclos por segundo se indican mediante la unidad de medida hertz (Hz).*

Friction   The force that resists motion between the surfaces of two objects or forms of matter.
Fricción.   *Fuerza que resiste el movimiento entre las superficies de dos objetos o formas de materia.*

Friction modifier   A class of materials used in brake linings to modify the final coefficient of friction of the linings.
Modificador de fricción   Clase de material que se usan en revestimientos de frenos para modificar el coeficiente final de fricción de los revestimientos.

Fulcrum   The pivot point of a lever.
Fulcro   Punto de pivote de una palanca.

Fuse   An electrical device used to protect a circuit against accidental overload or unit malfunction.
Fusible   Dispositivo eléctrico que se usa para proteger un circuito contra la sobrecarga accidental o mal funcionamiento de la unidad.

Fusible link   A type of fuse made of a special wire that melts to open a circuit when current draw is excessive.
Eslabón fusible   Tipo de fusible hecho de un alambre especial que se derrite para abrir un circuito cuando la llamada de corriente es excesiva.

Gas fade   Brake fade due to hot gas and dust that reduce friction between the drum or rotor during prolonged hard braking.
Pérdida de efectividad por gas   Pérdida de efectividad del freno debido al gas caliente y al polvo que reducen la fricción entre el tambor o rotor durante el frenado prolongado e intenso.

Geometric centerline   A static dimension represented by a line through the center of the vehicle from front to rear.
Línea central geométrica   Dimensión estática representada por una línea que atraviesa el centro del vehículo desde la parte delantera hasta la parte trasera.

Grade marks   Marks on fasteners that indicate strength.
Marcas de grado   Marcas en los dispositivos de fijación que indican fuerza.

Gravity bleeding   The process of letting old brake fluid and air drain from the brake hydraulic system through a wheel bleeder screw.
Purgado por gravedad   Proceso que permite el drenado de líquido de frenos anterior y de aire del sistema hidráulico de frenos mediante un tornillo purgador de la rueda.

Gross vehicle weight rating   Total weight of a vehicle plus its maximum rated payload.
Clasificación de peso bruto del vehículo   Peso total de un vehículo más su carga útil máxima valorada.

Hall effect switch   A device that produces a voltage pulse dependent on the presence of a magnetic field.
Interruptor de efecto Hall   Dispositivo que produce un pulso de voltaje que depende de la presencia de un campo magnético.

Hard code   A diagnostic trouble code that indicates a problem that is permanently present at the time of testing. A hard code might or might not keep the system from operating.
Código permanente   Código de diagnóstico de problemas que indica una dificultad que se encuentra presente de manera permanente al momento de realizar pruebas. Es posible que un código permanente pudiera evitar o no la operación del sistema.

Hard spots   Circular blue-gold glazed areas on drum or rotor surfaces where extreme heat has changed the molecular structure of the metal.
Manchas duras   Áreas circulares vidriadas en color azul y oro en las superficies del tambor o rotor donde el calor extremo ha cambiado la estructura molecular del metal.

Hardware   The electrical and mechanical parts of a computer system. Hardware includes resistors, diodes, capacitors, transistors, and other electronic parts mounted on a circuit board. Hardware also can include sensors and actuators in a system.
Hardware   Partes eléctricas y mecánicas de un sistema de computadora. El hardware incluye resistencias, diodos, capacitores, transistores y otras partes electrónicas montadas en una placa de circuitos. El hardware también puede incluir sensores y actuadores en un sistema.

Hazardous waste   Waste is considered hazardous if it has one or more of the following characteristics: ignitability, corrosivity, reactivity, and toxicity.
Desechos peligrosos   Los desechos se consideran peligrosos si tienen una o más de las características siguientes: tendencia a la inflamabilidad, a la corrosión, a la reactividad y a la toxicidad.

Heat checks   Small cracks on drum or rotor surfaces that usually can be machined away.
Comprobaciones de calor   Grietas pequeñas en las superficies del tambor o rotor que habitualmente pueden eliminarse mediante maquinado.

Height-sensing proportioning valve   A proportioning valve in which hydraulic pressure is adjusted automatically according to the vertical movement of the chassis in relation to the rear axle during braking.
Válvula proporcionadora detectora de altura   Válvula proporcionadora en la cual la presión hidráulica se ajusta automáticamente de acuerdo con el movimiento vertical del chasis en relación con el eje trasero durante el frenado.

HEPA filter   A high-efficiency particulate air filter.
Filtro HEPA   Filtro de partículas en el aire de alta eficacia.

Hertz   A unit of frequency equal to one cycle per second.
Hertz   Unidad de frecuencia igual a un ciclo por segundo.

Holddown springs   Small springs that hold drum brake shoes in position against the backing plate while providing flexibility for shoe application and release.
Resortes de anclaje   Pequeños resortes que sostienen en posición las zapatas de freno de tambor contra la placa de apoyo a la vez que proporcionan flexibilidad para la aplicación y liberación de la zapata.

Honing   A process whereby abrasive stones are rotated inside a cylinder to remove dirt, rust, or other slight corrosion and restore a uniform finish to the bore.
Rectificado   Proceso mediante el cual se hacen girar piedras abrasivas dentro de un cilindro para retirar impurezas, herrumbre u otro tipo de corrosión ligera y restaurar un acabado uniforme en el diámetro interior.

Hydro-Boost   A hydraulic power brake system that uses the power steering hydraulic system to provide boost for the brake system.
Hidrorefuerzo   *Sistema hidráulico de frenos de potencia que usa el sistema de dirección hidráulica para proporcionar refuerzo al sistema de frenos.*

Hydroplane   The action of a tire rolling on a layer of water on the road surface instead of staying in contact with the pavement.
Hidroplano   *Acción de una rueda que rueda sobre una capa de agua sobre la superficie de la carretera en lugar de permanecer en contacto con el pavimento.*

Hygroscopic   A descriptive term for a liquid's tendency to attract water and absorb moisture out of the air.
Higroscópico   *Término descriptivo para la tendencia que tiene un líquido de atraer agua y absorber humedad del aire.*

Impact sockets   Heavier walled sockets made of softer steel and designed for use with an impact wrench.
Casquillos de impacto   *Casquillos amurallados más pesados hechos de acero más suave y diseñados para usarse con una llave de impacto.*

Impedance   Refers to the operating resistance of a component or piece of equipment. The higher the impedance, the lower the operating amperage.
Impedancia   *Término que se refiere a la resistencia operativa de un componente o pieza de equipo. Cuanto mayor sea la impedancia, menor será el amperaje de operación.*

Induction   The process of producing electricity through magnetism rather than direct flow through a conductor.
Inducción   *Proceso de producción de electricidad mediante magnetismo en lugar de mediante flujo directo a través de un conductor.*

Inertia   The tendency of an object in motion to keep moving, and the tendency of an object at rest to remain at rest.
Inercia   *Tendencia de un objeto en movimiento a seguir en movimiento y tendencia de un objeto en reposo a permanecer en reposo.*

Insulators   Materials that have more than four electrons in their outer ring. With these materials, the force holding them in orbit is strong, and high voltages are needed to move them.
Aisladores   *Materiales que tienen más de cuatro electrones en el anillo exterior. Con estos materiales, la fuerza que los mantiene en órbita es fuerte y son necesarios voltajes altos para moverlos.*

Integral   The ABS that replaces the master cylinder and the power booster with a complete unit.
Integral   *Sistema de frenos antibloqueo que reemplaza el cilindro maestro y el reforzador de potencia con una unidad completa.*

Integrated ABS   An ABS in which the ABS hydraulic components, the standard brake hydraulic components, and a hydraulic power booster are joined in a single, integrated hydraulic system.
ABS integrado   *Sistema de frenos antibloqueo en el cual los componentes hidráulicos ABS, los componentes hidráulicos de*
frenos estándar y el reforzador de potencia hidráulico están unidos en un solo sistema hidráulico integrado.

Integrated circuit (IC)   A complete electronic circuit of many transistors and other devices, all formed on a single silicon semiconductor chip.
Circuito integrado (IC)   *Circuito electrónico completo de muchos transistores y otros dispositivos, todos formados en un solo chip semiconductor de silicio.*

Intermediate lever   Part of the parking brake linkage under the vehicle that increases application force and works with the equalizer to apply it equally to each wheel.
Palanca intermedia   *Pieza del mecanismo articulado del freno de estacionamiento bajo el vehículo que aumenta la aplicación de fuerza y funciona con el ecualizador para aplicarla de manera equitativa a cada rueda.*

ISO flare   A type of tubing flare connection in which a bubble-shaped end is formed on the tubing; also called a bubble flare.
Boquilla ISO   *Tipo de conexión de boquilla entubada en la cual se crea un extremo con forma de burbuja; también se conoce como boquilla de burbuja.*

Isolate, hold, and dump   The actions commanded by the microprocessor to regulate hydraulic pressures to an individual wheel brake.
Aislar, retener y vaciar   *Acciones ordenadas por el microprocesador para regular la presión hidráulica en un freno de rueda individual.*

Keep-alive memory (KAM)   Random-access memory that is retained by keeping a voltage applied to the circuits when the engine is off.
Memoria interna no borrable (KAM)   *Memoria de acceso aleatorio que se conserva al mantener aplicado un voltaje al circuito cuando el motor se apaga.*

Kinetic energy   The energy of mechanical work or motion.
Energía cinética   *Energía de movimiento o trabajo mecánico.*

Kinetic friction   Friction between two moving objects or between one moving object and a stationary surface.
Fricción cinética   *Fricción entre dos objetos en movimiento o entre un objeto en movimiento y una superficie estacionaria.*

Lands   The raised surfaces on a valve spool.
Superficies   *Las superficies elevadas en un carrete de válvula.*

Lateral accelerometer   A sensor used to measure vehicle cornering forces.
Acelerómetro lateral   *Sensor que se usa para medir la fuerza de un vehículo al dar una vuelta.*

Lateral runout   A side-to-side variation, or wobble, as the tire and wheel are rotated.
Descentramiento lateral   *Variación de lado a lado o bamboleo al girar el neumático y la rueda.*

Lathe-cut seal   A fixed seal for a caliper piston that has a square or irregular cross section; not round like an O-ring.
Sello de corte de torno   *Sello fijo para un pistón de mordaza que tiene una sección transversal cuadrada o irregular; no es redondo como un empaque anular.*

Leading shoe   The first shoe in the direction of drum rotation in a nonservo brake. When the vehicle is going forward, the forward shoe is the leading shoe, but the leading shoe can be the front or the rear shoe depending on whether the drum is rotating forward or backward and whether the wheel cylinder is at the top or the bottom of the backing plate.
Zapata principal   *Primera zapata en la dirección de giro del tambor en un freno noservo. Cuando el vehículo va hacia adelante, la zapata delantera es la zapata principal, pero la zapata principal puede ser la zapata delantera o trasera dependiendo de si el tambor gira hacia adelante o hacia atrás y si el cilindro de la rueda está en la parte superior o en la parte inferior de la placa de apoyo.*

LED   Light-emitting diode; a type of diode that gives off light as current passes through it.
LED   *Diodo emisor de luz; tipo de diodo que emite luz al pasarle corriente.*

Leverage   The use of a lever and fulcrum to create a mechanical advantage, usually to increase force applied to an object. The brake pedal is the first point of leverage in a vehicle brake system.
Apalancamiento   *El uso de una palanca y fulcro para crear una ventaja mecánica, habitualmente para aumentar la fuerza que se aplica a un objeto. El pedal del freno es el primer punto de apalancamiento en el sistema de frenos del vehículo.*

Limit pressure   The amount of pressure at which something occurs that will prevent or cause a buildup of pressure.
Presión límite   *La cantidad de presión a la cual algo ocurre que evita o causa acumulación de presión.*

Linearity   The expression of sensor accuracy throughout its dynamic range.
Linealidad   *Expresión de precisión de sensor a través de todo el rango dinámico.*

Lining fade   Brake fade due to a loss of brake lining coefficient of friction caused by excessive heat.
Pérdida de efectividad del revestimiento   *Pérdida de efectividad del freno debido a una pérdida de coeficiente de fricción del revestimiento de freno causado por calor excesivo.*

Loaded calipers   Pairs of calipers sold as service replacements that have been overhauled and are loaded with new pads. The complete caliper assembly is ready to install when purchased.
Mordazas cargadas   *Par de mordazas que se venden como reemplazo de servicio y que se han rehabilitado y cargado con pastillas nuevas. El ensamble de mordaza completo está listo para instalarse al momento de la compra.*

Loads   Devices, such as lights and motors, that use electricity to perform work.
Cargas   *Dispositivos, como las luces y los motores, que usan electricidad para realizar un trabajo.*

Mandrel   A tapered bar that acts as a template for the shaping or bending of metal.
Mandril   *Barra cónica que actúa como una plantilla para moldear o doblar metal.*

Manual bleeding   The process of using the brake pedal and master cylinder as a hydraulic pump to expel air and brake fluid from the system. Manual bleeding is a two-person operation.
Purgado manual   *Proceso en el que se usa el pedal del freno y el cilindro maestro como una bomba hidráulica para expulsar aire y líquido de frenos del sistema. Es un procedimiento que pueden realizar dos personas.*

Mass   The measure of the inertia of an object or form of matter or its resistance to acceleration; also the molecular density of an object.
Masa   *Medida de la inercia de un objeto o forma de materia o su resistencia a la aceleración; también, la densidad molecular de un objeto.*

Material safety data sheets (MSDS)   Information sheets containing chemical composition and precautionary information for all products that can present a health or safety hazard.
Hojas de datos de seguridad de materiales (MSDS)   *Hojas informativas que contienen la composición química y la información precautoria de los productos que pueden presentar peligros de seguridad o a la salud.*

Maxi-fuses   A heavy-duty circuit protection device that has the appearance of a large blade-type fuse.
Maxifusibles   *Dispositivo de protección de circuito para trabajo pesado que tiene la apariencia de un fusible grande de tipo álabe.*

Mechanical fade   Brake fade that occurs because of the limitations of the brake design.
Pérdida de efectividad mecánica   *Pérdida de efectividad de freno que ocurre debido a las limitaciones en el diseño de los frenos.*

Metallic lining   Brake friction material made from powdered metal that is formed into blocks by heat and pressure.
Revestimiento metálico   *Material de fricción de freno fabricada con metal en polvo al que se da forma de bloques por medio de calor y presión.*

Metering valve   A hydraulic control valve that delays pressure application to the front brakes until the rear drum brakes have started to operate.
Válvula de medición   *Válvula de control hidráulico que retrasa la aplicación de presión a los frenos delanteros hasta que los frenos de tambor traseros empiezan a operar.*

Micrometer   A precision measuring device used to measure small bores, diameters, and thicknesses; also called a mike.
Micrómetro   *Dispositivo de medición de precisión que se usa para medir pequeños orificios, diámetros y grosores; también se conoce como micro.*

Microprocessor   A digital computer, or processor, built on a single integrated circuit chip. A microprocessor can perform functions of arithmetic logic and control logic.
Microprocesador   *Computadora digital o procesador construido en un solo chip de circuito integrado. Un microprocesador puede realizar funciones de lógica aritmética y lógica de control.*

Mold-bonded lining   A pad assembly made by applying adhesive to the pad and then pouring the uncured lining material onto the pad in a mold. The assembly is cured at high temperature to fuse the lining and adhesive to the pad.
Revestimiento adherido al molde   *Ensamble de pastilla que se fabrica al aplicar adhesivo a la pastilla y luego dejar fluir el material de revestimiento no curado encima de la pastilla en un molde. El ensamble se cura a alta temperatura para fundir el revestimiento y el adhesivo a la pastilla.*

Momentum   The force of continuing motion. The momentum of a moving object equals its mass times its speed.
Impulso   *Fuerza de movimiento continuo. El impulso de un objeto en movimiento es igual a su masa multiplicada por su velocidad.*

MPMT switch   The common designation for a switch that has multiple throws and poles.
Interruptor MPMT   *La denominación común de un interruptor multipolar y multivanal.*

Multimeter   A tool that combines the voltmeter, ohmmeter, and ammeter together in one diagnostic instrument.
Multímetro   *Herramienta que combina juntos voltímetro, ohmiómetro y amperímetro en un instrumento de diagnóstico.*

Negative wheel slip   Wheel speed is less than vehicle speed during braking.
Deslizamiento negativo de la rueda   *La velocidad de la rueda es menor que la velocidad del vehículo durante el frenado.*

Network   The channel through which several computers share information.
Red   *Canal mediante el cual varias computadoras comparten información.*

NLGI   National Lubricating Grease Institute.
NLGI   *Instituto nacional de grasas lubricantes*

Nondirectional finish   A finish on the surface of a rotor that will not start a premature wear pattern on the pads.
Acabado no direccional   *Acabado en la superficie de un rotor que no iniciará un patrón de desgaste prematuro en las pastillas.*

Nonintegral   This is also called an add-on ABS unit and is separate from the master cylinder and power booster.
No integral   *También se conoce como unidad agregada de ABS y es independiente del cilindro maestro y el reforzador de potencia.*

Nonservo brake   A drum brake that develops self-energizing action only on the leading shoe.
Freno noservo   *Freno de tambor que desarrolla acción autoenergizante solo en la zapata principal.*

OBD II   On-board diagnostics—second generation.
OBD II   *Diagnóstico a bordo de segunda generación.*

Ohm   A unit of measured electrical resistance.
Ohmio   *Unidad de resistencia eléctrica medida.*

Ohmmeter   The meter used to measure electrical resistance.
Óhmetro   *Medidor que se usa para medir la resistencia eléctrica.*

Ohm's Law   A basic law of electricity expressing the relationship of current, resistance, and voltage in any electrical circuit.
Ley de Ohm   *Ley básica de la electricidad que expresa la relación de corriente, resistencia y voltaje en cualquier circuito eléctrico.*

Open   An electrical circuit that has a break in the wire or is not complete.
Abierto   *Circuito eléctrico que tiene una rotura en el alambre o que no está completo.*

Organic lining   Brake friction material made from non-metallic fibers bonded together in a composite material.
Revestimiento orgánico   *Material de fricción de freno fabricada con fibras no metálicas adheridas en un material compuesto.*

O-ring   A circular rubber seal shaped like the letter O.
Empaque anular   *Sello circular de caucho con forma de letra O.*

OSHA   The Occupational Safety and Health Administration, which is a government agency charged with ensuring safe work environments for all workers.
OSHA   *Administración de Salud y Seguridad Ocupacional, la cual es la agencia gubernamental que se encarga de asegurar entornos seguros de trabajo para todos los trabajadores.*

Output driver   Located in the processor and operated by the digital commands of the computer, an output driver is an electronic on/off switch used to control the ground circuit of a specific actuator.
Controlador de salida   *Localizado en el procesador y operado por los comandos digitales de la computadora, el controlador de salida es un interruptor electrónico de encendido y apagado que se usa para controlar el circuito a tierra de un actuador específico.*

Overload spring   A spring at the end of the cable in a cable-operated adjuster assembly that allows the cable to move without breaking if the pawl or starwheel is jammed.
Resorte de sobrecarga   *Resorte en el extremo de un cable en un ensamble ajustador operado por cable que permite que el cable se mueva y no se rompa si el trinquete o la rueda de estrella se atoran.*

Overvoltage Protection Relay (OVPR)   An electrical device used in ABS. It prevents damage of electronic components and circuits because of electrical surges or spikes.
Relé de protección de sobrevoltaje (OVPR)   *Dispositivo eléctrico que se usa en ABS. Evita el daño a componentes electrónicos y circuitos debido a sobretensión o descargas de corriente eléctrica.*

Ozone   An unstable molecule of oxygen with three atoms instead of the normal two. Ozone oxidizes other elements and compounds by giving up its extra oxygen atom.
Ozono   *Molécula inestable de oxígeno con tres átomos en lugar de dos, como es lo normal. El ozono oxida otros elementos y compuestos al ceder su átomo de oxígeno adicional.*

Pad hardware Miscellaneous small parts, such as antirattle clips, shims, and support clips, that hold brake pads in place and keep them from rattling.
Herrajes de pastilla *Varias piezas pequeñas, como clips antirruido, calces y clips de apoyo que sujetan las pastillas de freno y evitan que hagan ruido.*

Pad wear indicators Devices that warn the driver when disc brake linings have worn to the point where they need replacement. Wear indicators can be mechanical (audible) or electrical.
Indicadores de desgaste de pastillas *Dispositivos que alertan al conductor cuando se desgasta el revestimiento del freno de disco al punto en que es necesario reemplazarlo. Los indicadores de desgaste pueden ser mecánicos (audibles) o eléctricos.*

Parallel circuit An electrical circuit that has more than one path for the current to follow.
Circuito paralelo *Circuito eléctrico con más de una ruta por la cual sigue la corriente.*

Parallelism Thickness uniformity of a disc brake rotor. Both surfaces of a rotor must be parallel with each other within 0.001 inch or less.
Paralelismo *Uniformidad de grosor de un rotor de freno de disco. Ambas superficies de un rotor deben estar paralelas entre sí dentro de un límite de 0.001 pulgadas o menos.*

Parking brake control The pedal or lever used to apply the parking brakes.
Control de freno de estacionamiento *El pedal o palanca que se usa para aplicar el freno de estacionamiento.*

Parking brakes The disc or drum brakes that hold the vehicle stationary after the service brakes stop it.
Frenos de estacionamiento *Frenos de disco o de tambor que mantienen el vehículo estacionario después de que los frenos de servicio lo hacen detenerse.*

Pawl A hinged or pivoted component that engages a toothed wheel or rod to provide rotation or movement in one direction while preventing it in the opposite direction.
Trinquete *Componente articulado o de pivote que embraga una rueda dentada o varilla para proporcionar giro o movimiento en una dirección mientras la evita en la dirección contraria.*

Pedal free play The clearance between the brake pedal or booster pushrod and the primary piston in the master cylinder.
Juego libre de freno *Holgura entre el pedal de freno o varilla de empuje reforzadora y el pistón principal en el cilindro maestro.*

Permanent magnet (PM) generator A reluctance sensor. A sensor that generates a voltage signal by moving a conductor through a permanent magnetic field.
Generador de imán (PM) permanente *Sensor de reluctancia. Sensor que genera una señal de voltaje al mover un conductor a través de un campo magnético permanente.*

Phenolic plastic Plastic made primarily from phenol, a compound derived from benzene; also called carbolic acid.
Plástico fenólico *Plástico fabricado principalmente con fenol, un compuesto derivado de benceno; también se conoce como ácido carbónico.*

Phosgene A toxic gas formed by the decomposition of hydrocarbon solvents when they are subjected to heat.
Fosgeno *Gas tóxico que se forma por la descomposición de solventes de hidrocarburo cuando se someten al calor.*

Pickup coil sensor A reluctance sensor that generates a voltage signal by moving a conductor through a permanent magnetic field.
Sensor de bobina captadora *Sensor de reluctancia que genera una señal de voltaje al mover un conductor a través de un campo magnético permanente.*

Piston stop A metal part on a brake backing plate that keeps the wheel cylinder pistons from moving completely out of the cylinder bore.
Parada de pistón *Pieza metálica en una placa de apoyo de freno que evita que se salgan por completo los pistones del cilindro de la rueda del diámetro interior del cilindro.*

Plateaued finish A flattened finish to a cylinder bore in which all of the high spots or peaks are removed.
Acabado de meseta *Acabado aplanado de un diámetro interior en el cual se eliminan todos los puntos altos o picos.*

P-metric system The most common modern system to specify passenger car tire sizes.
Sistema p-métrico *El sistema moderno más común para especificar tamaños de neumático para automóviles de pasajeros.*

Polyglycol Polyalkylene-glycol-ether brake fluids that meet specifications for DOT 3 and DOT 4 brake fluids.
Poliglicol *Líquido de frenos éter glicólico polialcalino que cumple con las especificaciones DOT 3 y DOT 4 para líquido de frenos.*

Positive wheel slip Wheel speed is greater than vehicle speed during acceleration.
Deslizamiento positivo de la rueda *La velocidad de la rueda es mayor que la velocidad del vehículo durante la aceleración.*

POST An acronym for the power on system test used by the microprocessor at key on.
POST *Acrónimo para la prueba de alimentación de energía del sistema que realiza el microprocesador al momento de encenderlo.*

Potentiometer A variable resistor that acts as a voltage divider to produce a continuously variable output signal proportional to a mechanical position.
Potenciómetro *Resistencia variable que actúa como divisor de voltaje para producir una señal de salida variable continua proporcional a una posición mecánica.*

PowerMaster A self-contained hydraulic power brake system with its own hydraulic reservoir and independent electric pump; developed by General Motors for cars and trucks that could not economically use vacuum-assisted power brakes.
PowerMaster *Sistema hidráulico de frenos de potencia con su propio depósito hidráulico y bomba eléctrica independiente. Desarrollado por General Motors para automóviles y camiones que no podrían usar de manera económica frenos de potencia auxiliados por vacío.*

Press fit Forcing a part into an opening that is slightly smaller than the part itself in order to make a solid fit.
Ajuste a presión *Forzar una pieza en una abertura que es ligeramente más pequeña que la pieza misma para lograr un ajuste sólido.*

Pressure Force exerted on a given unit of surface area. Pressure equals force divided by area and is measured in pounds per square inch (psi) or kilopascals (kPa).
Presión *Fuerza que se ejerce sobre una unidad determinada de área de superficie. La presión es igual a la fuerza dividida por el área y se mide en libras por pulgada cuadrada o kilopascales (kPa).*

Pressure bleeding The process of using a tank filled with brake fluid and pressurized with compressed air to expel air and brake fluid from the system. Pressure bleeding is a one-person operation.
Purgado por presión *Proceso en el que se usa un tanque lleno de líquido de frenos y presurizado con aire comprimido para expulsar aire y líquido de frenos del sistema. Es un procedimiento que puede realizar una sola persona.*

Pressure differential The difference between two pressures on two surfaces or in two separate areas. The pressures can be either pneumatic (air) or hydraulic.
Diferencial de presión *La diferencia entre dos presiones en dos superficies o en dos áreas separadas. Las presiones pueden ser neumáticas (aire) o hidráulicas.*

Pressure differential valve A hydraulic valve that reacts to a difference in pressure between the halves of a split brake system. When a pressure differential exists, the valve turns on the brake warning lamp switch.
Válvula diferencial de presión *Válvula hidráulica que reacciona a diferencias de presión entre las mitades de un sistema de frenos dividido. Cuando existe una diferencia de presión, la válvula enciende el interruptor de la luz de advertencia del freno.*

Pressure modulation The rapid application and release of hydraulic pressure to one or more components.
Modulación de presión *Aplicación y liberación rápida de presión hidráulica en uno o más componentes.*

Primary shoe The leading shoe in a duo-servo brake. The primary shoe is self-energizing and applies servo action to the secondary shoe to increase its application force. Primary shoes have shorter linings than secondary shoes.
Zapata principal *Zapata principal en un freno duoservo. La zapata principal es autoenergizante y aplica servoacción a la zapata secundaria para aumentar su fuerza de aplicación. Las zapatas principales tienen revestimientos más cortos que las zapatas secundarias.*

Processor The term used to describe the metal box that houses the computer and its related components.
Procesador *Término que se usa para describir la caja metálica que alberga la computadora y sus componentes relacionados.*

Program The job instructions for a computer.
Programa *Instrucciones de trabajo para una computadora.*

Programmable read-only memory (PROM) A computer memory integrated circuit chip that can be programmed once to store the computer program and other data.
Memoria programable de solo lectura (PROM) *Chip de circuito integrado de memoria de computadora que se puede programar una vez para almacenar el programa de computadora y otros datos.*

Proportioning valve A hydraulic control valve that controls the pressure applied to rear drum brakes.
Válvula proporcionadora *Válvula de control hidráulico que controla la presión que se aplica a los frenos de tambor traseros.*

Pull up resistor A fixed resistor in a voltage divider circuit for a sensor input signal.
Resistencia de ajuste *Resistencia fija en un circuito divisor de voltaje para una señal de entrada de sensor.*

Pulse width The amount of time that an electromechanical device, such as a solenoid, is energized. Pulse width usually is measured in milliseconds.
Ancho de pulso. *La cantidad de tiempo en el cual se energiza un dispositivo electromecánico, como un solenoide. El ancho de pulso habitualmente se mide en milisegundos.*

Pulse-width modulation (PWM) The characteristic continuous on-and-off cycling of a solenoid for a fixed number of times per second. While the frequency of the cycles remains constant, the ratio of on-time to total cycle time varies, or is modulated.
Modulación de ancho de pulso (PWM) *Ciclo continuo característico de encendido y apagado de un solenoide para un número fijo de tiempos por segundo. Aunque la frecuencia del ciclo permanezca constante, la relación de tiempo encendido con el tiempo de ciclo total varía o se modula.*

Push (speed) nut A lightweight, stamped steel retainer that pushes onto a stud to hold two parts together temporarily.
Tuerca de empuje (velocidad) *Dispositivo de retención de acero troquelado y peso ligero que se empuja en un remache para sujetar dos piezas de manera temporal.*

Quick-takeup master cylinder A dual master cylinder that supplies a large volume of fluid to the front disc brakes on initial brake application, which takes up the clearance of low-drag calipers.
Cilindro maestro de recepción rápida *Cilindro maestro doble que proporciona un gran volumen de líquido a los frenos de disco delanteros al aplicar inicialmente el freno, lo cual ocupa el espacio libre de las mordazas de bajo arrastre.*

Quick-takeup valve The part of the quick-takeup master cylinder that controls fluid flow between the reservoir and the primary low-pressure chamber.
Válvula de recepción rápida *Parte del cilindro maestro de recepción rápida que controla el flujo de líquido entre el depósito y la cámara principal de presión baja.*

Radial ply tire Tire construction in which the cords in the body plies of the carcass run at an angle of 90 degrees to the steel beads in the inner rim of the carcass. Each cord is parallel to the radius of the tire circle.
Neumático de capa radial *Construcción de neumático en el cual las cuerdas en las capas del cuerpo de la carcasa corren en un ángulo de 90 grados hacia las cejas de acero en el rin*

interno de la carcasa. Cada cuerda es paralela al radio del círculo del neumático.

Radial runout An out-of-round condition in which the radius of the wheel or tire is not consistent from the wheel center to any point on the rim or the tread.
Descentramiento radial Condición ovalada en la cual el radio de la rueda o neumático no es consistente desde el centro de la rueda a cualquier punto en el rin o la banda de rodamiento.

Radio frequency interference (RFI) A narrow band of frequencies within the electromagnetic interference spectrum.
Interferencia de radiofrecuencia (RFI) Banda estrecha de frecuencias dentro del espectro de interferencia electromagnética.

Random-access memory (RAM) The permanent program memory of a computer. Instructions can be read from ROM, but nothing can be written into it and it cannot be changed.
Memoria de acceso aleatorio (RAM) Memoria permanente de programa de una computadora. Se pueden leer instrucciones de la memoria ROM pero no se puede escribir nada en ella y no se puede cambiar.

Reaction disc A vacuum power brake booster can be fitted with this, rather than a plate and levers, to provide pedal feel, or feedback, to the driver.
Disco de reacción Un reforzador de freno de potencia de vacío podría estar adaptado con esto en lugar de con una placa y palancas para proporcionar sensación de pedal o retroalimentación, al conductor.

Reaction plate and levers A vacuum power brake booster can be fitted with these, rather than a disc, to provide pedal feel, or feedback, to the driver.
Placa y palancas de reacción Un reforzador de freno de potencia de vacío podría estar adaptado con estos en lugar de con un disco para proporcionar sensación de pedal o retroalimentación, al conductor.

Read-only memory (ROM) A type of computer memory that is used to store programs.
Memoria de solo lectura (ROM) Tipo de memoria de computadora que se usa para almacenar programas.

Ream A process that cleans out or slightly enlarges the inside bore of something.
Abocardar Proceso que limpia o aumenta ligeramente el diámetro interior de algo.

Rear antilock brake system (RABS) A two-wheel Kelsey-Hayes ABS used on the rear wheels of Ford light-duty pickup trucks.
Sistema trasero de frenos antibloqueo (RABS) Sistema de frenos antibloqueo Kelsey-Hayes de dos ruedas que se usa en las ruedas traseras de las camionetas de carga para trabajo ligero Ford.

Rear wheel antilock (RWAL) One of the best-known Kelsey-Hayes systems. A two-wheel ABS used on the rear wheels of General Motors and Chrysler light-duty pickup trucks and some sport-utility vehicles.
Antibloqueo de ruedas traseras (RWAL) Uno de los sistemas Kelsey-Hayes más conocido. ABS Kelsey-Hayes de dos

ruedas que se usa en las ruedas traseras de las camionetas de carga para trabajo ligero y algunos vehículos utilitarios deportivos de General Motors y Chrysler.

Reference voltage A fixed voltage supplied to the sensor by a voltage regulator inside the computer or control module. As the sensor changes, the return voltage is altered and sent back to the computer for use. Most computer control systems operate with a 5-volt reference voltage.
Voltaje por referencia Voltaje fijo que se proporciona al sensor mediante un regulador de voltaje dentro de la computadora o módulo de control. Al cambiar el sensor, el voltaje de vuelta se altera y se envía de nuevo a la computadora para su uso. La mayoría de los sistemas de control de computadora operan con un voltaje de referencia de 5 voltios.

Reference voltage (Vref) sensors Sensors that provide input to the computer by modifying or controlling a constant, predetermined reference voltage signal.
Sensores de voltaje por referencia (Vref) Sensores que introducen información a la computadora al modificar o controlar una señal de voltaje de referencia constante y predeterminada.

Refractometer A test instrument that measures the deflection, or bending, of a beam of light.
Refractómetro Instrumento de prueba que mide el desvío o curvatura de un haz de luz.

Relay An electromagnetic device used to control a high-current circuit with a low-current control circuit.
Relé Dispositivo electromagnético que se usa para controlar un circuito de corriente alta con un circuito de control de corriente baja.

Reluctance sensor A magnetic pulse generator, or pickup coil, that sends a voltage signal in response to the varying reluctance of a magnetic field.
Sensor de reluctancia. Generador de pulso magnético o bobina captadora que envía una señal de voltaje en respuesta a la reluctancia variable de un campo magnético.

Reluctor A metal tooth ring used to influence the magnetic flux lines of the permanent magnet generator.
Reluctor Anillo dentado metálico que se usa para influir las líneas de flujo magnético del generador de imán permanente.

Repeatability The ability of a sensor to send the same signal voltage every time it measures the same value or quantity.
Repetición La capacidad de un sensor de enviar el mismo voltaje de señal cada vez que mide el mismo valor o cantidad.

Replenishing port The rearward port in the master cylinder bore. It lets fluid pass between each pressure chamber and its fluid reservoir during operation.
Puerto reabastecedor Puerto posterior en el diámetro interior del cilindro maestro que deja pasar líquido entre cada cámara de presión y su depósito de fluido durante la operación.

Residual pressure check valve A check valve used to retain slight pressure in the lines to the drum brakes when the brakes are not applied, which holds the wheel cylinder cup seals against the cylinder walls.
Válvula unidireccional de presión residual Válvula unidireccional que se usa para retener una presión ligera en las

líneas hacia los frenos de tambor cuando no se aplican los frenos, lo cual sostiene los sellos de copa del cilindro de la rueda contra las paredes del cilindro.

Return spring   A strong spring that retracts a drum brake shoe when hydraulic pressure is released.
Resorte de retorno   *Resorte resistente que retrae a una zapata de freno de tambor cuando se libera presión hidráulica.*

Rheostats   A type of variable resistor. Rheostats provide for varying amounts of voltage and current from their tap.
Reóstatos   *Tipo de resistencia variable. Proporcionan cantidades diversas de voltaje y corriente de su contacto.*

Right-to-know laws   Laws requiring employers to provide employees with a safe work place as it relates to hazardous materials.
Leyes sobre el derecho a saber   *Leyes que requieren que los empleadores proporcionen a los empleados un lugar seguro de trabajo en cuanto a materiales peligrosos.*

Riveted lining   Brake lining attached to the pad or shoe by copper or aluminum rivets.
Revestimiento remachado   *Revestimiento de freno que se une a la pastilla o zapata mediante remaches de cobre o aluminio.*

Rolled finish   The result of a manufacturing process in which a hardened metal is repeatedly rolled over, with great pressure, on the surface of the porous metal until the pores in the metal close.
Acabado laminado   *Resultado de un proceso de manufactura en el cual se hace pasar repetidamente un rodillo sobre un metal endurecido con gran presión sobre la superficie del metal poroso hasta que se cierran los poros en el metal.*

Rotor   The rotating part of a disc brake that is mounted on the wheel hub and contacted by the pads to develop friction to stop the vehicle. Also called a disc.
Rotor   *Pieza giratoria de un freno de disco montada en el cubo de la rueda y sobre la cual hacen contacto las pastillas para desarrollar la fricción que hace detener el vehículo. También se conoce como disco.*

Rotor lateral runout   Rotor side-to-side wobble as the rotor rotates.
Descentramiento lateral de rotor   *Giro excéntrico de lado a lado al girar el rotor.*

Schematics   Wiring diagrams used to show how circuits are constructed.
Dibujo esquemático   *Diagramas de cableado que se usan para mostrar cómo se construyen los circuitos.*

Screw pitch gauge   A tool used to check the threads per inch (pitch) of a fastener.
Medidor de roscas   *Herramienta que se usa para verificar el estriado por pulgada (espaciado) de un dispositivo de fijación.*

Scrub radius   The distance from the tire contact patch centerline to the point where the steering axis intersects the road.
Radio achaparrado   *Distancia de la línea central del parche de contacto al punto donde el eje de dirección cruza con la carretera.*

Secondary shoe   The trailing shoe in a duo-servo brake. The secondary shoe receives servo action from the primary shoe to increase its application force. Secondary shoes provide the greater braking force in a duo-servo brake and have longer linings than primary shoes do.
Zapata secundaria   *Zapata secundaria en un freno duoservo. La zapata secundaria recibe servoacción de la zapata principal para aumentar su fuerza de aplicación. Las zapatas secundarias proporcionan la fuerza de frenado mayor en un freno duoservo y tienen revestimientos más largos que las zapatas principales.*

Section width   The width of a tire across the widest point of its cross section, usually measured in millimeters.
Ancho de sección   *Ancho de un neumático de lado a lado del punto más ancho de la sección transversal, medido habitualmente en milímetros.*

Select low   The program used by the microprocessor to determine which wheel is beginning to lock up.
Seleccionar bajo   *Programa que usa un microprocesador para determinar cuál rueda se empieza a bloquear.*

Self-adjuster   A cable, lever, screw, strut, or other linkage part that provides automatic shoe adjustment and proper lining-to-drum clearance as a drum brake lining wears.
Autoajustador   *Cable, palanca, tornillo, puntal, u otra parte de unión que proporciona ajuste automático de la zapata y espacio libre apropiado del revestimiento al tambor al desgastarse el revestimiento del freno de tambor.*

Self-energizing   The action of a drum brake shoe when drum rotation increases the application force of the shoe by wedging it tightly against the drum.
Autoenergizante   *Acción de la zapata de un freno de tambor cuando el giro del tambor aumenta la fuerza de aplicación de la zapata al encajarla ajustadamente contra el tambor.*

Semiconductor   A material that is not a good conductor or a good insulator but can function as either when certain conditions exist.
Semiconductor   *Material que no es un buen conductor o buen aislante pero que funciona como cualquiera de los dos cuando existen ciertas condiciones.*

Semimetallic lining   Brake friction materials made from a mixture of organic or synthetic fibers and certain metals; they do not contain asbestos.
Revestimiento semimetálico   *Materiales de fricción de freno fabricados a partir de una mezcla de fibras orgánicas o sintéticas y ciertos metales; no contienen asbesto.*

Sensor   Any device that sends an input signal to a computer.
Sensor   *Cualquier dispositivo que envía una señal de salida a una computadora.*

Series circuit   An electrical circuit that has only one path for current to flow.
Circuito serie   *Circuito eléctrico que solo tiene una ruta para el flujo de corriente.*

Service brakes   The disc or drum brakes operated by the driver to stop the vehicle.
Frenos de servicio   *Frenos de disco o de tambor que opera el conductor para detener el vehículo.*

Servo action   The operation of a drum brake that uses the self-energizing operation of one shoe to apply mechanical force to the other shoe to assist its application. Broadly, servo action is any mechanical multiplication of force.
Servoacción   *Operación de un freno de tambor que usa la operación autoenergizante de una zapata para aplicar fuerza mecánica a la otra zapata para auxiliar en su aplicación. De forma amplia, la servoacción es cualquier multiplicación mecánica de fuerza.*

Setback   A difference in wheelbase from one side of a vehicle to the other.
Descuadre   *Diferencia de la distancia entre ejes de un lado a otro del vehículo.*

Shoe anchor   The large pin, or post, or block against which a drum brake shoe pivots or develops leverage.
Anclaje de zapata   *Pasador grande o poste o bloque contra el cual una zapata de freno de tambor gira o desarrolla apalancamiento.*

Short   An intentional or unintentional grounding of an electrical circuit.
Corto   *Puesta a tierra intencional o no intencional de un circuito eléctrico.*

Sliding caliper   A caliper that is mounted to its support on two fixed sliding surfaces, or ways. The caliper slides on the rigid ways and does not have the flexibility of a floating caliper.
Mordaza deslizante   *Mordaza que se monta en el soporte sobre dos superficies deslizantes fijas o guías. La mordaza se desliza sobre las guías rígidas y no tiene la flexibilidad de una mordaza flotante.*

Slope   The numerical ratio, or proportion, of rear drum brake pressure to full system pressure that is applied through a proportioning valve. If half of the system pressure is applied to the rear brakes, the slope is 1:2, or 50 percent.
Pendiente   *Índice numérico o proporción de la presión del freno de tambor trasero a la presión total del sistema que se aplica a través de una válvula proporcionadora. Si la mitad de la presión del sistema se aplica a los frenos traseros, la pendiente es de 1:2 ó 50 por ciento.*

Soft code   A diagnostic trouble code from a vehicle computer that indicates a problem that is not present at the time of testing.
Código flexible   *Código de diagnóstico de problemas de la computadora de un vehículo que indica un problema que no está presente al momento de realizar pruebas.*

Software   The various programs in RAM and ROM that provide a microprocessor with memory and operating instructions.
Software   *Diversos programas en RAM y ROM que le proporcionan a un microprocesador memoria e instrucciones de operación.*

Solenoid   An electromagnetic device similar in operation to a relay, but movement of the armature or iron core changes electrical energy into mechanical energy.
Solenoide   *Dispositivo electromagnético similar en operación a un relé, pero el movimiento de la armadura o núcleo de hierro convierte la energía eléctrica en energía mecánica.*

Solenoid valve   A mechanical valve operated by a solenoid to control the flow of liquid or gas.
Válvula solenoide   *Válvula mecánica operada por un solenoide para controlar el flujo de líquido o gas.*

Solid rotor   A rotor that is a solid piece of metal with a friction surface on each side.
Rotor sólido   *Rotor que es una pieza sólida de metal con una superficie de fricción en cada lado.*

Specific gravity   The weight of a volume of any liquid divided by the weight of an equal volume of water at equal temperature and pressure; the ratio of the weight of any liquid to the weight of water, which has a specific gravity of 1.000.
Peso específico   *El peso de un volumen o de cualquier líquido dividido entre el peso de un volumen igual de agua a temperatura y presión iguales; el índice de proporción del peso de cualquier líquido respecto al peso del agua, que tiene un peso específico de 1.000.*

Split point   The pressure at which a proportioning valve closes during brake application and reduces the rate at which further pressure is applied to rear drum brakes.
Punto de división   *Presión a la cual una válvula proporcionadora se cierra durante la aplicación del freno y reduce el índice mediante el cual se aplica más presión a los frenos de tambor traseros.*

Spool valve   A cylindrical sliding valve that uses lands and valleys around its circumference to control the flow of hydraulic fluid through the valve body.
Válvula de carrete   *Válvula cilíndrica deslizante que usa superficies y valles alrededor de su circunferencia para controlar el flujo de líquido hidráulico a través del cuerpo de la válvula.*

SPST switch   The common designation for a single-pole, single-throw switch.
Interruptor SPST   *La denominación común de un interruptor unipolar y univalvo.*

Square-cut seal   A fixed seal for a caliper piston that has a square cross section.
Sello de corte cuadrado   *Sello fijo para un pistón de mordaza que tiene una sección transversal cuadrada.*

Starwheel   A small wheel that is part of a drum brake adjusting link. Turning the starwheel lengthens or shortens the adjuster link to position the shoes for proper lining-to-drum clearance.
Rueda de estrella   *Rueda pequeña que es parte de una articulación de ajuste de freno de tambor. Hacer girar la rueda de estrella alarga o acorta la articulación del ajustador para colocar las zapatas con el fin de obtener el espacio libre apropiado del revestimiento al tambor.*

Static electricity   A charge of electricity that is not moving but will quickly move when given a path. Static electricity is normally caused by friction.
Electricidad estática   *Carga de electricidad que no se mueve pero que se mueve rápidamente cuando se le marca una ruta. La electricidad estática normalmente se debe a la fricción.*

Static friction   Friction between two stationary objects or surfaces.
Fricción estática   *Fricción entre dos objetos o superficies estacionarios.*

Static pressure   The pressure inside the hydraulic system when there is no fluid motion.
Presión estática   *Presión dentro del sistema hidráulico cuando no hay movimiento del líquido.*

Steering knuckle   The outboard part of the front suspension that pivots on the ball joints and lets the wheels turn for steering control.
Muñón de dirección   *Pieza en la parte exterior de la suspensión delantera que gira sobre las rótulas y permite que las ruedas den vuelta para controlar la dirección.*

Stroking seal   A seal, similar to a lip seal, that is installed on the piston and moves with it.
Sello de carrera   *Sello similar a un sello de pestaña que se instala en el pistón y se mueve con él.*

Surge bleeding   A supplementary bleeding method in which one person rapidly pumps the brake pedal to dislodge air pockets while another person opens the bleeder screw.
Purgado por incremento súbito   *Método de purgado suplementario en el cual una persona bombea rápidamente el pedal del freno para desprender bolsas de aire mientras otra persona abre el tornillo purgador.*

Swept area   The total area of the brake drum or rotor that contacts the friction surface of the brake lining.
Área de barrido   *El área total del tambor de freno o rotor que hace contacto con la superficie de fricción del revestimiento de freno.*

Synthetic lining   Brake friction materials made from nonorganic, nonmetallic, and nonasbestos materials; typically fiberglass and aramid fibers.
Revestimiento sintético   *Material de fricción de freno fabricada con materiales no orgánicos, no metálicos y no de asbesto. Habitualmente fibra de vidrio y fibra de aramida.*

Table   The outer surface of a brake shoe to which the lining is attached.
Tabla   *Superficie exterior de una zapata de freno al cual se une el revestimiento.*

Tandem booster   A power brake vacuum booster with two small diaphragms in tandem to provide additive vacuum force.
Reforzador en tándem   *Reforzador de vacío de frenos de potencia con dos pequeños diafragmas en tándem para proporcionar fuerza de vacío aditiva.*

Tap   A tap can cut and restore the threads in a bore.
Macho de roscar   *Herramienta que puede cortar y restaurar el estriado en un orificio.*

Tapered roller bearing   A specific kind of bearing that is based on tapered steel rollers held together in a cage.
Cojinete de rodillos cónicos   *Tipo específico de cojinete basado en rodillos cónicos de acero que se mantienen unidos en una jaula.*

Telescoping (snap) gauge   A tool used for measuring bore diameters and other clearances; also known as snap gauges.
Medidor exterior de extensión   *Herramienta que se usa para medir diámetros de orificios y otras tolerancias; también se conocen como calibradores exteriores.*

Tensile force   The moving force that slides or pulls an object over a surface.
Fuerza tensora   *Fuerza de movimiento que desliza o atrae a un objeto sobre una superficie.*

Test light   A test instrument that illuminates its lamp when there is the presence of voltage at one of its leads and it is connected to ground at the other lead.
Luz de prueba   *Instrumento de prueba que ilumina su bombilla cuando hay presencia de voltaje en uno de sus hilos conductores y el otro hilo conductor está conectado a tierra.*

Thermal energy   The energy of heat.
Energía térmica   *Energía de calor.*

Thermistor   A type of variable resistor. A thermistor is designed to change in value as its temperature changes.
Termistor   *Tipo de resistencia variable. El termistor está diseñado para cambiar de valor al elevarse su temperatura.*

Thread pitch   The number of threads in 1 inch of threaded bolt length. In the metric system, thread pitch is the distance in millimeters between two adjacent threads.
Espaciado de estriado   *El número de estrías en una pulgada de longitud de perno estriado. En el sistema métrico, espaciado de estriado es la distancia en milímetros entre dos estrías adyacentes.*

Torque stick   An extension for an impact wrench that includes the correct size socket for a wheel nut and that acts as a torsion bar to limit the torque applied by the impact wrench when installing a wheel.
Barra de torsión   *Extensión para llave de impacto que incluye el adaptador de tamaño correcto para una tuerca de rueda y que actúa como una barra de torsión para limitar la fuerza de torsión que aplica la llave de impacto al instalar una rueda.*

Torque wrenches   Torque wrenches measure how tight a nut or bolt is.
Llave de torsión   *Herramienta que mide la tensión de apriete de una tuerca o un perno.*

Traction control   A system that attempts to control wheel spin during acceleration on slick road surfaces.
Control de tracción   *Sistema que intenta controlar el giro de la rueda durante la aceleración sobre superficies resbaladizas en la carretera.*

Trailing shoe   The second shoe in the direction of drum rotation in a nonservo brake. The trailing shoe is non-self-energizing, and drum rotation works against shoe application.
Zapata secundaria   *Segunda zapata en la dirección de giro del tambor en un freno noservo. La zapata secundaria no es autoenergizante y el giro del tambor funciona contra la aplicación de la zapata.*

Transistor   A semiconductor device made of three primary elements. Transistors are typically used as switches in electronic circuits.
Transistor   *Dispositivo semiconductor fabricado con tres elementos principales. Los transistores se usan generalmente como interruptores en circuitos eléctricos.*

Tread   The layer of rubber on a tire that contacts the road and contains a distinctive pattern to provide traction.
Banda de rodamiento   *Capa de caucho en un neumático que hace contacto con la carretera y que contiene un patrón distintivo para proporcionar tracción.*

Tread contact patch   The area of the tire tread that contacts the road; determined by tire section width, diameter, and inflation pressure.
Parche de contacto de banda de rodamiento   *Área de la banda de rodamiento del neumático que hace contacto con la carretera; se determina por el ancho de la sección del neumático, por el diámetro y por la presión de inflado.*

Tread wear indicator   A continuous bar that appears across a tire tread when the tread wears down to the last ²/₃₂ (¹/₁₆) inch.
Indicador de desgaste de la banda de rodamiento   *Barra continua que aparece a lo ancho de la banda de rodamiento del neumático cuando la banda de rodamiento se desgasta a los últimos ²/₃₂ (¹/₁₆) de pulgada.*

Unidirectional rotor   A rotor with cooling fins that are curved or formed at an angle to the hub center to increase cooling airflow. Because the fins work properly only when the rotor turns in one direction, unidirectional rotors cannot be interchanged from right to left on a vehicle.
Rotor unidireccional   *Rotor con aletas de enfriamiento que están curvadas o formadas en ángulo hacia el centro del cubo para aumentar el flujo de aire de enfriamiento. Debido a que las aletas sólo funcionan apropiadamente cuando el rotor gira en una dirección, los rotores unidireccionales no se pueden intercambiar de derecha a izquierda en el vehículo.*

Vacuum   In automotive service, vacuum is generally considered to be air pressure lower than atmospheric pressure. A true vacuum, however, is a complete absence of air.
Vacío   *En servicio de automóviles, generalmente se considera como vacío a la presión de aire que es inferior a la presión atmosférica. Sin embargo, un vacío verdadero es la ausencia total de aire.*

Vacuum bleeding   The process of using a vacuum pump filled with brake fluid and attached to a wheel bleeder screw to draw old brake fluid and air from the system. Vacuum bleeding is a one-person operation.
Purgado al vacío   *Proceso en el que se usa una bomba de vacío llena de líquido de frenos y unida al tornillo purgador de la rueda para sacar el líquido de los frenos anteriores y el aire del sistema. Es un procedimiento que puede realizar una sola persona.*

Vacuum-suspended   A term that describes a power brake vacuum booster in which vacuum is present on both sides of the diaphragm when the brakes are released; the most common kind of vacuum booster.
Suspendido al vacío   *Término que describe un reforzador de vacío de frenos de potencia en el cual el vacío está presente en ambos lados del diafragma cuando los frenos se liberan; la forma más común de reforzador de vacío.*

Valleys   The annular grooves, or recessed areas, between the lands of a valve spool.
Valles   *Las ranuras anulares o áreas ahuecadas entre las superficies de un carrete de válvula.*

Vent port   The forward port in the master cylinder bore. It lets fluid pass between each pressure chamber and its fluid reservoir during operation.
Puerto de ventilación   *Puerto delantero en el diámetro interior del cilindro maestro que deja pasar líquido entre cada cámara de presión y su depósito de fluido durante la operación.*

Ventilated rotor   A rotor that has cooling fins cast between the braking surfaces to increase the cooling area of the rotor.
Rotor ventilado   *Rotor con aletas enfriadoras coladas entre las superficies de frenado para aumentar el área de enfriamiento del rotor.*

Vernier caliper   A vernier caliper is marked in both United States Customary System and metric divisions called a vernier scale, which consists of a stationary scale and a movable scale.
Pie de rey   *Un pie de rey está marcado con divisiones métricas y las correspondientes al sistema convencional de los Estados Unidos conocidas como escala vernier, el cual consiste de una escala estacionaria y una escala móvil.*

Volt   A unit of measurement of electromotive force. One volt of electromotive force applied steadily to a conductor of 1 ohm of resistance produces a current of 1 ampere.
Voltio   *Unidad de medida de la fuerza electromotriz. Un voltio de fuerza electromotriz aplicado uniformemente a un conductor de un ohmio de resistencia produce una corriente de un amperio.*

Voltage   Electrical pressure formed by the attraction of electrons to protons.
Voltaje   *Presión eléctrica que se forma por la atracción de electrones y protones.*

Voltage drop   The voltage lost by the passage of electrical current through resistance.
Caída de voltaje   *El voltaje que se pierde por el paso de corriente eléctrica a través de la resistencia.*

Voltage-generating sensors   Sensors that are capable of producing their own input voltage signal.
Sensores generadores de voltaje   *Sensores que son capaces de producir su propia señal de voltaje de entrada.*

Voltmeter   A tool used to measure the voltage at any point in an electrical system.
Voltímetro   *Herramienta que se usa para medir el voltaje en cualquier punto de un sistema eléctrico.*

Water fade   Brake fade that occurs when water is trapped between the brake linings and the drum or rotor and the coefficient of friction is reduced.
Pérdida de efectividad por agua   *Pérdida de efectividad del freno que ocurre cuando queda atrapada agua entre los revestimientos del freno y el tambor o rotor y el coeficiente de fricción se reduce.*

Ways   Machined surfaces on the caliper support on which a sliding caliper slides.
Guías   *Superficies maquinadas en el soporte de mordaza sobre el cual se desliza la mordaza deslizante.*

Web   The inner part of a brake shoe that is perpendicular to the table and to which all of the springs and other linkage parts attach.
Entramado   *Pieza interior de la zapata de frenos que es perpendicular a la tabla y a la cual se unen los resortes y otras piezas de enlace.*

Weight   The measure of the Earth's gravitational force, or pull, on an object.
Peso   *Medida de la fuerza gravitacional de la tierra o atracción, sobre un objeto.*

Wheel cylinder   The hydraulic slave cylinder mounted on the backing plate of a drum brake assembly. The wheel cylinders convert hydraulic pressure from the master cylinder to mechanical force that applies the brake shoes.
Cilindro de la rueda   *Cilindro esclavo hidráulico montado en la placa de apoyo de un ensamble de freno de tambor. Los cilindros de la rueda convierten la presión hidráulica del cilindro maestro en fuerza mecánica que se aplica a las zapatas de freno.*

Wheel speed sensors   Speed sensing units placed at the wheels of a vehicle to monitor the rotational speed of the individual wheels.
Sensores de velocidad de la rueda   *Unidades sensoras de velocidad colocadas en las ruedas de un vehículo para monitorear la velocidad de rotación de las ruedas individuales.*

Work   The amount of force applied to something and the distance over which it is applied.
Trabajo   *Cantidad de fuerza que se aplica a algo y la distancia sobre la cual se aplica.*

Yaw   The swinging motion to the left or right of the vertical centerline or rotation around the vertical centerline.
Derrape   *Se define como el movimiento de balanceo hacia la derecha o izquierda de la línea central vertical o rotación alrededor de la línea central vertical.*

Index